Theoretical and Observational Cosmology

NATO Science Series

A Series presenting the results of activities sponsored by the NATO Science Committee. The Series is published by IOS Press and Kluwer Academic Publishers, in conjunction with the NATO Scientific Affairs Division.

A. Life Sciences	IOS Press
B. Physics	Kluwer Academic Publishers
C. Mathematical and Physical Sciences	Kluwer Academic Publishers
D. Behavioural and Social Sciences	Kluwer Academic Publishers
E. Applied Sciences	Kluwer Academic Publishers
F. Computer and Systems Sciences	IOS Press
1. Disarmament Technologies	Kluwer Academic Publishers
2. Environmental Security	Kluwer Academic Publishers
3. High Technology	Kluwer Academic Publishers
4. Science and Technology Policy	IOS Press
5. Computer Networking	IOS Press

NATO-PCO-DATA BASE

The NATO Science Series continues the series of books published formerly in the NATO ASI Series. An electronic index to the NATO ASI Series provides full bibliographical references (with keywords and/or abstracts) to more than 50000 contributions from internatonal scientists published in all sections of the NATO ASI Series.
Access to the NATO-PCO-DATA BASE is possible via CD-ROM "NATO-PCO-DATA BASE" with user-friendly retrieval software in English, French and German (WTV GmbH and DATAWARE Technologies Inc. 1989).

The CD-ROM of the NATO ASI Series can be ordered from: PCO, Overijse, Belgium

Series C: Mathematical and Physical Sciences – Vol. 541

Theoretical and Observational Cosmology

edited by

Marc Lachièze-Rey

Service d'Astrophysique,
CE Saclay, Gif sur Yvette Cedex,
France

Kluwer Academic Publishers

Dordrecht / Boston / London

Published in cooperation with NATO Scientific Affairs Division

Proceedings of the NATO Advanced Study Institute on
Theoretical and Observational Cosmology
Cargèse, France
August 17–29, 1998

A C.I.P. Catalogue record for this book is available from the Library of Congress.

ISBN 0-7923-5945-3

Published by Kluwer Academic Publishers,
P.O. Box 17, 3300 AA Dordrecht, The Netherlands.

Sold and distributed in North, Central and South America
by Kluwer Academic Publishers,
101 Philip Drive, Norwell, MA 02061, U.S.A.

In all other countries, sold and distributed
by Kluwer Academic Publishers,
P.O. Box 322, 3300 AH Dordrecht, The Netherlands.

Printed on acid-free paper

Printed in the Netherlands.

Table of Contents

PREFACE

This volume contains the talks given at the NATO Advanced Study Institute summer school on *Theoretical and Observational Cosmology*. The school was held in Cargese on the island of Corsica (France), from august 17 to august 29, 1998. The lectures covered almost the whole field of theoretical and observational cosmology.

Fundamental techniques of general relativity were discussed in many of the lectures. G. Ellis presented the covariant kinetic and dynamical formalisms, which can be used to construct and interpret the different cosmological models, as well as to deal with their perturbations. J.-P. Luminet and B. Roukema presented the case of topology in general relativity. M. Lachièze-Rey presented mathematical considerations about the symmetries of the theory and their possible generalisations. Y. Mellier and F. Bernardeau derived the mathematical and observational aspects of the bending of light rays by gravitational fields.

G. Ellis introduced the different cosmological models and their associated parameters. The possibility, relevance and limitations of critical distinguishing observations were discussed. His introduction was followed by discussions about causality and horizons, symmetries of space and space-time and their mathematical representation (Lie groups and algebras, Killing vectors, isometries,...). These discussions were completed by the lecture of M. Lachièze-Rey, which considered the importance of spatial symmetries to approach the structure of space-time. G. Ellis also presented the main non-standard cosmological models. J.-P. Luminet and B. Roukema discussed those models with peculiar (multiconnected) topologies and their observational consequences. G. Ellis also considered relativistic perturbations of the cosmological models, with emphasis on the fundamental gauge problem.

Cosmic velocity fields were the subject of the lectures by J. Simmons. He detailed the difficult task of measuring the fields, while separating them from the cosmic expansion. He showed how the subject is related to the subjects of structure formation, of dark matter, and of the Cosmic Microwave Background, ...

A general presentation of the critical subject of the Cosmic Microwave Background (CMB) was given by N. Aghanim. She discussed its cosmological origin, as well as the recent observational results. The relativistic kinetic theory used to treat the CMB anisotropies (in a covariant fashion) was discussed in the lectures by G. Ellis. The use of CMB intrinsic anisotropies, for the determination of cosmological models and for testing structure formation models, were presented. N. Aghanim also discussed the non-intrinsic fluctuations due to the Sunyaev-Zeldovich effect, and their use for the determination of the cosmological parameters, for the study of galaxy clusters and for the large scale structure of the universe. She also presented work by herself and J L Puget, concerning a similar effect around quasars

R. Pain discussed the use of type Ia supernovae as standard candles for the determination of cosmological parameters. After an introduction to the physics of supernovae, he reviewed the reasons they are believed to be good cosmological distance indicators. He then presented the status and perspectives for their use in measuring the Hubble constant and the other cosmological parameters.

The existence of dark matter has important cosmological implications, not the last of which concerns the mean density in the universe. J. Rich reviewed the observational evidences for the existence of dark matter and discussed the various baryonic or non-baryonic candidates. The possibilities for the detection of non-baryonic particles as well as observation of (baryonic) dark matter through microlensing were discussed.

A very strong argument in favour of the big bang models is the agreement between the observed universal abundances of light elements and their predictions by the primordial nucleosynthesis calculations. K. Olive presented the subject, by introducing the nuclear reactions and the corresponding calculations, and by summarising the observational results. The limits of the baryonic component of the cosmic density were discussed, with the strong implications for dark matter, cosmological parameters, the age of the universe, etc.

Gravitational lensing was discussed in details, in the lectures by Y. Mellier and F. Bernardeau, who introduced the relevant equations and phenomenological parameters (Einstein radius, shear, amplification, critical lines and caustics,...). They distinguished the various cases of lensing with special attention payed to lensing by galaxy clusters, the three characteristic regimes being strong (giant arcs), medium (arclets) and weak (gravitational distortion) lensing. They discussed the different observational strategies, the present results and the perspectives. The results concerning galaxy clusters, combined with results from X-ray observations, and concerning the abundance and properties of dark matter, were emphasized. Introducing the concept of "astigmatic view of the distant universe" (gravitational optic deformation), they discussed the possibilities of investigating high redshift objects, and the large scale structure of space-time. Finally, J. Rich presented the results of searches for microlensing search in our Galaxy.

All lectures, even those oriented towards theory, emphasised the link with observations, to which some of them were more specifically devoted. D. Kunth discussed the field of primordial galaxies (related to the cosmic background radiations). G. Mathez discussed the nature and properties of quasars. Both classes of objects are relevant for the study of cosmic evolution, the knowledge of the very distant universe, and the measurement of the cosmological parameters.

In conclusion, the ASI presented a large and almost complete panorama of cosmology. The evidences for the big bang models were presented and discussed. The status of the cosmic parameters was reviewed with emphasis on the indications that the $\Omega=1$ and $\Lambda=0$ model seems excluded, while the evidence for a non zero cosmological constant is increasing. The school met its purpose, which was to offer a complete background for those students and researchers intending to work in the field of theoretical or observational cosmology.

I thank the NATO for providing the major part of the financial support, and also the Commissariat à l'Energie Atomique (Saclay, France). I thank also the Cargese scientific Institute, and its Director Elisabeth Dubois-Violette, for their hospitality and help. I am very grateful for all the practical help provided by Geneviève Thiery (from Saclay), Chantal Ariano and Brigitte Cassegrain, from Cargese, and all the local staff. Finally, it is a pleasure to thank all the lecturers and participants, who contributed to make this school attractive and pleasant.

Marc Lachièze-Rey

List of lecturers

Aghanim Nabila
Institut d'Astrophysique Spatiale, Université Paris XI, Bat. 121,
91405 Orsay Cedex, France

Bernardeau Francis
Service de Physique Théorique, CE Saclay,
91191 Gif sur Yvette Cedex, France

Bertchinger Edmund
Dept of Physics, Massachussets Institute of Technology,
room 6-207, Cambridge, MA02139, USA

Carr Bernard J.
Queen Mary College, Mile end road, London E14NS

Ellis George
Mathematics Department, University of Cape Town,
Rondebosch 7700, Cape Town, South Africa

Kunth Daniel
Institut d'Astrophysique de Paris, 98bis boulevard Arago,
75014 Paris, France

Lachièze-Rey Marc
Service d'Astrophysique, CE Saclay, Orme des Merisiers, 91191
Gif sur Yvette Cedex, France

Luminet Jean-Pierre
Observatoire de Meudon, 92195 Meudon cedex, France

Mathez Guy
Observatoire Midi-Pyrénées, LAT, 14 avenue Edouard Belin,
31400 Toulouse, France

Mellier Yannick
Institut d'Astrophysique de Paris, 98bis boulevard Arago,
75014 Paris, France

Olive Keith
Dept. of Physics, University of Minnesota,
Minneapolis MN 55416, USA

Pain Reynald
LPNHE, Université Paris VI et VII, 4 place Jussieu, Tour 33 Rdc,
75252 Paris Cedex 05, France

Puget Jean-Loup
Institut d'Astrophysique Spatiale, Université Paris XI, Bat. 121,
91405 Orsay Cedex, France

Rich Jim
SPP, CE Saclay, Orme des Merisiers,
91191 Gif sur Yvette Cedex, France

Simmons John
Dept. of Physics and Astronomy, Kelvin Building,
University of Glasgow, GLASGOW G12 8QQ, UK

Triay Roland
Centre de Physique Théorique, CNRS – Luminy Case 907,
13288 Marseille Cedex 5, France

List of students

Alibert Yann
CRAL-ENS, 46 allée d'Italie, 69364 Lyon cedex 07, France
Amram Philippe
IGRAP, Observatoire, 2 place Le Verrier,
13248 Marseille cedex 4, France
Athreya Ramana
Institut d'Astrophysique de Paris, 98 bis boulevard Arago,
75014 Paris, France
Banerjee Shyamal
Inter-University Centre for Astronomy & Astrophysics,
Post Bag 4, Ganeshkhind, Pune-411007, India
Barrett Richard
Dept.of Physics and Astronomy, University of Glasgow,
Glasgow, Scotland G12 8QQ, U.K
Beisbart Claus
Theoretische Physik, Universität München, Theresienst. 37 III,
D-80333 München,Germany
Blanc Guillaume
Observatoire Paris-Meudon, Place Jules Janssen,
92190 Meudon, France
Bleybel Ali
Observatoire de Meudon, 92195 Meudon principal cedex, France
Bogdanov Dimitri
Copenhagen University, Astronomical Observatory,
Juliane Maries Vej 30, DK-2100 Copenhagen, Denmark
Casas Rigoberto
Max-Planck-Institut für Astrophysik, Karl-Schwarzschild-
Strasse1, D-85740, Garching, Germany
Chaname Julio
Departamento de Astronomia y Astrofisica, Pontificia
Universidad Catolica de Chile, Faculdad de Fisica, Casilla 306,
Santiago de Chile, Chile
Chen Xuelei
Physics Department, Columbia University, NewYork,
NY 10027, USA
Clarkson Chris
University of Glasgow, University avenue, Glasgow, Scotland
Courty Stéphanie
Observatoire Paris-Meudon, Place Jules Janssen,
92190 Meudon, France
Czoske Oliver
Observatoire Midi-Pyrénées, 14 avenue Edouard Belin,
31400 Toulouse, France
Derue Frédéric
Laboratoire de l'accélérateur linéaire, bat. 200,
Université Paris Sud, 91405 Orsay, France
Di Pietro Elisa

Institut d'Astrophysique et de Geophysique, Université de Liege,
Avenue de Cointe, 5, B 4000 Liège (Belgium)
Diego José
Avda. Los Castros, Facultad de ciencias IFCA,
39005 Santander, Spain
Elbaz David
CE Saclay, SAp, Bat 141, 91191 Gif sur Yvette Cedex, France
Engineer Sunu
Research Scolar, Post bag 4, Inter-University Centre for
Astronomy & Astrophysics, Ganeshkhind, Pune-411007, India
Fadda Dario
Osservatorio astronomico di Trieste, Via Tiepolo 11,
I-34131 Trieste, Italy
Ferreras Ignacio
Instituto de fisica de Cantabria, Fac. de Ciencias,
av. Los Castro s/n, 39005 Santander, Spain
Galfard Christophe
CE Saclay, SAp, Bat 141, 91191 Gif sur Yvette Cedex, France
Godlowski Wlodzimierz
Astronomical Observatory,Jagellonian University, Orla 171;
Krakow, Poland
Gray Meghan
Institute of Astronomy, Cambridge University, Madingley Road,
Cambridge, CB3 0HA, UK
Grenacher Lukas
Institute of Theoretical Physics, University of Zurich,
Winterthurer Strasse 190, CH-8057 Zurich, Switzerland
Hanski Mikko
Tuorla Observatory, 21500 Piikkio, Finland
Hatzaminaoglou Evanthia
Observatoire Midi-Pyrénées, 14 avenue Edouard Belin,
31400 Toulouse, France
Kanekar Nissim
National Centre for Radio Astrophysics, Tata Institute of
Fundamental Research, University of Pune campus,
Ganeshkhind, Pune - 411007, India
Kannappan Sheila
Harvard-Smithsonian Center for Astrophysics, 60 Garden Street
MS-20, Cambridge, MA 02138, USA
Kolonko M.
ul. Krolewska 53/37, 30-081 Krakow, Poland
Kriwult Janusz
Kielce Pedagogical University, Institute of Physics,
ul. M. Konopnickiej 15, 25-406 Kielce, Poland
Lémonon Ludovic
Service d'Astrophysique, CE Saclay, Orme des Merisiers,
91191 Gif sur Yvette Cedex, France
Lepore Natasha

Center for Astrophysics, Harvard University, 60 Garden st.
Mailstop 10, Cambridge, MA 02138, USA
Lewin Alex
Theoretical Physics, Imperial college, Prince Consort Road,
London SW7 2BZ,UK
Maoli Roberto
Institut d'Astrophysique de Paris, 98 bis boulevard Arago,
75014 Paris, France
Miramonti Roberto
CE Saclay, SAp, Bat 141, 91191 Gif sur Yvette Cedex, France
Missonier Gilles
Institut d'Astrophysique de Paris, 98 bis boulevard Arago,
75014 Paris, France
Moeller Ole
Institute of Astronomy, Cambridge University, Madingley Road,
Cambridge, CB3 0HA, UK
Nayeri Ali
Inter-University Centre For Astronomy & Astrophysics (IUCAA),
Post Bag 4, Ganeshkhind, Pune 411 007, Maharashtra, India
Olsen Lisbeth
Astronomical Observatory Juliane Maries vej 30,
DK-2100 Copenhagen, Denmark
Orsini Michele
Universita di Bologna, Dip di Astronimia, v. Zamboni, 33,
40126 Bologna, Italy
Outram Philip
Institute of Astronomy, Cambridge University, Madingley Road,
Cambridge, CB3 0HA, UK
Perna Rosalba
Center for Astrophysics, Harvard University, 60 Garden st.
Mailstop 10, Cambridge, MA 02138, USA
Pissondes Jean-Claude
DAEC, Observatoire de Paris Meudon, Place Jules Janssen, 92195
Meudon Cedex, France
Raig Andreu
Dept. Astronomia i Meteorologia Marti i Franquée, 1,
Barcelone 08028, Spain
Régnault Nicolas
Laboratoire de l'accélérateur linéaire, bat. 200,
Université Paris Sud, 91405 Orsay, France
Ronen Shat
Institute of Astronomy, Cambridge University, Madingley Road,
Cambridge, CB3 0HA, UK
Rhodes Jason
Department of Physics, Jadwin hall, Princeton, NJ 08544
Roukema Boudewjin
Institut d'Astrophysique de Paris, 98bis boulevard Arago,
75014 Paris, France
Sanchez Sebastian

Instituto de fisica de Cantabria, Fac. de Ciencias,
av. Los Castro s/n, 39005 Santander, Spain
Schaye Joop
Institute of Astronomy, University of Cambridge
Madinglay Road, Cambridge, CB3 OHA, UK
Sicka Christian
Theresienstrasse 37 80333 München, ZI 341, Germany
Steer Daniele
D.A.M.P.T., Silver Street, Cambridge, CB3 9EW, UK
Szydtowski M.
Kielce Pedagogical University, Institute of Physics,
ul. M. Konopnickiej 15, 25-406 Kielce, Poland
Tesch Frank
Hamburger Sterwarte, Gojenbergsweg,
D-21029 Hamburg, Germany
Weller Jochen
Theoretical Physics, Imperial College, Prince Consort Road,
London SW7 2BZ, UK
Wichoski Ubirajara
Brown University, Department of Physics – Box 1843,
Providence RI02912, USA
Woszczyna Andrzej.
Astronomical Observatory, Jagellonian University, Orla 171;
Krakow, Poland
Wu Kelvin
Institute of Astronomy, University of Cambridge
Madinglay Road,
Cambridge, CB3 OHA, UK
Yvon Dominique
CE Saclay, SPP, Bat 141, 91191 Gif sur Yvette Cedex, France
Zlatev Ivaylo
Univ of Pennsilvania, USA

COSMOLOGICAL MODELS

Cargèse Lectures 1998

GEORGE F. R. ELLIS AND HENK VAN ELST
Cosmology Group
Department of Mathematics and Applied Mathematics
University of Cape Town
Rondebosch 7701
Cape Town
South Africa
Electronic addresses:
ellis@maths.uct.ac.za, henk@gmunu.mth.uct.ac.za

Abstract. The aim of this set of lectures is a systematic presentation of a $1+3$ covariant approach to studying the geometry, dynamics, and observational properties of relativistic cosmological models. In giving (i) the basic $1+3$ covariant relations for a cosmological fluid, the present lectures cover some of the same ground as a previous set of Cargèse lectures [7], but they then go on to give (ii) the full set of corresponding tetrad equations, (iii) a classification of cosmological models with exact symmetries, (iv) a brief discussion of some of the most useful exact models and their observational properties, and (v) an introduction to the gauge-invariant and $1+3$ covariant perturbation theory of almost-Friedmann–Lemaître–Robertson–Walker universes, with a fluid description for the matter and a kinetic theory description of the radiation.

1. Basic relations

A cosmological model represents the universe at a particular scale. We will assume that on large scales, space-time geometry is described by General Relativity (see e.g. d'Inverno [1], Wald [2], Hawking and Ellis [3], or Stephani [4]). Then a cosmological model is defined by specifying [5]–[7]:

M. Lachièze-Rey (ed.), Theoretical and Observational Cosmology, 1–116.

* the *space-time geometry* represented on some specific * averaging scale and determined by the metric $g_{ij}(x^k)$, which * — because of the requirement of compatibility with observations * — must either have some expanding Robertson–Walker ('RW') * geometries as a regular limit (see [8]), or else be * demonstrated to have observational properties compatible with the * major features of current astronomical observations of the * universe;

* the *matter present*, represented on the same averaging scale, and its *physical behaviour* (the energy-momentum tensor of each matter component, the equations governing the behaviour of each such component, and the interaction terms between them), which must represent physically plausible matter (ranging from early enough times to the present day, this will include most of the interactions described by present-day physics); and

* the *interaction of the geometry and matter* — how matter determines the geometry, which in turn determines the motion of the matter (see e.g. [9]). We assume this is through the *Einstein gravitational field equations* ('EFE') given by[1]

$$G_{ab} \equiv R_{ab} - \tfrac{1}{2} R\, g_{ab} = T_{ab} - \Lambda\, g_{ab} \;, \tag{1}$$

which, because of the *twice-contracted Bianchi identities*, guarantee the conservation of total energy-momentum

$$\nabla_b G^{ab} = 0 \quad \Rightarrow \quad \nabla_b T^{ab} = 0 \;, \tag{2}$$

provided the *cosmological constant* Λ satisfies the relation $\nabla_a \Lambda = 0$, i.e., it is constant in time and space.

Together, these determine the combined dynamical evolution of the model and the matter in it. The description must be sufficiently complete to determine

* the *observational relations* predicted by the model for both discrete sources and background radiation, implying a well-developed theory of *structure growth* for very small and for very large physical scales (i.e. for light atomic nuclei and for galaxies and clusters of galaxies), and of *radiation absorbtion and emission.*

To be useful in an explanatory role, a cosmological model must be easy to describe — that means they have symmetries or special properties of

[1]Throughout this review we employ geometrised units characterised by $c = 1 = 8\pi G/c^2$. Consequently, all geometrical variables occurring have physical dimensions that are integer powers of the dimension [length]. The index convention is such that space-time and spatial indices with respect to a general basis are denoted by $a, b, \ldots = 0, 1, 2, 3$ and $\alpha, \beta, \ldots = 1, 2, 3$, respectively, while space-time indices in a coordinate basis are $i, j, \ldots = 0, 1, 2, 3$.

some kind or other. The usual choices for the matter description will be some combination of

* a fluid with a physically well-motivated equation of state, for example a perfect fluid with specified equation of state (beware of imperfect fluids, unless they have well-defined and motivated physical properties);
* a mixture of fluids, usually with different 4-velocities;
* a set of particles represented by a kinetic theory description;
* a scalar field ϕ, with a given potential $V(\phi)$ (at early times);
* an electromagnetic field described by the Maxwell field equations.

As intimated above, the observational relations implied by cosmological models must be compared with astronomical observations. This determines those solutions that can usefully be considered as viable cosmological models of the real universe. A major aim of the present lectures is to point out that this class is wider than just the standard Friedmann–Lemaître–Robertson–Walker ('FLRW') cosmologies; indeed those models cannot be realistic on all scales of description, but they may also be inaccurate on large scales, or at very early and very late times. To examine this, we need to consider the subspace of the space of all cosmological solutions that contains models with observational properties like those of the real universe at some stage of their histories. Thus we are interested in the *full* state space of solutions, allowing us to see how realistic models are related to each other and to higher symmetry models, including particularly the FLRW models.

These lectures develop general techniques for examining this, and describe some specific models of interest. The first part looks at exact general relations valid in all cosmological models, the second part at exact cosmological solutions of the EFE, and the third part at approximate equations and solutions: specifically, 'almost-FLRW' models, linearised about a FLRW geometry.

2. $1+3$ covariant description and equations

Space-times can be described via

(a) the metric $g_{ij}(x^k)$ described in a particular set of coordinates, with its differential properties, as embodied by the connection, given through the Christoffel symbols;

(b) the metric described by means of particular tetrads, with its connection given through the Ricci rotation coefficients;

(c) 1+3 covariantly defined variables. In anisotropic cases, tetrad vectors can be uniquely defined in a $1+3$ covariant way and this approach merges into (b).

Here I will concentrate on the 1 + 3 covariant approach, based on [5, 10, 6, 7, 11], but dealing also with the tetrad approach which serves as a completion to the 1 + 3 covariant approach. The basic point here is that because we have complete coordinate freedom in General Relativity, it is preferable where possible to describe physics and geometry by tensor relations and quantities; these then remain valid whatever coordinate system is chosen.

2.1. VARIABLES

2.1.1. *Average 4-velocity of matter*

In a cosmological space-time $(\mathcal{M}, \mathbf{g})$, at late times there will be a family of preferred worldlines representing the average motion of matter at each point[2] (notionally, these represent the histories of clusters of galaxies, with associated 'fundamental observers'); at early times there will be uniquely defined notions of the average velocity of matter (at that time, interacting gas and radiation), and corresponding preferred worldlines. In each case their *4-velocity* is[3]

$$u^a = \frac{dx^a}{d\tau} \; , \qquad u_a u^a = -1 \; , \tag{3}$$

where τ is proper time measured along the fundamental worldlines. We assume this 4-velocity is unique: that is, there is a well-defined preferred motion of matter at each space-time event. At recent times this is taken to be the 4-velocity defined by the vanishing of the dipole of the cosmic microwave background radiation ('CBR'): for there is precisely one 4-velocity which will set this dipole to zero. It is usually assumed that this is the same as the average 4-velocity of matter in a suitably sized volume [6]; indeed this assumption is what underlies studies of large scale motions and the 'Great Attractor'.

Given u^a, there are defined unique *projection tensors*

$$U^a{}_b = -\,u^a\,u_b \;\; \Rightarrow \;\; U^a{}_c U^c{}_b = U^a{}_b \; , \; U^a{}_a = 1 \; , \; U_{ab}\,u^b = u_a \; , \tag{4}$$

$$h_{ab} = g_{ab} + u_a\,u_b \;\; \Rightarrow \;\; h^a{}_c h^c{}_b = h^a{}_b \; , \; h^a{}_a = 3 \; , \; h_{ab}\,u^b = 0 \; . \tag{5}$$

The first projects parallel to the velocity vector u^a, and the second determines the (orthogonal) metric properties of the instantaneous rest-spaces of observers moving with 4-velocity u^a. There is also defined a *volume element*

[2]We are here assuming a fluid description can be used on a large enough scale [5, 6]. The alternative is that the matter distribution is hierarchically structured at all levels or fractal (see e.g. [12] and refrences there), so that a fluid description does not apply. The success of the FLRW models encourages us to use the approach taken here.

[3]Merging from the one concept to the other as structure formation takes place.

for the rest-spaces:

$$\eta_{abc} = u^d \eta_{dabc} \quad \Rightarrow \quad \eta_{abc} = \eta_{[abc]} \ , \ \eta_{abc} u^c = 0 \ , \tag{6}$$

where η_{abcd} is the 4-dimensional volume element ($\eta_{abcd} = \eta_{[abcd]}$, $\eta_{0123} = \sqrt{|\det g_{ab}|}$).

Moreover, *two derivatives* are defined: the covariant time derivative $\dot{}$ along the fundamental worldlines, where for any tensor $T^{ab}{}_{cd}$

$$\dot{T}^{ab}{}_{cd} = u^e \nabla_e T^{ab}{}_{cd} \ , \tag{7}$$

and the fully orthogonally projected covariant derivative $\tilde{\nabla}$, where for any tensor $T^{ab}{}_{cd}$

$$\tilde{\nabla}_e T^{ab}{}_{cd} = h^a{}_f \, h^b{}_g \, h^p{}_c \, h^q{}_d \, h^r{}_e \nabla_r T^{fg}{}_{pq} \ , \tag{8}$$

with total projection on all free indices. The tilde serves as a reminder that if u^a has *non-zero* vorticity, $\tilde{\nabla}$ is *not* a proper 3-dimensional covariant derivative (see Eq. (27) below). Finally, following [11] (and see also [13]), we use angle brackets to denote orthogonal projections of vectors and the orthogonally projected symmetric trace-free part of tensors:

$$v^{\langle a \rangle} = h^a{}_b \, v^b \ , \quad T^{\langle ab \rangle} = [\, h^{(a}{}_c \, h^{b)}{}_d - \tfrac{1}{3} h^{ab} \, h_{cd} \,] T^{cd} \ ; \tag{9}$$

for convenience the angle brackets are also used to denote othogonal projections of covariant time derivatives along u^a (*'Fermi derivatives'*):

$$\dot{v}^{\langle a \rangle} = h^a{}_b \, \dot{v}^b \ , \quad \dot{T}^{\langle ab \rangle} = [\, h^{(a}{}_c \, h^{b)}{}_d - \tfrac{1}{3} h^{ab} \, h_{cd} \,] \dot{T}^{cd} \ . \tag{10}$$

Exercise: Show that the projected time and space derivatives of U_{ab}, h_{ab} and η_{abc} all vanish.

2.1.2. *Kinematical quantities*

We split the first covariant derivative of u_a into its irreducible parts, defined by their symmetry properties:

$$\nabla_a u_b = -\, u_a \, \dot{u}_b + \tilde{\nabla}_a u_b = -\, u_a \, \dot{u}_b + \tfrac{1}{3} \Theta \, h_{ab} + \sigma_{ab} + \omega_{ab} \ , \tag{11}$$

where the trace $\Theta = \tilde{\nabla}_a u^a$ is the *(volume) rate of expansion* of the fluid (with $H = \Theta/3$ the Hubble parameter); $\sigma_{ab} = \tilde{\nabla}_{\langle a} u_{b \rangle}$ is the trace-free symmetric *rate of shear* tensor ($\sigma_{ab} = \sigma_{(ab)}$, $\sigma_{ab} \, u^b = 0$, $\sigma^a{}_a = 0$), describing the rate of distortion of the matter flow; and $\omega_{ab} = \tilde{\nabla}_{[a} u_{b]}$ is the skew-symmetric *vorticity* tensor ($\omega_{ab} = \omega_{[ab]}$, $\omega_{ab} \, u^b = 0$),[4] describing the rotation

[4] The vorticity here is defined with respect to a right-handedly oriented spatial basis.

of the matter relative to a non-rotating (Fermi-propagated) frame. The stated meaning for these quantities follows from the evolution equation for a relative position vector $\eta^a_\perp = h^a{}_b\eta^b$, where η^a is a deviation vector for the family of fundamental worldlines, i.e. $u^b\nabla_b\eta^a = \eta^b\nabla_b u^a$. Writing $\eta^a_\perp = \delta\ell\, e^a$, $e_a e^a = 1$, we find the relative distance $\delta\ell$ obeys the propagation equation

$$\frac{(\delta\ell)^{\cdot}}{\delta\ell} = \tfrac{1}{3}\Theta + (\sigma_{ab}e^a e^b) \;, \tag{12}$$

(the generalised Hubble law), and the relative direction vector e^a the propagation equation

$$\dot{e}^{\langle a\rangle} = (\sigma^a{}_b - (\sigma_{cd}e^c e^d)\, h^a{}_b - \omega^a{}_b)\, e^b \;, \tag{13}$$

giving the observed rate of change of position in the sky of distant galaxies. Finally $\dot{u}^a = u^b\nabla_b u^a$ is the *relativistic acceleration* vector, representing the degree to which the matter moves under forces other than gravity plus inertia (which cannot be covariantly separated from each other in General Relativity: they are different aspects of the same effect). The acceleration vanishes for matter in free fall (i.e. moving under gravity plus inertia alone).

2.1.3. *Matter tensor*

The matter *energy-momentum tensor* T_{ab} can be decomposed relative to u^a in the form[5]

$$T_{ab} = \mu\, u_a\, u_b + q_a\, u_b + u_a\, q_b + p\, h_{ab} + \pi_{ab} \;, \tag{14}$$
$$q_a\, u^a = 0 \;,\; \pi^a{}_a = 0 \;,\; \pi_{ab} = \pi_{(ab)} \;,\; \pi_{ab}\, u^b = 0 \;,$$

where $\mu = (T_{ab}u^a u^b)$ is the *relativistic energy density* relative to u^a, $q^a = -T_{bc}\, u^b\, h^{ca}$ is the *relativistic momentum density*, which is also the energy flux relative to u^a, $p = \frac{1}{3}(T_{ab}h^{ab})$ is the *isotropic pressure*, and $\pi_{ab} = T_{cd}\, h^c{}_{\langle a}\, h^d{}_{b\rangle}$ is the trace-free *anisotropic pressure* (stress).

The *physics* of the situation is in the *equations of state* relating these quantities; for example, the commonly imposed restrictions

$$q^a = \pi_{ab} = 0 \quad \Leftrightarrow \quad T_{ab} = \mu\, u_a\, u_b + p\, h_{ab} \tag{15}$$

characterise a 'perfect fluid' with, in general, equation of state $p = p(\mu, s)$. If in addition we assume that $p = 0$, we have the simplest case: pressure-free matter ('dust' or 'Cold Dark Matter'). Otherwise we must specify an equation of state determining p from μ and possibly other thermodynamical

[5] We should really write $\mu = \mu(u^a)$, etc; but usually assume this dependence is understood.

variables. Whatever these relations may be, we usually require that various *energy conditions* hold: one or all of

$$\mu > 0 \,, \quad (\mu + p) > 0 \,, \quad (\mu + 3p) > 0 \,, \tag{16}$$

(the latter, however, being violated by scalar fields in inflationary universe models), and additionally demand the *adiabatic speed of sound* $c_s^2 = (\partial p/\partial \mu)_{s=\mathrm{const}}$ obeys

$$0 \leq c_s^2 \leq 1 \quad \Leftrightarrow \quad 0 \leq \left(\frac{\partial p}{\partial \mu}\right)_{s=\mathrm{const}} \leq 1 \,, \tag{17}$$

as required for local stability of matter (lower bound) and causality (upper bound), respectively.

2.1.4. *Maxwell field strength tensor*

The *Maxwell field strength tensor* F_{ab} of an electromagnetic field is split relative to u^a into *electric* and *magnetic field* parts by the relations (see [7])

$$E_a = F_{ab}\, u^b \quad \Rightarrow \quad E_a u^a = 0 \,, \tag{18}$$

$$H_a = \tfrac{1}{2}\, \eta_{abc}\, F^{bc} \quad \Rightarrow \quad H_a u^a = 0 \,. \tag{19}$$

2.1.5. *Weyl curvature tensor*

In analogy to F_{ab}, the *Weyl conformal curvature tensor* C_{abcd} is split relative to u^a into *'electric'* and *'magnetic' Weyl curvature* parts according to

$$E_{ab} = C_{acbd}\, u^c\, u^d \quad \Rightarrow \quad E^a{}_a = 0 \,, \; E_{ab} = E_{(ab)} \,, \; E_{ab}\, u^b = 0 \,, \tag{20}$$

$$H_{ab} = \tfrac{1}{2}\, \eta_{ade}\, C^{de}{}_{bc}\, u^c \quad \Rightarrow \quad H^a{}_a = 0 \,, \; H_{ab} = H_{(ab)} \,, \; H_{ab}\, u^b = 0 \,. \tag{21}$$

These represent the 'free gravitational field', enabling gravitational action at a distance (tidal forces, gravitational waves), and influence the motion of matter and radiation through the *geodesic deviation equation* for timelike and null vectors, respectively [14]–[18]. Together with the Ricci tensor R_{ab} (determined locally at each point by the matter tensor through the EFE (1)), these quantities completely represent the space-time *Riemann curvature tensor* R_{abcd}, which in fully $1+3$-decomposed form becomes[6]

$$R^{ab}{}_{cd} \;=\; R_P^{ab}{}_{cd} + R_I^{ab}{}_{cd} + R_E^{ab}{}_{cd} + R_H^{ab}{}_{cd} \,,$$

[6]Here P is the perfect fluid part, I the imperfect fluid part, E that due to the electric Weyl tensor, and H that due to the magnetic Weyl tensor. This obscures the similarities in these equations between E and π, and between H and q; however, this partial symmetry is broken by the field equations, so the splitting given here (due to M Shedden) is conceptually useful.

$$\begin{aligned}
R_{P\,cd}^{ab} &= \tfrac{2}{3}\,(\mu+3p-2\Lambda)\,u^{[a}\,u_{[c}\,h^{b]}{}_{d]} + \tfrac{2}{3}\,(\mu+\Lambda)\,h^{a}{}_{[c}\,h^{b}{}_{d]}\ , \\
R_{I\,cd}^{ab} &= -\,2\,u^{[a}\,h^{b]}{}_{[c}\,q_{d]} - 2\,u_{[c}\,h^{[a}{}_{d]}\,q^{b]} - 2\,u^{[a}\,u_{[c}\,\pi^{b]}{}_{d]} + 2\,h^{[a}{}_{[c}\,\pi^{b]}{}_{d]}\ , \\
R_{E\,cd}^{ab} &= 4\,u^{[a}\,u_{[c}\,E^{b]}{}_{d]} + 4\,h^{[a}{}_{[c}\,E^{b]}{}_{d]}\ , \\
R_{H\,cd}^{ab} &= 2\,\eta^{abe}\,u_{[c}\,H_{d]e} + 2\,\eta_{cde}\,u^{[a}\,H^{b]e}\ .
\end{aligned} \tag{22}$$

2.1.6. *Auxiliary quantities*

It is useful to define some associated kinematical quantities: the *vorticity vector*

$$\omega^a = \tfrac{1}{2}\,\eta^{abc}\,\omega_{bc} \quad\Rightarrow\quad \omega_a\,u^a = 0\ ,\ \omega_{ab}\,\omega^b = 0\ , \tag{23}$$

the magnitudes

$$\omega^2 = \tfrac{1}{2}\,(\omega_{ab}\omega^{ab}) \geq 0\ ,\quad \sigma^2 = \tfrac{1}{2}\,(\sigma_{ab}\sigma^{ab}) \geq 0\ , \tag{24}$$

and the *average length scale* S determined by

$$\frac{\dot{S}}{S} = \tfrac{1}{3}\,\Theta\ , \tag{25}$$

so the volume of a fluid element varies as S^3. Further it is helpful to define particular *spatial gradients* orthogonal to u^a, characterising the inhomogeneity of space-time:

$$X_a \equiv \tilde{\nabla}_a\mu\ ,\qquad Z_a \equiv \tilde{\nabla}_a\Theta\ . \tag{26}$$

The energy density μ (and also Θ) satisfies the important *commutation relation* for the $\tilde{\nabla}$-derivative [19]

$$\tilde{\nabla}_{[a}\tilde{\nabla}_{b]}\mu = \eta_{abc}\,\omega^c\,\dot{\mu}\ . \tag{27}$$

This shows that if $\omega^a\,\dot{\mu} \neq 0$ in an open set then $X_a \neq 0$ there, so non-zero vorticity implies anisotropic number counts in an expanding universe [20] (this is because there are then no 3-surfaces orthogonal to the fluid flow; see [6]).

2.2. 1+3 COVARIANT PROPAGATION AND CONSTRAINT EQUATIONS

There are three sets of equations to be considered, resulting from the EFE (1) and its associated integrability conditions.

2.2.1. *Ricci identities*

The first set arise from the *Ricci identities* for the vector field u^a, i.e.

$$2\,\nabla_{[a}\nabla_{b]}u^c = R_{ab}{}^{c}{}_{d}\,u^d\ . \tag{28}$$

On substituting in from (11), using (1), and separating out the orthogonally projected part into a trace, symmetric trace-free, and skew symmetric parts, and the parallel part similarly, we obtain three propagation equations and three constraint equations. The *propagation equations* are,

1. The *Raychaudhuri equation* [21]

$$\dot{\Theta} - \tilde{\nabla}_a \dot{u}^a = -\tfrac{1}{3}\Theta^2 + (\dot{u}_a \dot{u}^a) - 2\sigma^2 + 2\omega^2 - \tfrac{1}{2}(\mu + 3p) + \Lambda \,, \qquad (29)$$

which is the *basic equation of gravitational attraction* [5]–[7], showing the repulsive nature of a positive cosmological constant, leading to identification of $(\mu+3p)$ as the active gravitational mass density, and underlying the basic singularity theorem (see below).

2. The *vorticity propagation equation*

$$\dot{\omega}^{\langle a\rangle} - \tfrac{1}{2}\eta^{abc}\tilde{\nabla}_b \dot{u}_c = -\tfrac{2}{3}\Theta\omega^a + \sigma^a{}_b\,\omega^b \,; \qquad (30)$$

together with (38) below, showing how vorticity conservation follows if there is a perfect fluid with acceleration potential Φ [5, 7], since then, on using (27), $\eta^{abc}\tilde{\nabla}_b \dot{u}_c = \eta^{abc}\tilde{\nabla}_b \tilde{\nabla}_c \Phi = 2\omega^a \dot{\Phi}$,

3. The *shear propagation equation*

$$\begin{aligned} \dot{\sigma}^{\langle ab\rangle} - \tilde{\nabla}^{\langle a}\dot{u}^{b\rangle} &= -\tfrac{2}{3}\Theta\sigma^{ab} + \dot{u}^{\langle a}\dot{u}^{b\rangle} - \sigma^{\langle a}{}_c\,\sigma^{b\rangle c} - \omega^{\langle a}\omega^{b\rangle} \\ &\quad - (E^{ab} - \tfrac{1}{2}\pi^{ab}) \,, \end{aligned} \qquad (31)$$

the anisotropic pressure source term π_{ab} vanishing for a perfect fluid; this shows how the tidal gravitational field E_{ab} directly induces shear (which then feeds into the Raychaudhuri and vorticity propagation equations, thereby changing the nature of the fluid flow).

The *constraint equations* are,

1. The (0α)-*equation*

$$0 = (C_1)^a = \tilde{\nabla}_b \sigma^{ab} - \tfrac{2}{3}\tilde{\nabla}^a\Theta + \eta^{abc}[\,\tilde{\nabla}_b\omega_c + 2\,\dot{u}_b\,\omega_c\,] + q^a \,, \qquad (32)$$

showing how the momentum flux (zero for a perfect fluid) relates to the spatial inhomogeneity of the expansion;

2. The *vorticity divergence identity*

$$0 = (C_2) = \tilde{\nabla}_a\omega^a - (\dot{u}_a\omega^a) \,, \qquad (33)$$

3. The H_{ab}-*equation*

$$0 = (C_3)^{ab} = H^{ab} + 2\,\dot{u}^{\langle a}\omega^{b\rangle} + \tilde{\nabla}^{\langle a}\omega^{b\rangle} - (\mathrm{curl}\,\sigma)^{ab} \,, \qquad (34)$$

characterising the magnetic Weyl tensor as being constructed from the 'distortion' of the vorticity and the 'curl' of the shear, $(\mathrm{curl}\,\sigma)^{ab} = \eta^{cd\langle a}\,\tilde{\nabla}_c \sigma^{b\rangle}{}_d$.

2.2.2. *Twice-contracted Bianchi identities*
The second set of equations arise from the *twice-contracted Bianchi identities* which by the EFE (1) imply the conservation equations (2). Projecting parallel and perpendicular to u^a, we obtain the propagation equations

$$\dot{\mu} + \tilde{\nabla}_a q^a = -\,\Theta\,(\mu + p) - 2\,(\dot{u}_a q^a) - (\sigma^a{}_b \pi^b{}_a) \tag{35}$$

and

$$\dot{q}^{\langle a\rangle} + \tilde{\nabla}^a p + \tilde{\nabla}_b \pi^{ab} = -\,\tfrac{4}{3}\,\Theta\,q^a - \sigma^a{}_b\,q^b - (\mu + p)\,\dot{u}^a - \dot{u}_b\,\pi^{ab} - \eta^{abc}\,\omega_b\,q_c\ , \tag{36}$$

respectively. For perfect fluids, characterised by Eq. (15), these reduce to

$$\dot{\mu} = -\,\Theta\,(\mu + p)\ , \tag{37}$$

the *energy conservation equation*, and one constraint equation

$$0 = \tilde{\nabla}_a p + (\mu + p)\,\dot{u}_a\ , \tag{38}$$

the *momentum conservation equation*. This shows that $(\mu+p)$ is the inertial mass density, and also governs the conservation of energy. It is clear that if this quantity is zero (an effective cosmological constant) or negative, the behaviour of matter will be anomalous.

Exercise: Examine what happens in the two cases (i) $(\mu + p) = 0$, (ii) $(\mu + p) < 0$.

2.2.3. *Other Bianchi identities*
The third set of equations arise from the *Bianchi identities*

$$\nabla_{[a} R_{bc]de} = 0\ . \tag{39}$$

Double contraction gives Eq. (2), already considered. On using the splitting of R_{abcd} into R_{ab} and C_{abcd}, the above 1+3 splitting of those quantities, and the EFE, the once-contracted Bianchi identities give two further propagation equations and two further constraint equations, which are similar in form to the Maxwell field equations in an expanding universe (see [22, 7]).

The *propagation equations* are,

$$\begin{aligned}
&(\dot{E}^{\langle ab\rangle} + \tfrac{1}{2}\,\dot{\pi}^{\langle ab\rangle}) - (\mathrm{curl}\,H)^{ab} + \tfrac{1}{2}\,\tilde{\nabla}^{\langle a} q^{b\rangle} \\
&\quad = -\,\tfrac{1}{2}\,(\mu + p)\,\sigma^{ab} \\
&\qquad - \Theta\,(E^{ab} + \tfrac{1}{6}\,\pi^{ab}) + 3\,\sigma^{\langle a}{}_c\,(E^{b\rangle c} - \tfrac{1}{6}\,\pi^{b\rangle c}) - \dot{u}^{\langle a}\,q^{b\rangle} \\
&\qquad + \eta^{cd\langle a}\,[\,2\,\dot{u}_c\,H^{b\rangle}{}_d + \omega_c\,(E^{b\rangle}{}_d + \tfrac{1}{2}\,\pi^{b\rangle}{}_d)\,]\ ,
\end{aligned} \tag{40}$$

the $\dot{E}$-*equation*, and

$$\dot{H}^{\langle ab\rangle} + (\mathrm{curl}\,E)^{ab} - \tfrac{1}{2}(\mathrm{curl}\,\pi)^{ab}$$
$$= -\,\Theta\, H^{ab} + 3\,\sigma^{\langle a}{}_{c}\, H^{b\rangle c} + \tfrac{3}{2}\,\omega^{\langle a}\, q^{b\rangle}$$
$$-\ \eta^{cd\langle a}\,[\, 2\, \dot{u}_{c}\, E^{b\rangle}{}_{d} - \tfrac{1}{2}\,\sigma^{b\rangle}{}_{c}\, q_{d} - \omega_{c}\, H^{b\rangle}{}_{d}\,]\,, \qquad (41)$$

the $\dot{H}$-*equation*, where we have defined the 'curls'

$$(\mathrm{curl}\,H)^{ab} \;=\; \eta^{cd\langle a}\,\tilde{\nabla}_{c} H^{b\rangle}{}_{d}\,, \qquad (42)$$
$$(\mathrm{curl}\,E)^{ab} \;=\; \eta^{cd\langle a}\,\tilde{\nabla}_{c} E^{b\rangle}{}_{d}\,. \qquad (43)$$

These equations show how gravitational radiation arises: taking the time derivative of the $\dot{E}$-equation gives a term of the form $(\mathrm{curl}\,H)^{\cdot}$; commuting the derivatives and substituting from the $\dot{H}$-equation eliminates H, and results in a term in $\ddot{E}$ and a term of the form $(\mathrm{curl}\,\mathrm{curl}\,E)$, which together give the wave operator acting on E [22, 23]; similarly the time derivative of the $\dot{H}$-equation gives a wave equation for H.

The *constraint equations* are

$$0 \;=\; (C_{4})^{a} \;=\; \tilde{\nabla}_{b}(E^{ab} + \tfrac{1}{2}\,\pi^{ab}) - \tfrac{1}{3}\,\tilde{\nabla}^{a}\mu + \tfrac{1}{3}\,\Theta\, q^{a} - \tfrac{1}{2}\,\sigma^{a}{}_{b}\, q^{b} - 3\,\omega_{b}\, H^{ab}$$
$$-\ \eta^{abc}\,[\,\sigma_{bd}\, H^{d}{}_{c} - \tfrac{3}{2}\,\omega_{b}\, q_{c}\,]\,, \qquad (44)$$

the $(\mathrm{div}\,E)$-*equation* with source the spatial gradient of the energy density, which can be regarded as a vector analogue of the Newtonian Poisson equation [24], enabling tidal action at a distance, and

$$0 \;=\; (C_{5})^{a} \;=\; \tilde{\nabla}_{b} H^{ab} + (\mu + p)\,\omega^{a} + 3\,\omega_{b}\,(E^{ab} - \tfrac{1}{6}\,\pi^{ab})$$
$$+\ \eta^{abc}\,[\,\tfrac{1}{2}\,\tilde{\nabla}_{b} q_{c} + \sigma_{bd}\,(E^{d}{}_{c} + \tfrac{1}{2}\,\pi^{d}{}_{c})\,]\,, \qquad (45)$$

the $(\mathrm{div}\,H)$-*equation*, with source the fluid vorticity. These equations show respectively that scalar modes will result in a non-zero divergence of E_{ab} (and hence a non-zero E-field), and vector modes in a non-zero divergence of H_{ab} (and hence a non-zero H-field).

2.2.4. *Maxwell field equations*

Finally, we turn for completeness to the 1+3 decomposition of the *Maxwell field equations*

$$\nabla_{b} F^{ab} = j_{\mathrm{e}}^{a}\,, \qquad \nabla_{[a} F_{bc]} = 0\,. \qquad (46)$$

As shown in [7], the *propagation equations* can be written as

$$\dot{E}^{\langle a\rangle} - \eta^{abc}\,\tilde{\nabla}_{b} H_{c} \;=\; -\,j_{\mathrm{e}}^{\langle a\rangle} - \tfrac{2}{3}\,\Theta\, E^{a} + \sigma^{a}{}_{b}\, E^{b} \qquad (47)$$
$$+\ \eta^{abc}\,[\,\dot{u}_{b}\, H_{c} + \omega_{b}\, E_{c}\,]\,,$$
$$\dot{H}^{\langle a\rangle} + \eta^{abc}\,\tilde{\nabla}_{b} E_{c} \;=\; -\,\tfrac{2}{3}\,\Theta\, H^{a} + \sigma^{a}{}_{b}\, H^{b} \qquad (48)$$
$$-\ \eta^{abc}\,[\,\dot{u}_{b}\, E_{c} - \omega_{b}\, H_{c}\,]\,,$$

while the *constraint equations* assume the form

$$0 = (C_E) = \tilde{\nabla}_a E^a - 2(\omega_a H^a) - \rho_e , \tag{49}$$

$$0 = (C_H) = \tilde{\nabla}_a H^a + 2(\omega_a E^a) , \tag{50}$$

where $\rho_e = (-j_{e\,a} u^a)$.

2.3. PRESSURE-FREE MATTER ('DUST')

A particularly useful dynamical restriction is

$$p = 0 = q^a = \pi_{ab} \quad \Rightarrow \quad \dot{u}_a = 0 , \tag{51}$$

so the matter (often described as 'baryonic') is represented only by its 4-velocity u^a and its energy density $\mu > 0$. The implication follows from momentum conservation: (38) shows that the matter moves geodesically (as expected from the equivalence principle). This is the case of *pure gravitation*: it separates out the (non-linear) gravitational effects from all the fluid dynamical effects. The vanishing of the acceleration greatly simplifies the above set of equations.

2.4. IRROTATIONAL FLOW

If we have a barotropic perfect fluid:

$$q^a = \pi_{ab} = 0 , \quad p = p(\mu) , \quad \Rightarrow \quad \eta^{abc}\, \tilde{\nabla}_b \dot{u}_c = 0 , \tag{52}$$

then $\omega^a = 0$ is involutive: i.e.

$$\omega^a = 0 \text{ initially} \quad \Rightarrow \quad \dot{\omega}^{\langle a \rangle} = 0 \quad \Rightarrow \quad \omega^a = 0 \text{ at later times}$$

follows from the vorticity conservation equation (30) (and is true also in the special case $p = 0$). When the vorticity vanishes:

1. The fluid flow is hypersurface-orthogonal, and there exists a cosmic time function t such that $u_a = -g(x^b)\, \nabla_a t$; if additionally the acceleration vanishes, we can set $g = 1$;

2. The metric of the orthogonal 3-spaces is h_{ab},

3. From the Gauss equation and the Ricci identities for u^a, the Ricci tensor of these 3-spaces is given by [5, 6]

$$\begin{aligned} {}^3R_{ab} = & -\dot{\sigma}_{\langle ab \rangle} - \Theta\, \sigma_{ab} + \tilde{\nabla}_{\langle a} \dot{u}_{b \rangle} + \dot{u}_{\langle a} \dot{u}_{b \rangle} + \pi_{ab} \\ & + \tfrac{1}{3} h_{ab} \left[2\mu - \tfrac{2}{3}\Theta^2 + 2\sigma^2 + 2\Lambda \right] , \end{aligned} \tag{53}$$

which relates ${}^3R_{ab}$ to E_{ab} via (31), and their Ricci scalar is given by

$${}^3R = 2\mu - \tfrac{2}{3}\Theta^2 + 2\sigma^2 + 2\Lambda , \tag{54}$$

which is a generalised Friedmann equation, showing how the matter tensor determines the 3-space average curvature. These equations fully determine the curvature tensor ${}^3R_{abcd}$ of the orthogonal 3-spaces [6].

2.5. IMPLICATIONS

Altogether, in general we have six propagation equations and six constraint equations; considered as a set of evolution equations for the $1+3$ covariant variables, they are a first-order system of equations. This set is determinate once the fluid equations of state are given; together they then form a complete set of equations (the system closes up, but is essentially infinite dimensional because of the spatial derivatives that occur). The total set is normal hyperbolic at least in the case of a perfect fluid, although this is not obvious from the above form; it is shown by completing the equations to tetrad form (see the next section) and then taking combinations of the equations to give a symmetric hyperbolic normal form (see [25, 26]). We can determine many of the properties of specific solutions direct from these equations, once the nature of these solutions has been prescribed in $1+3$ covariant form (see for example the FLRW and Bianchi Type I cases considered below).

The *key issue* that arises is consistency of the constraints with the evolution equations. It is believed that they are *generally consistent* for physically reasonable and well-defined equations of state, i.e. they are consistent if no restrictions are placed on their evolution other than implied by the constraint equations and the equations of state (this has not been proved in general, but is very plausible; however, it has been shown for irrotational dust [11, 27]). It is this that makes consistent the overall hyperbolic nature of the equations with the 'instantaneous' action at a distance implicit in the Gauss-like equations (specifically, the (div E)-equation), the point being that the 'action at a distance' nature of the solutions to these equations is built into the initial data, which ensures that the constraints are satisfied initially, and are conserved thereafter because the time evolution preserves these constraints (cf. [28]). A particular aspect of this is that when $\omega^a = 0$, the generalised Friedmann-equation (54) is an integral of the Raychaudhuri equation (29) and energy equation (37).

One must be very cautious with imposing simplifying assumptions (e.g. vanishing shear) in order to obtain solutions: this can lead to major restrictions on the possible flows, and one can be badly misled if their consistency is not investigated carefully [29, 24]. Cases of particular interest are shear-free fluid motion (see [30]–[32]) and various restrictions on the Weyl tensor, including the 'silent universes', characterised by $H_{ab} = 0$ (and $p = 0$) [33, 34], or models with $\tilde{\nabla}_b H^{ab} = 0$ [35].

2.5.1. *Energy equation*

It is worth commenting here that, because of the equivalence principle, there is no agreed energy conservation equation for the gravitational field itself, nor is there a definition of its entropy (indeed some people — Freeman Dyson, for example [36] — claim it has no entropy). Thus the above set of equations does not contain expressions for gravitational energy[7] or entropy, and the concept of energy conservation does not play the major role for gravitation that it does in the rest of physics, neither is there any agreed view on the growth of entropy of the gravitational field.[8] However, energy conservation of the matter content of space-time, expressed by the divergence equation $\nabla_b T^{ab} = 0$, is of course of major importance.

If we assume a perfect fluid with a (linear) γ-law equation of state, then (37) shows that

$$p = (\gamma - 1)\,\mu \;,\; \dot{\gamma} = 0 \quad \Rightarrow \quad \mu = M/S^{3\gamma} \;,\; \dot{M} = 0 \;. \tag{55}$$

One can approximate ordinary fluids in this way with $1 \le \gamma \le 2$ in order that the causality and energy conditions are valid, with 'dust' and Cold Dark Matter ('CDM') corresponding to $\gamma = 1 \Rightarrow \mu = M/S^3$, and radiation to $\gamma = \frac{4}{3} \Rightarrow \mu = M/S^4$.

Exercise: Show how to generalise this to more realistic equations of state, taking account of entropy and of matter pressure (see e.g. [5]–[7]).

In the case of a mixture of non-interacting matter, radiation and CDM having the *same* 4-velocity, represented as a single perfect fluid, the total energy density is simply the sum of these components: $\mu = \mu_{\rm dust} + \mu_{\rm CDM} + \mu_{\rm radn}$. (NB: This is only possible in universes with spatially homogeneous radiation energy density, because the matter will move on geodesics which by the momentum conservation equation implies $\tilde{\nabla}_a p_{\rm radn} = 0 \Leftrightarrow \tilde{\nabla}_a \mu_{\rm radn} = 0$. This will not be true for a general inhomogeneous or perturbed FLRW model, but will be true in exact FLRW and orthogonal Bianchi models.)

Exercise: The pressure can still be related to the energy density by a γ-law as in (55) in this case of non-interacting matter and radiation, but

[7] There are some proposed 'super-energy' tensors, for example the Bel–Robinson tensor [37], but they do not play a significant role in the theory.

[8] Entropy is well understood in the case of black holes, but not for gravitational fields in the expanding universe.

γ will no longer be constant. What is the equation giving the variation of (i) γ, (ii) the speed of sound, with respect to the scale factor in this case? (See [38].)

A scalar field has a perfect fluid energy-momentum tensor if the surfaces $\{\phi = \text{const}\}$ are spacelike and we choose u^a normal to these surfaces. Then it approximates the equation of state (55) in the 'slow-rolling' regime, with $\gamma \approx 0$, and in the velocity-dominated regime, with $\gamma \approx 2$. In the former case the energy conditions are no longer valid, so 'inflationary' behaviour is possible, which changes the nature of the attractors in the space of space-times in an important way.

Exercise: Derive expressions for μ, p, $(\mu+p)$, $(\mu+3p)$ in this case. Under what conditions can a scalar field have (a) $(\mu+p)=0$, (b) $(\mu+3p)=0$, (c) $(\mu+3p)<0$?

2.5.2. *Basic singularity theorem*

Using the definition (25) of S, the Raychaudhuri equation can be rewritten in the form (cf. [21])

$$3\,\frac{\ddot{S}}{S} = -2\,(\sigma^2-\omega^2) + \tilde{\nabla}_a \dot{u}^a + (\dot{u}_a \dot{u}^a) - \tfrac{1}{2}\,(\mu+3p) + \Lambda\,, \tag{56}$$

showing how the curvature of the curve $S(\tau)$ along each worldline (in terms of proper time τ along that worldline) is determined by the kinematical quantities, the total energy density and pressure[9] in the combination $(\mu + 3p)$, and the cosmological constant Λ. This gives the basic

Singularity Theorem [21, 5, 6, 7]: In a universe where $(\mu+3p)>0$, $\Lambda \leq 0$, and $\dot{u}^a = \omega^a = 0$ at all times, at any instant when $H_0 = \frac{1}{3}\,\Theta_0 > 0$, there must have been a time $t_0 < 1/H_0$ ago such that $S \to 0$ as $t \to t_0$; a space-time singularity occurs there, where $\mu \to \infty$ and $t \to \infty$ for ordinary matter (with $(\mu+p)>0$).

The further singularity theorems of Hawking and Penrose (see [39, 3, 40]) utilize this result (and its null version) as an essential part of their proofs.

Closely related to this are two other results: the statements that (a) a static universe model containing ordinary matter requires $\Lambda > 0$ (Einstein's discovery of 1917), and (b) the Einstein static universe is unstable (Eddington's discovery of 1930). Proofs are left to the reader; they follow directly from (56).

[9]This form of the equation is valid for imperfect fluids also: the quantities q^a and π_{ab} do not directly enter this equation.

2.5.3. *Relations between important parameters*
Given the definitions

$$H_0 = \frac{\dot{S}_0}{S_0}\,, \quad q_0 = -\frac{1}{H_0^2}\frac{\ddot{S}_0}{S_0}\,, \quad \Omega_0 = \frac{\mu_0}{3H_0^2}\,, \quad w_0 = \frac{p_0}{\mu_0}\,, \quad \Omega_\Lambda = \frac{\Lambda}{3H_0^2}\,, \tag{57}$$

for the present-day values of the Hubble parameter ('constant'), deceleration parameter, density parameter, pressure to density ratio, and cosmological constant parameter, respectively, then from (56) we obtain

$$q_0 = 2\frac{(\sigma^2 - \omega^2)_0}{H_0^2} - \frac{(\tilde{\nabla}_a \dot{u}^a)_0 + (\dot{u}_a \dot{u}^a)_0}{3H_0^2} + \tfrac{1}{2}\,\Omega_0\,(1 + 3w_0) - \Omega_\Lambda\,. \tag{58}$$

Now CBR anisotropies let us deduce that the first two terms on the right are very small today, certainly less than 10^{-3}, and we can reasonably estimate from the nature of the matter that $p_0 \ll \mu_0$ and the third term on the right is also very small, so we estimate that in realistic universe models, at the present time

$$q_0 \approx \tfrac{1}{2}\,\Omega_0 - \Omega_\Lambda\,. \tag{59}$$

(Note we can estimate the magnitudes of the terms which have been neglected in this approximation.) This shows that a cosmological constant can cause an acceleration (negative q_0); if it vanishes, as commonly assumed, the expression simplifies:

$$\Lambda = 0 \;\Rightarrow\; q_0 \approx \tfrac{1}{2}\,\Omega_0\,, \tag{60}$$

expressing how the matter density present causes a deceleration of the universe. If we assume no rotation ($\omega^a = 0$), then from (54) we can estimate

$${}^3R_0 \approx 6\,H_0^2\,(\,\Omega_0 - 1 + \Omega_\Lambda\,)\,, \tag{61}$$

where we have dropped a term $(\sigma_0/H_0)^2$. If $\Lambda = 0$, then ${}^3R_0 \approx 6\,H_0^2\,(\,\Omega_0 - 1\,)$, showing that $\Omega_0 = 1$ is the critical value separating irrotational universes with positive spatial curvature ($\Omega_0 > 1 \Rightarrow {}^3R_0 > 0$) from those with negative spatial curvature ($\Omega_0 < 1 \Rightarrow {}^3R_0 < 0$).

Present day values of these parameters are almost certainly in the ranges [41]: baryon density: $0.01 \leq \Omega_0^{\text{baryons}} \leq 0.03$, total matter density: $0.1 \leq \Omega_0 \leq 0.3$ to 1 (implying that much matter may not be baryonic), Hubble constant: $45\,\text{km/sec/Mpc} \leq H_0 \leq 80\,\text{km/sec/Mpc}$, deceleration parameter: $-\,0.5 \leq q_0 \leq 0.5$, cosmological constant: $0 \leq \Omega_\Lambda \leq 1$.

2.6. NEWTONIAN CASE

Newtonian equations can be developed completely in parallel [42, 43, 6] and are very similar, but simpler; for example the Newtonian version of the Raychaudhuri equation is

$$\dot{\Theta} + \tfrac{1}{3}\Theta^2 + 2(\sigma^2 - \omega^2) - \mathrm{D}_\alpha a^\alpha + \tfrac{1}{2}\rho - \Lambda = 0\,, \tag{62}$$

where ρ is the matter density and $a_\alpha = \dot{v}_\alpha + \mathrm{D}_\alpha \Phi$ is the Newtonian analogue of the relativistic 'acceleration vector', with $\dot{}$ the convective derivative and Φ the Newtonian potential (with suitably generalised boundary conditions [44, 45]). The Newtonian analogue of E_{ab} is

$$E_{\alpha\beta} = \mathrm{D}_\alpha \mathrm{D}_\beta \Phi - \tfrac{1}{3}(\mathrm{D}_\gamma \mathrm{D}^\gamma \Phi)\, h_{\alpha\beta}\,, \tag{63}$$

where $h_{\alpha\beta}$ denotes the metric, and D_α the covariant derivative, of Euclidean space. For the latter $\mathrm{D}_\alpha h_{\beta\gamma} = 0$ and $[\mathrm{D}_\alpha, \mathrm{D}_\beta] = 0$. There is no analogue of H_{ab} in Newtonian theory [6], as shown by a strict limit process leading from relativistic to Newtonian solutions [46].

Exercise: Under what conditions will a relativistic cosmological solution allow a representation (63) for the electric part of the Weyl tensor? Will the potential Φ occurring here necessarily also relate to the acceleration of the reference timelike worldlines?

2.7. SOLUTIONS

Useful solutions are defined by considering appropriate restrictions on the kinematical quantities, Weyl tensor, or space-time geometry, for a specified plausible matter content. Given such restrictions,

(a) we need to understand the *dynamical evolution* that results, particularly fixed points, attractors, etc., in terms of suitable variables,

(b) we particularly seek to determine and characterise *involutive subsets* of the space of space-times: these are subspaces mapped into themselves by the dynamical evolution of the system, and so are left invariant by that evolution. The constraint and evolution equations must be consistent with each other on such subsets. A characterisation of these subspaces goes a long way to characterising the nature of self-consistent solutions of the full non-linear EFE.

As far as possible we aim to do this for the exact equations. We are also concerned with

(c) *linearisation* of the equations about known simple solutions, and determination of properties of the resulting linearised solutions, in particular considering whether they accurately represent the behaviour of the full non-linear theory in a neighbourhood of the background solution (the issue of *linearisation stability*),

(d) derivation of the *Newtonian limit* and its properties from the General Relativity equations, and understanding how accurately this represents the properties of the full relativistic equations (and of its linearised solutions); see [24] for a discussion.

3. Tetrad description

The $1+3$ covariant equations are immediately transparent in terms of representing relations between $1+3$ covariantly defined quantities with clear geometrical and/or physical significance. However, they do *not* form a complete set of equations guaranteeing the existence of a corresponding metric and connection. For that we need to use a tetrad description. The equations determined will then form a complete set, which will contain as a subset all the $1+3$ covariant equations just derived (albeit presented in a slightly different form). For completeness we will give these equations for a general dissipative relativistic fluid (recent presentations, giving the following form of the equations, are [47, 26]). First we briefly summarize a generic tetrad formalism, and then its application to cosmological models (cf. [30, 48]).

3.1. GENERAL TETRAD FORMALISM

A tetrad is a set of four orthogonal unit basis vector fields $\{\mathbf{e}_a\}$, $a = 0, 1, 2, 3$, which can be written in terms of a local coordinate basis by means of the *tetrad components* $e_a{}^i(x^j)$:

$$\mathbf{e}_a = e_a{}^i(x^j)\,\frac{\partial}{\partial x^i} \quad \Leftrightarrow \quad \mathbf{e}_a(f) = e_a{}^i(x^j)\,\frac{\partial f}{\partial x^i}\,, \quad e_a{}^i = \mathbf{e}_a(x^i)\,, \tag{64}$$

(the latter stating that the i-th component of the a-th tetrad vector is just the directional derivative of the i-th coordinate in the direction $\mathbf{e}_a$). This can be thought of as just a general change of vector basis, leading to a change of tensor components of the standard tensorial form: $T^{ab}{}_{cd} = e^a{}_i\, e^b{}_j\, e_c{}^k\, e_d{}^l\, T^{ij}{}_{kl}$ with obvious inverse, where the inverse components $e^a{}_i(x^j)$ (note the placing of the indices!) are defined by

$$e_a{}^i\, e^a{}_j = \delta^i{}_j \quad \Leftrightarrow \quad e_a{}^i\, e^b{}_i = \delta^b{}_a\,. \tag{65}$$

However, it is a change from an integrable basis to a non-integrable one, so non-tensorial relations (specifically: the form of the metric and connection components) are a bit different than when coordinate bases are used. A change of one tetrad basis to another will also lead to transformations of the standard tensor form for all tensorial quantities: if $\mathbf{e}_a =$

$\Lambda_{a}{}^{a}\,(x^i)\,\mathbf{e}_a$ is a change of tetrad basis with inverse $\mathbf{e}_a = \Lambda_a{}^{a}(x^i)\,\mathbf{e}_a$ (each of these matrices representing a Lorentz transformation), then $T^{ab}{}_{cd} = \Lambda_a{}^{a}\,\Lambda_b{}^{b}\,\Lambda_c{}^{c}\,\Lambda_d{}^{d}\,T^{a\,b}{}_{c\,d}$. Again the inverse is obvious.[10]

The metric tensor components in the tetrad form are given by

$$g_{ab} = g_{ij}\, e_a{}^i\, e_b{}^j = \mathbf{e}_a \cdot \mathbf{e}_b = \eta_{ab} \ , \tag{66}$$

where $\eta_{ab} = \mathrm{diag}\,(-1, +1, +1, +1)$, showing that the basis vectors are unit vectors orthogonal to each other (because the components g_{ab} are just the scalar products of these vectors with each other). The inverse equation

$$g_{ij}(x^k) = \eta_{ab}\, e^a{}_i(x^k)\, e^b{}_j(x^k) \tag{67}$$

explicitly constructs the coordinate components of the metric from the (inverse) tetrad components $e^a{}_i(x^j)$. We can raise and lower tetrad indices by use of the metric $g_{ab} = \eta_{ab}$ and its inverse $g^{ab} = \eta^{ab}$.

The *commutation functions* related to the tetrad are the quantities $\gamma^a{}_{bc}(x^i)$ defined by the *commutators* of the basis vectors:[11]

$$[\,\mathbf{e}_a,\, \mathbf{e}_b\,] = \gamma^c{}_{ab}(x^i)\, \mathbf{e}_c \quad \Rightarrow \quad \gamma^a{}_{bc}(x^i) = -\,\gamma^a{}_{cb}(x^i) \ . \tag{68}$$

It follows (apply this relation to the coordinate x^i) that in terms of the tetrad components,

$$\gamma^a{}_{bc}(x^i) = e^a{}_i\,(\,e_b{}^j\, \partial_j e_c{}^i - e_c{}^j\, \partial_j e_b{}^i\,) = -\,2\, e_b{}^i\, e_c{}^j\, \nabla_{[i} e^a{}_{j]} \ . \tag{69}$$

These quantities vanish iff the basis $\{\,\mathbf{e}_a\,\}$ is a coordinate basis: that is, there exist coordinates x^i such that $\mathbf{e}_a = \delta_a{}^i\, \partial/\partial x^i$, iff $[\,\mathbf{e}_a,\, \mathbf{e}_b\,] = 0 \Leftrightarrow \gamma^a{}_{bc} = 0$.

The *connection components* $\Gamma^a{}_{bc}$ for the tetrad ('Ricci rotation coefficients') are defined by the relations

$$\nabla_{\mathbf{e}_b} \mathbf{e}_a = \Gamma^c{}_{ab}\, \mathbf{e}_c \quad \Leftrightarrow \quad \Gamma^c{}_{ab} = e^c{}_i\, e_b{}^j\, \nabla_j e_a{}^i \ , \tag{70}$$

i.e. it is the c-component of the covariant derivative in the b-direction of the a-vector. It follows that all covariant derivatives can be written out in tetrad components in a way completely analogous to the usual tensor form,

[10] The tetrad components of any quantity are invariant when the coordinate basis is changed (for a fixed tetrad), and coordinate components are invariant when a change of tetrad basis is made (for a fixed set of coordinates); however, either change will alter the tetrad components relative to the given coordinates.

[11] Remember that the commutator of any two vectors X, Y is $[X, Y] = XY - YX$.

for example $\nabla_a T_{bc} = \mathbf{e}_a(T_{bc}) - \Gamma^d{}_{ba}\, T_{dc} - \Gamma^d{}_{ca}\, T_{bd}$, where for any function f, $\mathbf{e}_a(f) = e_a{}^i\, \partial f/\partial x^i$ is the derivative of f in the direction $\mathbf{e}_a$. In particular, because $\mathbf{e}_a(g_{bc}) = 0$ for $g_{ab} = \eta_{ab}$, applying this to the metric gives

$$\nabla_a g_{bc} = 0 \quad \Leftrightarrow \quad -\Gamma^d{}_{ba}\, g_{dc} - \Gamma^d{}_{ca}\, g_{bd} = 0 \quad \Leftrightarrow \quad \Gamma_{(ab)c} = 0\ , \qquad (71)$$

— the rotation coefficients are skew in their first two indices, when we raise and lower the first indices only. We obtain from this and the assumption of vanishing torsion the tetrad relations that are the analogue of the usual Christoffel relations:

$$\gamma^a{}_{bc} = -\left(\Gamma^a{}_{bc} - \Gamma^a{}_{cb}\right)\ , \qquad \Gamma_{abc} = \tfrac{1}{2}\left(g_{ad}\,\gamma^d{}_{cb} - g_{bd}\,\gamma^d{}_{ca} + g_{cd}\,\gamma^d{}_{ab}\right)\ . \quad (72)$$

This shows that the rotation coefficients and the commutation functions are each just linear combinations of the other.

Any set of vectors whatever must satisfy the *Jacobi identities*:

$$[X,[Y,Z]] + [Y,[Z,X]] + [Z,[X,Y]] = 0\ ,$$

which follows from the definition of a commutator. Applying this to the basis vectors $\mathbf{e}_a$, $\mathbf{e}_b$ and $\mathbf{e}_c$ gives the identities

$$\mathbf{e}_{[a}(\gamma^d{}_{bc]}) + \gamma^e{}_{[ab}\,\gamma^d{}_{c]e} = 0\ , \qquad (73)$$

which are the integrability conditions that the $\gamma^a{}_{bc}(x^i)$ are the commutation functions for the set of vectors $\mathbf{e}_a$.

If we apply the Ricci identities to the tetrad basis vectors $\mathbf{e}_a$, we obtain the Riemann curvature tensor components in the form

$$R^a{}_{bcd} = \mathbf{e}_c(\Gamma^a{}_{bd}) - \mathbf{e}_d(\Gamma^a{}_{bc}) + \Gamma^a{}_{ec}\,\Gamma^e{}_{bd} - \Gamma^a{}_{ed}\,\Gamma^e{}_{bc} - \Gamma^a{}_{be}\,\gamma^e{}_{cd}\ . \qquad (74)$$

Contracting this on a and c, one obtains the EFE (for $\Lambda = 0$) in the form

$$R_{bd} = \mathbf{e}_a(\Gamma^a{}_{bd}) - \mathbf{e}_d(\Gamma^a{}_{ba}) + \Gamma^a{}_{ea}\,\Gamma^e{}_{bd} - \Gamma^a{}_{de}\,\Gamma^e{}_{ba} = T_{bd} - \tfrac{1}{2}\,T\,g_{bd}\ . \quad (75)$$

It is not immediately obvious that this is symmetric, but this follows because (73) implies $R_{a[bcd]} = 0 \Rightarrow R_{ab} = R_{(ab)}$.

3.2. TETRAD FORMALISM IN COSMOLOGY

For a cosmological model we choose $\mathbf{e}_0$ to be the unit tangent of the matter flow, u^a. This fixing implies that the initial six-parameter freedom of using Lorentz transformations has been reduced to a three-parameter freedom of rotations of the spatial frame $\{\mathbf{e}_\alpha\}$. The 24 algebraically independent frame components of the space-time connection $\Gamma^a{}_{bc}$ can then be split into the set (see [30, 49, 26])

$$\Gamma_{\alpha 00} = \dot{u}_\alpha \tag{76}$$
$$\Gamma_{\alpha 0\beta} = \tfrac{1}{3}\Theta\,\delta_{\alpha\beta} + \sigma_{\alpha\beta} - \epsilon_{\alpha\beta\gamma}\,\omega^\gamma \tag{77}$$
$$\Gamma_{\alpha\beta 0} = \epsilon_{\alpha\beta\gamma}\,\Omega^\gamma \tag{78}$$
$$\Gamma_{\alpha\beta\gamma} = 2\,a_{[\alpha}\,\delta_{\beta]\gamma} + \epsilon_{\gamma\delta[\alpha}\,n^\delta{}_{\beta]} + \tfrac{1}{2}\,\epsilon_{\alpha\beta\delta}\,n^\delta{}_\gamma\,. \tag{79}$$

The first two sets contain the kinematical variables. In the third is the rate of rotation Ω^α of the spatial frame $\{\mathbf{e}_\alpha\}$ with respect to a *Fermi-propagated* basis. Finally, the quantities a^α and $n_{\alpha\beta} = n_{(\alpha\beta)}$ determine the 9 spatial rotation coefficients. The commutator equations (68) applied to any space-time scalar f take the form

$$[\,\mathbf{e}_0, \mathbf{e}_\alpha\,](f) = \dot{u}_\alpha\,\mathbf{e}_0(f) - [\,\tfrac{1}{3}\Theta\,\delta_\alpha{}^\beta + \sigma_\alpha{}^\beta + \epsilon_\alpha{}^\beta{}_\gamma\,(\omega^\gamma + \Omega^\gamma)\,]\,\mathbf{e}_\beta(f) \tag{80}$$

$$[\,\mathbf{e}_\alpha, \mathbf{e}_\beta\,](f) = 2\,\epsilon_{\alpha\beta\gamma}\,\omega^\gamma\,\mathbf{e}_0(f) + [\,2\,a_{[\alpha}\,\delta^\gamma{}_{\beta]} + \epsilon_{\alpha\beta\delta}\,n^{\delta\gamma}\,]\,\mathbf{e}_\gamma(f)\,; \tag{81}$$

The full set of equations for a gravitating fluid can be written as a set of constraints and a set of evolution equations, which include the tetrad form of the $1+3$ covariant equations given above, but complete them by giving all Ricci and Jacobi identities for the basis vectors. We now give these equations.

3.2.1. *Constraints*

The following set of relations does not contain any frame derivatives with respect to $\mathbf{e}_0$. Hence, we refer to these relations as 'constraints'. From the Ricci identities for u^a and the Jacobi identities we have the (0α)-equation $(C_1)^\alpha$, which, in Hamiltonian treatments of the EFE, is also referred to as the 'momentum constraint', the vorticity divergence identity (C_2) and the H_{ab}-equation $(C_3)^{\alpha\beta}$, respectively; the once-contracted Bianchi identities yield the $(\operatorname{div} E)$- and $(\operatorname{div} H)$-equations $(C_4)^\alpha$ and $(C_5)^\alpha$ [6, 47]; the constraint $(C_J)^\alpha$ again arises from the Jacobi identities while, finally, $(C_G)^{\alpha\beta}$ and (C_G) stem from the EFE. In detail,

$$\begin{aligned} 0 = (C_1)^\alpha = {} & (\mathbf{e}_\beta - 3\,a_\beta)\,(\sigma^{\alpha\beta}) - \tfrac{2}{3}\,\delta^{\alpha\beta}\,\mathbf{e}_\beta(\Theta) - n^\alpha{}_\beta\,\omega^\beta + q^\alpha \\ & + \epsilon^{\alpha\beta\gamma}\,[\,(\mathbf{e}_\beta + 2\,\dot{u}_\beta - a_\beta)\,(\omega_\gamma) - n_{\beta\delta}\,\sigma^\delta{}_\gamma\,] \end{aligned} \tag{82}$$

$$0 = (C_2) = (\mathbf{e}_\alpha - \dot{u}_\alpha - 2\,a_\alpha)\,(\omega^\alpha) \tag{83}$$

$$0 = (C_3)^{\alpha\beta} = H^{\alpha\beta} + (\delta^{\gamma\langle\alpha}\,\mathbf{e}_\gamma + 2\,\dot{u}^{\langle\alpha} + a^{\langle\alpha\rangle})\,(\omega^{\beta\rangle}) - \tfrac{1}{2}\,n^\gamma{}_\gamma\,\sigma^{\alpha\beta} \quad (84)$$
$$+ 3\,n^{\langle\alpha}{}_\gamma\,\sigma^{\beta\rangle\gamma} - \epsilon^{\gamma\delta\langle\alpha}\,[\,(\mathbf{e}_\gamma - a_\gamma)\,(\sigma^{\beta\rangle}{}_\delta) + n^{\beta\rangle}{}_\gamma\,\omega_\delta\,]$$

$$0 = (C_4)^\alpha = (\mathbf{e}_\beta - 3\,a_\beta)\,(E^{\alpha\beta} + \tfrac{1}{2}\,\pi^{\alpha\beta}) - \tfrac{1}{3}\,\delta^{\alpha\beta}\,\mathbf{e}_\beta(\mu) \quad (85)$$
$$+ \tfrac{1}{3}\,\Theta\,q^\alpha - \tfrac{1}{2}\,\sigma^\alpha{}_\beta\,q^\beta - 3\,\omega_\beta\,H^{\alpha\beta}$$
$$- \epsilon^{\alpha\beta\gamma}\,[\,\sigma_{\beta\delta}\,H^\delta{}_\gamma - \tfrac{3}{2}\,\omega_\beta\,q_\gamma + n_{\beta\delta}\,(E^\delta{}_\gamma + \tfrac{1}{2}\,\pi^\delta{}_\gamma)\,]$$

$$0 = (C_5)^\alpha = (\mathbf{e}_\beta - 3\,a_\beta)\,(H^{\alpha\beta}) + (\mu + p)\,\omega^\alpha \quad (86)$$
$$+ 3\,\omega_\beta\,(E^{\alpha\beta} - \tfrac{1}{6}\,\pi^{\alpha\beta}) - \tfrac{1}{2}\,n^\alpha{}_\beta\,q^\beta$$
$$+ \epsilon^{\alpha\beta\gamma}\,[\,\tfrac{1}{2}\,(\mathbf{e}_\beta - a_\beta)\,(q_\gamma) + \sigma_{\beta\delta}\,(E^\delta{}_\gamma + \tfrac{1}{2}\,\pi^\delta{}_\gamma)$$
$$- n_{\beta\delta}\,H^\delta{}_\gamma\,]$$

$$0 = (C_J)^\alpha = (\mathbf{e}_\beta - 2\,a_\beta)\,(n^{\alpha\beta}) + \tfrac{2}{3}\,\Theta\,\omega^\alpha + 2\,\sigma^\alpha{}_\beta\,\omega^\beta \quad (87)$$
$$+ \epsilon^{\alpha\beta\gamma}\,[\,\mathbf{e}_\beta(a_\gamma) - 2\,\omega_\beta\,\Omega_\gamma\,]$$

$$0 = (C_G)^{\alpha\beta} = {}^*S^{\alpha\beta} + \tfrac{1}{3}\,\Theta\,\sigma^{\alpha\beta} - \sigma^{\langle\alpha}{}_\gamma\,\sigma^{\beta\rangle\gamma} - \omega^{\langle\alpha}\,\omega^{\beta\rangle} \quad (88)$$
$$+ 2\,\omega^{\langle\alpha}\,\Omega^{\beta\rangle} - (E^{\alpha\beta} + \tfrac{1}{2}\,\pi^{\alpha\beta})$$

$$0 = (C_G) = {}^*R + \tfrac{2}{3}\,\Theta^2 - (\sigma^\alpha{}_\beta\sigma^\beta{}_\alpha) + 2\,(\omega_\alpha\omega^\alpha) \quad (89)$$
$$- 4\,(\omega_\alpha\Omega^\alpha) - 2\,\mu - 2\,\Lambda\,,$$

where

$$^*S_{\alpha\beta} = \mathbf{e}_{\langle\alpha}(a_{\beta\rangle}) + b_{\langle\alpha\beta\rangle} - \epsilon^{\gamma\delta}{}_{\langle\alpha}\,(\mathbf{e}_{|\gamma|} - 2\,a_{|\gamma|})\,(n_{\beta\rangle\delta}) \quad (90)$$

$$^*R = 2\,(2\,\mathbf{e}_\alpha - 3\,a_\alpha)\,(a^\alpha) - \tfrac{1}{2}\,b^\alpha{}_\alpha \quad (91)$$

$$b_{\alpha\beta} = 2\,n_{\alpha\gamma}\,n^\gamma{}_\beta - n^\gamma{}_\gamma\,n_{\alpha\beta}\,. \quad (92)$$

If $\omega^\alpha = 0$, so that u^a become the normals to a family of 3-spaces of constant time, the last two constraints in the set correspond to the symmetric trace-free and trace parts of the Gauss equation (53). In this case, one also speaks of (C_G) as the generalised Friedmann equation, alias the 'Hamiltonian constraint' or the 'energy constraint'.

3.2.2. *Evolution of spatial commutation functions*
The 9 spatial commutation functions a^α and $n_{\alpha\beta}$ are generally evolved by equations (40) and (41) given in [47]; these originate from the Jacobi identities. Employing each of the constraints $(C_1)^\alpha$ to $(C_3)^{\alpha\beta}$ listed in the previous paragraph, we can eliminate $\mathbf{e}_\alpha$ frame derivatives of the kinematical variables Θ, $\sigma_{\alpha\beta}$ and ω^α from their right-hand sides. Thus, we obtain the following equations for the evolution of the spatial connection components:

$$\begin{aligned} \mathbf{e}_0(a^\alpha) &= -\tfrac{1}{3}\,(\Theta\,\delta^\alpha{}_\beta - \tfrac{3}{2}\,\sigma^\alpha{}_\beta)\,(\dot{u}^\beta + a^\beta) + \tfrac{1}{2}\,n^\alpha{}_\beta\,\omega^\beta - \tfrac{1}{2}\,q^\alpha \\ &\quad - \tfrac{1}{2}\,\epsilon^{\alpha\beta\gamma}\,[\,(\dot{u}_\beta + a_\beta)\,\omega_\gamma - n_{\beta\delta}\,\sigma^\delta{}_\gamma - (\mathbf{e}_\beta + \dot{u}_\beta - 2\,a_\beta)\,(\Omega_\gamma)\,] \\ &\quad + \tfrac{1}{2}\,(C_1)^\alpha \end{aligned} \qquad (93)$$

$$\begin{aligned} \mathbf{e}_0(n^{\alpha\beta}) &= -\tfrac{1}{3}\,\Theta\,n^{\alpha\beta} - \sigma^{\langle\alpha}{}_\gamma\,n^{\beta\rangle\gamma} + \tfrac{1}{2}\,\sigma^{\alpha\beta}\,n^\gamma{}_\gamma - (\dot{u}^{\langle\alpha} + a^{\langle\alpha})\,\omega^{\beta\rangle} - H^{\alpha\beta} \\ &\quad + (\delta^{\gamma\langle\alpha}\,\mathbf{e}_\gamma + \dot{u}^{\langle\alpha})\,(\Omega^{\beta\rangle}) \\ &\quad - \tfrac{2}{3}\,\delta^{\alpha\beta}\,[\,2\,(\dot{u}_\gamma + a_\gamma)\,\omega^\gamma - \sigma^\gamma{}_\delta\,n^\delta{}_\gamma + (\mathbf{e}_\gamma + \dot{u}_\gamma)\,(\Omega^\gamma)\,] \\ &\quad - \epsilon^{\gamma\delta\langle\alpha}\,[\,(\dot{u}_\gamma + a_\gamma)\,\sigma^{\beta\rangle}{}_\delta - (\omega_\gamma + 2\,\Omega_\gamma)\,n^{\beta\rangle}{}_\delta\,] \\ &\quad - \tfrac{2}{3}\,\delta^{\alpha\beta}\,(C_2) + (C_3)^{\alpha\beta}\,. \end{aligned} \qquad (94)$$

3.2.3. *Evolution of kinematical variables*
The evolution equations for the 9 kinematical variables Θ, ω^α and $\sigma_{\alpha\beta}$ are provided by the Ricci identities for u^a, i.e.

$$\begin{aligned} \mathbf{e}_0(\Theta) - \mathbf{e}_\alpha(\dot{u}^\alpha) &= -\tfrac{1}{3}\,\Theta^2 + (\dot{u}_\alpha - 2\,a_\alpha)\,\dot{u}^\alpha - (\sigma^\alpha{}_\beta\sigma^\beta{}_\alpha) \\ &\quad + 2\,(\omega_\alpha\omega^\alpha) - \tfrac{1}{2}\,(\mu + 3p) + \Lambda \end{aligned} \qquad (95)$$

$$\begin{aligned} \mathbf{e}_0(\omega^\alpha) - \tfrac{1}{2}\,\epsilon^{\alpha\beta\gamma}\,\mathbf{e}_\beta(\dot{u}_\gamma) &= -\tfrac{2}{3}\,\Theta\,\omega^\alpha + \sigma^\alpha{}_\beta\,\omega^\beta - \tfrac{1}{2}\,n^\alpha{}_\beta\,\dot{u}^\beta \\ &\quad - \tfrac{1}{2}\,\epsilon^{\alpha\beta\gamma}\,[\,a_\beta\,\dot{u}_\gamma - 2\,\Omega_\beta\,\omega_\gamma\,] \end{aligned} \qquad (96)$$

$$\begin{aligned} \mathbf{e}_0(\sigma^{\alpha\beta}) - \delta^{\gamma\langle\alpha}\,\mathbf{e}_\gamma(\dot{u}^{\beta\rangle}) &= -\tfrac{2}{3}\,\Theta\,\sigma^{\alpha\beta} + (\dot{u}^{\langle\alpha} + a^{\langle\alpha})\,\dot{u}^{\beta\rangle} - \sigma^{\langle\alpha}{}_\gamma\,\sigma^{\beta\rangle\gamma} \\ &\quad - \omega^{\langle\alpha}\,\omega^{\beta\rangle} - (E^{\alpha\beta} - \tfrac{1}{2}\,\pi^{\alpha\beta}) \\ &\quad + \epsilon^{\gamma\delta\langle\alpha}\,[\,2\,\Omega_\gamma\,\sigma^{\beta\rangle}{}_\delta - n^{\beta\rangle}{}_\gamma\,\dot{u}_\delta\,]\,. \end{aligned} \qquad (97)$$

3.2.4. *Evolution of matter and Weyl curvature variables*
Finally, we have the equations for the 4 matter variables μ and q^α and the 10 Weyl curvature variables $E_{\alpha\beta}$ and $H_{\alpha\beta}$, which are obtained from the twice-contracted and once-contracted Bianchi identities, respectively:

$$\mathbf{e}_0(\mu) + \mathbf{e}_\alpha(q^\alpha)$$

$$= -\Theta\,(\mu + p) - 2\,(\dot{u}_\alpha - a_\alpha)\,q^\alpha - (\sigma^\alpha{}_\beta \pi^\beta{}_\alpha) \tag{98}$$

$$\begin{aligned}
&\mathbf{e}_0(q^\alpha) + \delta^{\alpha\beta}\,\mathbf{e}_\beta(p) + \mathbf{e}_\beta(\pi^{\alpha\beta}) \\
&\quad = -\tfrac{4}{3}\,\Theta\,q^\alpha - \sigma^\alpha{}_\beta\,q^\beta - (\mu + p)\,\dot{u}^\alpha - (\dot{u}_\beta - 3\,a_\beta)\,\pi^{\alpha\beta} \\
&\qquad - \epsilon^{\alpha\beta\gamma}\,[\,(\omega_\beta - \Omega_\beta)\,q_\gamma - n_{\beta\delta}\,\pi^\delta{}_\gamma\,]
\end{aligned} \tag{99}$$

$$\begin{aligned}
&\mathbf{e}_0(E^{\alpha\beta} + \tfrac{1}{2}\,\pi^{\alpha\beta}) - \epsilon^{\gamma\delta\langle\alpha}\,\mathbf{e}_\gamma(H^{\beta\rangle}{}_\delta) + \tfrac{1}{2}\,\delta^{\gamma\langle\alpha}\,\mathbf{e}_\gamma(q^{\beta\rangle}) \\
&\quad = -\tfrac{1}{2}\,(\mu + p)\,\sigma^{\alpha\beta} - \Theta\,(E^{\alpha\beta} + \tfrac{1}{6}\,\pi^{\alpha\beta}) + 3\,\sigma^{\langle\alpha}{}_\gamma\,(E^{\beta\rangle\gamma} - \tfrac{1}{6}\,\pi^{\beta\rangle\gamma}) \\
&\qquad + \tfrac{1}{2}\,n^\gamma{}_\gamma\,H^{\alpha\beta} - 3\,n^{\langle\alpha}{}_\gamma\,H^{\beta\rangle\gamma} - \tfrac{1}{2}\,(2\,\dot{u}^{\langle\alpha} + a^{\langle\alpha})\,q^{\beta\rangle} \\
&\qquad + \epsilon^{\gamma\delta\langle\alpha}\,[\,(2\,\dot{u}_\gamma - a_\gamma)\,H^{\beta\rangle}{}_\delta + (\omega_\gamma + 2\,\Omega_\gamma)\,(E^{\beta\rangle}{}_\delta + \tfrac{1}{2}\,\pi^{\beta\rangle}{}_\delta) \\
&\qquad\qquad + \tfrac{1}{2}\,n^{\beta\rangle}{}_\gamma\,q_\delta\,]
\end{aligned} \tag{100}$$

$$\begin{aligned}
&\mathbf{e}_0(H^{\alpha\beta}) + \epsilon^{\gamma\delta\langle\alpha}\,\mathbf{e}_\gamma(E^{\beta\rangle}{}_\delta - \tfrac{1}{2}\,\pi^{\beta\rangle}{}_\delta) \\
&\quad = -\Theta\,H^{\alpha\beta} + 3\,\sigma^{\langle\alpha}{}_\gamma\,H^{\beta\rangle\gamma} + \tfrac{3}{2}\,\omega^{\langle\alpha}\,q^{\beta\rangle} \\
&\qquad - \tfrac{1}{2}\,n^\gamma{}_\gamma\,(E^{\alpha\beta} - \tfrac{1}{2}\,\pi^{\alpha\beta}) + 3\,n^{\langle\alpha}{}_\gamma\,(E^{\beta\rangle\gamma} - \tfrac{1}{2}\,\pi^{\beta\rangle\gamma}) \\
&\qquad + \epsilon^{\gamma\delta\langle\alpha}\,[\,a_\gamma\,(E^{\beta\rangle}{}_\delta - \tfrac{1}{2}\,\pi^{\beta\rangle}{}_\delta) - 2\,\dot{u}_\gamma\,E^{\beta\rangle}{}_\delta + \tfrac{1}{2}\,\sigma^{\beta\rangle}{}_\gamma\,q_\delta \\
&\qquad\qquad + (\omega_\gamma + 2\,\Omega_\gamma)\,H^{\beta\rangle}{}_\delta\,]\,.
\end{aligned} \tag{101}$$

Exercise: (a) Show how most of these equations are the tetrad version of corresponding $1+3$ covariant equations. For which of the tetrad equations is this not true? (b) Explain why there are no equations for $\mathbf{e}_0(\Omega^\alpha)$ and $\mathbf{e}_0(\dot{u}^\alpha)$. [Hint: What freedom is there in choosing the tetrad?]

3.3. COMPLETE SET

For a prescribed set of matter equations of state, this gives the complete set of tetrad relations, which can be used to characterise particular families of solutions in detail. It clearly contains all the $1+3$ covariant equations above, plus others required to form a complete set. It can be recast into a symmetric normal hyperbolic form [26], showing the hyperbolic nature of the equations and determining their characteristics. Detailed studies of exact solutions will need a coordinate system and vector basis, and usually it will be advantageous to use tetrads for this purpose, because the tetrad vectors can be chosen in physically preferred directions (see [30, 50] for the use of tetrads to study locally rotationally symmetric space-times, and [49, 51] for Bianchi universes; these cases are both discussed below).

Finally it is important to note that when tetrad vectors are chosen uniquely in an invariant way (e.g. as eigenvectors of a non-degenerate shear tensor or of the electric Weyl tensor), then — because they are uniquely defined from 1+3 covariant quantities — all the rotation coefficients above are in fact covariantly defined scalars, so all these equations are invariant equations. The only times when it is not possible to define unique tetrads in this way is when the space-times are isotropic or locally rotationally symmetric, as discussed below.

4. FLRW universes and observational relations

A particularly important involutive subspace is that of the Friedmann–Lemaître ('FL') universes, based on the everywhere-isotropic Robertson–Walker ('RW') geometry. It is characterised by a perfect fluid matter tensor and the condition that *local isotropy* holds everywhere:

$$0 = \dot{u}^a = \sigma_{ab} = \omega^a \Leftrightarrow 0 = E_{ab} = H_{ab} \Rightarrow 0 = X_a = Z_a = \tilde{\nabla}_a p \; , \qquad (102)$$

the first conditions stating the kinematical quantities are locally isotropic, the second that these universes are conformally flat, and the third that they are spatially homogeneous.

Exercise: Show that the implications in this relation follow from the 1+3 covariant equations in the previous section when $p = p(\mu)$, thus showing that isotropy everywhere implies spatial homogeneity in this case.

4.1. COORDINATES AND METRIC

It follows then that (see [52]):

1. Comoving coordinates can be found[12] so that the metric takes the form

$$ds^2 = -\,dt^2 + S^2(t)\,(\,dr^2 + f^2(r)\,d\Omega^2\,)\;, \quad u^a = \delta^a{}_0\;, \qquad (103)$$

where $d\Omega^2 = d\theta^2 + \sin^2\theta\,d\phi^2$, $u_a = -\nabla_a t$, and $\dot{S}/S = \frac{1}{3}\Theta$, characterising $S(t)$ as the scale factor for distances between any pair of fundamental observers. The expansion of matter depends only on one scale length, so it is isotropic (there is no distortion or rotation).

2. The Ricci tensor ${}^3R_{ab}$ is isotropic, so the 3-spaces $\{t = \text{const}\}$ are 3-spaces of constant curvature k/S^2 where k can be normalised to ± 1, if it is non-zero. Using the geodesic deviation equation in these 3-spaces, one finds that (see [52, 53])

$$f(r) = \sin r\;,\; r\;,\; \sinh r \quad \text{if} \quad k = +1\;,\; 0\;,\; -1\;. \qquad (104)$$

[12]There are many other coordinate systems in use, for example with different definitions of the radial distance r.

Thus when $k = +1$ the surface area $4\pi S^2(t) f^2(r)$ of a geodesic 2-sphere in these spaces, centred on the (arbitrary) point $r = 0$, increases to a maximum at $r = \pi/2$ and then decreases to zero again at the antipodal point $r = \pi$; hence the point at $r = 2\pi$ has to be the same point as $r = 0$, and these 3-spaces are necessarily closed, with finite total volume. In the other cases the 3-spaces are usually unbounded and the surface areas of these 2-spaces increase without limit; however, unusual topologies still allow the spatial sections to be closed [54].

Exercise: Find the obvious orthonormal tetrad associated with these coordinates, and determine their commutators and Ricci rotation coefficients.

4.2. DYNAMICAL EQUATIONS

The remaining non-trivial equations are the energy equation (37), the Raychaudhuri equation (29), which now takes the form

$$3\,\frac{\ddot{S}}{S} + \tfrac{1}{2}\,(\mu + 3p) = 0\ , \tag{105}$$

and the Friedmann equation that follows from (54):

$${}^3R = 2\,\mu - \tfrac{2}{3}\,\Theta^2 = \frac{6\,k}{S^2}\ , \tag{106}$$

where k is a constant. Any two of these equations imply the third if $\dot{S} \neq 0$ (the latter equation being a first integral of the other two). All one has to do then to determine the dynamics is to solve the Friedmann equation. The solution depends on what form is assumed for the matter: Usually it is taken to be a perfect fluid with equation of state $p = p(\mu)$, or as a sum of such fluids, or as a scalar field with given potential $V(\phi)$. For the γ-law discussed above, the energy equation integrates to give (55), which can then be used to represent μ in the Friedmann equation.

Exercise: Show that on using the tetrad found above, all the other $1+3$ covariant and tetrad equations are identically true when these equations are satisfied.

4.2.1. *Basic parameters*

As well as the parameters H_0, Ω_0, Ω_Λ and q_0, the FLRW models are characterised by the spatial curvature parameter $K_0 = k/S_0^2 = {}^3R_0/6$. These parameters are related by the equations (59) and (61), which are now exact rather than approximate relations.

4.2.2. *Singularity and ages*

The existence of the big bang, and age limits on the universe, follow directly from the Raychaudhuri equation, together with the energy assumption $(\mu + 3p) > 0$ (true at least when quantum fields do not dominate), because the universe is expanding today ($\Theta_0 > 0$). That is, the singularity theorem above applies in particular to FLRW models. Furthermore, from the Raychaudhuri equation, in any FLRW model, the fundamental age relation holds (see e.g. [52]):

> **Age Theorem:** In an expanding FLRW universe with vanishing cosmological constant and satisfying the active gravitational mass density energy condition, ages are strictly constrained by the Hubble expansion rate: namely, at every instant, the age t_0 of the universe (the time since the big bang) is less than the inverse Hubble constant at that time:

$$(\mu + 3p) > 0 \ , \ \Lambda = 0 \quad \Rightarrow \quad t_0 < 1/H_0 \ . \tag{107}$$

More precise ages $t_0(H_0, \Omega_0)$ can be determined for any specific cosmological model from the Friedmann equation (106); in particular, in a matter-dominated early universe the same result will hold with a factor 2/3 on the right-hand side, while in a radiation dominated universe the factor will be 1/2. Note that this relation applies in the early universe when the expansion rate was much higher, and, hence, shows that the hot early epoch ended shortly after the initial singularity [52].

The age limits are one of the central issues in modern cosmology [55, 41]. Hipparchos satellite measurements suggest a lowering of the age estimates of globular clusters to about 1.2×10^9 years, together with a decrease in the estimate of the Hubble constant to about $H_0 \approx 50$ km/sec/Mpc. This corresponds to a Hubble time $1/H_0$ of about 1.8×10^9 years, implying there is no problem, but red giant and Cepheid measurements suggest $H_0 \approx 72 - 77$ km/sec/Mpc [56], implying the situation is very tight indeed. However, recent supernovae measurements [57] suggest a positive cosmological constant, allowing violation of the age constraint, and hence easing the situation. All these figures should still be treated with caution; the issue is fundamental to the viability of the FLRW models, and still needs resolution.

4.3. EXACT AND APPROXIMATE SOLUTIONS

If $\Lambda = 0$ and the energy conditions are satisfied, FLRW models expand forever from a big bang if $k = -1$ or $k = 0$, and recollapse in the future if $k = +1$. A positive value of Λ gives a much wider choice for behaviours [58, 59].

4.3.1. *Simplest models*

a) *Einstein static*: $S(t) = \text{const}$, $k = +1$, $\Lambda = \frac{1}{2}(\mu + 3p) > 0$, where everything is constant in space and time, and there is no redshift. This model is unstable (see above).

b) *de Sitter*: $S(t) = \exp(H\,t)$, $H = \text{const}$, $k = 0$, a steady state solution in a constant curvature space-time: it is empty, because $(\mu + p) = 0$, i.e. it does not contain ordinary matter, but rather a cosmological constant,[13] or a scalar field in the strict 'no-rolling' case. It has ambiguous redshift because the choice of families of worldlines and space sections is not unique in this case; see [60].

c) *Milne*: $S(t) = t$, $k = -1$. This is flat, empty space-time in expanding coordinates (again $(\mu + p) = 0$).

d) *Einstein–de Sitter*: the simplest non-empty expanding model, with

$$k = 0 = \Lambda : \quad S(t) = a\,t^{2/3} \;, \quad a = \text{const} \;\; \text{if} \;\; p = 0 \;.$$

$\Omega = 1$ is always identically true in this case (this is the critical density case that just manages to expand forever). The age of such a universe is $t_0 = 2/(3H_0)$; if the cosmological constant vanishes, higher density universes ($\Omega_0 > 1$) will have ages less than this, and lower density universes ($0 < \Omega_0 < 1$) ages between this value and (107). This is the present state of the universe if the standard inflationary universe theory is correct, the high value of Ω then implying that most of the matter in the universe is invisible (the 'dark matter' issue; see [41] for a summary of ways of estimating the matter content of the universe, leading to estimates that the detected matter in the universe in fact corresponds to $\Omega_0 \approx 0.2$ to 0.3). It is thus difficult to reconcile this model with observations (the universe could have flat space sections and a large cosmological constant; but then that is not the Einstein–de Sitter model).

4.3.2. *Parametric solutions*

Use conformal time $\tau = (1/S_0) \int dt/S(t)$ and rescale $S \to y = S(t)/S_0$. Then for a non-interacting mixture of pressure-free matter and radiation, we find in the three cases $k = +1, 0, -1$,

$$k = +1: \quad y = \alpha(1 - \cos\tau) + \beta \sin\tau \;, \tag{108}$$

$$k = 0: \quad y = \alpha\tau^2/2 + \beta\tau \;, \tag{109}$$

$$k = -1: \quad y = \alpha(\cosh\tau - 1) + \beta \sinh\tau \;, \tag{110}$$

[13] A fluid with $(\mu + p) = 0$ is equivalent to a cosmological constant.

where $\alpha = S_0^2 H_0^2 \Omega_m/2$, $\beta = (S_0^2 H_0^2 \Omega_r)^{1/2}$, and, on setting $t = \tau = 0$ when $S = 0$,

$$k = +1: \quad t = S_0[\,\alpha(\tau - \sin\tau) + \beta(1 - \cos\tau)\,]\,, \tag{111}$$
$$k = 0: \quad t = S_0[\,\alpha\tau^3/6 + \beta\tau^2/2\,]\,, \tag{112}$$
$$k = -1: \quad t = S_0[\,\alpha(\sinh\tau - \tau) + \beta(\cosh\tau - 1)\,]\,. \tag{113}$$

It is interesting how in this parametrization the dust and radiation terms decouple; this solution includes as special cases the pure dust solutions, $\beta = 0$, and the pure radiation solution, $\alpha = 0$. The general case represents a smooth transition from a radiation dominated early era to a matter dominated later era, and (if $k \neq 0$) on to a curvature dominated era, recollapsing if $k = +1$.

4.3.3. *Early-time solutions*
At early times, when matter is relativistic or negligible compared with radiation, the equation of state is $p = \frac{1}{3}\mu$ and the curvature term can be ignored. The solution is

$$S(t) = c\,t^{1/2}\,, \quad c = \text{const}\,, \quad \mu = \tfrac{3}{4}t^{-2}\,, \quad T = \left(\frac{3}{4a}\right)^{1/4} \frac{1}{t^{1/2}}\,, \tag{114}$$

which determines the expansion time scale during nucleosynthesis and so the way the temperature T varies with time (and hence determines the element fractions produced), and has no adjustable parameters. Consequently the degree of agreement attained between nucleosynthesis theory based on this time scale and element abundance observations [61]–[63] may be taken as supporting both a FLRW geometry and the validity of the EFE at that epoch.

The standard thermal history of the hot early universe (e.g. [61]) follows; going back in time, the temperature rises indefinitely (at least until an inflationary or quantum-dominated epoch occurs), so that the very early universe is an opaque near-equilibrium mixture of elementary particles that combine to form nuclei, atoms, and then molecules after pair production ends and the mix cools down as the universe expands, while various forms of radiation (gravitational radiation, neutrinos, electromagnetic radiation) successively decouple and travel freely through the universe that has become transparent to them. This picture is very well supported by the detection of the extremely accurate black body spectrum of the CBR, together with the good agreement of nucleosynthesis observations with predictions based on the FLRW time scales (114) for the early universe.

Exercise: The early universe was radiation dominated but later became matter dominated (as at the present day). Determine at what values S_{equ} of the scale factor $S(t)$ matter–radiation equality occurs, as a function of Ω_0. For what values of Ω_0 does this occur before decoupling of matter and radiation? (Note that if the universe is dominated by Cold Dark Matter ('CDM') then equality of baryon and radiation density occurs after this time.) When does the universe become curvature dominated?

4.3.4. *Scalar field*

The inflationary universe models use many approximations to model a FLRW universe with a scalar field ϕ as the dominant contribution to the dynamics, so allowing accelerating models that expand quasi-exponentially through many efoldings at a very early time [64, 65], possibly leading to a very inhomogeneous structure on very large (super-particle-horizon) scales [66]. This then leads to important links between particle physics and cosmology, and there is a very large literature on this subject. If an inflationary period occurs in the very early universe, the matter and radiation densities drop very close to zero while the inflaton field dominates, but is restored during 'reheating' at the end of inflation when the scalar field energy converts to radiation.

This will not be pursued further here, except to make one point: because the potential $V(\phi)$ is unspecified (the nature of the inflaton is not known) and the initial value of the 'rolling rate' $\dot{\phi}$ can be chosen at will, it is possible to specify a precise procedure whereby any desired evolutionary history $S(t)$ is attained by appropriate choice of the potential $V(\phi)$ and the initial 'rolling rate' (see [67] for details). Thus, inflationary models may be adjusted to give essentially any desired results in terms of expansion history.

4.3.5. *Kinetic theory*

While a fluid description is used most often, it is also of interest to use a kinetic theory description of the matter in the universe [68]. The details of collisionless isotropic kinetic models in a FLRW geometry are given by Ehlers, Geren and Sachs [69]; this is extended to collisions in [70]. Curiously it is also possible to obtain exact anisotropic collisionless solutions in FLRW geometries; details are given in [71].

4.4. PHASE PLANES

From these equations, as well as finding simple exact solutions, one can determine evolutionary phase planes for this family of models; see Refsdal and Stabell [59] for (Ω_m, q_0), Ehlers and Rindler [72] for $(\Omega_m, \Omega_r, q_0)$, Wainwright and Ellis [51] for (Ω_0, H_0), and Madsen and Ellis [38] for (Ω, S). The latter are based on the phase-plane equation

$$\frac{d\Omega}{dS} = -(3\gamma - 2)\frac{\Omega}{S}(1 - \Omega) . \qquad (115)$$

This equation is valid for any γ, i.e. for arbitrary relations between μ and p, but gives a (Ω, S) phase plane flow if $\gamma = \gamma(\Omega, S)$, and in particular if $\gamma = \gamma(S)$ or $\gamma =$ const. Non-static solutions can be followed through turnaround points where $\dot{S} = 0$ (and so Ω is infinite). This enables one to attain complete (time-symmetric) phase planes for models with and without inflation; see [38] and [73] for details.

4.5. OBSERVATIONS

Astronomical observations are based on radiation travelling to us on the geodesic null rays that generate our past light cone. In the case of a FLRW universe, we may consider only radial null rays as these are generic (because of spatial homogeneity, we can choose the origin of coordinates on any light ray of interest; because of isotropy, light rays travelling in any direction are equivalent to those travelling in any other direction). Thus we may consider geodesic light rays travelling in the FLRW metric (103) such that $ds^2 = 0 = d\theta = d\phi$; then it follows that $0 = -dt^2 + S^2(t)\, dr^2$ on these geodesics. Hence, radiation emitted at E and received at O obeys the basic relations

$$r = \int_E^O dr = \int_{t_E}^{t_0} \frac{dt}{S(t)} = \int_{S_E}^{S_0} \frac{dS}{S(t)\,\dot{S}(t)} , \qquad (116)$$

where the term $\dot{S}$ may be found from the Friedmann equation (106), once a suitable matter description has been chosen.

4.5.1. *Redshift*

The first fundamental quantity is *redshift*. Considering two successive pulses sent from E to O, each remaining at the same comoving coordinate position, it follows from (116) that the cosmological redshift in a FLRW model is given by

$$(1 + z_c) = \frac{\lambda_0}{\lambda_E} = \frac{\Delta T_0}{\Delta T_E} = \frac{S(t_0)}{S(t_E)} , \qquad (117)$$

and so directly measures the expansion of the universe between when light was emitted and when it is received. Two comments are in order. First, redshift is essentially a time-dilation effect, and will be apparent in all observations of a source, not just in its spectra; this characterisation has the important consequences that (i) redshift is achromatic — the fractional shift in wavelength is independent of wavelength, (ii) the width of any emitted frequency band $d\nu_E$ is altered proportional to the redshift when it reaches the observer, i.e. the observed width of the band is $d\nu_0 = (1+z)\, d\nu_E$, and (iii) the observed rate of emission of radiation and the rate of any time variation in its intensity will both also be proportional to $(1+z)$. Second, there can be local gravitational and Doppler contributions z_0 at the observer, and z_E at the emitter; observations of spectra tell us the overall redshift z, given by

$$(1+z) = (1+z_0)\,(1+z_c)\,(1+z_E)\ , \tag{118}$$

but cannot tell us what part is cosmological and what part is due to local effects at the source and the observer. The latter can be determined from the CBR anisotropy, but the former can only be estimated by identifying cluster members and subtracting off the mean cluster motion. The essential problem is in identifying which sources should be considered members of the same cluster. This is the source of the controversies between Arp et al. and the rest of the observational community (see e.g. Field et al [74]).

4.5.2. *Areas*

The second fundamental issue is apparent size. Considering light rays converging to the observer at time t_0 in a solid angle $d\Omega = \sin\theta\, d\theta\, d\phi$, from the metric form (103) the corresponding null rays[14] will be described by constant values of θ and ϕ and at the time t_E will encompass an area $dA = S^2(t_E) f^2(r) d\Omega$ orthogonal to the light rays, where r is given by (116). Thus, on defining the *observer area distance* $r_0(z)$ by the standard area relation, we find

$$dA = r_0^2\, d\Omega \quad \Rightarrow \quad r_0^2 = S^2(t_E)\, f^2(r)\ . \tag{119}$$

Because these models are isotropic about each point, the *same* distance will relate the observed angle α corresponding to a linear length scale ℓ orthogonal to the light rays:

$$\ell = r_0\, \alpha\ . \tag{120}$$

One can now calculate r_0 from this formula together with (116) and the Friedmann equation, or from the geodesic deviation equation (see [53]), to obtain for a non-interacting mixture of matter and radiation [75],

$$\begin{aligned} r_0(z) &= \frac{1}{H_0 q_0 (q_0 + \beta - 1)} \\ &\quad \times \frac{\left[\, (q_0 - 1)\left\{1 + 2q_0 z + q_0 z^2 (1-\beta)\right\}^{1/2} - (q_0 - q_0\beta z - 1)\,\right]}{(1+z)^2}\ , \end{aligned} \tag{121}$$

[14] Bounded by geodesics located at (ϕ_0, θ_0), $(\phi_0 + d\phi, \theta_0)$, $(\phi_0, \theta_0 + d\theta)$, $(\phi_0 + d\phi, \theta_0 + d\theta)$.

where β represents the matter to radiation ratio: $(1-\beta)\,\rho_{m0} = 2\,\beta\,\rho_{r0}$. The standard Mattig relation for pressure-free matter is obtained for $\beta = 1$ [76], and the corresponding radiation result for $\beta = 0$.

An important consequence of this relation is refocusing of the past light cone: the universe as a whole acts as a gravitational lens, so that there is a redshift z_* such that the area distance reaches a maximum there and then decreases for larger z; correspondingly, the apparent size of an object of fixed size would reach a minimum there and then increase as the object was moved further away [77]. As a specific example, in the simplest (Einstein–de Sitter) case with $p = \Lambda = k = 0$, we find

$$\beta = 1\ ,\ q_0 = \tfrac{1}{2} \quad \Rightarrow \quad r_0(z) = \frac{2}{H_0}\,\frac{1}{(1+z)^{3/2}}\,(\sqrt{1+z} - 1)\ , \qquad (122)$$

which refocuses at $z_* = 5/4$ [78]; objects further away will look the same size as much closer objects. For example, an object at a redshift $z_1 = 1023$ (i.e. at about last scattering) will appear the same angular size as an object of identical size at redshift $z_2 = 0.0019$ (which is very close — it corresponds to a speed of recession of about 570 km/sec). In a low density universe, refocusing takes place further out, at redshifts up to $z \approx 4$, depending on the density, and with apparent sizes depending on possible source size evolution [79].

The predicted (*angular size, distance*)–relations are difficult to test observationally because objects of more or less fixed size (such as spherical galaxies) do not have sharp edges that can be used for measuring angular size and so one has rather to measure isophotal diameters (see e.g. [80]), while objects with well-defined linear dimensions, such as double radio sources, are usually rapidly evolving and so one does not know their intrinsic size. Thus, these tests, while in principle clean, are in fact difficult to use in practice.

4.5.3. *Luminosity and reciprocity theorem*

There is a remarkable relation between upgoing and downgoing bundles of null geodesics connecting the source at t_E and the observer at t_0. Define *galaxy area distance* r_G as above for observer area distance but for the upgoing rather than downgoing bundle of null geodesics. The expression for this distance will be exactly the same as (119) except that the times t_E and t_0 will be interchanged. Consequently, on using the redshift relation (117),

Reciprocity Theorem: The observer area distance and galaxy area distance are identical up to redshift factors:

$$\frac{r_0^2}{r_G^2} = \frac{1}{(1+z)^2} \ . \tag{123}$$

This is true in any space-time as a consequence of the standard first integral of the geodesic deviation equation [81, 6].

Now from photon conservation, the *flux of light* received from a source of *luminosity* L at time t_E will be measured to be

$$F = \frac{L(t_E)}{4\pi} \frac{1}{(1+z)^2} \frac{1}{r_G^2} \ ,$$

with r given by (116), and the two factors $(1+z)$ coming from photon redshift and time dilation of the emission rate, respectively. On using the reciprocity result this becomes

$$F = \frac{L(t_E)}{4\pi} \frac{1}{(1+z)^4} \frac{1}{r_0^2} \ , \tag{124}$$

where $r_0(q_0, z)$ is given by (121). On taking logarithms, this gives the standard (*luminosity, redshift*)–relation of observational cosmology [77]. Observations of this Hubble relation basically agree with these predictions, but are not accurate enough to distinguish between the various FLRW models. The hopes that this relation would determine q_0 from galaxy observations have faded away because of the major problem of *source evolution*: we do not know what the source luminosity would have been at the time of evolution. We lack standard candles of known luminosity (or equivalently, rigid objects of known linear size, from which apparent size measurements would give the answer). Various other distance estimators such as the Tully–Fisher relation have helped considerably, but not enough to give a definitive answer. Happily it now seems that Type Ia supernovae may provide the answer in the next decade, because their luminosity can be determined from their light curves, which should depend only on local physics rather than their evolutionary history. This is an extremely promising development at the present time (see e.g. [57]).

4.5.4. *Specific intensity*

In practice, we measure (a) in a limited waveband rather than over all wavelengths as the 'bolometric' calculation above suggests; and (b) real detectors measure specific intensity (radiation received per unit solid angle) at each point of an image, rather than total source luminosity. Putting these together, we see that if the *source spectrum* is $\mathcal{I}(\nu_E)$, i.e. a fraction $\mathcal{I}(\nu_E)\,d\nu_E$ of the source radiation is emitted in the frequency range $d\nu_E$, then the observed *specific intensity* at each image point is given by[15]

$$I_\nu \, d_\nu = \frac{B_E}{(1+z)^3}\,\mathcal{I}(\nu(1+z))\,d\nu \; , \tag{125}$$

where B_E is the *surface brightness* of the emitting object, and the area distance $r_0(z)$ has canceled out (because of the reciprocity theorem). This tells us the apparent intensity of radiation detected in each direction — which is independent of (area) distance, and dependent only on the source redshift, spectrum, and surface brightness. Together with the angular diameter relation (120), this determines what is actually measured by a detector [80].

An immediate application is black body radiation: if any radiation is emitted as black body radiation at temperature T_E, it follows (*Exercise!*) from the black body expression $\mathcal{I}_\nu = \nu^3\, b(\nu/T_E)$ that the received radiation will also be black body (i.e. have this same black body form) but with a measured temperature of

$$T_0 = \frac{T_E}{(1+z)} \; . \tag{126}$$

Note this is true in all cosmologies: the result does not depend on the FLRW symmetries. The importance of this, of course, is that it applies to the observed CBR.

4.5.5. *Number counts*

If we observe sources in a given solid angle $d\Omega$ in the distance range $(r, r + dr)$, the corresponding volume is $dV = S^3(t_E)\, r_0^2\, dr\, d\Omega$, so if the source density is $n(t_E)$ and the probability of detection is p, the number of sources observed will be

$$dN = p\, n(t_E)\, dV = p\left[\frac{n(t_E)}{(1+z)^3}\right] S^3(t_0)\, f^2(r)\, dr\, d\Omega \; , \tag{127}$$

with r given by (116). This is the *basic number count relation*, where dr can be expressed in terms of observable quantities such as dz; the quantity in brackets is constant if source numbers are conserved in a FLRW model, that is

$$n(t_E) = n(t_0)(1+z)^3 \, . \tag{128}$$

The FLRW predictions agree with observations only if we allow for source number and/or luminosity evolution (cf. the discussion of spherically symmetric models in the next section); but we have no good theory for source

[15] Absorption effects will modify this if there is sufficient absorbing matter present; see [6] for relevant formulae.

evolution.

The additional problem is that there are many undetectable objects in the sky, including entire galaxies, because they lie below the detection threshold; thus we face the problem of *dark matter*, which is very difficult to detect by cosmological observations except by its lensing effects (if it is clustered) and its effects on the age of the universe (if it is smoothly distributed). The current view is that there is indeed such dark matter, detected particularly through its dynamical effects in galaxies and clusters of galaxies (see [41] for a summary), with the present day total matter density most probably in the range $0.1 \leq \Omega_0 \leq 0.3$, while the baryon density is of the order of $0.01 \leq \Omega_0^{\text{baryons}} \leq 0.03$ (from nucleosynthesis arguments). Thus, most of the dark matter is probably non-baryonic.

To properly deal with source statistics in general, and number counts in particular, one should have a reasonably good model of detection limits. It is highly misleading to represent such limits as depending on source apparent magnitude alone (see Disney [82]); this does not take into account the possible occurrence of low-surface brightness galaxies. A useful model based on both source apparent size and magnitude is presented in Ellis, Perry and Sievers [83], summarized in [52]. One should note from this particularly that if there is an evolution in source size, this has a more important effect on source detectability than an evolution in surface brightness.

Exercise: Explain why an observer in a FLRW model may at late times of its evolution see a situation that looks like an island universe (see [84]).

4.6. OBSERVATIONAL LIMITS

The first basic observational limit is that we cannot observe anything outside our past light cone, given by (116). Combined with the finite age of the universe, this leads to a maximum (comoving) coordinate distance from the origin for matter with which we can have had any causal connection: namely

$$r_{ph}(t_0) = \int_0^{t_0} \frac{dt}{S(t)}\,, \tag{129}$$

which converges for any ordinary matter. Matter outside is not visible to us, indeed we cannot have had any causal contact with it. Consequently (see Rindler [85]), the particles at this comoving coordinate value define the *particle horizon*: they separate that matter which can have had any causal contact with us since the origin of the universe from that which cannot. This is most clearly seen by using Penrose's conformal diagrams, obtained on using as coordinates the comoving radius and conformal time; see Penrose [86] and Tipler, Clarke and Ellis [40]. The present day distance to the horizon is

$$D_{ph}(t_0) = S(t_0)\, r_{ph} = S(t_0) \int_0^{t_0} \frac{dt}{S(t)} \,. \tag{130}$$

From (117), this is a sphere corresponding to infinite measured redshift (because $S(t) \to 0$ as $t \to 0$).

Exercise: Show that once comoving matter has entered the particle horizon, it cannot leave it (i.e. once causal contact has been established in a FLRW universe, it cannot cease).

Actually we cannot even see as far as the particle horizon: on our past light cone information rapidly fades with redshift (because of (125)); and because the early universe is opaque, we can only see (by means of any kind of electromagnetic radiation) to the *visual horizon* (Ellis and Stoeger [87]), which is the sphere at comoving coordinate distance

$$r_{vh}(t_0) = \int_{t_d}^{t_0} \frac{dt}{S(t)} \,, \tag{131}$$

where t_d is the time of *decoupling of matter and radiation*l, when the universe became transparent (at about a redshift of $z = 1100$). The matter we see at that time is the matter which emitted the CBR we measure today with a present temperature of $2.73\,^\circ$K; its present distance from us is

$$D_{vh}(t_0) = S(t_0)\, r_{vh} \,. \tag{132}$$

If we evaluate these quantities in an Einstein–de Sitter universe, we find an interesting paradox: (re-establishing the fundamental constant c,) the present day distance to the particle horizon is $D_{ph}(t_0) = 3\,ct_0$. The question is how can this be bigger than ct_0 (the distance corresponding to the age of the universe). This suggests that the matter comprising the particle horizon has been moving away from us at faster than the speed of light in order to reach that distance. How can this be? To investigate this (Rothman and Ellis [78]), note that the *proper distance* from the origin to a galaxy at comoving coordinate r at time t is $D(t,r) = S(t)\, r$. Its velocity away from us is thus given by the Hubble law

$$v = \dot{D} = \dot{S}\, r = \frac{\dot{S}}{S} D = H\, D \,. \tag{133}$$

Thus, at any time t, $v = c$ when $D_c(t) = c/H(t)$; this is the *speed of light sphere*, where galaxies are (at the present time) receding away from us at the speed of light; those galaxies at a larger distance will be (instantaneously) moving away at a speed greater than c. In the case of an Einstein–de Sitter universe, it occurs when $D_c = 3\,ct_0/2$. This is precisely half the present distance to the horizon, which is thus *not* the distance where points are moving away from us at the speed of light (however, it *is* the surface of infinite redshift).

To see that this is compatible with local causality, change from Lagrangian coordinates (t, r) to Eulerian coordinates (t, D), where D is the instantaneous proper distance, as above. Then we find the (non-comoving) metric form

$$\begin{aligned} ds^2 &= -(1 - \frac{(\dot{S}/c)^2}{S^2} D^2)\,(dct)^2 - 2\,\frac{\dot{S}/c}{S}\,D^2\,dct\,dD + dD^2 \\ &\quad + S^2(t)\,f^2(r(t,D))\,d\Omega^2 \,. \end{aligned} \tag{134}$$

It follows that the local light cones are given by

$$\frac{dD_\pm}{dt} = \frac{\dot{S}}{S}\,D \pm c\,. \tag{135}$$

It is easily seen than that there is no violation of local causality. We also find from this that the past light cone of $t = t_0$ intersects the family of speed of light spheres at its maximum distance from the origin (the place where the past light cone starts refocusing), i.e. at

$$t_* = \frac{8}{27}\,t_0\,,\ D_* = \frac{4}{9}\,ct_0 = \frac{8}{27}\,\frac{c}{H_0}\,,\ S(t_*) = \frac{4}{9}\,S(t_0)\,,\ z_* = 1.25\,. \tag{136}$$

At that intersection, $dD_-/dt = 0$ (maximum distance!), $dD_+/dt = 2c$, so there is no causality violation by the matter moving at speed c relative to the central worldline. That matter is presently at a distance ct_0 from us. By contrast, the matter comprising the visual horizon was moving away from us at a speed $v = 61c$ when it emitted the CBR, and was at a distance of about 10^7 light years from our past worldline at that time. Hence, it is the fastest moving matter we shall ever see, but was not at the greatest proper distance to which we can see (which is D_*, see (136)). For a full investigation of these matters see [78].

Finally it should be noted that an early inflationary era will move the particle horizon out to very large distances, thus [64, 65] solving the causal problem presented by the isotropy of CBR arriving here from causally disconnected regions (see [87, 78] for the relevant causal diagrams), but it will have no effect on the visual horizon. Thus, it changes the causal limitations, but does not affect the visual limits on the part of the universe we can see.

Exercise: Determine the angular size seen today for the horizon distance $D_{ph}(t_d)$ at the time of decoupling. What is the physical significance of this distance? How might this relate to CBR anisotropies?

4.6.1. *Small universes*

The existence of visual horizons represent absolute limits on what we can ever know; because of them, we can only hope to investigate a small fraction of all the matter in the universe. Furthermore, they imply we do not in fact have the data needed to predict to the future, for at any time gravitational radiation from as yet unseen objects (e.g. domain walls in a chaotic inflationary universe) may cross the visual horizon and undermine any predictions we may have made [88]. However, there is one exceptional situation: it is possible we live in a small universe, with a spatially closed topology on such a length scale (say, 300 to 800 Mpc) that we have already seen around the universe many times, thus already having seen all the matter there is in the universe. The effect is like being in a room with mirrors on the floor, ceiling, and all walls; images from a finite number of objects seem to stretch to infinity. There are many possible topologies, whatever the sign of k [54]; the observational result — best modelled by considering many identical copies of a basic cell attached to each other in an infinitely repeating pattern[16] — can be very like the real universe (Ellis and Schreiber [89]). In this case we would be able to see our own galaxy many times over, thus being able to observationally examine its historical evolution once we had identified which images of distant galaxies were in fact repeated images of our own galaxy.

It is possible the real universe is like this. Observational tests can be carried out by trying to identify the same cluster of galaxies, QSO's [90], or X-ray sources in different directions in the sky [91]; or by detecting circles of identical temperature variation in the CBR sky (Cornish et al [92]). If no such circles are detected, this will be a reasonably convincing proof that we do not live in such a small universe — which has various philosophical advantages over the more conventional models with infinite spatial sections [88]. Inter alia they give some degree of mixing of CBR modes so giving a potentially powerful explanation of the low degree of CBR anisotropy (but this effect is not as strong as some have claimed; see [93]).

[16] In mathematical terms, the universal covering space.

4.7. FLRW UNIVERSES AS COSMOLOGICAL MODELS

These models are very successful in explaining the major features of the observed universe — its expansion from a hot big bang leading to the observed galactic redshifts and remnant black body radiation, tied in well with element abundance predictions and observations (Peebles et al [94]). However, these models do not describe the real universe well in an essential way, in that the highly idealized degree of symmetry does not correspond to the lumpy real universe. Thus, they can serve as basic models giving the largest-scale smoothed out features of the observable physical universe, but one needs to perturb them to get realistic ('almost-FLRW') universe models that can be used to examine the inhomogeneities and anisotropies arising during structure formation, and that can be compared in detail with observations. This is the topic of the last sections.

However, there is a major underlying issue: because of their high symmetry, these models are infinitely improbable in the space of all possible universes. This high symmetry represents a very high degree of fine tuning of initial conditions, which is extraordinarily improbable, unless we can show physical reasons why it should develop from much more general conditions. In order to examine that question, one needs to look at much more general models and see if they do indeed evolve towards the FLRW models because of physical processes (this is the chaotic cosmology programme, initiated by Misner [95, 96], and taken up much later by the inflationary universe proposal of Guth [64]). Additionally, while the FLRW models seem good models for the observed universe at the present time, one can ask (a) are they the only possible models that will fit the observations? (b) does the universe necessarily have the same symmetries on very large scales (outside the particle horizon), or at very early and/or very late times?

To study these issues, we need to look at more general models, developing some understanding of their geometry and dynamics. This is the topic of the next section. We will find there is a range of models in addition to the FLRW models that can fulfill all present day observational requirements. Nevertheless, it is important to state that the family of perturbed FLRW models can meet all present observational requirements, *provided* we allow suitable evolution of source properties back in the past. They also provide a powerful theoretical framework for considering the nature of and effects of cosmic evolution. Hence, they are justifiably the standard models of cosmology. No evidence stands solidly against them.

Exercise: Apart from the detection of major anisotropy, there are a series of other observations which could, if they were ever observed, decisively disprove this family of standard models. What are these observations? (See [97] for some.)

4.8. GENERAL OBSERVATIONAL RELATIONS

Before moving to that section, we briefly consider general observational relations. The present section has focussed on the observational relations holding in FLRW universe models. However, corresponding generic relations can be found determining observations in arbitrary cosmologies (see Kristian and Sachs [98] and Ellis [6]). The essential points are as follows.

In a general model, observations take place on our *past light cone*, which will develop many cusps and caustics at early times because of gravitational lensing, but is still locally generated by geodesic null rays. The information we receive comes to us along these null rays, with tangent vector

$$k^a = \frac{dx^a}{dv} \; , \; k_a k^a = 0 \; , \; k_a = \nabla_a \phi \; \Rightarrow \; k^b \nabla_b k^a = 0 \; , \; k^a \nabla_a \phi = 0 \; . \quad (137)$$

The phase factor ϕ determines the local light cone $\{\phi = \text{const}\}$. Relative to an observer with 4-velocity u^a the null vector k^a determines a redshift factor $(-k_a u^a)$ and a direction e_a:

$$k^a = (-k_b u^b)\,(u^a + e^a) \; , \quad e_a u^a = 0 \; , \quad e_a e^a = 1 \; . \quad (138)$$

Considering the observed variation $\dot{\phi} = u^a \nabla_a \phi$ of the phase ϕ, we see that the observed cosmological redshift z for comoving matter[17] is given by

$$(1 + z) = \frac{\lambda_O}{\lambda_E} = \frac{(k_a u^a)_E}{(k_b u^b)_O} \; . \quad (139)$$

Taking the derivative of this equation along k^a, we get the fundamental equation [5, 6]

$$\frac{d\lambda}{\lambda} = -\,\frac{d(k_a u^a)}{(k_b u^b)} = (\,\tfrac{1}{3}\Theta + (\sigma_{ab} e^a e^b) + \dot{u}_a e^a\,)\,dl \; , \quad (140)$$

where $dl = (-k_a u^a)\,dv$. This shows directly the isotropic and anisotropic contributions to redshift from the expansion and shear, respectively, and the gravitational redshift contribution from the acceleration.[18] In a FLRW universe, the last two contributions will vanish.

[17] Cf. the comment on cosmological and local sources of redshift above.

[18] In a static gravitational field, this will be given by an acceleration potential: $\dot{u}_a = \tilde{\nabla}_a \Phi$; see [5].

Area distances are defined as before, and because of the geodesic deviation equation, the reciprocity theorem holds unchanged [6].[19] Consequently, the same surface brightness results as discussed above hold generically; specifically, Eq. (125) holds in any anisotropic or inhomogeneous cosmology. The major difference from the isotropic case is that due to the effect of the electric and magnetic Weyl tensors in the geodesic deviation equation, distortions occur in bundles of null geodesics which then cause focusing, resulting both in strong lensing (multiple images, Einstein rings, and arcs related to cusps and caustics in the past light cone) and weak lensing (systematic distortion of images in an observed area); see the lectures by Y Mellier and F Bernardeau, the book by Schneider, Ehlers and Falco [99], and the work by Holz and Wald [100].

Power series equations showing how the kinematical quantities and the electric and magnetic Weyl tensors affect cosmological observations have been given in a beautiful paper by Kristian and Sachs [98]. The generalisation of those relations to generic cosmologies has been investigated by Ellis et al [101], showing how in principle cosmological observations can directly determine the space-time structure on the past null cone, and thence off it. Needless to say, major practical observational difficulties make this a formidable task, but some progress in this direction is possible (see e.g. [88], [102] and [103]).

Exercise: Apart from area distances, distortions, matter densities, and redshifts, a crucial data set needed to completely determine the space-time geometry from the EFE is the transverse velocities of the matter we observe on the past light cone [101]. Consider how one might try to measure these velocity components, and what are the best limits one might place on them by practical measurement techniques. [Hint: One possible route is by solar system interferometry. Another is by the Sunyaev–Zel'dovich effect [104].]

5. Solutions with symmetries

5.1. SYMMETRIES OF COSMOLOGIES

Symmetries of a space or a space-time (generically, 'space') are transformations of the space into itself that leave the metric tensor and all physical and geometrical properties invariant. We deal here only with continuous symmetries, characterised by a continuous group of transformations and associated vector fields [105].

[19]Because of the first integrals of the geodesic deviation equation; this result can also be shown from use of Liouville's theorem in kinetic theory.

5.1.1. *Killing vectors*

A space or space-time *symmetry*, or *isometry*, is a transformation that drags the metric along a certain congruence of curves into itself. The generating vector field ξ_i of such curves is called a *Killing vector (field)* (or 'KV'), and obeys Killing's equations,

$$(L_\xi g)_{ij} = 0 \quad \Leftrightarrow \quad \nabla_{(i}\xi_{j)} = 0 \quad \Leftrightarrow \quad \nabla_i\xi_j = -\nabla_j\xi_i \ , \tag{141}$$

where L_X is the *Lie derivative*. By the Ricci identities for a KV, this implies the curvature equation:

$$\nabla_i\nabla_j\xi_k = R^m{}_{ijk}\,\xi_m \ , \tag{142}$$

and so the infinite series of further equations that follows by taking covariant derivatives of this one, e.g.

$$\nabla_l\nabla_i\nabla_j\xi_k = (\nabla_l R^m{}_{ijk})\,\xi_m + R^m{}_{ijk}\,\nabla_l\xi_m \ . \tag{143}$$

The set of all KV's forms a Lie algebra with a basis $\{\xi_a\}$, $a = 1, 2, \ldots, r$, of dimension $r \le \frac{1}{2}n\,(n-1)$. ξ^i_a denote the components with respect to a local coordinate basis, a, b, c label the KV basis, and i, j, k the coordinate components. Any KV can be written in terms of this basis, with *constant coefficients*. Hence: if we take the commutator $[\xi_a, \xi_b]$ of two of the basis KV's, this is also a KV, and so can be written in terms of its components relative to the KV basis, which will be constants. We can write the constants as $C^c{}_{ab}$, obtaining[20]

$$[\xi_a, \xi_b] = C^c{}_{ab}\,\xi_c \ , \qquad C^a{}_{bc} = C^a{}_{[bc]} \ . \tag{144}$$

By the Jacobi identities for the basis vectors, these structure constants must satisfy

$$C^a{}_{e[b}C^e{}_{cd]} = 0 \ , \tag{145}$$

(which is just equation (73) specialized to the case of a set of vectors with constant commutation functions). These are the integrability conditions that must be satisfied in order that the Lie algebra exist in a consistent way. The transformations generated by the Lie algebra form a Lie group of the same dimension (see Eisenhart [105] or Cohn [106]).

Arbitrariness of the basis: We can change the basis of KV's in the usual way;

$$\xi_{a'} = \Lambda_{a'}{}^a\,\xi_a \quad \Leftrightarrow \quad \xi^i_{a'} = \Lambda_{a'}{}^a\,\xi^i_a \ , \tag{146}$$

[20] Cf. equation (68).

where the $\Lambda_a{}^{a}$ are constants with $\det(\Lambda_a{}^{a}) \neq 0$, so unique inverse matrices $\Lambda^{a}{}_{a}$ exist. Then the structure constants transform as tensors:

$$C^{c}{}_{ab} = \Lambda^{c}{}_{c}\Lambda_{a}{}^{a}\Lambda_{b}{}^{b}C^{c}{}_{ab} \,. \tag{147}$$

Thus the possible equivalence of two Lie algebras is not obvious, as they may be given in quite different bases.

5.1.2. *Groups of isometries*

The isometries of a space of dimension n must be a group, as the identity is an isometry, the inverse of an isometry is an isometry, and the composition of two isometries is an isometry. Continuous isometries are generated by the Lie algebra of KV's. The group structure is determined locally by the Lie algebra, in turn characterised by the structure constants [106]. The action of the group is characterised by the nature of its orbits in space; this is only partially determined by the group structure (indeed the same group can act as a space-time symmetry group in quite different ways).

5.1.3. *Dimensionality of groups and orbits*

Most spaces have no KV's, but special spaces (with symmetries) have some. The group action defines orbits in the space where it acts, and the dimensionality of these orbits determines the kind of symmetry that is present.

The *orbit* of a point p is the set of all points into which p can be moved by the action of the isometries of a space. Orbits are necessarily homogeneous (all physical quantities are the same at each point). An *invariant variety* is a set of points moved into itself by the group. This will be bigger than (or equal to) all orbits it contains. The orbits are necessarily invariant varieties; indeed they are sometimes called *minimum invariant varieties*, because they are the smallest subspaces that are always moved into themselves by all the isometries in the group. *Fixed points* of a group of isometries are those points which are left invariant by the isometries (thus the orbit of such a point is just the point itself). These are the points where all KV's vanish (however, the derivatives of the KV's there are non-zero; the KV's generate isotropies about these points). *General points* are those where the dimension of the space spanned by the KV's (that is, the dimension of the orbit through the point) takes the value it has almost everywhere; *special points* are those where it has a lower dimension (e.g. fixed points). Consequently, the dimension of the orbits through special points is lower than that of orbits through general points. The dimension of the orbit and isotropy group is the same at each point of an orbit, because of the equivalence of the group action at all points on each orbit.

The group is *transitive on a surface S* (of whatever dimension) if it can move any point of S into any other point of S. Orbits are the largest surfaces through each point on which the group is transitive; they are therefore sometimes referred to as *surfaces of transitivity.* We define their dimension as follows, and determine limits from the maximal possible initial data for KV's:
dim surface of transitivity $= s$, where in a space of dimension n, $s \leq n$.

At each point we can also consider the dimension of the isotropy group (the group of isometries leaving that point fixed), generated by all those KV's that vanish at that point:
dim of isotropy group $= q$, where $q \leq \frac{1}{2}\, n\,(n-1)$.

The *dimension r of the group of symmetries* of a space of dimension n is $r = s + q$ (translations plus rotations). From the above limits , $0 \leq r \leq n + \frac{1}{2}\, n\,(n-1) = \frac{1}{2}\, n\,(n+1)$ (the maximal number of translations and of rotations). This shows the Lie algebra of KV's is finite dimensional.

Maximal dimensions: If $r = \frac{1}{2}\, n\,(n+1)$, we have a space(-time) of constant curvature (maximal symmetry for a space of dimension n). In this case,

$$R_{ijkl} = K\,(\,g_{ik}\,g_{jl} - g_{il}\,g_{jk}\,)\;, \tag{148}$$

with K a constant; and K necessarily *is* a constant if this equation is true and $n \geq 3$. One cannot get $q = \frac{1}{2}\, n\,(n-1) - 1$ so $r \neq \frac{1}{2}\, n\,(n+1) - 1$.

A group is *simply transitive* if $r = s \;\Leftrightarrow\; q = 0$ (no redundancy: dimensionality of group of isometries is just sufficient to move each point in a surface of transitivity into each other point). There is no continuous isotropy group.

A group is *multiply transitive* if $r > s \;\Leftrightarrow\; q > 0$ (there is redundancy in that the dimension of the group of isometries is larger than is needed to move each point in an orbit into each other point). There exist non-trivial isotropies.

5.2. CLASSIFICATION OF COSMOLOGICAL SYMMETRIES

We consider non-empty perfect fluid models, i.e. (15) holds with $(\mu{+}p) > 0$.

For a cosmological model, because space-time is 4-dimensional, the possibilities for the dimension of the surface of transitivity are $s = 0, 1, 2, 3, 4$. As to isotropy, we assume $(\mu + p) \neq 0$; then $q = 3$, 1, or 0 because u^a is invariant and so the isotropy group at each point has to be a sub-group of the rotations acting orthogonally to u^a (and there is no 2-dimensional

Dimension Isotropy Group	Dim invariant variety s = 2	s = 3	s = 4
	inhomogeneous	spatially homogeneous	space-time homogeneous
q = 0 aniso- tropic	generic metric form known: Spatially self-similar, Abelian G_2 on 2-d spacelike surfaces, non-Abelian G_2	Bianchi: orthogonal, tilted	Osvath/Kerr
q = 1 LRS	Lemaitre-Tolman-Bondi family	Kantowski-Sachs, LRS Bianchi	G"odel
q = 3 iso- tropic	none (cannot happen)	Friedmann	Einstein static
	two non-ignorable coordinates	one non-ignorable coordinate	algebraic EFE (no redshift)

Dimension Isotropy Group	Dim invariant variety s = 0	s = 1
	inhomogeneous	inhomogeneous/no isotropy group
q = 0	Szekeres-Szafron, Stephani-Barnes, Oleson type N The real universe!	General metric form independent of one coord; KV h.s.o./not h.s.o.

Figure 1. Classification of cosmological models (with $(\mu + p) > 0$) by isotropy and homogeneity.

subgroup of $O(3)$.) The dimension q of the isotropy group can vary over the space (but not over an orbit): it can be greater at special points (e.g. an axis centre of symmetry) where the dimension s of the orbit is less, but r (the dimension of the total symmetry group) must stay the same everywhere. Thus the possibilities for isotropy at a general point are:

a) Isotropic: $q = 3$, the Weyl tensor vanishes, kinematical quantities vanish except Θ. All observations (at every point) are isotropic. This is the FLRW family of geometries;

b) Local Rotational Symmetry ('LRS'): $q = 1$, the Weyl tensor is of algebraic Petrov type D, kinematical quantities are rotationally symmetric about a preferred spatial direction. All observations at every general point are rotationally symmetric about this direction. All metrics are known in the case of dust [30] and a perfect fluid (see [50] and also [107]).

c) Anisotropic: $q = 0$; there are no rotational symmetries. Observations in each direction are different from observations in each other direction.

Putting this together with the possibilities for the dimensions of the surfaces of transitivity, we have the following possibilities (see Figure 1):

5.2.1. *Space-time homogeneous models*

These models with $s = 4$ are unchanging in space and time, hence μ is a constant, so by the energy conservation equation (37) they cannot expand: $\Theta = 0$. Thus by (139) they cannot produce an almost isotropic redshift, and are not useful as models of the real universe. Nevertheless they are of some interest.

The *isotropic case* $q = 3$ ($\Rightarrow r = 7$) is the Einstein static universe, the non-expanding FLRW model (briefly mentioned above) that was the first relativistic cosmological model found. It is not a viable cosmology inter alia because it has no redshifts, but it laid the foundation for the discovery of the expanding FLRW models.

Exercise: What other features make this space-time problematic as a cosmological model?

The *LRS case* $q = 1$ ($\Rightarrow r = 5$) is the Gödel stationary rotating universe [108], also with no redshifts. This model was important because of the new understanding it brought as to the nature of time in General Relativity (see [3, 40, 109]). Inter alia, it is a model in which causality is violated (there exist closed timelike lines through each space-time point) and there exists no cosmic time function whatsoever.

The anisotropic models $q = 0$ ($\Rightarrow r = 4$) are all known, [110], but are interesting only for the light they shed on Mach's principle; see [111].

5.2.2. *Spatially homogeneous universes*
These models with $s = 3$ are the major models of theoretical cosmology, because they express mathematically the idea of the 'cosmological principle': all points of space at the same time are equivalent to each other [112].

The *isotropic case* $q = 3$ ($\Rightarrow r = 6$) is the family of FLRW models, the standard models of cosmology discussed above that have the comoving metric form (103).

The *LRS case* $q = 1$ ($\Rightarrow r = 4$) is the family of Kantowski–Sachs universes [113]–[115] plus the LRS orthogonal [49] and tilted [116] Bianchi models. The simplest are the Kantowski–Sachs family, with comoving metric form

$$ds^2 = -\,dt^2 + A^2(t)\,dr^2 + B^2(t)\,(\,d\theta^2 + f^2(\theta)\,d\phi^2\,)\;, \qquad (149)$$

where $f(\theta)$ is given by (104).

The *anisotropic case* $q = 0$ ($\Rightarrow r = 3$) is the family of Bianchi universes with a group of isometries G_3 acting simply transitively on spacelike surfaces. They can be orthogonal or tilted; the simplest class is the Bianchi Type I family, discussed later in this section. The family as a whole has quite complex properties; these models are discussed in the following section.

5.2.3. *Spatially inhomogeneous universes*
These models have $s \leq 2$.

The *LRS cases* ($q = 1 \Rightarrow s = 2, r = 3$) are the spherically symmetric family with metric form

$$ds^2 = -\,C^2(t,r)\,dt^2 + A^2(t,r)\,dr^2 + B^2(t,r)\,(\,d\theta^2 + f^2(\theta)\,d\phi^2\,)\;, \qquad (150)$$

where $f(\theta)$ is given by (104). In the dust case, we can set $C(t,r) = 1$ and can integrate the EFE analytically; for $k = +1$, these are the Lemaître–Tolman–Bondi ('LTB') spherically symmetric models [117]–[119], discussed later in this section. They may have a centre of symmetry (a timelike worldline), and can even allow two such centres, but they cannot be isotropic about a general point (because isotropy everywhere implies spatial homogeneity; see the discussion of FLRW models).

The *anisotropic cases* ($q = 0 \Rightarrow s \leq 2, r \leq 2$) include solutions admitting an Abelian or non-Abelian group of isometries G_2, and spatially self-similar models (see e.g. [51]).

Solutions with no symmetries at all have $r = 0 \Rightarrow s = 0, q = 0$. The real universe, of course, belongs to this class; all the others are intended as approximations to this unique universe. Remarkably, we know some exact solutions without symmetries, specifically (a) the Szekeres quasi-spherical models [120, 121], that are in a sense non-linear FLRW perturbations [122], with metric

$$ds^2 = -\,dt^2 + e^{2A}\,dx^2 + e^{2B}(dy^2 + dz^2)\ , \tag{151}$$

$$A = A(t,x,y,z)\ , \qquad B = B(t,x,y,z)\ ,$$

(b) Stephani's conformally flat models [123, 124], and (c) Oleson's type N solutions (for a discussion of these and all the other inhomogeneous models, see Krasiński [8] and Kramer et al [125]). One further interesting family without global symmetries are the Swiss-Cheese models made by cutting and pasting segments of spherically symmetric models. These are discussed below.

We now discuss the simplest useful anisotropic and inhomogeneous models, before turning to the Bianchi models in the next section.

5.3. BIANCHI TYPE I UNIVERSES ($S = 3$)

These are the simplest anisotropically expanding universe models. The metric can be given in the form [126]

$$ds^2 = -\,dt^2 + X^2(t)\,dx^2 + Y^2(t)\,dy^2 + Z^2(t)\,dz^2\ , \qquad u^a = \delta^a{}_0\ . \tag{152}$$

This is the simplest generalisation of the flat FLRW models to allow for different expansion factors in three orthogonal directions; the corresponding *average expansion scale factor* is $S(t) = (XYZ)^{1/3}$. They are spatially homogeneous, being invariant under an Abelian group of isometries G_3 simply transitive on spacelike surfaces $\{t = \text{const}\}$, so $s = 3$; in general $q = 0 \Rightarrow r = 3$, but there are LRS and isotropic subcases (the latter being the Einstein–de Sitter universe). The space sections $\{t = \text{const}\}$ are flat (when $t = t_0$, all the metric coefficients are constant), and all invariants depend only on the time coordinate t. The fluid flow (orthogonal to these homogeneous surfaces) is necessarily geodesic and irrotational. Thus these models obey the restrictions

$$0 = \dot{u}^a = \omega^a\ , \qquad 0 = X_a = Z_a = \tilde{\nabla}_a p\ , \qquad 0 = {}^3R_{ab}\ . \tag{153}$$

The $1+3$ covariant equations obeyed by these models follow from the $1+3$ covariant equations in the previous section, on making these restrictions.

We can find a tetrad in the obvious way from the above coordinates ($e_1{}^i = X(t)^{-1}\,\delta_1{}^i$, etc.); then the tetrad equations of the previous section hold with

$$0 = \dot{u}^\alpha = \omega^\alpha = \Omega^\alpha \ , \qquad 0 = a^\alpha = n_{\alpha\beta} \ ,$$
$$0 = \mathbf{e}_\alpha(\Theta) = \mathbf{e}_\alpha(\sigma_{\beta\gamma}) \ , \qquad 0 = \mathbf{e}_\alpha(\mu) = \mathbf{e}_\alpha(p) \ . \tag{154}$$

It follows that the (0α)-equation (32), which is $(C_1)^\alpha$ in the tetrad form, is identically satisfied, and also that $H_{ab} = 0$ and $\tilde{\nabla}_b E^{ab} = 0$. From the Gauss equation (53), the shear obeys

$$(S^3\sigma_{ab})^{\cdot} = 0 \quad \Rightarrow \quad \sigma_{ab} = \frac{\Sigma_{ab}}{S^3} \ , \quad (\Sigma_{ab})^{\cdot} = 0 \ , \tag{155}$$

which implies

$$\sigma^2 = \frac{\Sigma^2}{S^6} \ , \quad \Sigma^2 = \tfrac{1}{2}\,\Sigma_{ab}\Sigma^{ab} \ , \quad (\Sigma^2)^{\cdot} = 0 \ . \tag{156}$$

All the EFE will then be satisfied if the conservation equation (37), the Raychaudhuri equation (29), and the Friedmann-like equation (54) are satisfied. As in the FLRW case, the latter is the first integral of the other two. Assuming a γ-law equation of state, (55) will be satisfied and, using (156), equation (54) becomes the generalised Friedmann equation,

$$3\,\frac{\dot{S}^2}{S^2} = \frac{\Sigma^2}{S^6} + \frac{M}{S^{3\gamma}} \ . \tag{157}$$

This shows that no matter how small the shear today, it will (for ordinary matter) dominate the very early evolution of the universe, which will then approximate the Kasner vacuum solution [125].

On writing out the tetrad components of the shear equation (155), using the commutator relations (80) to determine the shear components, one finds that the individual length scales are given by,

$$X(t) = S(t)\,\exp(\Sigma_1\,W(t)) \ ,$$
$$Y(t) = S(t)\,\exp(\Sigma_2\,W(t)) \ ,$$
$$Z(t) = S(t)\,\exp(\Sigma_3\,W(t)) \ ,$$

where

$$W(t) = \int \frac{dt}{S^3(t)} \ , \tag{158}$$

and the constants Σ_α satisfy

$$\Sigma_1 + \Sigma_2 + \Sigma_3 = 0 \ , \quad \Sigma_1^2 + \Sigma_2^2 + \Sigma_3^2 = 2\Sigma^2 \ .$$

These relations can be satisfied by setting

$$\Sigma_\alpha = (2\Sigma/3)\sin\alpha_\alpha \,, \qquad \alpha_1 = \alpha \,, \; \alpha_2 = \alpha + \frac{2\pi}{3} \,, \; \alpha_3 = \alpha + \frac{4\pi}{3} \,, \qquad (159)$$

and α is a constant. Thus the solution is given by choosing a value for γ, and then integrating successively (157) and (158).

Exercise: Show that, on using the obvious tetrad associated with the coordinates above, all the tetrad (and $1+3$ covariant) equations are then satisfied.

For example, in the case of dust ($\gamma = 1$):

$$S(t) = (\frac{9}{2}Mt^2 + \sqrt{3}\Sigma t)^{1/3} \,, \qquad W(t) = \frac{1}{\sqrt{3}\Sigma}\ln\left(\frac{t}{\frac{3}{4}Mt + \sqrt{3}\Sigma}\right) \,,$$

so

$$X(t) = S(t)\left(\frac{t^2}{S(t)^3}\right)^{\frac{2}{3}\sin\alpha_1} \,,$$

$$Y(t) = S(t)\left(\frac{t^2}{S(t)^3}\right)^{\frac{2}{3}\sin\alpha_2} \,,$$

$$Z(t) = S(t)\left(\frac{t^2}{S(t)^3}\right)^{\frac{2}{3}\sin\alpha_3} \,.$$

The generic case is anisotropic; LRS cases occur when $\alpha = \pi/6$ and $\alpha = \pi/2$ in (159), and isotropy when $\Sigma = 0$.

At late times this isotropizes to give the Einstein–de Sitter model, and, hence, as mentioned above, can be a good model of the real universe if Σ is chosen appropriately. However, at early times, the situation is quite different. As $t \to 0$, provided $\Sigma \neq 0$, then $S(t) \to (\sqrt{3}\Sigma)^{1/3} t^{1/3}$ and

$$X(t) \to X_0\, t^{\frac{1}{3}(1+2\sin\alpha_1)} \,,$$

$$Y(t) \to Y_0\, t^{\frac{1}{3}(1+2\sin\alpha_2)} \,,$$

$$Z(t) \to Z_0\, t^{\frac{1}{3}(1+2\sin\alpha_3)} \,.$$

Plotting the function $f(\alpha) = \frac{2}{3}\,(\frac{1}{2}+\sin\alpha)$, we see that the generic behaviour occurs for $\alpha \neq \pi/2$; in this case two of the powers are positive but one is negative, so going backwards in time, the collapse along the preferred axis reverses and changes to a (divergent) expansion, while collapse continues (divergently) along the two orthogonal direction; the singularity is a *cigar singularity*. Going forward in time, a collapse along the preferred axis stops and reverses to become an expansion. However when $\alpha = \pi/2$, one exponent is positive but the other two are zero. Hence, going back in time, collapse continues divergently along the preferred direction in these LRS solutions back to the singularity, but in the orthogonal directions it slows down and halts; this is a *pancake singularity*. An important consequence in this special case is that horizons are broken in the preferred direction — communication is possible to arbitrary distance in a cylinder around this axis [3].

One can work out detailed observational relations in these models. Because of the high symmetry, the null geodesics can be found explicitly; those along the three preferred axes are particularly simple. Redshift along each of these axes simply scales with the expansion ratio in that direction. Area distances can be found explicitly [127, 128]. An interesting feature is that all observations will show an eight-fold discrete isotropy symmetry about the preferred axes. One can also work out helium production and CBR anisotropy in these models, following the pioneering paper by Thorne [129]. Because the shear can dominate the dynamics at nucleosynthesis or baryosynthesis time, causing a speeding up of the expansion, one can get quite different results than in the FLRW models. Consequently, one can use the nucleosynthesis observations to limit the shear constant Σ, but still allowing extra freedom at the time of baryosynthesis. The CBR quadrupole anisotropy will directly measure the difference in expansion along the three principal axes since last scattering, and, hence, may also be used to limit the anisotropy parameter Σ. Nucleosynthesis gives stronger limits, because it probes to earlier times. These models have also been investigated in the case of viscous fluids and kinetic theory solutions (Misner [130]), with electromagnetic fields, and also the effects of 'reheating' on the CBR anisotropy and spectrum have been examined; see Rees [131].

Thus these models can have arbitrarily small shear at the present day, thus can be arbitrarily close to an Einstein–de Sitter universe since decoupling, but can be quite different early on.

Exercise: Show how the solutions will be altered by (i) a fluid with simple viscosity: $\pi_{ab} = -\eta\,\sigma_{ab}$ with constant viscocity coefficient η, (ii) freely propagating neutrinos [130].

5.4. LEMAÎTRE–TOLMAN–BONDI FAMILY ($S = 2$)

The simplest inhomogeneous models are those that are spherically symmetric. In general they are time-dependent, with 2-dimensional spherical surfaces of symmetry: $s = 2$, $q = 1 \Rightarrow r = 3$. The geometry of this family (including the closely related models with flat and negatively curved 2-surfaces of symmetry), is examined in a $1+3$ covariant way by van Elst and Ellis [107], and a tetrad analysis is given by Ellis [30] (the pressure-free case) and Stewart and Ellis [50] (for perfect fluids). Here we only consider the dust case, because then a simple analytic solution is possible; the perfect fluid case includes spherical stellar models and collapse solutions (see e.g. Misner, Thorne and Wheeler [132]).

The general spherically symmetric metric for an irrotational dust matter source in synchronous comoving coordinates is the Lemaître–Tolman–Bondi ('LTB') metric [117]–[119]

$$ds^2 = -\,dt^2 + X^2(t,r)\,dr^2 + Y^2(t,r)\,d\Omega^2\;, \qquad u^a = \delta^a{}_0\;. \tag{160}$$

The function $Y = Y(t,r)$ is the areal radius, since the proper area of a sphere of coordinate radius r on a time slice of constant t is $4\pi\,Y^2$ (upon re-establishing factors of π). Solving the EFE [119] shows

$$ds^2 = -\,dt^2 + \frac{[\,Y'(t,r)\,]^2}{1+2E(r)}\,dr^2 + Y^2(t,r)\,d\Omega^2\;, \tag{161}$$

where $Y'(t,r) = \partial Y(t,r)/\partial r$, and $d\Omega^2 = d\theta^2 + \sin^2\theta\,d\phi^2$, with $Y(t,r)$ obeying a generalised Friedmann equation,

$$\dot{Y}(t,r) = \pm\sqrt{\frac{2\,M(r)}{Y(t,r)} + 2\,E(r)}\;, \tag{162}$$

and the density given by

$$4\pi\,\mu(t,r) = \frac{M'(r)}{Y^2(t,r)\,Y'(t,r)}\;. \tag{163}$$

Equation (162) can be solved in terms of a parameter $\eta = \eta(t,r)$:

$$Y(t,r) = \frac{M(r)}{\mathcal{E}(r)}\,\phi_0(t,r)\;, \qquad \xi(t,r) = \frac{[\,\mathcal{E}(r)\,]^{3/2}\,(t - t_B(r))}{M(r)}\;, \tag{164}$$

where[21]

$$\mathcal{E}(r) = \begin{cases} 2E(r), \\ 1, \\ -2E(r), \end{cases} \qquad \phi_0 = \begin{cases} \cosh\eta - 1, \\ (1/2)\eta^2, \\ 1 - \cos\eta, \end{cases} \qquad \xi = \begin{cases} \sinh\eta - \eta, \\ (1/6)\eta^3, \\ \eta - \sin\eta, \end{cases}$$

$$\text{when} \begin{cases} E > 0 \\ E = 0 \\ E < 0 \end{cases} , \tag{165}$$

for hyperbolic, parabolic and elliptic solutions, respectively.

The LTB model is characterised by three arbitrary functions of comoving coordinate radius r. $E = E(r) \geq -1$ has a geometrical role, determining the local 'embedding angle' of spatial slices, and also a dynamical role, determining the local energy per unit mass of the dust particles, and, hence, the type of evolution of Y. $M = M(r)$ is the effective gravitational mass with coordinate radius r. $t_B = t_B(r)$ is the local time at which $Y = 0$, i.e., the local time of the big bang — we have a non-simultaneous bang surface. Specification of these three arbitrary functions — $M(r)$, $E(r)$ and $t_B(r)$ — fully determines the model, and whilst all have some type of physical or geometrical interpretation, they admit a freedom to choose the radial coordinate, leaving two physically meaningful choices, e.g. $r = r(M)$, $E = E(M)$, $t_B = t_B(M)$. For particular choices of this initial data, one obtains FLRW models, which of course are special cases of these spherical models with very specific initial data. In fact, the FLRW models are obtained if one sets

$$2\,E(r) = -\,k\,r^2 \ , \quad Y(t,r) = S(t)\,r \ , \quad M(r) = \frac{4\pi}{3}\,\mu(t)\,Y^3 \ . \tag{166}$$

The LTB models have been used in a number of interesting ways in cosmology:

* to give simple models of structure formation [133, 134], for example by looking at evolution of a locally open region in a closed universe [135] and evolution of density contrast [136],

* to give universe models that are inhomogeneous on a cosmological scale [137, 138],

* to examine inhomogeneous big bang structures [139],

* to examine CBR anisotropies [140]–[142],

[21] Strictly speaking, the hyperbolic, parabolic and elliptic solutions obtain when $YE/M > 0$, $= 0$ and < 0, respectively, since $E = 0$ at a spherical origin in both hyperbolic and elliptic models.

* to investigate observational conditions for spatial homogeneity [143]–[147],
* to trace the effect of averaging on spatial inhomogeneities [148], and
* to look at the relationship between cosmic evolution and closure of the universe [149].

These aspects are discussed in Krasiński's book [8], Part III. Here I will only summarize one interesting result: namely, regarding observational tests of whether the real universe is more like a LTB inhomogeneous model, or a FLRW model. In Mustapha, Hellaby and Ellis [144], the following result is shown:

Isotropic Observations Theorem (1): Any given isotropic set of source observations $n(z)$ and $m(z)$, together with any given source luminosity and number evolution functions $L(z)$ and $N(z)$, can be fitted by a spherically symmetric dust cosmology — a LTB model — in which observations are spherically symmetric about us because we are located near the central worldline.

This shows that any spherically symmetric observations we may eventually make can be accommodated by appropriate inhomogeneities in a LTB model — irrespective of what source evolution may occur. In particular, one can find such a model that will fit the observations if there is no source evolution. that The following result also holds:

Isotropic Observations Theorem (2): Given any spherically symmetric geometry and any spherically symmetric set of observations, we can find evolution functions that will make the model compatible with the observations. This applies in particular if we want to fit observations to a FLRW model.

The point of the first result is that it shows that these models — spherical inhomogeneous generalisations of the FLRW models — are viable models of the real universe since decoupling, because they cannot be observationally disproved (at least not in any simple way). The usual response to this is: but the FLRW models are confirmed by observations. The second result clarifies this: yes they are, provided you allow an evolution function to be chosen specifically so that the initially discrepant number counts fit the FLRW model predictions. That is, we *assume* the FLRW geometry, and then determine what source evolution makes this assumption compatible with observations [150]. Without this freedom, the FLRW models are contradicted by radio source observations, for example, which without evolution are better fitted by a flat (Euclidean) model. Thus the FLRW model fit is obtained only because of this freedom, allowed because we do not understand source luminosity and number evolution. The inhomogeneous LTB models provide an alternative understanding of the data; the observations do not contradict them.

Exercise: Suppose (a) observations are isotropic, and (b) we knew the source evolution function and selection function, and (c) we were to observationally show that after taking them into account, the area distance relation has precisely the FLRW form (121) for $\beta = 1$ and the number count relation implies the FLRW form (128). Assuming the space-time matter content is dust, (i) prove from this that the space-time is a FLRW space-time. Now (ii) explain the observational difficulties that prevent us using this exact result to prove spatial homogeneity in practice.

An alternative approach to proving homogeneity is via the Postulate of Uniform Thermal Histories ('PUTH') — that is, the assumption that because we see similar kinds of objects at great distances and nearby, they must have had similar thermal histories. One might then hope that from this one could deduce spatial homogeneity of the space-time geometry (for otherwise the thermal histories would have been different). Unfortunately the argument here is not watertight, as can be shown by a counterexample based on the LTB models (Bonnor and Ellis [143]). Proving — rather than assuming — spatial homogeneity remains elusive. We cannot observationally disprove spatial inhomogeneity. However, we can give a solid argument for it via the EGS theorem discussed below.

5.5. SWISS-CHEESE MODELS

Finally, an interesting family of inhomogeneous models is the Swiss-Cheese family of models, obtained by repeatedly cutting out a spherical region from a FLRW model and filling it in with another spherical model: Schwarzschild or LTB, for example. This requires:

(i) locating the 3-dimensional timelike *junction surfaces* $\Sigma_\pm$ in each of the two models;

(ii) defining a proposed *identification* Φ between Σ_+ and Σ_-;

(iii) determining the *junction conditions* that (a) the 3-dimensional metrics of Σ_+ and Σ_- (the first fundamental forms of these surfaces) be isometric under this identification, so that there be no discontinuity when we glue them together — we arrive at the same metric from both sides — and (b) the second fundamental forms of these surfaces (i.e. the covariant first derivatives of the 3-dimensional metrics along the spacelike normal directions) must also be isometric when we make this identification, so that they too are continuous in the resultant space-time — equivalently, there is no discontinuity in the direction of the spacelike unit normal vector as we cross the junction surface Σ (this is the condition that there be no surface layer on Σ once we make the join; see Israel [151]).

Satisfying these junction conditions involves deciding how the 3-dimensional junction surfaces $\Sigma_\pm$ should be placed in the respective background space-times. It follows from them that 4 of the 10 components of T_{ab} must be continuous: if n^a denotes the spacelike (or, in some other matching problems, timelike) unit normal to Σ and $p^a{}_b$ the tensor projecting orthogonal to n^a, then $(T_{ab}n^a n^b)$ and $T_{bc}\, n^b\, p^c{}_a$ must be continuous, but the other 6 components $T_{cd}\, p^c{}_a\, p^d{}_b$ can be discontinuous (thus at the surface of a star, in which case n^a will be spacelike, the pressure is continuous but the energy density can be discontinuous). Conservation of the energy-momentum tensor across the junction surface Σ will then be satisfied by the constraints $0 = (G_{ab} - T_{ab})\, n^a n^b$ and $0 = (G_{bc} - T_{bc})\, n^b\, p^c{}_a$.

(iv) Having determined that these junction conditions can be satisfied for some particular identification of points, one can then proceed to identify these corresponding points in the two surfaces Σ_+ and Σ_-, thus gluing an interior Schwarzschild part to an exterior FLRW part, for example. Because of the reciprocal nature of the junction conditions, it is then clear we could have joined them the other way also, obtaining a well-matched FLRW interior and Schwarzschild exterior.

(v) One can continue in this way, obtaining a family of holes of different sizes in a FLRW model with different interior fillings, with further FLRW model segments fitted into the interiors of some of these regions, obtaining a Swiss-Cheese model. One can even obtain a hierarchically structured family of spherically symmetric vacuum and non-vacuum regions in this way.

It is important to note that one *cannot* match arbitrary masses. It follows from the junction conditions that the Schwarzschild mass in the interior of a combined FLRW–Schwarzschild solution must be the same as the mass that has been removed: $M_{\mathrm{Schw}} = (4\pi/3)\,(\mu\, S^3\, r^3)_{\mathrm{FLRW}}$. If the masses were wrongly matched, there would be an excess gravitational field from the mass in the interior that would not fit the exterior gravitational field, and the result would be to distort the FLRW geometry in the exterior region — which then would no longer be a FLRW model. Alternatively viewed, the reason this matching of masses is needed is that otherwise we will have fitted the wrong background geometry to the inhomogeneous Swiss-Cheese model — averaging the masses in that model will not give the correct background average [152], and they could not have arisen from rearranging uniformly distributed masses in an inhomogeneous way (this is the content of the Traschen integral constraints [153]). Consequently, there can be *no* long-range effects of such matching: the Schwarzschild mass cannot cause large-scale motions of matter in the FLRW region.

These models were originally developed by Einstein and Strauss [154] (see also Schücking [155]) to examine the effect of the expansion of the universe on the solar system (can we measure the expansion of the universe by laser ranging within the solar system?). Their matching of a Schwarzschild interior to a FLRW exterior showed that this expansion has no effect on the motion of planets in the Schwarzschild region. It does not, however, answer the question as to *where* the boundary between the regions should be placed — which determines which regions are affected by the universal expansion. Subsequent uses of these models have included:

* examining Oppenheimer–Snyder collapse in an expanding universe [156]–[158],
* examining gravitational lensing effects on area distances [159],
* investigating CBR anisotropies [160]–[162],
* modelling voids in large-scale structure [163, 164], perhaps using surface-layers [165],
* modelling the universe as a patchwork of domains of different curvature $k = 0, \pm 1$ [166].

Exercise: Show how appropriate choice of initial data in a LTB model can give an effective Swiss-Cheese model with one centre surrounded by a series of successive FLRW and non-FLRW spherical regions. Can you include (i) flat, (ii) vacuum (Schwarzschild) regions in this construction?

One of the most intriguing questions is what non-spherically symmetric models can be joined regularly onto a FLRW model. Bonnor has shown that some Szekeres anisotropic and inhomogeneous models can be matched to a dust FLRW model across a comoving spherical junction surface [167]. Dyer et al [168] have shown that one can match FLRW and LRS Kasner (anisotropic vacuum Bianchi Type I) models across a flat 3-dimensional timelike junction surface. Optical properties of these Cheese-slice models have been investigated in depth [169].

6. Bianchi universes ($s = 3$)

These are the models in which there is a group of isometries G_3 simply transitive on spacelike surfaces, so they are spatially homogeneous. There is only *one* essential dynamical coordinate (the time t), and the EFE reduce to ordinary differential equations, because the inhomogeneous degrees of freedom have been 'frozen out'. They are thus quite special in geometrical terms; nevertheless, they form a rich set of models where one can study the exact dynamics of the full non-linear field equations. The solutions to the EFE will depend on the matter in the space-time. In the case of a fluid (with uniquely defined flow lines), we have two different kinds of models:

Orthogonal models, with the fluid flow lines orthogonal to the surfaces of homogeneity (Ellis and MacCallum [49], see also [51]);
Tilted models, with the fluid flow lines not orthogonal to the surfaces of homogeneity; the components of the fluid peculiar velocity enter as further variables (King and Ellis [116], see also [170]).

Rotating models *must* be tilted (cf. Eq. (27)), and are much more complex than non-rotating models.

6.1. CONSTRUCTING BIANCHI UNIVERSES

There are essentially three direct ways of constructing the orthogonal models, all based on properties of a tetrad of vectors $\mathbf{e}_a$ that commute with the basis of KV's $\{ \xi_\alpha \}$, and usually with the timelike basis vector chosen parallel to the unit normal $n_a = - \nabla_a t$ to the surfaces of homogeneity, i.e. $\mathbf{e}_0 = \mathbf{n}$.

The *first approach* (Taub [171], Heckmann and Schücking [126]) puts all the time variation in the metric components:

$$ds^2 = - dt^2 + \gamma_{\alpha\beta}(t)\, (e^\alpha{}_i(x^\gamma)\, dx^i)\, (e^\beta{}_j(x^\delta)\, dx^j) \ , \tag{167}$$

where $e^\alpha{}_i(x^\gamma)$ are 1-forms inverse to the spatial vector triad $e_\alpha{}^i(x^\gamma)$, which have the same commutators $C'^\alpha{}_{\beta\gamma}$, $\alpha, \beta, \gamma, \ldots = 1, 2, 3$, as the structure constants of the group of isometries and commute with the unit normal vector $\mathbf{e}_0$ to the surfaces of homogeneity; that is, $\mathbf{e}_\alpha = e_\alpha{}^i\, \partial/\partial x^i$, $\mathbf{e}_0 = \partial/\partial t$ obey the commutator relations

$$[\,\mathbf{e}_\alpha, \mathbf{e}_\beta\,] = C^\gamma{}_{\alpha\beta}\, \mathbf{e}_\gamma \ , \qquad [\,\mathbf{e}_0, \mathbf{e}_\alpha\,] = 0 \ , \tag{168}$$

where the $C^\gamma{}_{\alpha\beta}$ are the Lie algebra structure constants satisfying the Jacobi identities (145). The EFE (1) become ordinary differential equations for the metric functions $\gamma_{\alpha\beta}(t)$.

The *second approach* is based on use of the automorphism group of the symmetry group with time-dependent parameters. We will not consider it further here (see Collins and Hawking [172], Jantzen [173, 174] and Wainwright and Ellis [51] for a discussion).

The *third approach* (Ellis and MacCallum [49]), which is in my view the preferable one, uses an orthonormal tetrad based on the normals to the surfaces of homogeneity (i.e. $\mathbf{e}_0 = \mathbf{n}$, the unit normal vector to these surfaces). The tetrad is chosen to be invariant under the group of isometries, i.e. the tetrad vectors commute with the KV's, and the metric components in the tetrad are space-time constants, $g_{ab} = \eta_{ab}$; now the dynamical time variation is in the commutation functions for the basis vectors, which then determine the time-(and space-)dependence in the basis vectors themselves. Thus we have an orthonormal basis $\mathbf{e}_a$, $a = 0, 1, 2, 3$, such that

$$[\,\mathbf{e}_a, \mathbf{e}_b\,] = \gamma^c{}_{ab}(t)\,\mathbf{e}_c\ . \tag{169}$$

The commutation functions $\gamma^a{}_{bc}(t)$, together with the matter variables, are then treated as the dynamical variables. The EFE (1) are first-order equations for these quantities, supplemented by the Jacobi identities for the $\gamma^a{}_{bc}(t)$, which are also first-order equations. It is sometimes useful to introduce also the Weyl tensor components as auxiliary variables, but this is not necessary in order to obtain solutions. Thus the equations needed are just the tetrad equations given in section 3, for the case

$$\dot{u}^\alpha = \omega^\alpha = 0 = \mathbf{e}_\alpha(\gamma^a{}_{bc})\ . \tag{170}$$

The spatial commutation functions $\gamma^\alpha{}_{\beta\gamma}(t)$ can be decomposed into a time-dependent matrix $n_{\alpha\beta}(t)$ and vector $a^\alpha(t)$ (see (79)), and are equivalent to the structure constants $C^\alpha{}_{\beta\gamma}$ of the symmetry group at each point.[22] In view of (170), the Jacobi identities (73) now take the simple form

$$n^{\alpha\beta}\,a_\beta = 0\ . \tag{171}$$

The tetrad basis can be chosen to diagonalise $n_{\alpha\beta}$ to attain $n_{\alpha\beta} = \mathrm{diag}\,(n_1, n_2, n_3)$ and to set $a^\alpha = (a, 0, 0)$, so that the Jacobi identities are then simply $n_1\,a = 0$. Consequently we define two major classes of structure

[22] That is, they can be brought to the canonical forms of the $C^\alpha{}_{\beta\gamma}$ by a suitable change of group-invariant basis (the final normalisation to ± 1 may require changing from normalised basis vectors); the transformation to do so is different at each point and at each time.

Class	Type	n_1	n_2	n_3	a	
A	I	0	0	0	0	Abelian
	II	+ve	0	0	0	
	VI_0	0	+ve	-ve	0	
	VII_0	0	+ve	+ve	0	
	VIII	-ve	+ve	+ve	0	
	IX	+ve	+ve	+ve	0	
B	V	0	0	0	+ve	
	IV	0	0	+ve	+ve	
	VI_h	0	+ve	-ve	+ve	h < 0
	III	0	+ve	-ve	n2n3	same as VI_1
	VII_h	0	+ve	+ve	+ve	h > 0

Figure 2. Canonical structure constants for different Bianchi types. The parameter $h = a^2/n_2n_3$.

constants (and so Lie algebras):

Class A: $a = 0$,
Class B: $a \neq 0$.

Following Schücking, the adaptation of the Bianchi classification of G_3 group types used is as in Figure 2. Given a specific group type at one instant, this type will be preserved by the evolution equations for the quantities $n_\alpha(t)$ and $a(t)$. This is a consequence of a generic property of the EFE: they will always preserve symmetries in initial data (within the Cauchy development of that data); see Hawking and Ellis [3].

In some cases, the Bianchi groups allow higher symmetry subcases: isotropic (FLRW) or LRS models. Figure 3 gives the Bianchi symmetry groups admitted by FLRW and LRS solutions [49], i.e. these are the simply transitive 3-dimensional subgroups allowed by the full G_6 of isometries (in the FLRW case) and the G_4 of isometries (in the LRS case). The only LRS models not allowing a simply transitive subgroup G_3 are the Kantowski–Sachs models for $k = 1$.

Tilted models can be constructed similarly, as discussed below, or by

```
==========================================================
                  Isotropic Bianchi models

 FLRW k = +1: Bianchi IX [two commuting groups]
 FLRW k =  0: Bianchi  I, Bianchi VII_0
 FLRW k = -1: Bianchi  V, Bianchi VII_h
==========================================================
                     LRS Bianchi models

Orthogonal              c = 0                 c \neq 0

   Taub-NUT I        [KS +1: no subgroup]   Bianchi IX
   Taub-NUT 3        Bianchi I, VII_0       Bianchi II
   Taub-NUT 2        Bianchi III [KS -1]    Bianchi VII_h,
                                            III
Tilted

                     Bianchi V, VII_h
                     Farnsworth,
                     Collins-Ellis
==========================================================
```

Figure 3. The Bianchi models permitting higher symmetry subcases. The parameter c is zero iff the preferred spatial vector is hypersurface-orthogonal.

using non-orthogonal bases in various ways [116]; those possibilities will not be pursued further here.

6.2. DYNAMICS OF BIANCHI UNIVERSES

The set of tetrad equations (section 3) with these restrictions will determine the evolution of all the commutation functions and matter variables, and, hence, determine the metric and also the evolution of the Weyl tensor (these are regarded as auxiliary variables). In the case of orthogonal models — the fluid 4-velocity u^a is *parallel* to the normal vectors n^a — the matter variables will be just the fluid density and pressure [49]; in the case of tilted models — the fluid 4-velocity is *not parallel* to the normals — we also need the peculiar velocity of the fluid relative to the normal vectors [116], determining the fluid energy-momentum tensor decomposition relative to the normal vectors (a perfect fluid will appear as an imperfect fluid in that frame). Various papers relate these equations to variational principles and Hamiltonian formalisms, thus expressing them in terms of a potential formalism that gives an intuitive feel for what the evolution will be like [48, 175]. There have also been many numerical investigations of these dynamical equations and the resulting solutions. I will briefly consider three specific aspects here, then the relation to observations, and finally the related dynamical systems approach.

6.2.1. *Chaos in these universes?*
An ongoing issue since Misner's discovery of the 'Mixmaster' behaviour of the Type IX universes has been whether or not these solutions show chaotic behaviour as they approach the initial singularity (see [176] and Hobill in [51]). The potential approach represents these solutions as bouncing in an expanding approximately triangular shaped potential well, with three deep troughs attached to the corners. The return map approximation (a series of Kasner-like epochs, separated by collisions with the potential walls and consequent change of the Kasner parameters) suggests the motion is chaotic, but the question is whether this map represents the solutions of the differential equations well enough to reach this conclusion (for example, the potential walls are represented as flat in this approximation; and there are times when the solution moves up one of the troughs and then reflects back, but the return map does not represent this part of the motion). Part of the problem is that the usual definitions of chaos in terms of a Lyapunov parameter depend on the definition of time variable used, and there is a good case for changing to conformal Misner time in these investigations.

The issue may have been solved now by an analysis of the motion in terms of the attractors in phase space given by Cornish and Levin [177], suggesting that the motion is indeed chaotic, independent of the definition of time used. There may also be chaos in Type XIII solutions. Moreover, chaotic behaviour near the initial singularity was observed in solutions when a source-free magnetic Maxwell field is coupled to fluid space-times of Type I [178] and Type VI_0 [179].

6.2.2. *Horizons and whimper singularities*
In tilted Class B models, it is possible for there to be a dramatic change in the nature of the solution. This occurs where the surfaces of homogeneity change from being spacelike (at late times) to being timelike (at early times), these regions being separated by a null surface $\mathcal{H}$, the horizon associated with this change of symmetry. At earlier times the solution is no longer spatially homogeneous — it is inhomogeneous and stationary.[23] Associated with the horizon is a singularity where all scalar quantities are

[23]This kind of change happens also in the maximally extended Schwarzschild solution at the event horizon.

finite but components of the matter energy-momentum tensor diverge when measured in a parallelly propagated frame as one approaches the boundary of space-time (this happens because the parallelly propagated frame gets infinitely rescaled in a finite proper time relative to a family of KV's, which in the limit have this singularity as a fixed point). The matter itself originates at an anisotropic big bang singularity at the origin of the universe in the stationary inhomogeneous region.

Details of how this happens are given in Ellis and King [180], and phase plane diagrams for the simplest models in which this occurs — tilted LRS Type V models — in Collins and Ellis [170]. These models isotropise at late times, and can be arbitrarily similar to a low density FLRW model at the present day.

6.2.3. *Isotropisation properties*

An issue of importance is whether these models tend to isotropy at early or late times. An important paper by Collins and Hawking [172] shows that for ordinary matter many Bianchi models become anisotropic at very late times, even if they are very nearly isotropic at present. Thus isotropy is unstable in this case. On the other hand, a paper by Wald [181] showed that Bianchi models will tend to isotropise at late times if there is a positive cosmological constant present, implying that an inflationary era can cause anisotropies to die away. The latter work, however, while applicable to models with non-zero tilt angle, did not show this angle dies away, and indeed it does not do so in general (Goliath and Ellis [182]). Inflation also only occurs in Bianchi models if there is not too much anisotropy to begin with (Rothman and Ellis [183]), and it is not clear that shear and spatial curvature are in fact removed in all cases [184]. Hence, some Bianchi models isotropise due to inflation, but not all.

To study this kinds of question properly needs the use of phase planes. These will be discussed after briefly considering observations.

6.3. OBSERVATIONAL RELATIONS

Observational relations in these universes have been examined in detail.

(a) Redshift, area distance, and galaxy observations ((M, z) and (N, z) relations) are considered in MacCallum and Ellis [128]. Anisotropies can occur in all these relations, but many of the models will display discrete isotropies in the sky.

(b) The effect of tilt is to make the universe look inhomogeneous, even though it is spatially homogeneous (King and Ellis [116]). This will be reflected in particular in a dipole anisotropy in number counts, which will thus occur in rotating universes [20].[24]

(c) Element formation will be altered primarily through possible changes in the expansion time scale at the time of nucleosynthesis ([129, 186, 187]). This enables us to put limits on anisotropy from measured element abundances in particular Bianchi types. This effect could in principle go either way, so a useful conjecture [188] is that in fact the effect of anisotropy will always — despite the possible presence of rotation — be to speed up the expansion time scale in Bianchi models.

(d) CBR anisotropies will result in anisotropic universe models, for example many Class B Bianchi models will show a hot-spot and associated spiral pattern in the CBR sky [189]–[191].[25] This enables us to put limits on anisotropy from observed CBR anisotropy limits (Collins and Hawking [189], Bunn et al [192]). If 'reheating' takes place in an anisotropic universe, this will mix anisotropic temperatures from different directions, and hence distort the CBR spectrum [131].

Limits on present-day anisotropy from the CBR and element abundance measurements are very stringent: $|\sigma_0|/\Theta_0 \leq 10^{-6}$ to 10^{-12}, depending on the model. However, because of the anisotropies that can build up in both directions in time, this does not imply that the very early universe (before nucleosynthesis) or late universe will also be isotropic. The conclusion applies back to last scattering (CBR measurements) and to nucleosynthesis (element abundances). In both cases the conclusion is quite model dependent. Although very strong limits apply to some Bianchi models, they are much weaker for other types. Hence, one should be a bit cautious in what one claims in this regard.

6.4. DYNAMICAL SYSTEMS APPROACH

The most illuminating description of the evolution of families of Bianchi models is a dynamical systems approach based on the use of orthonormal tetrads, presented in detail in Wainwright and Ellis [51]. The main vari-

[24] They will also occur in FLRW models seen from a reference frame that is not comoving; hence, they should occur in the real universe if the standard interpretation of the CBR anisotropy as due to our motion relative to a FLRW universe is correct; see Ellis and Baldwin [185].

[25] This result is derived in a gauge-dependent way; it would be useful to have a gauge-invariant version.

ables used are essentially the commutation functions mentioned above, but rescaled by a common time dependent factor.

6.4.1. *Reduced differential equations*

The basic idea (Collins [193], Wainwright [194]) is to write the EFE in a way that enables one to study the evolution of the various physical and geometrical quantities *relative to the overall rate of expansion of the universe*, as described by the rate of expansion scalar Θ, or equivalently *the Hubble parameter* $H = \frac{1}{3}\Theta$. The remaining freedom in the choice of orthonormal tetrad needs to be eliminated by specifying the variables Ω^α implicitly or explicitly (for example by specifying them as functions of the $\sigma_{\alpha\beta}$). This also simplifies the other quantities (for example choice of a shear eigenframe will result in the tensor $\sigma_{\alpha\beta}$ being represented by two diagonal terms). One so obtains a reduced set of variables, consisting of H and the remaining commutation functions, which we denote symbolically by

$$\mathbf{x} = \left(\gamma^a{}_{bc}|_{\text{reduced}}\right) . \tag{172}$$

The physical state of the model is thus described by the vector $(H, \mathbf{x})$. The details of this reduction differ for the Class A and B models, and in the latter case there is an algebraic constraint of the form

$$g(\mathbf{x}) = 0 , \tag{173}$$

where g is a homogeneous polynomial.

The idea is now to normalise $\mathbf{x}$ with the Hubble parameter H. We denote the resulting variables by a vector $\mathbf{y} \in R^n$, and write:

$$\mathbf{y} = \frac{\mathbf{x}}{H} . \tag{174}$$

These new variables are *dimensionless*, and will be referred to as *expansion-normalised variables*. It is clear that each dimensionless state $\mathbf{y}$ determines a 1-parameter family of physical states $(\mathbf{x}, H)$. The evolution equations for the $\gamma^a{}_{bc}$ lead to evolution equations for H and $\mathbf{x}$ and hence for $\mathbf{y}$. In deriving the evolution equations for $\mathbf{y}$ from those for $\mathbf{x}$, the *deceleration parameter* q plays an important role. The Hubble parameter H can be used to define a scale factor S according to (25)

$$H = \frac{\dot{S}}{S} , \tag{175}$$

where $\cdot$ denotes differentiation with respect to t. The deceleration parameter, defined (see (57)) by $q = -\ddot{S}\,S/\dot{S}^2$, is related to $\dot{H}$ according to

$$\dot{H} = -(1+q)\,H^2 . \tag{176}$$

In order that the evolution equations define a flow, it is necessary, in conjunction with the rescaling (174), to introduce a *dimensionless time variable* τ according to

$$S = S_0 \, e^{\tau} \, , \tag{177}$$

where S_0 is the value of the scale factor at some arbitrary reference time. Since S assumes values $0 < S < +\infty$ in an ever-expanding model, τ assumes all real values, with $\tau \to -\infty$ at the initial singularity and $\tau \to +\infty$ at late times. It follows from equations (175) and (177) that

$$\frac{dt}{d\tau} = \frac{1}{H} \, , \tag{178}$$

and the evolution equation (176) for H can be written

$$\frac{dH}{d\tau} = -\,(1+q)\,H \, . \tag{179}$$

Since the right-hand side of the evolution equations for the $\gamma^a{}_{bc}$ are homogeneous of degree 2 in the $\gamma^a{}_{bc}$, the change (178) of the time variable results in H canceling out of the evolution equation for $\mathbf{y}$, yielding an autonomous differential equation (DE):

$$\frac{d\mathbf{y}}{d\tau} = \mathbf{f}(\mathbf{y}) \, , \quad \mathbf{y} \in R^n \, . \tag{180}$$

The constraint $g(\mathbf{x}) = 0$ translates into a constraint

$$g(\mathbf{y}) = 0 \, , \tag{181}$$

which is preserved by the DE. The functions $\mathbf{f} : R^n \to R^n$ and $g : R^n \to R$ are polynomial functions in $\mathbf{y}$. An essential feature of this process is that the evolution equation for H, namely (179), decouples from the remaining equations (180) and (181). In other words, the DE (180) describes the evolution of the non-tilted Bianchi cosmologies, the transformation (174) essentially scaling away the effects of the overall expansion. An important consequence is that the new variables are bounded near the initial singularity.

6.4.2. *Equations and orbits*

The first step in the analysis is to formulate the EFE, using expansion-normalised variables, as a DE (180) in R^n, possibly subject to a constraint (181). Thus one uses the tetrad equations presented above, now adapted to apply to the variables rescaled in this way. Since τ assumes all real values (for models which expand indefinitely), the solutions of (180) are defined for all τ and hence define a *flow* $\{\phi_\tau\}$ on R^n. The evolution of the cosmological models can thus be analyzed by studying the orbits of this flow in the physical region of state space, which is a subset of R^n defined by the requirement that the matter energy density μ be non-negative, i.e.

$$\Omega(\mathbf{y}) = \frac{\mu}{3H^2} \geq 0 , \tag{182}$$

where the density parameter Ω (see (57)) is a dimensionless measure of μ.

The *vacuum boundary*, defined by $\Omega(\mathbf{y}) = 0$, describes the evolution of vacuum Bianchi models, and is an invariant set which plays an important role in the qualitative analysis because vacuum models can be asymptotic states for perfect fluid models near the big-bang or at late times. There are other invariant sets which are also specified by simple restrictions on $\mathbf{y}$ which play a special role: the subsets representing each Bianchi type (Figure 2), and the subsets representing higher symmetry models, specifically the FLRW models and the LRS Bianchi models (according to Figure 3).

It is desirable that the dimensionless state space D in R^n is a *compact set*. In this case each orbit will have non-empty future and past limit sets, and hence there will exist a past attractor and a future attractor in state space. When using expansion-normalised variables, compactness of the state space has a direct physical meaning for ever-expanding models: if the state space is compact, then at the big-bang no physical or geometrical quantity diverges more rapidly than the appropriate power of H, and at late times no such quantity tends to zero less rapidly than the appropriate power of H. This will happen for many models; however, the state space for Bianchi Type VII_0 and Type VIII models is non-compact. This lack of compactness manifests itself in the behaviour of the Weyl tensor at late times.

6.4.3. *Equilibrium points and self-similar cosmologies*

Each ordinary orbit in the dimensionless state space corresponds to a one-parameter family of physical universes, which are conformally related by a constant rescaling of the metric. On the other hand, for an equilibrium point $\mathbf{y}^*$ of the DE (180) (which satisfies $\mathbf{f}(\mathbf{y}^*) = \mathbf{0}$), the deceleration parameter q is a constant, i.e. $q(\mathbf{y}^*) = q^*$, and we find

$$H(\tau) = H_0\, e^{(1+q\ )\tau} .$$

In this case, however, the parameter H_0 is no longer essential, since it can be set to unity by a translation of τ, $\tau \to \tau + \mathrm{const}$; then (178) implies that

$$H\,t = \frac{1}{1+q^*} , \tag{183}$$

so that by (172) and (174) the commutation functions are of the form (const)$\times t^{-1}$. It follows that the resulting cosmological model is *self-similar*. It then turns out that *to each equilibrium point of the DE (180) there corresponds a unique self-similar cosmological model.* In such a model the physical states at different times differ only by an overall change in the length scale. Such models are expanding, but in such a way that their dimensionless state does not change. They include the flat FLRW model ($\Omega = 1$) and the Milne model ($\Omega = 0$). All vacuum and non-tilted perfect fluid self-similar Bianchi solutions have been given by Hsu and Wainwright [195]. The equilibrium points determine the asymptotic behaviour of other more general models.

6.4.4. *Phase planes*

Many phase planes can be constructed explicitly. The reader is referred to Wainright and Ellis [51] for a comprehensive presentation and survey of results attained so far. Several interesting points emerge:

* *Relation to lower dimensional spaces*: it seems that the lower dimensional spaces, delineating higher symmetry models, can be skeletons guiding the development of the higher dimensional spaces (the more generic models). This is one reason why study of the exact higher symmetry models is of significance. The way this occurs is the subject of ongoing investigation (the key issue being how the finite dimensional dynamical systems corresponding to models with symmetry are imbedded in and relate to the infinite dimensional dynamical system describing the evolution of models without symmetry).

* *Identification of models in state space*: the analysis of the phase planes for Bianchi models shows that the procedure sometimes adopted of identifying all points in state space corresponding to the same model is not a good idea. For example the Kasner ring that serves as a framework for evolution of many other Bianchi models contains multiple realizations of the same Kasner model. To identify them as the same point in state space would make the evolution patterns very difficult to follow. It is better to keep them separate, but to learn to identify where multiple realizations of the same model occur (which is just the *equivalence problem* for cosmological models).

* *Isotropisation* is a particular issue that can be studied by use of these planes [196, 182]. It turns out that even in the classes of non-inflationary Bianchi models that contain FLRW models as special cases, not all models isotropise at some period of their evolution; and of those that do so, most

become anisotropic again at late times. Only an inflationary equation of state will lead to such isotropisation for a fairly general class of models (but in the tilted case it is not clear that the tilt angle will die away [182]); once it has turned off, anisotropic modes will again occur.

An important idea that arises out of this study is that of *intermediate isotropisation*: namely, models that become very like a FLRW model for a period of their evolution but start off and end up quite unlike these models. It turns out that many Bianchi types allow intermediate isotropisation, because the FLRW models are saddle points in the relevant phase planes. This leads to the following two interesting results:

> **Bianchi Evolution Theorem (1)**: Consider a family of Bianchi models that allow intermediate isotropisation. Define an ϵ-neighbourhood of a FLRW model as a region in state space where all geometrical and physical quantities are closer than ϵ to their values in a FLRW model. Choose a time scale L. Then no matter how small ϵ and how large L, there is an open set of Bianchi models in the state space such that each model spends longer than L within the corresponding ϵ-neighbourhood of the FLRW model.

(This follows because the saddle point is a fixed point of the phase flow; consequently the phase flow vector becomes arbitrarily close to zero at all points in a small enough open region around the FLRW point in state space.)

Consequently, although these models are quite unlike FLRW models at very early and very late times, there is an open set of them that are observationally indistinguishable from a FLRW model (choose L long enough to encompass from today to last coupling or nucleosynthesis, and ϵ to correspond to current observational bounds). Thus there exist many such models that are viable as models of the real universe in terms of compatibility with astronomical observations.

> **Bianchi Evolution Theorem (2)**: In each set of Bianchi models of a type admitting intermediate isotropisation, there will be spatially homogeneous models that are linearisations of these Bianchi models about FLRW models. These perturbation modes will occur in any almost-FLRW model that is generic rather than fine-tuned; however, the exact models approximated by these linearisations will be quite unlike FLRW models at very early and very late times.

(Proof is by linearising the equations above (see the following section) to obtain the Bianchi equations linearised about the FLRW models that occur at the saddle point leading to the intermediate isotropisation. These modes will be the solutions in a small neighbourhood about the saddle point permitted by the linearised equations (given existence of solutions to the non-linear equations, linearisation will not prevent corresponding linearised solutions existing).)

The point is that these modes can exist as linearisations of the FLRW model; if they do not occur, then initial data has been chosen to set these modes precisely to zero (rather than being made very small), which requires very special initial conditions. Thus these modes will occur in almost all almost-FLRW universes. Hence, if one believes in generality arguments, they will occur in the real universe. When they occur, they will at early and late times grow until the model is very far from a FLRW geometry (while being arbitrarily close to an FLRW model for a very long time, as per the previous theorem).

Exercise: Most studies of CBR anisotropies and nucleosynthesis are carried out for the Bianchi types that allow FLRW models as special cases (see Figure 3). Show that Bianchi models can approximate FLRW models for extended periods even if they do not belong to those types. What kinds of CBR anisotropies can occur in these models? (See e.g. [51].)

7. Almost-FLRW universes

The real universe is *not* FLRW because of all the structure it contains, and (because of the non-linearity of the EFE) the other exact solutions we can attain have higher symmetry than the real universe. Thus in order to obtain realistic models we can compare with detailed observations we need to approximate, aiming to obtain 'almost-FLRW' models representing a universe that is FLRW-like on a large scale but allowing for generic inhomogeneities on a small scale.

7.1. GAUGE PROBLEM

The major problem in studying perturbed models is the *gauge problem*, due to the fact that there is *no* identifiable fixed background model in General Relativity. One can start with a unique FLRW universe model with metric $\overline{g}_{ab}$ in some coordinate system, and perturb it to obtain a more realistic model: $\overline{g}_{ab} \to g_{ab} = \overline{g}_{ab} + \delta g_{ab}$, but then the process has no unique inverse: the background model $\overline{g}_{ab}$ is not uniquely determined by the lumpy universe model g_{ab} (no unique tensorial averaging process has been defined that will recover $\overline{g}_{ab}$ from g_{ab}). Many choices can be made. However, the usual variables describing perturbations depend on the way the (fictitious) background model $\overline{g}_{ab}$ is fitted to the real universe g_{ab}; these variables can be given any values one wants by changing this correspondence.[26] For example, the *dimensionless density contrast* δ representing a density perturbation is usually defined by

$$\delta(x) = \frac{\mu(x) - \overline{\mu}(x)}{\mu(x)}\ , \tag{184}$$

where $\mu(x)$ is the actual value of the density at the point x, while $\overline{\mu}(x)$ is the (fictitious) background value there, determined by the chosen mapping of the background model into the realistic lumpy model (Ellis and Bruni [197]). This quantity can be given any value we desire by altering that map; we can for example set it to zero by choosing the real surfaces of constant density to be the background surfaces of constant time and, hence, of constant density. Consequently, perturbation equations written in terms of this variable have as solution both physical modes and gauge modes, the latter corresponding to variation of gauge choice rather than to physical variation.

One way to solve this is by very carefully keeping track of the gauge used and the resulting gauge freedom; see Ma and Bertschinger [198] and Prof. Bertschinger's lectures here. The alternative is to use gauge-invariant variables. A widely used and fundamentally important set of such variables are those introduced by Bardeen [199], and used for example by Bardeen, Steinhardt and Turner [200]. Another possibility is use of Gauge-Invariant and 1+3 Covariant ('GIC') variables, i.e. variables that are gauge-invariant and also $1+3$ covariantly defined so that they have a clear geometrical meaning, and can be examined in any desired coordinate system. That is what will be pursued here.

In more detail: our aim is to examine perturbed models by using $1+3$ covariant variables defined in the real space-time (not the background), deriving exact equations for these variables in that space-time, and then approximating by linearising about a RW geometry to get the linearised equations describing the evolution of density inhomogeneities in almost-FLRW universes. How do we handle gauge invariance in this approach? We rely on the

> **Gauge Invariance Lemma** (Stewart and Walker [201]): If a quantity $T^{\cdots}{}_{\cdots}$ vanishes in the background space-time, then it is gauge-invariant (to <u>all</u> orders).

[26]This is often represented implicitly rather than explicitly, by assuming that points with the same coordinate values in the background space and more realistic model map to each other; then the gauge freedom is contained in the coordinate freedom available in the realistic universe model.

[The proof is straightforward : If $\overline{T}^{\cdots}{}_{\cdots} = 0$, then $\delta T^{\cdots}{}_{\cdots} = T^{\cdots}{}_{\cdots} - \overline{T}^{\cdots}{}_{\cdots} = T^{\cdots}{}_{\cdots}$, which is manifestly independent of the mapping Φ from $\overline{S}$ to S (it does not matter how we map $\overline{T}^{\cdots}{}_{\cdots}$ from $\overline{S}$ to S when $\overline{T}^{\cdots}{}_{\cdots}$ vanishes).] The application to almost-FLRW models follows (see Ellis and Bruni [197]), where we use an order-of-magnitude notation as follows: Given a smallness parameter ϵ, $\mathcal{O}[n]$ denotes $\mathcal{O}(\epsilon^n)$, and $A \approx B$ means $A - B = \mathcal{O}[2]$ (i.e. these variables are equivalent to $\mathcal{O}[1]$). When $A \approx 0$ we shall regard A as vanishing when we linearise (for it is zero to the accuracy of relevant first-order calculations). Then,

- Zero-order variables are μ, p, Θ, and their time derivatives, $\dot{\mu}$, $\dot{p}$, $\dot{\Theta}$,
- First-order variables are $\dot{u}^a$, σ_{ab}, ω^a, q^a, π_{ab}, E_{ab}, H_{ab}, X_a, Z_a, and their time and space derivatives.

As these first-order variables all vanish in exact FLRW universes, provided u^a is uniquely defined in the realistic (lumpy) almost-FLRW universe model, they are all uniquely defined GIC variables. Thus this set of variables provides what we wanted: $1+3$ covariant variables characterising departures from a FLRW geometry (and, in particular, the spatial inhomogeneity of a universe) that are gauge-invariant when the universe is almost-FLRW. Because they are tensors defined in the real space-time, we can evaluate them in any coordinate system we like in that space-time.

7.1.1. *Key variables*

Two simple gauge-invariant quantities give us the information we need to discuss the time evolution of density fluctuations. The basic quantities we start with are the orthogonal projections of the energy density gradient, i.e. the vector $X_a \equiv \tilde{\nabla}_a \mu$, and of the expansion gradient, i.e. the vector $Z_a \equiv \tilde{\nabla}_a \Theta$. The first can be determined (a) from virial theorem estimates and large-scale structure observations (as, e.g., in the POTENT programme), (b) by observing gradients in the numbers of observed sources and estimating the mass-to-light ratio (Kristian and Sachs [98], Eq. (39)), and (c) by gravitational lensing observations. However, these do not directly correspond to the quantities usually calculated; but two closely related quantities do. The first is the *comoving fractional density gradient*:

$$\mathcal{D}_a \equiv S \, \frac{X_a}{\mu} \, , \tag{185}$$

which is gauge-invariant and dimensionless, and represents the spatial density variation over a fixed comoving scale. Note that S, and so $\mathcal{D}_a$, is defined only up to a constant by equation (25); this allows it to represent the density variation between any neighbouring worldlines. The vector $\mathcal{D}_a$ can be separated into a magnitude $\mathcal{D}$ and direction e_a

$$\mathcal{D}_a = \mathcal{D} \, e_a \ , \ e_a e^a = 1 \ , \ e_a u^a = 0 \quad \Rightarrow \quad \mathcal{D} = (\mathcal{D}_a \mathcal{D}^a)^{1/2} \, . \tag{186}$$

The magnitude $\mathcal{D}$ is the gauge-invariant variable[27] that most closely corresponds to the intention of the usual $\delta = \delta\mu/\mu$ given in (184). The crucial difference from the usual definition is that $\mathcal{D}$ represents a (real) spatial fluctuation, rather than a (fictitious) time fluctuation, and does so in a GIC manner. An important auxiliary variable in what follows is the *comoving spatial expansion gradient*:

$$\mathcal{Z}_a \equiv S\, Z_a \ . \tag{187}$$

The issue now is, can we find a set of equations determining how these variables evolve? Yes we can; they follow from the $1+3$ covariant exact equations of the previous sections.

7.2. DYNAMICAL EQUATIONS

We can determine exact propagation equations along the fluid flow lines for the quantities defined in the previous section, and then linearise these to the almost-FLRW case. The basic linearised equations are given by Hawking [22] (see his equations (13) to (19)); we add to them the linearised propagation equations for the gauge-invariant spatial gradients defined above [197].

7.2.1. *Growth of inhomogeneity*

Taking the spatial gradient of the equation of energy conservation (37) (for the case of a perfect fluid), we find [197]

$$\tilde{\nabla}_a(\dot{\mu}) + \Theta\, \tilde{\nabla}_a(\mu+p) + (\mu+p)\, \tilde{\nabla}_a\Theta = 0 \ ,$$

i.e.

$$h^b{}_a \nabla_b(u^c \nabla_c \mu) + \Theta\,(X_a + \tilde{\nabla}_a p) + (\mu+p)\, Z_a = 0 \ .$$

Using Leibniz' Rule and changing the order of integration in the second-derivative term (and noting that the pressure-gradient term cancels on using the momentum conservation equation (38)), we obtain the *fundamental equation for the growth of inhomogeneity*:

$$\dot{X}^{\langle a\rangle} + \tfrac{4}{3}\,\Theta\, X^a \;=\; -\,(\mu+p)\, Z^a - \sigma^a{}_b\, X^b + \eta^{abc}\,\omega_b\, X_c \ , \tag{188}$$

with source term Z^a. On taking the gradient of the Raychaudhuri equation (29) we find the companion equation for that source term:

$$\begin{aligned}\dot{Z}^{\langle a\rangle} + \Theta\, Z^a \;=&\; -\tfrac{1}{2}\, X^a - \sigma^a{}_b\, Z^b + \eta^{abc}\,\omega_b\, Z_c + \dot{u}^a\, \mathcal{R} - 2\,\tilde{\nabla}^a(\sigma^2 - \omega^2) \\ &+ \tilde{\nabla}^a(\tilde{\nabla}_b \dot{u}^b + \dot{u}_b \dot{u}^b) \ , \end{aligned} \tag{189}$$

[27] Or, equivalently in the linear case, the spatial divergence of $\mathcal{D}_a$.

where

$$\mathcal{R} = -\tfrac{1}{3}\Theta^2 + \tilde{\nabla}_a \dot{u}^a + (\dot{u}_a \dot{u}^a) - 2\sigma^2 + 2\omega^2 + \mu + \Lambda\,. \qquad (190)$$

These equations contain no information not implied by the others already given; nevertheless, they are useful in that they are exact equations directly giving the rate of growth of inhomogeneity in the generic (perfect fluid) case, the second, together with the evolution equations above, giving the rate of change of all the source terms in the first.

The procedure now is to systematically approximate all the dynamical equations, and, in particular, the structure growth equations given above, by dropping all terms of second order or higher in the implicit[28] expansion variable ϵ. Thus we obtain the linearised equations as approximations to the exact equations above by noting that in the almost-FLRW case,

$$\dot{X}^{\langle a \rangle} + \tfrac{4}{3}\Theta X^a = -(\mu + p)\, Z^a + \mathcal{O}[2] \qquad (191)$$

$$\dot{Z}^{\langle a \rangle} + \Theta Z^a = -\tfrac{1}{2} X^a + \dot{u}^a \mathcal{R} + \tilde{\nabla}^a(\tilde{\nabla}_b \dot{u}^b) + \mathcal{O}[2]\,, \qquad (192)$$

where

$$\mathcal{R} = -\tfrac{1}{3}\Theta^2 + \tilde{\nabla}_a \dot{u}^a + \mu + \Lambda + \mathcal{O}[2]\,. \qquad (193)$$

Then we linearise the equations by dropping the terms $\mathcal{O}[2]$, *so from now on in this section '=' means equal up to terms of order* ϵ^2.

7.3. DUST

In the case of dust, $p = 0 \Rightarrow \dot{u}^a = 0$, and the equations (191) and (192) for growth of inhomogeneity become

$$S^{-4}\, h^a{}_b (S^4 X^b)^{\cdot} = -\mu Z^a \qquad (194)$$

$$S^{-3}\, h^a{}_b (S^3 Z^b)^{\cdot} = -\tfrac{1}{2} X^a\,, \qquad (195)$$

This closes up to give a second-order equation (take the time derivative of the first and substitute from the second and the energy conservation equation (37)). To compare with the usual equations, change to the variables $\mathcal{D}_a$ and $\mathcal{Z}_a$ (see (185) and (187)). Then the equations become

$$\dot{\mathcal{D}}^{\langle a \rangle} = -\mathcal{Z}^a \qquad (196)$$

$$\dot{\mathcal{Z}}^{\langle a \rangle} = -\tfrac{2}{3}\Theta\, \mathcal{Z}^a - \tfrac{1}{2}\mu\, \mathcal{D}^a\,. \qquad (197)$$

These directly imply the second-order equation (take the time derivative of the first equation!)

$$0 = \ddot{\mathcal{D}}_{\langle a \rangle} + \tfrac{2}{3}\Theta\, \dot{\mathcal{D}}_{\langle a \rangle} - \tfrac{1}{2}\mu\, \mathcal{D}_a\,, \qquad (198)$$

[28] One could make this variable explicit but there does not seem to be much gain in doing so.

which is the usual equation for growth of density inhomogeneity in dust universes, and has the usual solutions: when $k = 0$, then[29] $S(t) \propto t^{2/3}$, and we obtain

$$\mathcal{D}_a(t, x^\alpha) = d_{+a}(x^\alpha)\, t^{2/3} + d_{-a}(x^\alpha)\, t^{-1} \ , \qquad \dot{d}_{ia} = 0 \ , \tag{199}$$

(where t is proper time along the flow lines).[30] This shows the growing mode that leads to structure formation and the decaying mode that dissipates previously existing inhomogeneities. It has been obtained in a GIC way: all the first-order variables, including in particular those in this equation, are gauge-invariant, and there are no gauge modes. Furthermore, we have available the fully non-linear equations, Eqs. (188) and (189), and so can estimate the errors in the neglected terms, and set up a systematic higher-order approximation scheme for solutions of these equations. Solutions for other background models (with $k = \pm 1$ or $\Lambda \neq 0$) can be obtained by substituting the appropriate values in (198) for the background variables Θ and μ, possibly changing to conformal time to simplify the calculation.

Exercise: What is the growth rate at late times in a low-density universe, when the expansion is curvature dominated and so is linear: $S(t) = c_0\, t$? What if there is a cosmological constant, so the late time expansion is exponential: $S(t) = c_0 \exp(H_0\, t)$, where c_0 and H_0 are constants?

7.3.1. *Other quantities*

We have concentrated here on the growth of inhomogeneities. However, all the $1 + 3$ covariant equations of the previous sections apply and can be linearised in a straightforward way to the almost-FLRW case (and one can find suitable coordinates and tetrads in order to employ the complete set of tetrad equations in this context, too). Doing so, one can in particular study vorticity perturbations and gravitational wave perturbations; see the pioneering paper by Hawking [22]. We will not consider these further here; however, a series of interesting issues arise. The GIC approach to gravitational waves is examined in Hogan and Ellis [202] and Dunsby, Bassett and Ellis [23], and the possibility of longitudinal waves besides the familiar transverse ones in van Elst and Ellis [26].

A further important issue is the effect of perturbations on observations (apart from the CBR anisotropies, discussed below). It has been known for

[29] This is the background rather than the real value of this quantity, which is what should really be used here; it is determined in the real space-time by $\dot{S}/S = \frac{1}{3}\,\Theta$. However, as we are linearising, the difference makes only a second-order change in the coefficients in the equations, which we can neglect.

[30] We can also see that if $\Theta = 0$, there will be an exponential rather than power-law growth.

a long time that anisotropies can affect area distances as well as redshifts (see Bertotti [203], Kantowski [204]); the Dyer–Roeder formula ([205], see also [99]) can be used at any redshift for those many rays that propagate in the lower-density regions between inhomogeneities; however, this formula is not accurate for those ray bundles that pass very close to matter, where shearing becomes important. This is closely related to the *averaging problem* (see e.g. Ellis [88] or Boersma [206]): how can dynamics and observations of a universe which is basically empty almost everywhere average out correctly to give the same dynamics and observations as a universe which is exactly spatially homogeneous? What differences will there be from the FLRW case? We will not pursue this further here except to state that it is believed this does in fact work out OK: in the fully inhomogeneous case it is the Weyl tensor that causes distortions in the empty spaces between astrophysical objects (as in gravitational lensing), and hence causes convergence of both timelike and null geodesics; these, however, average out to give zero average distortion and the same convergence effect as a FLRW space-time with zero Weyl tensor, but with the Ricci tensor causing focusing of these curves (see [207] for a discussion of the null case; however, some subtleties arise here in terms of the way areas are defined when strong lensing takes place [208, 145, 209]).

7.4. PERFECT FLUIDS

A similar GIC analysis to the dust case has been given by Ellis, Hwang and Bruni determining FLRW perturbations for the perfect fluid case [210, 19]. This gives the single-fluid equation for growth of structure in the universe (again, derived in a GIC manner), and includes as special cases a fully $1+3$ covariant derivation of the Jeans length and of the speed of sound for barotropic perfect fluids. To evaluate the last two terms in (192) when $p \neq 0$, using (38) we see that, to first order,

$$\tilde{\nabla}_a(\tilde{\nabla}_b \dot{u}^b) = -\frac{\tilde{\nabla}_a(\tilde{\nabla}_b \tilde{\nabla}^b p)}{(\mu + p)} \,. \tag{200}$$

But, for simplicity considering only the case of vanishing vorticity,[31] we have

$$\tilde{\nabla}_b \tilde{\nabla}^b \tilde{\nabla}_a p = \tilde{\nabla}_b \tilde{\nabla}_a \tilde{\nabla}^b p \,, \tag{201}$$

and, on using the Ricci identities for the $\tilde{\nabla}$-derivatives and the zero-order relation ${}^3R_{ab} = \frac{1}{3}\,{}^3R\, h_{ab}$ for the 3-dimensional Ricci tensor, we obtain

$$\tilde{\nabla}_a \tilde{\nabla}_b \tilde{\nabla}^b p = \tilde{\nabla}_b \tilde{\nabla}_a \tilde{\nabla}^b p - \tfrac{1}{3}\,{}^3R\, \tilde{\nabla}_a p \,. \tag{202}$$

[31] When it is non-zero, (27) must be taken into account when commuting derivatives; see [19].

Thus, on using ${}^3R = \mathcal{K} = 6\,k/S^2$, we find

$$\tfrac{1}{2}\mathcal{K}\,\dot{u}_a = -\frac{1}{(\mu+p)}\frac{3\,k}{S^2}\tilde{\nabla}_a p\,,$$

$$\tilde{\nabla}_a(\tilde{\nabla}_b \dot{u}^b) = \frac{1}{(\mu+p)}\left(\frac{2k}{S^2}\tilde{\nabla}_a p - \tilde{\nabla}^2\tilde{\nabla}_a p\right)\,, \tag{203}$$

introducing the notation $\tilde{\nabla}^2\tilde{\nabla}_a p \equiv \tilde{\nabla}_b\tilde{\nabla}^b\tilde{\nabla}_a p$. In performing this calculation, note that there will *not* be 3-spaces orthogonal to the fluid flow if $\omega^a \neq 0$, but still we can calculate the 3-dimensional orthogonal derivatives as usual (by using the projection tensor $h^a{}_b$); the difference from when $\omega^a = 0$ will be that the quantity we calculate as a curvature tensor, using the usual definition from commutation of second derivatives, will not have all the usual curvature tensor symmetries. Nevertheless, the zero-order equations, representing the curvature of the 3-spaces orthogonal to the fluid flow in the background model, will agree with the linearised equations up to the required accuracy.

Now if $p = p(\mu, s)$, where s is the entropy per particle, we find

$$\tilde{\nabla}_a p = \left(\frac{\partial p}{\partial \mu}\right)_{s=\text{const}}\tilde{\nabla}_a \mu + \left(\frac{\partial p}{\partial s}\right)_{\mu=\text{const}}\tilde{\nabla}_a s\,. \tag{204}$$

We assume we can ignore the second term (pressure variations caused by spatial entropy variations) relative to the first (pressure variations caused by energy density variations) and spatial variations in the scale function S (which would at most cause second-order variations in the propagation equations). Then (ignoring terms due to the spatial variation of $\partial p/\partial\mu$, which will again cause second-order variations) we find in the zero-vorticity case,

$$S\left[\,\tfrac{1}{2}\mathcal{K}\,\dot{u}_a + \tilde{\nabla}_a(\tilde{\nabla}_b\dot{u}^b)\,\right] = -\frac{1}{(1+p/\mu)}\left(\frac{\partial p}{\partial\mu}\right)\left(\frac{k}{S^2}\mathcal{D}_a + \tilde{\nabla}^2\mathcal{D}_a\right)\,. \tag{205}$$

This is the result that we need in proceeding with (192).

7.4.1. *Second-order equations*

The equations for propagation can now be used to obtain second-order equations for $\mathcal{D}_a$.[32] For easy comparison, we follow Bardeen [199] by defining

$$w = \frac{p}{\mu}\,, \quad c_s^2 = \frac{\partial p}{\partial \mu} \quad \Rightarrow \quad \left(\frac{p}{\mu}\right)^{\cdot} \equiv \dot{w} = -\,\Theta\,(1+w)\,(c_s^2 - w)\,. \tag{206}$$

[32] And the variable $\Phi_a = \mu\, S^2\, \mathcal{D}_a$ that corresponds more closely to Bardeen's variable; see [19].

Now differentiation of (191), projection orthogonal to u^a, and linearisation gives a second-order equation for $\mathcal{D}_a$ (we use Eqs. (29), (192), (206) and (205) in the process). We find

$$\begin{aligned} 0 &= \ddot{\mathcal{D}}_{\langle a\rangle} + (\tfrac{2}{3} - 2w + c_s^2)\,\Theta\,\dot{\mathcal{D}}_{\langle a\rangle} \\ &\quad - [\,(\tfrac{1}{2} + 4w - \tfrac{3}{2}w^2 - 3c_s^2)\,\mu + (c_s^2 - w)\,\frac{12\,k}{S^2}\,]\,\mathcal{D}_a \\ &\quad + c_s^2\,[\,\frac{2\,k}{S^2}\,\mathcal{D}_a - \tilde{\nabla}^2\mathcal{D}_a\,]\,. \end{aligned} \tag{207}$$

This equation is the *basic result of this subsection*; the rest of the discussion examines its properties and special cases. It is a second-order equation determining the evolution of the GIC density variation variable $\mathcal{D}_a$ along the fluid flow lines. It has the form of a *wave equation* with extra terms due to the expansion of the universe, gravity and the spatial curvature.[33] We bracket the last two terms together, because when we make a harmonic decomposition these terms together give the harmonic eigenvalues n^2.

This form of the equations allows for a variation of $w = p/\mu$ with time. However, if $w = \text{const}$, then from (206) $c_s^2 = w$, and the equation simplifies to

$$\begin{aligned} 0 &= \ddot{\mathcal{D}}_{\langle a\rangle} + (\tfrac{2}{3} - w)\,\Theta\,\dot{\mathcal{D}}_{\langle a\rangle} \\ &\quad - \tfrac{1}{2}\,(1-w)\,(1+3w)\,\mu\,\mathcal{D}_a + w\,[\,\frac{2\,k}{S^2}\,\mathcal{D}_a - \tilde{\nabla}^2\mathcal{D}_a\,]\,. \end{aligned} \tag{208}$$

The matter source term vanishes if $w = 1$ (the case of 'stiff matter' $\Leftrightarrow p = \mu$) or $w = -\frac{1}{3}$ (the case $p = -\frac{1}{3}\mu$, corresponding to matter with no active gravitational mass). Between these two limits ('ordinary matter'), the matter term is positive and tends to cause the density gradient to increase ('gravitational aggregation'); outside these limits, the term is negative and tends to cause the density gradient to decrease ('gravitational smoothing'). Also the sign of the damping term (giving the adiabatic decay of inhomogeneities) is positive if $\frac{2}{3} > w$ (that is, $2\mu > 3p$) but negative otherwise (they adiabatically grow rather than decay in this case). The equation reduces correctly to the corresponding dust equation in the case $w = 0$. Two other cases of importance are:

* *Speed of sound*: when Θ, μ, and k/S^2 can be neglected, we see directly from (207) that c_s introduced above *is* the speed of sound (and that imaginary values of c_s, that is, negative values of $\partial p/\partial\mu$, then lead to exponential

[33] We have dropped Λ in these equations; it can be represented by setting $w = -1$.

growth or decay rather than oscillations).

* *Radiation*: In the case of pure radiation, $\gamma = \frac{4}{3}$ and $w = \frac{1}{3} = c_s^2$. Then we find from (208)

$$0 = \ddot{\mathcal{D}}_{\langle a\rangle} + \tfrac{1}{3}\Theta\,\dot{\mathcal{D}}_{\langle a\rangle} - \tfrac{2}{3}\,(\tfrac{1}{3}\Theta^2 + \frac{3\,k}{S^2})\,\mathcal{D}_a + \tfrac{1}{3}\,[\,\frac{2\,k}{S^2}\,\mathcal{D}_a - \tilde{\nabla}^2\mathcal{D}_a\,]\,. \qquad (209)$$

7.4.2. *Harmonic decomposition*

It is standard (see, e.g., [22] and [199]) to decompose the variables harmonically, thus effectively separating out the time and space variations by turning the differential equations for time variation of the perturbations as a whole into separate time variation equations for each component of spatial variation, characterised by comoving wavenumber. This conveniently represents the idea of a comoving wavelength for the matter inhomogeneities. In our case we do so by writing $\mathcal{D}_a$ in terms of harmonic vectors $Q_a^{(n)}$, from which the background expansion has been factored out.

We start with the defining equations of the $1+3$ covariant *scalar harmonics* $Q^{(n)}$,

$$\dot{Q}^{(n)} = 0\;, \qquad \tilde{\nabla}^2 Q^{(n)} = -\,\frac{n^2}{S^2}\,Q^{(n)}\;, \qquad (210)$$

corresponding to Bardeen's scalar Helmholtz equation (2.7) [199], but expressed $1+3$ covariantly following Hawking [22]. From these quantities we define the $1+3$ covariant *vector harmonics* (cf. [199], equations (2.8) and (2.10); we do not divide by the wavenumber, however, so our equations are valid even if $n = 0$)

$$Q_a^{(n)} \equiv S\,\tilde{\nabla}_a Q^{(n)} \Rightarrow Q_a^{(n)}\,u^a = 0\;,\; \dot{Q}_{\langle a\rangle}^{(n)} \approx 0\;,\; \tilde{\nabla}^2 Q_a^{(n)} = -\,\frac{(n^2 - 2k)}{S^2}\,Q_a^{(n)}\;, \qquad (211)$$

(the factor S ensuring these vector harmonics are approximately covariantly constant along the fluid flow lines in the almost-FLRW case). We write $\mathcal{D}_a$ in terms of these harmonics:

$$\mathcal{D}_a = \sum_n \mathcal{D}^{(n)}\,Q_a^{(n)}\;, \qquad \tilde{\nabla}_a\mathcal{D}^{(n)} \approx 0\;, \qquad (212)$$

where $\mathcal{D}^{(n)}$ is the harmonic component of $\mathcal{D}_a$ corresponding to the *comoving wavenumber* n, containing the time variation of that component; to first order, $\mathcal{D}^{(n)} \equiv \mu^{(n)}/\mu$. Putting this decomposition in the linearised equations (207) and (208), the harmonics decouple. Thus, for example we obtain from

(208) the n-th harmonic equation

$$0 = \ddot{\mathcal{D}}^{(n)} + (\tfrac{2}{3} - w)\,\Theta\,\dot{\mathcal{D}}^{(n)} - [\ \tfrac{1}{2}(1-w)\,(1+3w)\,\mu - w\,\frac{n^2}{S^2}\ ]\ \mathcal{D}^{(n)}\ , \quad (213)$$

(valid for each $n \geq 0$), where one can if one wishes substitute for μ from the Friedmann equation in terms of Θ and k/S^2. This equation shows how the growth of the inhomogeneity depends on the comoving wavelength. For the case of radiation, $w = \frac{1}{3}$, this is

$$0 = \ddot{\mathcal{D}}^{(n)} + \tfrac{1}{3}\,\Theta\,\dot{\mathcal{D}}^{(n)} - [\ \tfrac{2}{3}\,\mu - \tfrac{1}{3}\,\frac{n^2}{S^2}\ ]\ \mathcal{D}^{(n)}\ . \quad (214)$$

7.5. IMPLICATIONS

To determine the solutions explicitly, we have to substitute for μ, Θ and S from the zero-order equations. The most important issue is which terms dominate.

7.5.1. *Jeans instability*

Jeans' criterion is that *gravitational collapse* will tend to occur if the combination of the matter term and the term containing the Laplace operator in (207) or (208) is *positive* [211]; that is, if

$$\tfrac{1}{2}(1-w)\,(1+3w)\,\mu\,\mathcal{D}_a > w\,[\ \frac{2\,k}{S^2}\,\mathcal{D}_a - \tilde{\nabla}^2\mathcal{D}_a\ ]\ , \quad (215)$$

when $c_s^2 = w$. Using the harmonic decomposition, this can be expressed in terms of an equivalent scale: from (213) gravitational collapse tends to occur for a mode $\mathcal{D}^{(n)}$ if

$$\tfrac{1}{2}(1-w)\,(1+3w)\,\mu > w\,\frac{n^2}{S^2}\ , \quad (216)$$

that is, if

$$n_J \equiv \left[\ (1-w)\left(\frac{1}{w}+3\right)\frac{\mu(t)}{2}\ \right]^{1/2} S(t) > n\ . \quad (217)$$

In terms of wavelengths the *Jeans length* is defined by

$$\lambda_J \equiv \frac{2\pi\,S(t)}{n_J} = c_s\,c\,\sqrt{\frac{\pi}{G\mu(t)}\ \frac{1}{(1-w)(1+3w)}}\ , \quad (218)$$

where we have re-established the fundamental constants c and G ($w = (c_s/c)^2$). Thus, *gravitational collapse* will occur for small n (wavelengths longer than λ_J), but not for large n (wavelengths less than λ_J), for the pressure gradients are then large enough to resist the collapse and lead to *oscillations* instead.

For *non-relativistic matter*, $|w| \ll 1$ and $\mu = \rho_m c^2$, where ρ_m is the mass density of the matter, so

$$\lambda_J = c_s \sqrt{\frac{\pi}{G\rho_m(t)}} \; . \tag{219}$$

Then the Jeans mass will be

$$M_J = \frac{4\pi}{3}\,\rho_m\,\lambda_J^3 = \frac{4\pi}{3}\,c_s^3 \left(\frac{\pi}{G}\right)^{3/2} \rho_m^{-1/2} \; . \tag{220}$$

For *radiation*, where $w = \frac{1}{3}$ and $\mu = \rho_r c^2$, collapse will occur if

$$(2\mu)^{1/2} > \frac{n_J}{S} \quad \Leftrightarrow \quad \lambda < \lambda_J = c_s \sqrt{\frac{3\pi}{4G\rho_r(t)}} \; . \tag{221}$$

The corresponding Jeans mass for matter coupled to the radiation will be

$$M_J = \frac{4\pi}{3}\,\rho_m\,\lambda_J^3 = \frac{4\pi}{3}\,c_s^3 \left(\frac{3\pi}{4G\rho_r}\right)^{3/2} \rho_m \; . \tag{222}$$

7.5.2. *Short-wavelength solutions*

For wavelengths much shorter than the Jeans length, equation (213) becomes the damped harmonic equation

$$0 = \ddot{\mathcal{D}}^{(n)} + (\tfrac{2}{3} - w)\,\Theta\,\dot{\mathcal{D}}^{(n)} + w\,\frac{n^2}{S^2}\,\mathcal{D}^{(n)} \; , \tag{223}$$

giving oscillations. In the early universe, during radiation dominated expansion before decoupling, the tight coupling of the dominant radiation and matter leads to a fluid with $w = \frac{1}{3}$; then the short-wavelength equation becomes

$$0 = \ddot{\mathcal{D}}^{(n)} + \tfrac{1}{3}\,\Theta\,\dot{\mathcal{D}}^{(n)} + \tfrac{1}{3}\,\frac{n^2}{S^2}\,\mathcal{D}^{(n)} \; , \tag{224}$$

giving the *acoustic oscillations* during that era for modes such that $\lambda < \lambda_J = c_s c\,(3\pi/4G\mu)^{1/2}$.

7.5.3. *Long-wavelength solutions*

For wavelengths much longer than the Jeans length, we can drop the Laplace operator terms in (213) to obtain

$$0 = \ddot{\mathcal{D}}^{(n)} + (\tfrac{2}{3} - w)\,\Theta\,\dot{\mathcal{D}}^{(n)} - \tfrac{1}{2}\,(1 - w)\,(1 + 3w)\,\mu\,\mathcal{D}^{(n)} \; . \tag{225}$$

Thus, the second-order propagation equations become ordinary differential equations along the fluid flow lines, easily solved for particular equations of state. In the case of radiation ($w = \frac{1}{3}$) we find

$$0 = \ddot{\mathcal{D}}^{(n)} + \tfrac{1}{3}\,\Theta\,\dot{\mathcal{D}}^{(n)} - \tfrac{2}{3}\left(\tfrac{1}{3}\,\Theta^2 + \frac{3\,k}{S^2}\right)\mathcal{D}^{(n)}\;, \tag{226}$$

when $\lambda > \lambda_J = c_s c\,(3\pi/4G\mu)^{1/2}$. When $k = 0$, then $S(t) \propto t^{1/2}$, and we obtain in this long-wavelength limit

$$0 = \mathcal{D}_a = d_{+a}(x^\alpha)\;t + d_{-a}(x^\alpha)\;t^{-1/2}\;, \qquad \dot{d}_{ia}\;, \tag{227}$$

(where t is proper time along the flow lines). The corresponding standard results in the synchronous and comoving proper time gauges differ, being modes proportional to t and to $t^{1/2}$; we obtain the same growth law as derived in the comoving time orthogonal gauge and equivalent gauges. As our variables are GIC, we believe they show the latter gauges represent the physics more accurately than any other. Note that, moreover, we obtain no fictitious modes (proportional to t^{-1}) because we are using GIC variables.

7.5.4. *Change of behaviour with time*

Any particular inhomogeneity will have a constant comoving size and, hence, constant comoving wavelength λ and constant comoving wavenumber n as defined above. However, the Jeans length will *vary with time*.

During the radiation era,[34] $S \propto t^{1/2}$ and $\mu = \frac{3}{4}t^{-2}$ (see (114)), so (dropping the dimensional constants) $\rho_m \propto t^{-3/2}$, and the comoving Jeans length

$$\lambda_J = \sqrt{\tfrac{1}{3}\,\pi\,t^2} \propto t \tag{228}$$

will steadily grow to a value $\lambda_J^{\max}$ at matter–radiation equality, while the Jeans mass of coupled matter will grow as

$$M_J = \frac{4\pi}{3}\,\rho_m\,\lambda_J^3 \propto t^{-3/2}\,t^3 = t^{3/2}\;. \tag{229}$$

Thereafter, until recombination, the Jeans mass stays constant: the matter and radiation are still tightly coupled but now the universe is matter dominated and the speed of sound of the coupled fluid depends on the matter density: $c_s = (c/\sqrt{3})\,(\,1 + (3\rho_m/4\rho_r)\,)$ (see Rees [212]). After recombination the Jeans length and mass will rapidly die away because as the matter and radiation decouple, leading to $c_s \to 0$ and so $\lambda_J \to 0$. Each wavelength

[34]This is early enough that we can ignore the curvature term in the Friedmann equation.

λ longer than λ_J^{max} will have a growing mode as in (227) until the Jeans wavelength becomes greater than λ; it will then stop growing and undergo acoustic oscillations which will last until decoupling when the Jeans length drops towards zero and matter-dominated growth starts according to (199). Growth of small perturbations eventually slows down when the universe becomes curvature dominated at late times (when this happens depends on Ω_0; in a critical density universe, it never occurs).

Thus, the *key times* for any wavelength after the initial perturbations have been seeded[35] are (i) t_J, when they become smaller than the Jeans length (if they are small enough that this occurs), (ii) t_{equ}, when the matter dominated era starts (which will be before decoupling of matter and radiation, because $\Omega_0 \geq 0.1$), (iii) t_{dec}, when decoupling takes place. The acoustic oscillations have constant amplitude in the radiation dominated era from t_J until t_{equ}, and then die away as $t^{-1/6}$ in the matter dominated era until they end at t_{dec} [212]. Baryonic inhomogeneities 'freeze out' at that time; they then start growing by damped gravitational attraction. If they grow large enough, changing to non-linear collapse and ultimately star formation, then local energy generation starts.

Exercise: Establish these behaviours from the equations given above.

By contrast CDM freezes out at t_{equ} and starts growth at that time (Rees [97]). Thus in a CDM dominated universe, as is often supposed, the CDM fluctuations that govern structure formation start gravitational growth earlier than the baryons. They then govern the growth of inhomogeneities, attracting the baryons into their potential wells; a 2-fluid description representing the separate average velocities and their relative motion (see below) is needed to examine this.

This picture has to be modified, however, by allowing for diffusion effects. Kinetic theory is the best way to tackle this. The result is damping of perturbations below diffusion scales which depend on whether or not Hot Dark Matter ('HDM') is present; baryonic fluctuations on small scales are attenuated by photon viscosity and free-streaming of neutrinos [212, 97].

[35]The usual assumption is that perturbations are essentially unaffected by all the strong interactions in the early universe after the end of inflation, including the ending of pair production (when matter ceases to be relativistic), decoupling of neutrinos, and the irreversible interactions during baryosynthesis and nucleosynthesis, and decoupling with Hot Dark Matter ('HDM') or CDM.

7.6. OTHER MATTER

Many other cases have been examined in this GIC formalism. We list them with major references.

7.6.1. *Scalar fields*

The case of scalar fields is dealt with in a GIC way by Bruni, Ellis and Dunsby [213]. This analysis leads to the usual conserved quantities and theory of growth of inhomogeneities in an inflationary era. A *key element* here is choice of 4-velocity; for small perturbations there is a unique obvious choice, namely, choosing u^a orthogonal to the surfaces on which the scalar field ϕ is constant. The energy-momentum tensor then has the form of a perfect fluid, but with density and pressure depending on both kinetic and potential energy terms for ϕ.

7.6.2. *Multi-fluids and imperfect fluids*

The physically important case of multi-fluids is dealt with by Dunsby, Bruni and Ellis [214]; for example, enabling modelling of perturbations that include a matter–radiation interaction. The *key element* again is choice of 4-velocity. Each component has a separate 4-velocity $u^a_{(i)}$, and there are various options now for the reference 4-velocity u^a. When linear changes are made in this choice of 4-velocity, the essential effect is to alter the measured momentum in the u^a frame (Maartens et al [215]). The equations are simplified most by choosing u^a as the centre of mass 4-velocity for the sum of all components, and the 4-velocity $u^a_{(i)}$ of the i-th component as its centre of mass 4-velocity. One must then carefully check the separate momentum and energy equations for each component, as well as for the matter as a whole. These determine the evolution of the $1+3$ covariantly defined *relative velocities*: $V^a_{ij} = u^a_{(i)} - u^a_{(j)}$, and the separate matter densities $\mu_{(i)}$.

Exercise: Establish the equations for the relative velocity and the density inhomogeneities in the 2-fluid case.

The case of imperfect fluids is closely related, and the same issue of choice of 4-velocity arises. As pointed out earlier, it is essential to use realistic equations of state in studying perturbations of an imperfect fluid, such as described by the Müller–Israel–Stewart theory [216, 217] (see Maartens and Triginier [218] for a detailed GIC analysis of such imperfect fluids).

7.6.3. *Magnetic fields*

These have been examined in a GIC way by Barrow and Tsagas [219], using the $1+3$ covariant splitting of the electromagnetic field and the Maxwell field equations [7].

7.6.4. *Newtonian version*

A Newtonian version of the analysis can be developed fully in parallel to the relativistic version [220], based on the Newtonian analogue of the approach to cosmology presented in [6], and including derivations of the Newtonian Jeans length and Newtonian formulae for the growth of inhomogeneities.

Exercise: Establish these equations, and hence determine the main differences between the Newtonian and relativistic versions of structure formation.

7.6.5. *Alternative gravity*

The same GIC approach can be used to analyze higher-derivative gravitational theories; details are in [221].

7.7. RELATION TO OTHER FORMALISMS

The relation between the GIC approach to perturbations and the very influential Bardeen gauge-invariant formalism [199] has been examined in depth [222]. The essential points are that

* as might be expected, the implications of both approaches for structure formation are the same,
* the implications of the GIC formalism can be worked out in any desired coordinate system, including the Bardeen coordinates (which are incorporated into that approach in an essential way from the start),
* the Bardeen approach is essentially based on the linearised equations, while the GIC starts with the full non-linear equations and linearises them, as explained above. This enables an estimate of the errors involved, and a systematic n-th order approximation scheme,
* the GIC formalism does not use a *non-local* splitting [223] into scalar, vector and tensor modes, and only uses a harmonic splitting (into wavelengths) at a late stage of the analysis; these are both built into the Bardeen approach *ab initio*.

Both approaches have the advantage over gauge approaches that they do not involve gauge modes, and the differential equations are of minimal order needed to characterise the physics of the problem. Many papers on the use of various gauges are rather confused; however, the major paper by Ma and Bertschinger [198] clarifies the relations between important gauges in a clear way. As shown there, the answers obtained for large-scale growth of inhomogeneities is indeed gauge-dependent, and this becomes significant particularly on very large scales. The GIC formalism obviates this problem. However, whatever formalism is used, the issue is how the results relate to observations. The perturbation theory predicts structure growth, gravitational wave emission, gravitational lensing, and background radiation anisotropies. The latter are one of the most important tests of the geometry and physics of perturbed models, and are the topic of the final section.

8. CBR anisotropies

Central to present day cosmology is the study of the information obtainable from measurements of CBR anisotropies. A GIC version of the pioneering Sachs–Wolfe paper [224], based on photon path integration and calculation of the redshift along these paths (cf. the integration in terms of Bardeen's variables by Panek [225]), is given by Dunsby [226] and Challinor and Lasenby [227]. However, a kinetic theory approach enables a more in-depth study of the photons' evolution and interactions with the matter inhomogeneities, and so is the dominant way of analyzing CBR anisotropies.

8.1. COVARIANT KINETIC THEORY

Relativistic kinetic theory (see e.g. [68] and [228]–[230]) provides a self-consistent microscopically based treatment where there is a natural unifying framework in which to deal with a gas of particles in circumstances ranging from hydrodynamical to free-streaming behaviour. The photon gas undergoes a transition from hydrodynamical tight coupling with matter, through the process of decoupling from matter, to non-hydrodynamical free-streaming. This transition is characterised by the evolution of the photon mean free path from effectively zero to effectively infinity. This range of behaviour can appropriately be described by kinetic theory with non-relativistic classical *Thomson scattering* (see e.g. Jackson [231] or Feynman I [232]), and the baryonic matter with which radiation interacts can reasonably be described hydrodynamically during these times.

In this approach, the single-particle *photon distribution function* $f(x^i, p^a)$ represents the number of photons measured in the 3-volume element dV at the event x^i that have momenta in the momentum volume element π about the momentum p^a through the equation

$$dN = f(x^i, p^a)\,(-p_a u^a)\,\pi\, dV \;, \tag{230}$$

where u^a is the observer's 4-velocity and the redshift factor $(-p_a u^a)$ makes f into a (observer-independent) scalar. The rate of change of f is determined by the *Boltzmann equation*

$$L(f) = C \;, \tag{231}$$

where the *Liouville operator*

$$L(f) = \frac{\partial f}{\partial x^i}\frac{dx^i}{dv} + \frac{\partial f}{\partial p^a}\frac{dp^a}{dv} \tag{232}$$

gives the change of f in parameter distance dv along the geodesics that characterise the particle motions. The collision term C determines the rate of change of f due to emission, absorption and scattering processes; it can represent Thomson scattering, binary collisions, etc. Over the period of importance for CBR anisotropies, i.e. considerably after electron–positron annihilation, the average photon energy is much less than the electron rest mass and the electron thermal energy may be neglected, so that the quantised Compton interaction between photons and electrons (the dominant interaction between radiation and matter) may reasonably be described in the non-quantised Thomson limit.[36] After decoupling, there is very little interaction between matter and the CBR so we can use the *Liouville equation*:

$$C = 0 \;\Rightarrow\; L(f) = 0\,. \tag{233}$$

The energy-momentum tensor of the photons is

$$T^{ab} = \int_{T_x} p^a p^b f \pi\,, \tag{234}$$

where π is the volume element in the momentum tangent space T_x. This satisfies the conservation equations (35) and (36), and is part of the total T^{ab} tensor determining the space-time curvature.

Exercise: The same theory applies to particles with non-zero rest-mass. (i) What is the form of the Boltzmann collision term for binary particle collisions? (ii) Show that energy-momentum conservation in individual collisions will lead to conservation of the particle stress-tensor T^{ab}. (iii) Using the appropriate integral definition of entropy, determine a H-theorem for this form of collision. (iv) Under what conditions can equilibrium exist for such a gas of particles, and what is the equilibrium form of the distribution function? (See [5].)

[36]The $1+3$ covariant treatment described in this section also neglects polarisation effects.

8.2. ANGULAR HARMONIC DECOMPOSITION

In the $1+3$ covariant approach of [233, 234],[37] the photon 4-momentum p^a (where $p_a p^a = 0$) is split relative to an observer moving with 4-velocity u^a as

$$p^a = E\,(u^a + e^a)\,, \qquad e_a e^a = 1\,, \qquad e_a u^a = 0\,, \tag{235}$$

where $E = (-p_a u^a)$ is the *energy* and $e^a = E^{-1}\,h^a{}_b\,p^b$ is the *spatial propagation direction*, as measured by a comoving (fundamental) observer. Then the *photon distribution function is decomposed into* $1+3$ *covariant harmonics* via the expansion [233, 238]

$$\begin{aligned} f(x,p) &= f(x,E,e) = \sum_{\ell\geq 0} F_{A_\ell}(x,E)\,e^{A_\ell} \\ &= F + F_a e^a + F_{ab} e^a e^b + \cdots = \sum_{\ell\geq 0} F_{A_\ell} e^{\langle A_\ell\rangle}\,, \end{aligned} \tag{236}$$

where $e^{A_\ell} \equiv e^{a_1} e^{a_2} \cdots e^{a_\ell}$ and, $e^{\langle A_\ell\rangle}$ is the symmetric trace-free part of e^{A_ℓ}. The $1+3$ covariant *distribution function anisotropy multipoles* F_{A_ℓ} are irreducible since they are Projected, Symmetric, and Trace-Free ('PSTF'), i.e.

$$F_{a\cdots b} = F_{\langle a\cdots b\rangle} \quad \Leftrightarrow \quad F_{a\cdots b} = F_{(a\cdots b)}\,, \; F_{a\cdots b}\,u^b = 0 = F_{a\cdots bc}\,h^{bc}\,.$$

They encode the anisotropy structure of the distribution function in the same way as the usual spherical harmonic expansion

$$f = \sum_{\ell\geq 0} \sum_{m=-\ell}^{+\ell} f_\ell^m(x,E)\,Y_\ell^m(\vec{e})\,,$$

but with two major advantages: (a) the F_{A_ℓ} are $1+3$ covariant, and thus independent of any choice of coordinates in momentum space, unlike the f_ℓ^m; (b) F_{A_ℓ} is a rank-ℓ tensor field on space-time for each *fixed* E, and directly determines the ℓ-multipole of radiation anisotropy after integration over E. The multipoles can be recovered from the distribution function via

$$F_{A_\ell} = \Delta_\ell^{-1} \int f(x,e)\,e_{\langle A_\ell\rangle}\,d\Omega\,, \quad \text{with} \quad \Delta_\ell = 4\pi\,\frac{2^\ell\,(\ell!)^2}{(2\ell+1)!}\,, \tag{237}$$

[37] Based on the Ph.D. Thesis of R Treciokas, Cambridge University, 1972, combined with the covariant formalism of F A E Pirani [235] (see also K S Thorne [236]) by G F R Ellis when in Hamburg (1st Institute of Theoretical Physics) in 1972; see also [69, 70]. A similar formalism has been developed by M L Wilson [237].

where $d\Omega = d^2e$ is a solid angle in momentum space. A further useful identity is [233]

$$\int e^{A_\ell}\, d\Omega = \frac{4\pi}{\ell+1}\begin{cases} 0 & \ell \text{ odd}\,, \\ h^{(a_1a_2}\,h^{a_3a_4}\ldots h^{a_{\ell-1}a_\ell)} & \ell \text{ even}\,. \end{cases} \tag{238}$$

The first three multipoles determine the radiation energy-momentum tensor, which is (from (234) and (14))

$$T_R^{ab}(x) = \int p^a\, p^b\, f(x,p)\, d^3p = \mu_R\,(u^a\,u^b + \tfrac{1}{3}\,h^{ab}) + 2\,q_R^{(a}\,u^{b)} + \pi_R^{ab}\,, \tag{239}$$

where $d^3p = E\,dE\,d\Omega$ is the covariant volume element on the future null cone at the event x. It follows from (236) and (239) that the dynamical quantities of the radiation (in the u^a-frame) are:

$$\begin{aligned} \mu_R &= 4\pi\int_0^\infty E^3\,F\,dE \\ q_R^a &= \frac{4\pi}{3}\int_0^\infty E^3\,F^a\,dE \\ \pi_R^{ab} &= \frac{8\pi}{15}\int_0^\infty E^3\,F^{ab}\,dE\,. \end{aligned} \tag{240}$$

We extend these dynamical quantities to all multipole orders by defining the 1 + 3 covariant *brightness anisotropy multipoles*[38] [238]

$$\Pi_{a_1\cdots a_\ell} = \int_0^\infty E^3\,F_{a_1\cdots a_\ell}\,dE\,, \tag{241}$$

so that $\Pi = \mu_R/4\pi$, $\Pi^a = 3\,q_R^a/4\pi$ and $\Pi^{ab} = 15\,\pi_R^{ab}/8\pi$. Writing the Boltzmann equation in the form

$$\frac{df}{dv} \equiv p^a\,\frac{\partial f}{\partial x^a} - \Gamma^a{}_{bc}\,p^b\,p^c\,\frac{\partial f}{\partial p^a} = C[f]\,, \tag{242}$$

the *collision term is also decomposed into* 1 + 3 *covariant harmonics:*

$$C[f] = \sum_{\ell\geq 0} b_{A_\ell}(x,E)\,e^{A_\ell} = b + b_a e^a + b_{ab}e^a e^b + \cdots\,, \tag{243}$$

[38]Because photons are massless, we do not need the complexity of the moment definitions used in [233]. From now on, all energy integrals will be understood to be over the range $0 \leq E \leq \infty$.

where the 1+3 covariant *scattering multipoles* $b_{A_\ell} = b_{\langle A_\ell \rangle}$ encode irreducible properties of the particle interactions. Then the Boltzmann equation is equivalent to an *infinite hierarchy of* $1+3$ *covariant multipole equations*

$$L_{A_\ell}(x,E) = b_{A_\ell}[F_{A_m}](x,E) ,$$

where L_{A_ℓ} are the multipoles of df/dv, and will be given in the next section. These multipole equations are tensor field equations on space-time for each value of the photon energy E (but note that energy changes along each photon path). Given the solutions $F_{A_\ell}(x,E)$ of the equations, the relation (236) then determines the full photon distribution $f(x,E,e)$ as a scalar field over phase space.

8.3. NON-LINEAR $1+3$ COVARIANT MULTIPOLE EQUATIONS

The full Boltzmann equation in photon phase space contains more information than necessary to analyze radiation anisotropies in an inhomogeneous universe. For that purpose, when the radiation is close to black body, we do not require the full spectral behaviour of the distribution multipoles, but only the *energy-integrated multipoles*. The monopole leads to the average (all-sky) temperature, while the higher-order multipoles determine the temperature anisotropies. The GIC definition of the *average temperature* T is given according to the Stefan–Boltzmann law by

$$\mu_R(x) = 4\pi \int E^3\, F(x,E)\, dE = r\, T(x)^4 , \tag{244}$$

where r is the radiation constant. If f is close to a Planck distribution, then T is the thermal black body average temperature. But note that *no* notion of background temperature is involved in this definition. There is an all-sky average implied in (244). Fluctuations across the sky are measured by integrating the higher multipoles (a precise definition is given below), i.e. the fluctuations are determined by the $\Pi_{a_1\cdots a_\ell}$ ($\ell \geq 1$) defined in (241).

The form of $C[f]$ shows that $1+3$ covariant equations for the temperature fluctuations arise from decomposing the *energy-integrated Boltzmann equation*

$$\int E^2 \frac{df}{dv}\, dE = \int E^2\, C[f]\, dE \tag{245}$$

into $1+3$ covariant multipoles. We begin with the right-hand side, which requires the $1+3$ covariant form of the Thomson scattering term. Defining the $1+3$ covariant *energy-integrated scattering multipoles* [215]

$$K_{A_\ell} = \int E^2\, b_{A_\ell}\, dE ,$$

we find that

$$
\begin{aligned}
K &= n_E\sigma_T\left[\,\tfrac{4}{3}\Pi v_B^2 - \tfrac{1}{3}\Pi^a v_{Ba}\,\right] + \mathcal{O}[3]\,, && (246)\\
K^a &= -\,n_E\sigma_T\left[\,\Pi^a - 4\,\Pi v_B^a - \tfrac{2}{5}\Pi^{ab}v_{Bb}\,\right] + \mathcal{O}[3]\,, && (247)\\
K^{ab} &= -\,n_E\sigma_T\left[\,\tfrac{9}{10}\Pi^{ab} - \tfrac{1}{2}\Pi^{\langle a}v_B^{b\rangle} - \tfrac{3}{7}\Pi^{abc}v_{Bc} - 3\Pi v_B^{\langle a}v_B^{b\rangle}\,\right] && (248)\\
&\quad + \mathcal{O}[3]\,,\\
K^{abc} &= -\,n_E\sigma_T\left[\,\Pi^{abc} - \tfrac{3}{2}\Pi^{\langle ab}v_B^{c\rangle} - \tfrac{4}{9}\Pi^{abcd}v_{Bd}\,\right] + \mathcal{O}[3]\,, && (249)
\end{aligned}
$$

and, for $\ell > 3$,

$$
K^{A_\ell} = -\,n_E\sigma_T\left[\,\Pi^{A_\ell} - \Pi^{\langle A_{\ell-1}}v_B^{a_\ell\rangle} - \left(\frac{\ell+1}{2\ell+3}\right)\Pi^{A_\ell a}v_{Ba}\,\right] + \mathcal{O}[3]\,, \quad (250)
$$

where the expansion is in terms of the peculiar velocity v_B^a of the baryons relative to the reference frame u^a. Parameters are the *free electron number density* n_E and the *Thomson scattering cross section* σ_T, the latter being proportional to the square of the classical electron radius [231, 232]. The first three multipoles are affected by Thomson scattering differently than the higher multipoles.

Equations (246)–(250) [215] are a non-linear generalisation of the results given by Challinor and Lasenby [239]. They show the *coupling of baryonic bulk velocity to the radiation multipoles, arising from* $1+3$ *covariant non-linear effects in Thomson scattering.* If we linearise fully, i.e. neglect all terms containing v_B except the $\mu_R v_B^a$ term in the dipole K^a, which is first-order, then our equations reduce to those in [239]. The generalised non-linear equations apply to the analysis of $1+3$ covariant second-order effects on an FLRW background, to first-order effects on a spatially homogeneous but anisotropic background, and more generally, to any situation where the baryonic frame is in non-relativistic motion relative to the fundamental u^a-frame.

Next we require the multipoles of df/dv. These can be read off directly from the general expressions in [233], which are exact, $1+3$ covariant and also include the case of massive particles. For clarity and completeness, we outline an alternative, $1+3$ covariant derivation (the derivation in [233] uses tetrads). We require the rate of change of the photon energy along the null geodesics, given by [69, 70]

$$
\frac{dE}{dv} = -\left[\,\tfrac{1}{3}\Theta + (\sigma_{ab}e^a e^b) + (\dot{u}_a e^a)\,\right]E^2\,, \quad (251)
$$

which follows directly from $E = (-p_a u^a)$, $p^b \nabla_b p^a = 0$, and Eq. (11). Then

$$
\begin{aligned}
&\frac{d}{dv}[\, F_{a_1 \cdots a_\ell}(x,E)\, e^{a_1} \cdots e^{a_\ell}\,] \\
&= \frac{d}{dv}\left[\, E^{-\ell} F_{a_1 \cdots a_\ell}(x,E)\, p^{a_1} \cdots p^{a_\ell}\,\right] \\
&= E\Big\{ [\, \tfrac{1}{3}\Theta + \dot{u}_b e^b + \sigma_{bc} e^b e^c\,] \left(\ell\, F_{a_1 \cdots a_\ell} - E\, F'_{a_1 \cdots a_\ell}\right) e^{a_1} \cdots e^{a_\ell} \\
&\qquad + (u^{a_1} + e^{a_1}) \cdots (u^{a_\ell} + e^{a_\ell})\, [\, \dot{F}_{a_1 \cdots a_\ell} + e^b \nabla_b F_{a_1 \cdots a_\ell}\,]\Big\} ,
\end{aligned}
$$

where a prime denotes $\partial/\partial E$. The first term is readily put into irreducible PSTF form using identities in [233], p470. In the second term, when the round brackets are expanded, only those terms with at most one u^{a_r} survive, and

$$
u^a \dot{F}_{a\cdots} = -\,\dot{u}^a F_{a\cdots}\, , \qquad u^b \nabla^a F_{b\cdots} = -\,(\,\tfrac{1}{3}\Theta\, h^{ab} + \sigma^{ab} + \eta^{abc}\omega_c\,)\, F_{b\cdots}\, .
$$

Thus the $1+3$ covariant *anisotropy multipoles* b_{A_ℓ} of df/dv are

$$
\begin{aligned}
E^{-1} b_{A_\ell} = {}& \dot{F}_{\langle A_\ell \rangle} - \tfrac{1}{3}\Theta\, E\, F'_{A_\ell} + \tilde{\nabla}_{\langle a_\ell} F_{A_{\ell-1}\rangle} + \frac{(\ell+1)}{(2\ell+3)} \tilde{\nabla}^a F_{aA_\ell} \\
& - \frac{(\ell+1)}{(2\ell+3)} E^{-(\ell+1)} \left[E^{\ell+2} F_{aA_\ell}\right]' \dot{u}^a \\
& - E^\ell \left[E^{1-\ell} F_{\langle A_{\ell-1}}\right]' \dot{u}_{a_\ell\rangle} + \ell\, \omega^b\, \eta_{bc\langle a_\ell} F_{A_{\ell-1}\rangle}{}^c \\
& - \frac{(\ell+1)(\ell+2)}{(2\ell+3)(2\ell+5)} E^{-(\ell+2)} \left[E^{\ell+3} F_{abA_\ell}\right]' \sigma^{ab} \\
& - \frac{2\ell}{(2\ell+3)} E^{-1/2} \left[E^{3/2} F_{b\langle A_{\ell-1}}\right]' \sigma_{a_\ell\rangle}{}^b \\
& - E^{\ell-1} \left[E^{2-\ell} F_{\langle A_{\ell-2}}\right]' \sigma_{a_{\ell-1} a_\ell\rangle}\, .
\end{aligned} \tag{252}
$$

This regains the result of [233] (equation (4.12)) in the massless case. The form given here benefits from the streamlined version of the $1+3$ covariant formalism. We re-iterate that this result is exact and holds for any photon or (massless) neutrino distribution in *any* space-time.

We now multiply (252) by E^3 and integrate over all energies, using integration by parts and the fact that $E^n F_{a\cdots} \to 0$ as $E \to \infty$ for any positive n. We obtain the multipole equations that determine the brightness anisotropies multipoles Π_{A_ℓ}:

$$
\begin{aligned}
K_{A_\ell} = \dot{\Pi}_{\langle A_\ell \rangle} + \tfrac{4}{3}\,\Theta\,\Pi_{A_\ell} + \tilde{\nabla}_{\langle a_\ell}\Pi_{A_{\ell-1}\rangle} + \frac{(\ell+1)}{(2\ell+3)}\,\tilde{\nabla}^b\Pi_{bA_\ell} \\
- \frac{(\ell+1)(\ell-2)}{(2\ell+3)}\,\dot{u}^b\,\Pi_{bA_\ell} + (\ell+3)\,\dot{u}_{\langle a_\ell}\Pi_{A_{\ell-1}\rangle} \\
+ \ell\,\omega^b\,\eta_{bc(a_\ell}\Pi_{A_{\ell-1})}{}^c - \frac{(\ell-1)(\ell+1)(\ell+2)}{(2\ell+3)(2\ell+5)}\,\sigma^{bc}\,\Pi_{bcA_\ell} \\
+ \frac{5\ell}{(2\ell+3)}\,\sigma^b{}_{\langle a_\ell}\Pi_{A_{\ell-1}\rangle b} - (\ell+2)\,\sigma_{\langle a_\ell a_{\ell-1}}\,\Pi_{A_{\ell-2}\rangle}\,. \qquad (253)
\end{aligned}
$$

Once again, this is an exact result, and it holds also for any collision term, i.e. any K_{A_ℓ}. For decoupled neutrinos, we have $K_N^{A_\ell} = 0$ in this equation. For photons undergoing Thomson scattering, the left-hand side of (253) is given by (250), which is exact in the kinematical and dynamical quantities, but first-order in the relative baryonic velocity. The equations (250) and (253) thus constitute a non-linear generalisation of the FLRW-linearised case given by Challinor and Lasenby [239].

The monopole and dipole of equation (253) give the evolution equations of the energy and momentum densities:

$$
\begin{aligned}
K &= \dot{\Pi} + \tfrac{4}{3}\,\Theta\,\Pi + \tfrac{1}{3}\,\tilde{\nabla}_a\Pi^a + \tfrac{2}{3}\,\dot{u}_a\Pi^a + \tfrac{2}{15}\,\sigma_{ab}\Pi^{ab}\,, \qquad (254)\\
K^a &= \dot{\Pi}^{\langle a\rangle} + \tfrac{4}{3}\,\Theta\,\Pi^a + \tilde{\nabla}^a\Pi + \tfrac{2}{5}\,\tilde{\nabla}_b\Pi^{ab} \\
&\quad + \tfrac{2}{5}\,\dot{u}_b\,\Pi^{ab} + 4\,\Pi\,\dot{u}^a + \eta^{abc}\,\omega_b\,\Pi_c + \sigma^{ab}\,\Pi_b\,, \qquad (255)
\end{aligned}
$$

(these are thus the equations giving the divergence of T^{ab} in (234)). For *photons*, the equations are:

$$
\begin{aligned}
\dot{\mu}_R + \tfrac{4}{3}\,\Theta\,\mu_R + \tilde{\nabla}_a q_R^a + 2\,\dot{u}_a q_R^a + \sigma_{ab}\pi_R^{ab} &= n_E\sigma_T\left(\tfrac{4}{3}\,\mu_R v_B^2 - q_R^a v_{Ba}\right) \\
&\quad + \mathcal{O}[3]\,, \qquad (256)\\
\dot{q}_R^{\langle a\rangle} + \tfrac{4}{3}\,\Theta\,q_R^a + \tfrac{4}{3}\,\mu_R\,\dot{u}^a + \tfrac{1}{3}\,\tilde{\nabla}^a\mu_R + \tilde{\nabla}_b\pi_R^{ab} & \\
+ \sigma^a{}_b\,q_R^b + \eta^{abc}\,\omega_b\,q_{Rc} + \dot{u}_b\,\pi_R^{ab} &= n_E\sigma_T\left(\tfrac{4}{3}\,\mu_R\,v_B^a - q_R^a\right. \qquad (257)\\
&\quad \left. + \pi_R^{ab}\,v_{Bb}\right) + \mathcal{O}[3]\,,
\end{aligned}
$$

(the present versions of the fluid energy and momentum conservation equations (35) and (36)).

The non-linear dynamical equations are completed by the integrated Boltzmann multipole equations. For *photons*, the quadrupole evolution equation is

$$\dot{\pi}_{R}^{\langle ab\rangle} + \tfrac{4}{3}\,\Theta\,\pi_{R}^{ab} + \tfrac{8}{15}\,\mu_{R}\,\sigma^{ab} + \tfrac{2}{5}\,\tilde{\nabla}^{\langle a}q_{R}^{b\rangle} + \frac{8\pi}{35}\,\tilde{\nabla}_{c}\Pi^{abc}$$
$$+\,2\,\dot{u}^{\langle a}\,q_{R}^{b\rangle} + 2\,\omega^{c}\,\eta_{cd}{}^{\langle a}\,\pi_{R}^{b\rangle d} + \tfrac{2}{7}\,\sigma_{c}{}^{\langle a}\,\pi_{R}^{b\rangle c} - \frac{32\pi}{315}\,\sigma_{cd}\,\Pi^{abcd}$$
$$= -\,n_{E}\sigma_{T}\,\big(\tfrac{9}{10}\,\pi_{R}^{ab} - \tfrac{1}{5}\,q_{R}^{\langle a}\,v_{B}^{b\rangle} - \frac{8\pi}{35}\,\Pi^{abc}\,v_{Bc} - \tfrac{2}{5}\,\rho_{R}\,v_{B}^{\langle a}\,v_{B}^{b\rangle}\big)$$
$$+\,\mathcal{O}[3]\,, \qquad (258)$$

(a fluid description gives no analogue of this and the following equations). The higher multipoles ($\ell > 3$) evolve according to

$$\dot{\Pi}^{\langle A_\ell\rangle} + \tfrac{4}{3}\,\Theta\,\Pi^{A_\ell} + \tilde{\nabla}^{\langle a_\ell}\Pi^{A_{\ell\,1}\rangle} + \frac{(\ell+1)}{(2\ell+3)}\,\tilde{\nabla}_{b}\Pi^{bA_\ell}$$
$$-\;\frac{(\ell+1)(\ell-2)}{(2\ell+3)}\,\dot{u}_{b}\,\Pi^{bA_\ell} + (\ell+3)\,\dot{u}^{\langle a_\ell}\,\Pi^{A_{\ell\,1}\rangle}$$
$$+\;\ell\,\omega^{b}\,\eta_{bc}{}^{\langle a_\ell}\,\Pi^{A_{\ell\,1}\rangle c} - \frac{(\ell-1)(\ell+1)(\ell+2)}{(2\ell+3)(2\ell+5)}\,\sigma_{bc}\,\Pi^{bcA_\ell}$$
$$+\;\frac{5\ell}{(2\ell+3)}\,\sigma_{b}{}^{\langle a_\ell}\,\Pi^{A_{\ell\,1}\rangle b} - (\ell+2)\,\sigma^{\langle a_\ell a_{\ell\,1}}\,\Pi^{A_{\ell\,2}\rangle}$$
$$= -\,n_{E}\sigma_{T}\left[\,\Pi^{A_\ell} - \Pi^{\langle A_{\ell\,1}}\,v_{B}^{a_\ell\rangle} - \left(\frac{\ell+1}{2\ell+3}\right)\Pi^{A_\ell a}\,v_{Ba}\,\right]$$
$$+\,\mathcal{O}[3]\,. \qquad (259)$$

For $\ell = 3$, the $\Pi^{\langle A_{\ell\,1}}\,v_{B}^{a_\ell\rangle}$ term on the right-hand side of equation (259) must be multiplied by $\frac{3}{2}$. For neutrinos, the equations are the same except without the Thomson scattering terms. Note that these equations link angular multipoles of order $\ell-2$, $\ell-1$, ℓ, $\ell+1$, $\ell+2$, i.e they link *five* successive harmonic terms. This is the source of the *harmonic mixing* that occurs as the radiation propagates.

These equations for the radiation (and neutrino) multipoles generalise the equations given by Challinor and Lasenby [239], to which they reduce when we remove all terms $\mathcal{O}(\epsilon v_{I})$ and $\mathcal{O}(\epsilon^{2})$. In this case, on introducing a *FLRW-linearisation*, there is major simplification of the equations:

$$\dot{\mu}_{R} + \tfrac{4}{3}\,\Theta\,\mu_{R} + \tilde{\nabla}_{a}q_{R}^{a} \approx 0\,, \qquad (260)$$

$$\dot{q}_{R}^{\langle a\rangle} + \tfrac{4}{3}\,\Theta\,q_{R}^{a} + \tfrac{4}{3}\,\mu_{R}\,\dot{u}^{a} + \tfrac{1}{3}\,\tilde{\nabla}^{a}\mu_{R} + \tilde{\nabla}_{b}\pi_{R}^{ab} \approx n_{E}\sigma_{T}\,\big(\tfrac{4}{3}\,\mu_{R}\,v_{B}^{a} - q_{R}^{a}\big)\,, \qquad (261)$$

$$\dot{\pi}_{R}^{\langle ab\rangle} + \tfrac{4}{3}\,\Theta\,\pi_{R}^{ab} + \tfrac{8}{15}\,\mu_{R}\,\sigma^{ab} + \tfrac{2}{5}\,\tilde{\nabla}^{\langle a}q_{R}^{b\rangle} + \frac{8\pi}{35}\,\tilde{\nabla}_{c}\Pi^{abc} \approx -\,\tfrac{9}{10}\,n_{E}\sigma_{T}\,\pi_{R}^{ab}\,, \qquad (262)$$

and, for $\ell \geq 3$,

$$\dot{\Pi}^{\langle A_\ell \rangle} + \tfrac{4}{3}\,\Theta\,\Pi^{A_\ell} + \tilde{\nabla}^{\langle a_\ell}\Pi^{A_{\ell\ 1}\rangle} + \frac{(\ell+1)}{(2\ell+3)}\,\tilde{\nabla}_b\Pi^{bA_\ell} \approx -\,n_{_E}\sigma_{_T}\,\Pi^{A_\ell}\ . \tag{263}$$

Note that these equations now link only angular multipoles of order $\ell - 1$, ℓ, $\ell + 1$, i.e they link *three* successive terms. This is a major qualitative difference from the full non-linear equations.

8.4. TEMPERATURE ANISOTROPY MULTIPOLES

Finally, we return to the definition of temperature anisotropies. As noted above, these are determined by the Π_{A_ℓ}. We define the *temperature fluctuation* $\tau(x,e)$ via the directional temperature which is determined by the directional bolometric brightness:

$$T(x,e) = T(x)\,[\,1 + \tau(x,e)\,] = \left[\frac{4\pi}{r}\int E^3\,f(x,E,e)\,dE\right]^{1/4}\ . \tag{264}$$

This is a GIC definition which is also exact. We can rewrite it explicitly in terms of the Π_{A_ℓ}:

$$\tau(x,e) = \left[\,1 + \frac{4\pi}{\mu_R}\sum_{\ell\geq 1}\Pi_{A_\ell}e^{A_\ell}\right]^{1/4} - 1 = \tau_a e^a + \tau_{ab}e^a e^b + \cdots\ . \tag{265}$$

In principle, we can extract the 1+3 covariant irreducible PSTF *temperature anisotropy multipoles* by using the inversion (237):

$$\tau_{A_\ell}(x) = \Delta_\ell^{-1}\int \tau(x,e)\,e_{\langle A_\ell\rangle}\,d\Omega\ . \tag{266}$$

In the almost-FLRW case, when τ is $\mathcal{O}[1]$, we regain from (265) the linearised definition given in [240]:

$$\tau_{A_\ell} \approx \left(\frac{\pi}{\mu_R}\right)\Pi_{A_\ell}\ , \tag{267}$$

where $\ell \geq 1$. In particular, the dipole and quadrupole are

$$\tau^a \approx \frac{3\,q_R^a}{4\,\mu_R} \quad \text{and} \quad \tau^{ab} \approx \frac{15\,\pi_R^{ab}}{2\,\mu_R}\ . \tag{268}$$

We can normalise the radiation dynamical multipoles Π^{A_ℓ} to define the dimensionless $1+3$ covariant *brightness temperature anisotropy multipoles* ($\ell \geq 1$)

$$\mathcal{T}^{A_\ell} = \left(\frac{\pi}{r\,T^4}\right)\Pi^{A_\ell} \approx \tau^{A_\ell}\ .$$

Thus the $\mathcal{T}^{A_\ell}$ are equal to the temperature fluctuation multipoles plus non-linear corrections. In terms of these quantities, the *hierarchy of radiation multipoles becomes*:

$$\begin{aligned}
\frac{\dot{T}}{T} &= -\tfrac{1}{3}\Theta - \tfrac{1}{3}\tilde{\nabla}_a\mathcal{T}^a - \tfrac{4}{3}\mathcal{T}^a\,\frac{\tilde{\nabla}_a T}{T} - \tfrac{2}{3}\,\dot{u}_a\,\mathcal{T}^a - \tfrac{2}{15}\,\sigma_{ab}\,\mathcal{T}^{ab} \\
&\quad + \tfrac{1}{3}\,n_E\,\sigma_T\,v_{Ba}\,(\,v_B^a - \mathcal{T}^a\,) + \mathcal{O}[3]\,, \qquad (269)\\
\dot{\mathcal{T}}^{\langle a\rangle} &= -4\left(\frac{\dot{T}}{T} + \tfrac{1}{3}\Theta\right)\mathcal{T}^a - \frac{\tilde{\nabla}^a T}{T} - \dot{u}^a - \tfrac{2}{5}\,\tilde{\nabla}_b\mathcal{T}^{ab} \\
&\quad + n_E\sigma_T\,(\,v_B^a - \mathcal{T}^a\,) + \tfrac{2}{5}\,n_E\sigma_T\,\mathcal{T}^{ab}\,v_{Bb} - \sigma^a{}_b\,\mathcal{T}^b - \tfrac{2}{5}\,\dot{u}_b\,\mathcal{T}^{ab} \\
&\quad - \eta^{abc}\,\omega_b\,\mathcal{T}_c - \tfrac{8}{5}\,\mathcal{T}^{ab}\,\frac{\tilde{\nabla}_b T}{T} + \mathcal{O}[3]\,, \qquad (270)\\
\dot{\mathcal{T}}^{\langle ab\rangle} &= -4\left(\frac{\dot{T}}{T} + \tfrac{1}{3}\Theta\right)\mathcal{T}^{ab} - \sigma^{ab} - \tilde{\nabla}^{\langle a}\mathcal{T}^{b\rangle} - \tfrac{3}{7}\,\tilde{\nabla}_c\mathcal{T}^{abc} \\
&\quad - \tfrac{9}{10}\,n_E\sigma_T\,\mathcal{T}^{ab} + n_E\sigma_T\,(\,\tfrac{1}{2}\,\mathcal{T}^{\langle a}\,v_B^{b\rangle} + \tfrac{3}{7}\,\mathcal{T}^{abc}\,v_{Bc} + \tfrac{3}{4}v_B^{\langle a}v_B^{b\rangle}) \\
&\quad - 5\,\dot{u}^{\langle a}\,\mathcal{T}^{b\rangle} - \tfrac{4}{21}\,\sigma_{cd}\,\mathcal{T}^{abcd} - 2\,\omega^c\,\eta_{cd}{}^{\langle a}\,\mathcal{T}^{b\rangle d} \\
&\quad - \tfrac{10}{7}\,\sigma_c{}^{\langle a}\,\mathcal{T}^{b\rangle c} - \tfrac{12}{7}\,\mathcal{T}^{abc}\,\frac{\tilde{\nabla}_c T}{T} + \mathcal{O}[3]\,, \qquad (271)
\end{aligned}$$

and, for $\ell > 3$:

$$\begin{aligned}
\dot{\mathcal{T}}^{\langle A_\ell\rangle} &= -4\left(\frac{\dot{T}}{T} + \tfrac{1}{3}\Theta\right)\mathcal{T}^{A_\ell} - \tilde{\nabla}^{\langle a_\ell}\mathcal{T}^{A_{\ell-1}\rangle} - \frac{(\ell+1)}{(2\ell+3)}\,\tilde{\nabla}_b\mathcal{T}^{bA_\ell} \\
&\quad - n_E\sigma_T\,\mathcal{T}^{A_\ell} + n_E\sigma_T\left[\,\mathcal{T}^{\langle A_{\ell-1}}\,v_B^{a_\ell\rangle} + \left(\frac{\ell+1}{2\ell+3}\right)\mathcal{T}^{A_\ell b}\,v_{Bb}\,\right] \\
&\quad + \frac{(\ell+1)\,(\ell-2)}{(2\ell+3)}\,\dot{u}_b\,\mathcal{T}^{bA_\ell} - (\ell+3)\,\dot{u}^{\langle a_\ell}\,\mathcal{T}^{A_{\ell-1}\rangle} \\
&\quad - \ell\,\omega^b\,\eta_{bc}{}^{\langle a_\ell}\,\mathcal{T}^{A_{\ell-1}\rangle c} + (\ell+2)\,\sigma^{\langle a_\ell a_{\ell-1}}\,\mathcal{T}^{A_{\ell-2}\rangle} \\
&\quad + \frac{(\ell-1)\,(\ell+1)\,(\ell+2)}{(2\ell+3)\,(2\ell+5)}\,\sigma_{bc}\,\mathcal{T}^{bcA_\ell} - \frac{5\ell}{(2\ell+3)}\,\sigma_b{}^{\langle a_\ell}\,\mathcal{T}^{A_{\ell-1}\rangle b} \\
&\quad - 4\,\frac{(\ell+1)}{(2\ell+3)}\,\mathcal{T}^{A_\ell b}\,\frac{\tilde{\nabla}_b T}{T} + \mathcal{O}[3]\,. \qquad (272)
\end{aligned}$$

For $\ell = 3$, the Thomson scattering term $\mathcal{T}^{\langle A_{\ell-1}}\,v_B^{a_\ell\rangle}$ must be multiplied by a factor $\frac{3}{2}$.

The $1+3$ covariant non-linear multipole equations given in this form show more clearly the evolution of temperature anisotropies (including the monopole, i.e. the average temperature T), in general linking five successive harmonics. Although the $\mathcal{T}^{A_\ell}$ only determine the actual temperature fluctuations τ^{A_ℓ} to linear order, they are a useful dimensionless measure of anisotropy. Furthermore, equations (269)–(272) apply as the evolution equations for the temperature fluctuation multipoles when the radiation anisotropy is small (i.e. $\mathcal{T}^{A_\ell} \approx \tau^{A_\ell}$), but the space-time inhomogeneity and anisotropy are not restricted. This includes the particular case of small CBR anisotropies in general Bianchi universes, or in perturbed Bianchi universes.

FLRW-linearisation, i.e. the case when only first-order effects relative to the FLRW limit are considered, reduces the above equations to the linearised form, generically linking three successive harmonics:

$$\frac{\dot{T}}{T} \approx -\tfrac{1}{3}\Theta - \tfrac{1}{3}\tilde{\nabla}_a\tau^a , \tag{273}$$

$$\dot{\tau}^{\langle a\rangle} \approx -\frac{\tilde{\nabla}^a T}{T} - \dot{u}^a - \tfrac{2}{5}\tilde{\nabla}_b\tau^{ab} + n_E\sigma_T\left(v_B^a - \tau^a\right) , \tag{274}$$

$$\dot{\tau}^{\langle ab\rangle} \approx -\sigma^{ab} - \tilde{\nabla}^{\langle a}\tau^{b\rangle} - \tfrac{3}{7}\tilde{\nabla}_c\tau^{abc} - \tfrac{9}{10}n_E\sigma_T\,\tau^{ab} , \tag{275}$$

and, for $\ell \geq 3$:

$$\dot{\tau}^{\langle A_\ell\rangle} \approx -\tilde{\nabla}^{\langle a_\ell}\tau^{A_{\ell-1}\rangle} - \frac{(\ell+1)}{(2\ell+3)}\tilde{\nabla}_b\tau^{bA_\ell} - n_E\sigma_T\,\tau^{A_\ell} . \tag{276}$$

These are GIC multipole equations leading to the Fourier mode formulation of the integrated Boltzmann equations used in the standard literature (see e.g. [241] and the references therein) when they are decomposed into spatial harmonics and associated wavelengths.[39] This then allows examination of diffusion effects, which are wavelength-dependent.

These linearised equations, together with the linearised equations governing the kinematical and gravitational quantities, may be 1+3 *covariantly split* into scalar, vector and tensor modes, as described in [222, 239]. The modes can then be expanded in 1+3 covariant eigentensors of the comoving Laplacian, and the Fourier coefficients obey ordinary differential equations.

Exercise: Determine the resulting hierarchy of mode equations. Show from these equations that there is a wavelength λ_S, such that for shorter wavelengths the perturbations are heavily damped (the physical reason is photon diffusion).

Numerical integration of these equations are performed for scalar modes by Challinor and Lasenby [239], with further analytic results given in [242].

[39]They contain the free-streaming subcase for $n_E = 0$.

These are the Sachs–Wolfe family of integrations, of fundamental importance in determining CBR anisotropies, as discussed in many places; see e.g. Hu and Sugiyama [241], Hu and White [243] (the GIC version will be given in the series of papers by Gebbie, Maartens, Dunsby and Ellis). However, they assume an almost-FLRW geometry. We turn now to justifying that assumption.

8.5. ALMOST-EGS-THEOREM AND ITS APPLICATIONS

One of our most important understandings of the nature of the universe is that at recent times it is well-represented by the standard spatially homogeneous and isotropic FLRW models. The basic reason for this belief is the observed high degree of isotropy of the CBR, together with a fundamental result of Ehlers, Geren and Sachs [69] (hereafter 'EGS'), taken nowadays (see e.g. [3]) to establish that the universe is almost-FLRW at recent times (i.e. since decoupling of matter and radiation).

The EGS programme can be summarized as follows: Using (a) the measured high isotropy of the CBR at our space-time position, and (b) the Copernican assumption that we are not at a privileged position in the universe, the aim is to deduce that the universe is accurately FLRW. EGS gave an exact theorem of this kind: if a family of freely-falling observers measure self-gravitating background radiation to be everywhere *exactly* isotropic in the case of non-interacting matter and radiation, then the universe is *exactly* FLRW. This is taken to establish the desired conclusion, in view of the measured near-isotropy of the radiation at our space-time location. However, of course the CBR is not exactly isotropic. Generally, we want to show stability of arguments we use [244]; in this case, we wish to do so by showing the EGS result remains *nearly* true if the radiation is *nearly* isotropic, thus providing the foundation on which further analysis, such as that in the famous Sachs–Wolfe paper [224], are based. More precisely, we aim to prove the following theorem [245].

> **Almost-EGS-Theorem:** *If* the Einstein–Liouville equations are satisfied in an expanding universe, where there is present pressure-free matter with 4-velocity vector field u^a ($u_a u^a = -1$) such that (freely-propagating) background radiation is everywhere almost-isotropic relative to u^a in some domain U, *then* space-time is almost-FLRW in U.

This description is intended to represent the situation since decoupling to the present day. The pressure-free matter represents the galaxies on which fundamental observers live, who measure the radiation to be almost isotropic.

8.5.1. *Assumptions*

In detail, we consider matter and radiation in a space-time region U. In our application, we consider U to be the region within and near our past light cone from decoupling to the present day (this is the observable space-time region where we would like to prove the universe is almost-FLRW; before decoupling a different analysis is needed, for collisions dominate there, and we do not have sufficient data to comment on the situation far from our light cone). Our assumptions are that, in the region considered,

(1) The EFE are satisfied, with the total energy-momentum tensor T_{ab} composed of *non-interacting* matter and radiation components:

$$T^{ab} = T_M^{ab} + T_R^{ab} \,, \qquad \nabla_b T_M^{ab} = 0 \,, \qquad \nabla_b T_R^{ab} = 0 \,. \tag{277}$$

The independent conservation equations express the decoupling of matter from radiation; the matter energy-momentum tensor is $T_M^{ab} = \rho\, u^a\, u^b$, ($\mu_M = \rho$ is the matter density) and, at each point, the radiation energy-momentum tensor is $T_R^{ab} = \int f\, p^a\, p^b\, \pi$, where π is the momentumspace volume element. The only non-zero energy-momentum tensor contribution from the matter, relative to the 4-velocity u^a, is the energy density; so the total energy density is

$$\mu = \mu_R + \rho = \mathcal{O}[0] \,, \tag{278}$$

while all other energy-momentum tensor components are simply equal to the radiation contributions. Without confusion we will *omit* a subscript 'R' on these terms.

(2) The matter 4-velocity field u^a is geodesic and expanding:

$$\dot{u}^a = 0 \,, \qquad \Theta = \mathcal{O}[0] > 0 \,, \tag{279}$$

(the first requirement in fact follows from momentum conservation (38) for the pressure-free matter). The assumption of an expanding universe is essential to what follows, for otherwise there are counter-examples to the result [246].

(3) The radiation obeys Liouville's equation (233): $L(f) = 0$, where L is the Liouville operator. This means that there is no entropy production, so that q^a and π_{ab} are *not* dissipative quantities, but measure the extent to which f deviates from isotropy.

(4) Relative to u^a — i.e. for all matter-comoving observers — the distribution function is almost isotropic everywhere in the region U. Formally, in this region $F = F(x^i, E)$ and its time derivatives are zero-order, while

$F_{A_l} = F_{A_l}(x^i, E)$ plus their space and time derivatives are at most first-order for $l > 0$. In brief:

$$F,\ \dot{F} = \mathcal{O}[0]\ , \quad F_{A_l},\ \dot{F}_{\langle A_l\rangle},\ \tilde{\nabla}_a F_{A_l} = \mathcal{O}[1]\ , \tag{280}$$

(and we assume all higher derivatives of F_{A_l} are also $\mathcal{O}[1]$ for $l > 0$).

It immediately follows, under reasonable assumptions about the phase space integrals of the harmonic components, that the scalar moments are zero-order but the tensor moments and their derivatives are first-order. We assume the conditions required to ensure this are true: i.e. we additionally suppose

$$\mu_R,\ p_R = \mathcal{O}[0]\ , \quad q_a,\ \dot{q}_{\langle a\rangle},\ \tilde{\nabla}_a q_b = \mathcal{O}[1]\ , \quad \pi_{ab},\ \dot{\pi}_{\langle ab\rangle},\ \tilde{\nabla}_a \pi_{bc} = \mathcal{O}[1] \tag{281}$$

are satisfied as a consequence (280), and that the same holds for the higher space and time derivatives of q^a, π_{ab} and for all higher moments (specifically, those defined in (289) below). The *first aim* now is to show that

(a) the kinematical quantities and Weyl tensor are almost-FLRW, i.e. σ_{ab} and ω^a are first-order, which then implies that E_{ab} and H_{ab} are also first-order. The *second aim* is to show that then,

(b) there are coordinates such that the metric tensor takes a perturbed RW form.

8.5.2. *Proving almost-FLRW kinematics*

Through an appropriate integration over momentum space, the zeroth harmonic of the Liouville equation (233) gives the energy conservation equation for the radiation (cf. (256)),

$$\dot{\mu}_R + \tfrac{4}{3}\,\Theta\,\mu_R + \sigma_{ab}\pi^{ab} + \tilde{\nabla}_a q^a = 0\ , \tag{282}$$

while the first harmonic gives the momentum conservation equation (cf. (257)),

$$\dot{q}^{\langle a\rangle} + \tfrac{4}{3}\,\Theta\,q^a + \tfrac{1}{3}\,\tilde{\nabla}^a\mu_R + \tilde{\nabla}_b\pi^{ab} + \sigma^a{}_b\,q^b + \eta^{abc}\,\omega_b\,q_c = 0\ , \tag{283}$$

which implies by (281),

$$\tilde{\nabla}_a\mu_R = \mathcal{O}[1]\ . \tag{284}$$

Taking spatial derivatives of (283), the same result follows for the higher spatial derivatives of μ_R; in particular, its second derivatives are at most first-order.

Now the definition of the $\tilde{\nabla}$-derivative leads to the identity (27), giving

$$(\tilde{\nabla}_a\tilde{\nabla}_b - \tilde{\nabla}_b\tilde{\nabla}_a)\mu_R = 2\,\eta_{abc}\,\omega^c\,\dot{\mu}_R\ . \tag{285}$$

In an expanding universe, by (281) the energy conservation equation (282) for the radiation shows

$$\dot{\mu}_R + \tfrac{4}{3}\Theta\,\mu_R = \mathcal{O}[1]\ . \tag{286}$$

Thus, because the universe is expanding, $\dot{\mu}_R$ is zero-order. However, the left-hand side of (285) is first-order; consequently, by the higher-derivative version of (284),

$$\omega^a = \mathcal{O}[1]\ . \tag{287}$$

The second harmonic of the Liouville equation (233) leads, after an appropriate integration over momentum space, to an evolution equation for the anisotropic stress tensor π_{ab} (which is the present version of (258)):

$$\begin{aligned} 0 \;=\;& \dot{\pi}^{\langle ab\rangle} + \tfrac{4}{3}\,\Theta\,\pi^{ab} + \tfrac{8}{15}\,\mu_R\,\sigma^{ab} + \tfrac{2}{7}\,\sigma_c{}^{\langle a}\,\pi^{b\rangle c} + 2\,\omega^c\,\eta_{cd}{}^{\langle a}\,\pi^{b\rangle d} \\ &+ \tfrac{2}{5}\,\tilde{\nabla}^{\langle a}q^{b\rangle} + J^{ab} - \frac{32\pi}{315}\,\sigma_{cd}\,\Pi^{abcd}\ , \end{aligned} \tag{288}$$

where (cf. (241))

$$J^{ab} = \frac{8\pi}{35}\int_0^\infty E^3\,\tilde{\nabla}_c F^{abc}\,dE\ , \qquad \Pi^{abcd} = \int_0^\infty E^3\,F^{abcd}\,dE\ . \tag{289}$$

Consequently, by (287), (280) and (281),

$$\sigma_{ab} = \mathcal{O}[1]\ , \tag{290}$$

and (on taking derivatives of the above equation) the same is true for its space and time derivatives. Then, to first order, the evolution equations become

$$\dot{\mu}_R + \tfrac{4}{3}\,\Theta\,\mu_R + \tilde{\nabla}_a q^a \approx 0\ , \tag{291}$$

$$\dot{q}^{\langle a\rangle} + \tfrac{4}{3}\,\Theta\,q^a + \tfrac{1}{3}\,\tilde{\nabla}^a\mu_R + \tilde{\nabla}_b\pi^{ab} \approx 0\ , \tag{292}$$

$$\dot{\pi}^{\langle ab\rangle} + \tfrac{4}{3}\,\Theta\,\pi^{ab} + \tfrac{8}{15}\,\mu_R\,\sigma^{ab} + \tfrac{2}{5}\,\tilde{\nabla}^{\langle a}q^{b\rangle} + J^{ab} \approx 0\ , \tag{293}$$

showing the equations can *only* close at first order *if* we can argue that $J^{ab} = \mathcal{O}[2]$. *Any such approximation must be done with utmost care* because of the radiation multipole truncation theorem (see below).

It now follows from the shear propagation equation (31) and the Ricci identity (34) involving the magnetic Weyl tensor H_{ab}, that all the Weyl tensor components are also at most first-order:

$$E_{ab} = \mathcal{O}[1]\ , \qquad H_{ab} = \mathcal{O}[1]\ . \tag{294}$$

Consequently, the Bianchi identity (44) shows that

$$X_a \equiv \tilde{\nabla}_a \mu = \mathcal{O}[1] \quad \Rightarrow \quad \tilde{\nabla}_a \rho = \mathcal{O}[1] \, , \tag{295}$$

by (278). Finally, the (0α)-equations (32), or spatial derivatives of (286), show that

$$Z_a \equiv \tilde{\nabla}_a \Theta = \mathcal{O}[1] \, . \tag{296}$$

This establishes that the kinematical quantities for the matter flow are almost-FLRW: all the quantities that vanish in the FLRW case, are at most first-order here. Thus the only zero-order $1+3$ covariantly defined quantities are those that are non-vanishing in FLRW universes.

8.5.3. *Proving almost-FLRW dynamics*

It follows that the zero-order equations governing the dynamics are just those of a FLRW universe. Because the kinematical quantities are precisely those we expect in a perturbed FLRW universe, we can linearise the $1+3$ covariant equations about the FLRW values in the usual way, in which the background model will — as we have just seen — obey the usual FLRW equations. Hence, the usual linearised $1+3$ covariant FLRW perturbation analyses can be applied [197, 222], leading to the usual results for growth of inhomogeneities in almost-FLRW universes.

8.5.4. *Finding an almost-RW metric*

Given that the kinematical quantities take an almost-FLRW form, the *key issue* in proving existence of an almost-RW metric is choice of a time function to use as a cosmic time (in the realistic, inhomogeneous universe model). The problem is that although we have shown the vorticity will be small, it will in general not be zero; hence, there will be no time surfaces orthogonal to the matter flow lines [5, 6]. The problem can equivalently be viewed as the need to find a vorticity-free (that is, hypersurface-orthogonal) congruence of curves to use as a kinematical reference frame, which, if possible, one would like to also be geodesic. As we cannot assume the matter 4-velocity fulfills this condition, we need to introduce another congruence of curves, say $\hat{u}^a$, that *is* hypersurface-orthogonal and that does not differ too greatly from u^a, so the matter is moving slowly relative to that frame. Then we have to change to that frame, using a 'hat' to denote its kinematical quantities.

8.5.5. *Result*

This gives the result we want; we have shown that freely propagating almost-isotropic background radiation everywhere in a region U implies the universe is an almost-FLRW universe in that region. Thus we have proved the stability of the Ehlers, Geren and Sachs [69] result. This result is the foundation for the important analysis of Sachs and Wolfe [224] and all related analyses, determining the effect of inhomogeneities on the CBR by integration from last scattering till today in an almost-FLRW universe, for these papers *start off* with the assumption that the universe is almost-FLRW since decoupling, and build on that basis.

It should also be noted that the assumption that radiation is isotropic about distant observers can be partially checked by testing how close the CBR spectrum is to black body in those directions where we detect the Sunyaev–Zel'dovich effect [104] (see also Goodman [247]). Since that effect mixes and scatters incoming radiation, substantial radiation anisotropies relative to those clusters of galaxies inducing the effect would result in a significant distortion of the outgoing spectrum.

Exercise: How good are the limits we might obtain on the CBR anisotropy at distance points on our past null cone by this method? Are there any other ways one can obtain such limits?

8.6. OTHER CBR CALCULATIONS

One can extend the above analysis to obtain model-independent limits on the CBR anisotropy (Stoeger et al [240]). However, there are also a whole series of model-dependent analyses available.

8.6.1. *Sachs–Wolfe and related effects*

I will not pursue here the issue of Sachs–Wolfe [224] type calculations, integrating up the redshift from the surface of last scattering to the observer and so determining the CBR anisotropy, and, hence, cosmological parameters [248], because they are so extensively covered in the literature. The GIC version is covered by Challinor and Lasenby [239] and developed in depth in the series of papers by Gebbie, Dunsby, Maartens and Ellis ([215, 238, 242]). However, a few comments are in order.

One can approach this topic $1+3$ covariantly by a direct generalisation of the photon type of calculation (see [226, 227]), or by use of the kinetic formalism developed above, as in the papers just cited. Two things need to be very carefully considered. These are,

(i) The issue of putting *correct limits* on the Sachs–Wolfe integral. This integral needs to be taken to the *real* surface of last scattering, not the background surface of last scattering (which is fictitious). If this is not done, the results may be gauge-dependent. One should place the limits on this integral properly, which can be done quite easily by calculating when the optical depth due to Thomson scattering is unity [225, 249].

(ii) In order to obtain *solutions*, one has somehow to cut off the harmonic series, for otherwise there is an infinite regress whereby higher-order harmonics determine the evolution of lower-order ones, as is clear from the multipole equations above. However, truncation of harmonics in a kinetic theory description needs to be approached with great caution, because of the following theorem for collision-free radiation [234]:

> **Radiation Multipole Truncation Theorem**: In the exact Einstein–Liouville theory, truncation of the angular harmonics seen by a family of geodesic observers *at any order whatever* leads to the vanishing of the shear of the family of observers; hence, this can only occur in highly restricted spaces.

The proof is given in [234]. This is an exact result of the full theory. However, *it remains true in the linear theory* if the linearisation is done carefully, *even though the term responsible for the conclusion is second-order* and so is omitted in most linearised calculations.

The point is that in that equation at the relevant order, there are *only* second-order terms in the equation, so one cannot drop the terms responsible for this conclusion (one can only drop them in an equation where there are also linear terms, so the second-order terms are negligible compared with the linear terms; that is not the case here). Thus, a proper justification for the acceptability of some effective truncation procedure relies on showing how a specific approximation procedure gives an acceptable approximation to the results of the exact theory, despite this disjuncture.

Exercise: Give such a justification.

8.6.2. *Other models*

As mentioned previously in these lectures, there is an extensive literature on the CBR anisotropy

- in Bianchi (exact spatially homogeneous) models; see e.g. [190]–[192],
- in Swiss-Cheese (exact inhomogeneous) models; see e.g. [161] and [142],
- in small universes (FLRW models with compact spatial topology that closes up on a small scale, so that we have seen round the universe already since decoupling [89]); see e.g. [250] and [92].

These provide important parametrized sets of models that one can use to test and exploit the restrictions the CBR observations place on alternative universe models that are not necessarily close to the standard models at early or late time, but are still potentially compatible with observations, and deserve full exploration.

In each case, whether carrying out Sachs–Wolfe type or more specific model calculations, we get tighter limits than in the EGS case but they are also more model-dependent. In each case they can be carried out using the $1+3$ covariantly defined harmonic variables and used to put limits on the $1+3$ covariantly defined quantities that are the theme of this article.

The overall conclusion is that — given the Copernican assumption underlying use of the almost-FLRW models — the high degree of observed CBR isotropy confirms use of these models for the observable universe and puts limits on the amplitude of anisotropic and inhomogeneous modes in this region (i.e. within the horizon and since decoupling). Other models remain viable as representations of the universe at earlier and later times and outside the horizon.

9. Conclusion and open issues

The sections above have presented the $1+3$ covariant variables and equations, together with the tetrad equations, a series of interesting exact cosmological models, and a systematic procedure for obtaining approximate almost-FLRW models and examining observations in these models. It has been shown how a number of anisotropic and inhomogeneous models are useful in studying observational limits on the geometry of the real universe, and how interesting dynamical and observational issues arise in considering these models; indeed issues such as whether inflation in fact succeeds in making the universe isotropic or not cannot be tackled without examining such models [251].

9.1. CONCLUSION

The standard model of cosmology is vindicated in the following sense (cf. [94, 41]):

- The universe is expanding and evolving, as evidenced by the (*magnitude, redshift*)-relation for galaxies and other sources together with number count observations and a broad compatibility of age estimates;
- It started from a hot big bang early stage which evolved to the presently observed state, this early era being evidenced by the CBR spectrum and concordance with nucleosynthesis observations;

- Given a[40] Copernican assumption, the high degree of isotropy of the CBR and other observations support an almost-FLRW (nearly homogeneous and isotropic) model within the observable region (inside our past light cone) since decoupling and probably back to nucleosynthesis times;
- There are globally inhomogeneous spherically symmetric models that are also compatible with the observations if we do not introduce the Copernican assumption [144], and there are inhomogeneous and anisotropic modes that suggest the universe is not almost-FLRW at very early and very late times [51, 196], and on very large scales (outside the horizon) [66].

The issue of cosmological parameters may be resolved in the next decade due to the flood of new data coming in — from deep space number counts and redshift measurements, observations of supernovae in distant galaxies, measurement of strong and weak gravitational lensing, and CBR anisotropy measurements (see Coles and Ellis [41] for a discussion).

The further extensions of this model proposed from the particle physics side are at present mainly compatible but not yet as compelling from an observational viewpoint, namely:

- An early inflationary era helps resolve some philosophical puzzles about the structure of the universe related specifically to the homogeneity and isotropy of the universe [64]–[66], but the link to particle physics will not be compelling until a specific inflaton candidate is identified;
- Structure formation initiated by quantum fluctuations at very early stages in an inflationary model are a very attractive idea, with the proposal of a CDM-dominated late universe and associated predictions for the CBR anisotropy giving a strong link to observations that may be confirmed in the coming decade; however, theoretical details of this scheme, and particularly the associated issues of biassing and the normalisation of matter to radiation, are still to be resolved;
- The density of matter in the universe is probably below the critical value predicted by the majority of inflationary models [41]; compatibility with inflation may be preserved by either moving to low density inflationary models[41] or by confirmation of a cosmological constant that is dynamically dominant at the present time;[42] the latter pro-

[40] Necessarily philosophically based [150].
[41] For an early proposal, see [252] and [73].
[42] But way below that predicted by present field theory [253].

posal needs testing relative to number counts, lensing observations, and structure formation scenarios;
- The vibrant variety of pre-inflationary proposals involve a huge extrapolation of presently known physics to way beyond the testable domain and, given their variety, do not as yet give a compelling unique view of that era.

Hence, it is suggested [41] that these proposals should not be regarded as part of the standard model but rather as interesting avenues under investigation.

9.2. OPEN ISSUES

Considered from a broader viewpoint, substantial issues remain unresolved. Among them are,

(1) The Newtonian theory of cosmology is not yet adequately resolved. Newtonian theory is only a good theory of gravitation when it is a good approximation to General Relativity; obtaining this limit in non-linear cosmological situations raises a series of questions and issues that still need clarifying [24], particularly relating to boundary conditions in realistic Newtonian cosmological models.

(2) We have some understanding of how the evolution of families of inhomogeneous models relates to that of families of higher symmetry models. It has been indicated that a skeleton of higher symmetry models seems to guide the evolution of lower symmetry models in the state space (the space of space-times) [51]. This relation needs further elucidation. Also anisotropic and inhomogeneous inflationary models are relatively little explored and problems remain [184, 254];

(3) We need to find a suitable measure of probability in the full space of cosmological space-times, and in its involutive subspaces. The requirement is a natural measure that is plausible. Some progress has been made in the FLRW subcase [255]–[258], but even here is not definitive. Closely related to this is the issue of the stability of the results we derive from cosmological modelling [244];

(4) We need to be able to relate descriptions of the same space-time on different scales of description. This leads to the issue of averaging and the resulting effective (polarization) contributions to the energy-momentum tensor, arising because averaging does not commute with calculating the field equations for a given metric [88], and we do not have a good proce-

dure for fitting a FLRW model to a lumpy realistic universe model [259]. It includes the issue (discussed briefly above) of how the almost-everywhere empty real universe can have dynamical and observational relations that average out to high precision to the FLRW relations on a large scale.

(5) Related to this is the question of definition of entropy for gravitating systems in general,[43] and cosmological models in particular. This may be expected to imply a coarse-graining in general, and so is strongly related to the averaging question. It is an important issue in terms of its relation to the spontaneous formation of structure in the early universe, and also relates to the still unresolved arrow of time problem [260], which in turn relates to a series of further issues concerned with the effect on local physics of boundary conditions at the beginning of the universe (see e.g. [28]).

(6) One can approach relating a model cosmology to astronomical observations by a strictly observational approach [98, 101], as opposed to the more usual model-based approach as envisaged in the main sections above and indeed in most texts on cosmology. Intermediate between them is a best-fitting approach [259]. Use of such a fitting approach is probably the best way to tackle modelling the real universe [261], but it is not yet well-developed.

(7) Finally, underlying all these issues is the series of problems arising because of the uniqueness of the universe, which is what particularly gives cosmology its unique character and underlies the special problems in cosmological modelling and application of probability theory to cosmology [262]. Proposals to deal with this by considering an ensemble of universes realized in one or other of a number of possible ways are in fact untestable and, hence, of a metaphysical rather than physical nature; but this needs further exploration.

There is interesting work still to be done in all these areas. It will be important in tackling these issues to as far as possible use gauge-invariant and 1+3 covariant methods, because coordinate methods can be misleading.

Acknowledgements

GFRE and HvE thank Roy Maartens, John Wainwright, Peter Dunsby, and Tim Gebbie for work done together, some of which is presented in these lectures, Charles Hellaby for useful comments and references, Tim

[43] In the case of black holes, there is a highly developed theory; but there is no definition for a general gravitational field; see e.g. [260].

Gebbie for helpful comments, and Anthony Challinor and Roy Maartens for corrections to some equations. We are grateful to the Foundation for Research and Development (South Africa) and the University of Cape Town for financial support. HvE acknowledges the support by a grant from the Deutsche Forschungsgemeinschaft (Germany).

References

1. R d'Inverno, *Introducing Einstein's Relativity* (Oxford Univerity Press, Oxford, 1992).
2. R M Wald, *General Relativity* (University of Chicago Press, Chicago, 1984).
3. S W Hawking and G F R Ellis, *The Large Scale Structure of Space-Time* (Cambridge University Press, Cambridge, 1973).
4. H Stephani, *General Relativity* (Cambridge University Press, Cambridge, 1990).
5. J Ehlers, *Akad. Wiss. Lit. Mainz, Abhandl. Math.-Nat. Kl.* **11**, 793 (1961). Translation: J Ehlers, *Gen. Rel. Grav.* **25**, 1225 (1993).
6. G F R Ellis, in *General Relativity and Cosmology, Proceedings of the XLVII Enrico Fermi Summer School*, Ed. R K Sachs (Academic Press, New York, 1971).
7. G F R Ellis, in *Cargèse Lectures in Physics, Vol. 6*, Ed. E Schatzman (Gordon and Breach, New York, 1973).
8. A Krasiński, *Physics in an Inhomogeneous Universe* (Cambridge University Press, Cambridge, 1996).
9. J A Wheeler, *Einsteins Vision* (Springer, Berlin, 1968).
10. W Kundt and M Trümper, *Akad. Wiss. Lit. Mainz, Abhandl. Math.-Nat. Kl.* **12**, 1 (1962).
    M Trümper, *J. Math. Phys.* **6**, 584 (1965).
11. R Maartens, *Phys. Rev.* D **55**, 463 (1997).
12. M B Ribeiro and A Y Miguelote, *Braz. Journ. Phys.* **28**, 132 (1998).
13. R Maartens, G F R Ellis and S T C Siklos, *Class. Quantum Grav.* **14**, 1927 (1997).
14. T Levi-Civita, *Math. Ann.* **97**, 291 (1926).
15. F A E Pirani, *Acta Phys. Polon.* **15**, 389 (1956).
16. F A E Pirani, *Phys. Rev.* **105**, 1089 (1957).
17. P Szekeres, *J. Math. Phys.* **6**, 1387 (1965).
18. P Szekeres, *J. Math. Phys.* **7**, 751 (1966).
19. G F R Ellis, M Bruni and J C Hwang, *Phys. Rev.* D **42**, 1035 (1990).
20. K Gödel, *Proc. Int. Cong. Math.* (Am. Math. Soc.) **175**, (1952).
21. A Raychaudhuri, *Phys. Rev.* **98**, 1123 (1955).
22. S W Hawking, *Astrophys. J.* **145**, 544 (1966).
23. P K S Dunsby, B A C Bassett and G F R Ellis, *Class. Quantum Grav.* **14**, 1215 (1996).
24. H van Elst and G F R Ellis, *Class. Quantum Grav.* **15**, 3545 (1998).
25. H Friedrich, *Phys. Rev.* D **57**, 2317 (1998).
26. H van Elst and G F R Ellis, *Phys. Rev.* D **59**, 024013 (1999).
27. T Velden, *Diplomarbeit*, Universität Bielefeld/Albert–Einstein–Institut, Potsdam, (1997).
28. G F R Ellis and D W Sciama, in *General Relativity* (Synge Festschrift), Ed. L O'Raifeartaigh (Oxford University Press, Oxford, 1972).
29. G F R Ellis, in *Gravitation and Cosmology* (Proceedings of ICGC95), Eds. S Dhurandhar and T Padmanabhan (Kluwer, Dordrecht, 1997).
30. G F R Ellis, *J. Math. Phys.* **8**, 1171 (1967).
31. C B Collins, *J. Math. Phys.* **26**, 2009 (1985).
32. J M Senovilla, C Sopuerta and P Szekeres, *Gen. Rel. Grav.* **30**, 389 (1998).
33. S Matarrese, O Pantano and D Saez, *Phys. Rev. Lett.* **72**, 320 (1994).

34. H van Elst, C Uggla, W M Lesame, G F R Ellis and R Maartens, *Class. Quantum Grav.* **14**, 1151 (1997).
35. C F Sopuerta, R Maartens, G F R Ellis and W M Lesame, Nonperturbative gravitomagnetic fields are not transverse. Preprint *gr-qc/9809085*, (1998).
36. F J Dyson, *Sci. Am.*, September (1971).
37. See C W Misner, K S Thorne and J A Wheeler, *Gravitation* (Freeman and Co., New York, 1973), Exercise 15.2 (p. 382), and references given there.
38. M S Madsen and G F R Ellis, *Mon. Not. Roy. Astr. Soc.* **234**, 67 (1988).
39. S W Hawking and R Penrose, *Proc. R. Soc. London A* **314**, 529 (1970).
40. F J Tipler, C J S Clarke and G F R Ellis, in *General Relativity and Gravitation: One Hundred Years after the Birth of Albert Einstein, Vol. 2*, Ed. A Held (Plenum Press, New York, 1980).
41. P Coles and G F R Ellis, *The Density of Matter in the Universe* (Cambridge University Press, Cambridge, 1997).
42. A Raychaudhuri, *Zeits. f. Astroph.* **43**, 161 (1957).
43. O Heckmann, *Astronom. Journ.* **66**, 205 (1961).
44. O Heckmann and E Schücking, *Zeits. f. Astrophys.* **38**, 95 (1955).
45. O Heckmann and E Schücking, *Zeits. f. Astrophys.* **40**, 81 (1956).
46. J Ehlers and T Buchert, On the Newtonian limit of the Weyl-tensor. Preprint AEI/MPI für Gravitationsphysik, (1996).
47. H van Elst and C Uggla, *Class. Quantum Grav.* **14**, 2673 (1997).
48. M A H MacCallum, in *Cargèse Lectures in Physics, Vol. 6*, Ed. E Schatzman (Gordon and Breach, New York, 1973).
49. G F R Ellis and M A H MacCallum, *Commun. Math. Phys.* **12**, 108 (1969).
50. J M Stewart and G F R Ellis, *J. Math. Phys.* **9**, 1072 (1968).
51. J Wainwright and G F R Ellis (Eds.), *Dynamical Systems in Cosmology* (Cambridge University Press, Cambridge, 1997).
52. G F R Ellis, in *Vth Brazilian School on Cosmology and Gravitation*, Ed. M Novello (World Scientific, Singapore, 1987).
53. G F R Ellis and H van Elst, in *On Einstein's Path — Essays in Honor of Engelbert Schücking*, Ed. A Harvey, (Springer, New York, 1999). Available as preprint *gr-qc/9709060*.
54. G F R Ellis, *Gen. Rel. Grav.* **2**, 7 (1971).
55. G F R Ellis et al, in *Dahlem Workshop Report ES19, The Evolution of the Universe*, Eds. S Gottlöber and G Börner (Wiley, New York, 1996).
56. W E Harris, R P Durrell, M J Pierce, and J Seckers, *Nature* **395**, 45 (1998). B F Madore et al, *Nature* **395**, 47 (1998).
57. S Perlmutter et al, *Nature* **391**, 51 (1998).
58. H P Robertson, *Rev. Mod. Phys.* **5**, 62 (1933).
59. R Stabell and S Refsdal, *Mon. Not. Roy. Astr. Soc.* **132**, 379 (1966).
60. E Schrödinger, *Expanding Universes* (Cambridge University Press, Cambridge, 1956).
61. S Weinberg, *Gravitation and Cosmology* (Wiley, New York, 1972).
62. R M Barnett et al, *Rev. Mod. Phys.* **68**, 708 (1996).
63. D N Schramm and M S Turner, *Rev. Mod. Phys.* **70**, 303 (1998).
64. A H Guth, *Phys. Rev.* D **23**, 347 (1981).
65. E W Kolb and M S Turner, *The Early Universe* (Wiley, New York, 1990).
66. A D Linde, *Particle Physics and Inflationary Cosmology* (Harwood, Chur, 1990).
67. G F R Ellis and M Madsen, *Class. Quantum Grav.* **8**, 667 (1991).
68. J Ehlers, in *General Relativity and Cosmology, Proceedings of the XLVII Enrico Fermi Summer School*, Ed. R K Sachs (Academic Press, New York, 1971).
69. J Ehlers, P Geren and R K Sachs, *J. Math. Phys.* **9**, 1344 (1968).
70. R Treciokas and G F R Ellis, *Commun. Math. Phys.* **23**, 1 (1971).
71. G F R Ellis, D R Matravers and R Treciokas, *Gen. Rel. Grav.* **15**, 931 (1983).
72. J Ehlers and W Rindler, *Mon. Not. Roy. Astr. Soc.* **238**, 503 (1989).

73. G F R Ellis, in *Gravitation, Proceedings of the Banff Summer Research Institute on Gravitation*, Eds. R Mann and P Wesson (World Scientific, Singapore, 1991).
74. G B Field, H Arp and J N Bahcall, *The Redshift Controversy* (Benjamin, Reading, MA, 1973).
75. D R Matravers and A M Aziz, *Mon. Not. Astr. Soc. S.A.* **47**, 124 (1988).
76. W Mattig, *Astr. Nach.* **284**, 109 (1958).
77. A Sandage, *Astrophys. J.* **133**, 355 (1961).
78. G F R Ellis and T Rothman, *Am. J. Phys.* **61**, 93 (1993).
79. G F R Ellis and G Tivon, *Observatory* **105**, 189 (1985).
80. G F R Ellis and J J Perry, *Mon. Not. Roy. Astr. Soc.* **187**, 357 (1979).
81. I M H Etherington, *Phil. Mag.* **15**, 761 (1933).
82. M Disney, *Nature* **263**, 573 (1976).
83. G F R Ellis, J J Perry and A Sievers, *Astronom. Journ.* **89**, 1124 (1984).
84. T Rothman and G F R Ellis, *Observatory* **107**, 24 (1987).
85. W Rindler, *Mon. Not. Roy. Astr. Soc.* **116**, 662 (1956).
86. R Penrose, in *Relativity Groups and Topology*, Eds. C M DeWitt and B S DeWitt (Gordon and Breach, New York, 1963).
87. G F R Ellis and W R Stoeger, *Class. Quantum Grav.* **5**, 207 (1988).
88. G F R Ellis, in *General Relativity and Gravitation*, Eds. B Bertotti, F de Felice and A Pascolini (Reidel, Dordrecht, 1984).
89. G F R Ellis and G Schreiber, *Phys. Lett.* **115A**, 97 (1986).
90. B F Roukema, *Mon. Not. Roy. Astr. Soc.* **283**, 1147 (1996).
91. B Roukema and A C Edge, *Mon. Not. Roy. Astr. Soc.* **292**, 105 (1997).
92. N J Cornish, D N Spergel and G D Starkman, *Phys. Rev. Lett.* **77**, 215 (1996).
93. G Ellis and R Tavakol, *Class. Quantum Grav.* **11**, 675 (1994).
94. P J E Peebles, D N Schramm, E L Turner and R G Kron, *Nature* **352**, 769 (1991).
95. C W Misner, *Astrophys. J.* **151**, 431 (1968).
96. C W Misner, *Phys. Rev. Lett.* **22**, 1071 (1969).
97. M J Rees, *Perspectives in Astrophysical Cosmology* (Cambridge University Press, Cambridge, 1995).
98. J Kristian and R K Sachs, *Astrophys. J.* **143**, 379 (1966).
99. P Schneider, J Ehlers and E Falco, *Gravitational Lenses* (Springer, Berlin, 1992).
100. D E Holz and R M Wald, *Phys. Rev.* D **58**, 063501 (1998).
101. G F R Ellis, S D Nel, W Stoeger, R Maartens and A P Whitman, *Phys. Rep.* **124**, 315 (1985).
102. G F R Ellis, in *Galaxies and the Young Universe*, Eds. H von Hippelein, K Meisenheimer and J H Roser (Springer, Berlin, 1995).
103. W Stoeger (Ed.), *Theory and Observational Limits in Cosmology*, Specola Vaticana (The Vatican Observatory), (1987).
104. R A Sunyaev and Ya B Zel'dovich, *Astrophys. Space Sci.* **9**, 368 (1970).
105. L P Eisenhart, *Continuous Groups of Transformations* (Princeton University Press, Princeton, 1933). Reprinted: (Dover, New York, 1961).
106. P M Cohn, *Lie Algebras* (Cambride University Press, Cambridge, 1961).
107. H van Elst and G F R Ellis, *Class. Quantum Grav.* **13**, 1099 (1996).
108. K Gödel, *Rev. Mod. Phys.* **21**, 447 (1949).
109. G F R Ellis, in *Gödel 96*, Lecture Notes in Logic 6, Ed. P Hajek (Springer, Berlin, 1996).
110. I Oszvath, *J. Math. Phys.* **6**, 590 (1965); *J. Math. Phys.* **9**, 2871 (1970).
111. I Oszvath and E Schücking, *Nature* **193**, 1168 (1962).
112. H Bondi, *Cosmology* (Cambridge University Press, Cambridge, 1960).
113. A S Kompaneets and A S Chernov, *Sov. Phys. JETP* **20**, 1303 (1965).
114. R Kantowski and R K Sachs, *J. Math. Phys.* **7**, 443 (1966).
115. C B Collins, *J. Math. Phys.* **18**, 2116 (1977).
116. A R King and G F R Ellis, *Commun. Math. Phys.* **31**, 209 (1973).
117. G Lemaître, *Ann. Soc. Sci. Bruxelles I* **A 53**, 51 (1933).

Translation: G Lemaître, *Gen. Rel. Grav.* **29**, 641 (1997).
118. R C Tolman, *Proc. Nat. Acad. Sci. U.S.* **20**, 69 (1934).
119. H Bondi, *Mon. Not. Roy. Astr. Soc.* **107**, 410 (1947).
120. P Szekeres, *Commun. Math. Phys.* **41**, 55 (1975).
121. P Szekeres, *Phys. Rev.* D **12**, 2941 (1975).
122. S W Goode and J Wainwright, *Phys. Rev.* D **26**, 3315 (1982).
123. H Stephani, *Class. Quantum Grav.* **4**, 125 (1987).
124. A Krasiński, *Gen. Rel. Grav.* **15**, 673 (1983).
125. D Kramer, H Stephani, M A H MacCallum and E Herlt, *Exact Solutions of Einstein's Field Equations* (Cambridge University Press, Cambridge, 1980).
126. O Heckmann and E Schücking, in *Gravitation*, Ed. L Witten (Wiley, New York, 1962).
127. K Tomita, *Prog. Theor. Phys.* **40**, 264 (1968).
128. M A H MacCallum and G F R Ellis, *Commun. Math. Phys.* **19**, 31 (1970).
129. K S Thorne, *Astrophys. J.* **148**, 51 (1967).
130. C W Misner, *Phys. Rev. Lett.* **19**, 533 (1967).
131. M J Rees, *Astrophys. J.* **153**, L1 (1968).
132. C W Misner, K S Thorne and J A Wheeler, *Gravitation* (Freeman and Co., New York, 1973).
133. W B Bonnor, *Mon. Not. Roy. Astr. Soc.* **159**, 261 (1972).
134. W B Bonnor, *Mon. Not. Roy. Astr. Soc.* **167**, 55 (1974).
135. Ya B Zel'dovich and L P Grishchuk, *Mon. Not. Roy. Astr. Soc.* **207**, 23P (1984).
136. F C Mena and R Tavakol, *Class. Quantum Grav.* **16**, 435 (1999).
137. J Silk, *Astron. Astrophys.* **59**, 53 (1977).
138. R Kantowski, *Astrophys. J.* **155**, 1023 (1969).
139. C Hellaby and K Lake, *Astrophys. J.* **282**, 1 (1984).
140. D J Raine and E G Thomas, *Mon. Not. Roy. Astr. Soc.* **195**, 649 (1981).
141. B Paczynski and T Piran, *Astrophys. J.* **364**, 341 (1990).
142. M Panek, *Astrophys. J.* **388**, 225 (1992).
143. W B Bonnor and G F R Ellis, *Mon. Not. Roy. Astr. Soc.* **218**, 605 (1986).
144. N Mustapha, C Hellaby and G F R Ellis, *Mon. Not. Roy. Astr. Soc.* **292**, 817 (1998).
145. N Mustapha, B A C C Bassett, C Hellaby and G F R Ellis, *Class. Quantum Grav.* **15**, 2363 (1998).
146. R Maartens, N P Humphreys, D R Matravers and W Stoeger, *Class. Quantum Grav.* **13**, 253 (1996).
147. N P Humphreys, R Maartens and D R Matravers, Regular spherical dust spacetimes, Preprint *gr-qc/9804023*, (1998).
148. C Hellaby, *Gen. Rel. Grav.* **20**, 1203 (1988).
149. C Hellaby and K Lake, *Astrophys. J.* **290**, 381 (1985); *Astrophys. J.* **300**, 461 (1986).
150. G F R Ellis, *Qu. Journ. Roy. Astr. Soc.* **16**, 245 (1975).
151. W Israel, *Il Nuovo Cimento Serie X* **44**, 1 (1966); Corrections: **48**B, N2 (1967).
152. G F R Ellis and M Jaklitsch, *Astrophys. J.* **346**, 601 (1989).
153. J Traschen, *Phys. Rev.* D **29**, 1563 (1984);
Phys. Rev. D **31**, 283 (1985).
154. A Einstein and E G Strauss, *Rev. Mod. Phys.* **17**, 120 (1945);
Rev. Mod. Phys. **18**, 148 (1945).
155. E Schücking, *Z. Physik* **137**, 595 (1954).
156. K Lake, *Astrophys. J.* **240**, 744 (1980);
Astrophys. J. **242**, 1238 (1980).
157. C Hellaby and K Lake, *Astrophys. J.* **251**, 429 (1981).
158. C Hellaby and K Lake, *Astrophys. Lett.* **23**, 81 (1983).
159. R Kantowski, *Astrophys. J.* **155**, 89 (1969).
160. M J Rees and D W Sciama, *Nature* **217**, 511 (1968).

161. C C Dyer, *Mon. Not. Roy. Astr. Soc.* **175**, 429 (1976).
162. A Meszaros and Z Molner, *Astrophys. J.* **470**, 49 (1996).
163. W B Bonnor and A Chamorro, *Astrophys. J.* **361**, 21 (1990); *Astrophys. J.* **378**, 461 (1991).
164. A Chamorro, *Astrophys. J.* **383**, 51 (1991).
165. K Lake, in *Vth Brazilian school of Cosmology and Gravitation*, Ed. M Novello (World Scientific, Singapore, 1987).
166. M Harwit, *Astrophys. J.* **392**, 394 (1992).
167. W B Bonnor, *Commun. Math. Phys.* **51**, 191 (1976).
168. C C Dyer, S Landry and E G Shaver, *Phys. Rev.* D **47**, 1404 (1993).
169. S Landry and C C Dyer, *Phys. Rev.* D **56**, 3307 (1997).
170. C B Collins and G F R Ellis, *Phys Rep.* **56**, 63 (1979).
171. A H Taub, *Ann. Math.* **53**, 472 (1951).
172. C B Collins and S W Hawking, *Astrophys. J.* **180**, 317 (1973).
173. R T Jantzen, *Commun. Math. Phys.* **64**, 211 (1979).
174. R T Jantzen, in *Cosmology of the Early Universe*, Ed. L Z Fang and R Ruffini (World Scientific, Singapore, 1984).
175. M A H MacCallum, in *General Relativity, An Einstein Centenary Survey*, Eds. S W Hawking and W Israel (Cambridge University Press, Cambridge, 1979).
176. D W Hobill, A Burd and A A Coley (Eds.), *Deterministic Chaos in General Relativity* (Plenum Press, New York, 1994).
177. N J Cornish and J J Levin, *Phys. Rev. Lett.* **78**, 998 (1997).
178. V G LeBlanc, *Class. Quantum Grav.* **14**, 2281 (1997).
179. V G LeBlanc, D Kerr and J Wainwright, *Class. Quantum Grav.* **12**, 513 (1995).
180. G F R Ellis and A R King, *Commun. Math. Phys.* **38**, 119 (1974).
181. R M Wald, *Phys. Rev.* D **28**, 2118 (1983).
182. M Goliath and G F R Ellis, Homogeneous cosmologies with cosmological constant. Preprint, (1998). Accepted by *Phys. Rev.* D . Available at *gr-qc/9811068*.
183. A Rothman and G F R Ellis, *Phys. Lett.* **180B**, 19 (1986).
184. A Raychaudhuri and B Modak, *Class. Quantum Grav.* **5**, 225 (1988).
185. G F R Ellis and J Baldwin, *Mon. Not. Roy. Astr. Soc.* **206**, 377 (1984).
186. J D Barrow, *Mon. Not. Roy. Astr. Soc.* **175**, 359 (1976); *Mon. Not. Roy. Astr. Soc.* **211**, 221 (1984).
187. A Rothman and R Matzner, *Phys. Rev.* D **30**, 1649 (1984).
188. R Matzner, A Rothman and G F R Ellis, *Phys. Rev.* D **34**, 2926 (1986).
189. C B Collins and S W Hawking, *Mon. Not. Roy. Astr. Soc.* **162**, 307 (1973).
190. J D Barrow, R Juszkiewicz and D H Sonoda, *Nature* **305**, 397 (1983); *Mon. Not. Roy. Astr. Soc.* **213**, 917 (1985).
191. R Bajtlik, M Justkiewicz, M Prozszynski and P Amsterdamski, *Astrophys. J.* **300**, 463 (1986).
192. E F Bunn, P Ferreira and J Silk, *Phys. Rev. Lett.* **77**, 2883 (1996).
193. C B Collins, *Commun. Math. Phys.* **23**, 137 (1971).
194. J Wainwright, in *Relativity Today*, Ed. Z Perjes (World Scientific, Singapore, 1988).
195. L Hsu and J Wainwright, *Class. Quantum Grav.* **3**, 1105 (1986).
196. J Wainright, A A Coley, G F R Ellis and M Hancock, *Class. Quantum Grav.* **15**, 331 (1998).
197. G F R Ellis and M Bruni, *Phys. Rev.* D **40**, 1804 (1989).
198. C P Ma and E Bertschinger, *Astrophys. J.* **455**, 7 (1996).
199. J Bardeen, *Phys. Rev.* D **22**, 1882 (1980).
200. J Bardeen, P Steinhardt and M S Turner, *Phys. Rev.* D **28**, 679 (1983).
201. J M Stewart and M Walker, *Proc. R. Soc. London A* **341**, 49 (1974).
202. P Hogan and G F R Ellis, *Class. Quantum Grav.* **14**, A171 (1997).
203. B Bertotti, *Proc. R. Soc. London A* **294**, 195 (1966).
204. R Kantowski, *Astrophys. J.* **155**, 89 (1969).
205. C C Dyer and R C Roeder, *Astrophys. J. Lett.* **180**, L31 (1973);

Gen. Rel. Grav. **13**, 1157 (1981).
206. J P Boersma, *Phys. Rev.* D **57**, 798 (1998).
207. S. Weinberg, *Astrophys. J.* **208**, L1 (1976).
208. G F R Ellis, B A C C Bassett and P K S Dunsby, *Class. Quantum Grav.* **15**, 2345 (1998).
209. G F R Ellis and D M Solomons, *Class. Quantum Grav.* **15**, 2381 (1998).
210. G F R Ellis, J Hwang and M Bruni (1989), *Phys. Rev.* D **40**, 1819 (1989).
211. J H Jeans, *Phil. Trans.* **129**, 587 (1902).
212. M J Rees, in *General Relativity and Cosmology, Proceedings of the XLVII Enrico Fermi Summer School*, Ed. R K Sachs (Academic Press, New York, 1971).
213. M Bruni, G F R Ellis and P K S Dunsby, *Class. Quantum Grav.* **9**, 921 (1992).
214. P K S Dunsby, M Bruni and G F R Ellis, *Astrophys. J.* **395**, 54 (1992).
215. R Maartens, T Gebbie and G F R Ellis, Cosmic microwave background anisotropies: Nonlinear dynamics. Preprint, (1998). Accepted by *Phys. Rev.* D . Available at *astro-ph/9808163*.
216. I Müller, *Z. Physik* **198**, 329 (1967).
217. W Israel and J M Stewart, *Ann. Phys. (N.Y.)* **118**, 341 (1979).
218. R Maartens and J Triginier, *Phys. Rev.* D **56**, 4640 (1997); *Phys. Rev.* D **58**, 123507 (1998).
219. C G Tsagas and J D Barrow, *Class. Quantum Grav.* **14**, 2539 (1997); *Class. Quantum Grav.* **15**, 3523 (1998).
220. G F R Ellis, *Mon. Not. Roy. Astr. Soc.* **243**, 509 (1990).
221. T Hirai and K Maeda, *Astrophys. J.* **431**, 6 (1994).
222. M Bruni, P K S Dunsby and G F R Ellis, *Astrophys. J.* **395**, 34 (1992).
223. J M Stewart, *Class. Quantum Grav.* **7**, 1169 (1990).
224. R K Sachs and A M Wolfe, *Astrophys. J.* **147**, 73 (1967).
225. M Panek, *Phys. Rev.* D **34**, 416 (1986).
226. P K S Dunsby, *Class. Quantum Grav.* **14**, 3391 (1997).
227. A Challinor and A Lasenby, *Phys. Rev.* D **58**, 023001 (1998).
228. R W Lindquist, *Ann. Phys.* **37**, 487 (1966).
229. W Israel, in *General Relativity*, Ed. O'Raifeartaigh (Oxford University Press, Oxford, 1972).
230. J M Stewart, *Non Equilibrium Relativistic Kinetic Theory*, Springer Lecture Notes in Physics 10 (Springer, Berlin, 1971).
231. J D Jackson, *Classical Electrodynamics* (Wiley, New York, 1962).
232. R P Feynman, R B Leighton and M Sands, *The Feynman Lectures on Physics*, Vol. I (Addison–Wesley, Reading, MA, 1963).
233. G F R Ellis, D R Matravers and R Treciokas, *Ann. Phys.* **150**, 455 (1983).
234. G F R Ellis, R Treciokas and D R Matravers, *Ann. Phys.* **150**, 487 (1983).
235. F A E Pirani, in *Lectures in General Relativity*, Eds. S Deser and K Ford (Prentice–Hall, Englewood Cliffs, 1964).
236. K S Thorne, *Rev. Mod. Phys.* **52**, 299 (1980).
237. M L Wilson, *Astrophys. J.* **273**, 2 (1983).
238. T Gebbie and G F R Ellis, Covariant cosmic microwave background anisotropies. I: Algebraic relations for mode and multipole representations. Preprint *astro-ph/9804316*, (1998).
239. A D Challinor and A N Lasenby, Cosmic microwave background anisotropies in the CDM model: A covariant and gauge-invariant approach. Preprint *astro-ph/9804301*, (1998).
240. R Maartens, G F R Ellis and W J Stoeger, *Phys. Rev.* D **51**, 1525 (1995); *Phys. Rev.* D **51**, 5942 (1995).
241. W Hu and N Sugiyama, *Phys. Rev.* D **51**, 2599 (1995).
242. T Gebbie and G F R Ellis: Covariant cosmic microwave background anisotropies. II: Almost-FLRW standard model. Preprint, (1998).
243. W Hu and M White, *Astrophys. J.* **471**, 30 (1996).

244. R K Tavakol and G F R Ellis, *Phys. Lett.* **130A**, 217 (1988).
245. W Stoeger, R Maartens and G F R Ellis, *Astrophys. J.* **443**, 1 (1995).
246. G F R Ellis, R Maartens and S D Nel, *Mon. Not. Roy. Astr. Soc.* **184**, 439 (1978).
247. J Goodman, *Phys. Rev.* D **52**, 1821 (1995).
248. G Jungman, M Kamionkowski, A Kosowsky and D N Spergel, *Phys. Rev.* D **54**, 1332 (1996).
249. W R Stoeger, C Xu, G F R Ellis and M Katz, *Astrophys. J.* **445**, 17 (1995).
250. D Stevens, D Scott and J Silk, *Phys. Rev. Lett.* **71**, 20 (1993).
251. P Anninos, R A Matzner, T Rothman and M P Ryan, *Phys. Rev.* D **43**, 3821 (1991).
252. G F R Ellis, D H Lyth and M B Mijić, *Phys. Lett.* **271B**, 52 (1991).
253. S Weinberg, *Rev. Mod. Phys.* **61**, 1 (1989).
254. R Penrose, in *Proc. 14th Texas Symp. on Relativistic Astrophysics*, Ed. E J Fergus, (New York Academy of Sciences, New York, 1989).
255. G W Gibbons, S W Hawking and J M Stewart, *Nucl. Phys.* B **281**, 736 (1987).
256. D N Page, *Phys. Rev.* D **36**, 1607 (1987).
257. G Evrard and P Coles, *Class. Quantum Grav.* **12**, L93 (1995).
258. D H Coule, *Class. Quantum Grav.* **12**, 455 (1995).
259. G F R Ellis and W R Stoeger, *Class. Quantum Grav.* **4**, 1679 (1987).
260. R Penrose, *The Emperor's New Mind: Concerning Computers, Minds, and the Laws of Physics* (Oxford University Press, Oxford, 1989).
261. D R Matravers, G F R Ellis and W R Stoeger, *Qu. J. Roy. Astr. Soc.* **36**, 29 (1995).
262. G F R Ellis, *Mem. Ital. Astr. Soc.* **62**, 553 (1991).

TOPOLOGY OF THE UNIVERSE: THEORY AND OBSERVATION

JEAN-PIERRE LUMINET[1] AND BOUDEWIJN F. ROUKEMA[2,3]
[1] *DARC, Observatoire de Paris-Meudon, 5 place Jules Janssen, F-92195 Meudon Cedex, France (Jean-Pierre.Luminet@obspm.fr)*

AND

[2] *Institut d'Astrophysique de Paris, 98bis Bd Arago, F-75.014, Paris, France*
[3] *Inter-University Centre for Astronomy and Astrophysics, Post Bag 4*
Ganeshkhind, Pune, 411 007, India (boud@iucaa.ernet.in)

Abstract. "One could imagine that as a result of enormously extended astronomical experience, the entire universe consists of countless identical copies of our Milky Way, that the infinite space can be partitioned into cubes each containing an exactly identical copy of our Milky Way. Would we really cling on to the assumption of infinitely many identical repetitions of the same world? ... We would be much happier with the view that these repetitions are illusory, that in reality space has peculiar connection properties so that if we leave any one cube through a side, then we immediately reenter it through the opposite side." (Schwarzschild 1900, translation 1998)

Developments in the theoretical and observational sides of cosmic topology were slow for most of the century, but are now progressing rapidly at a scale of most interest, with the hope of detecting the topological parameters of the universe within the next ten years.

The historical, mathematical and observational aspects of the subject are briefly reviewed in this course.

1. Introduction

Think of a right triangle drawn on a transparency. The transparency is a piece of an *infinite* Euclidean plane whose flat metric is expressed by

M. Lachièze-Rey (ed.), Theoretical and Observational Cosmology, 117–156.

Pythagoras' Theorem. Bend the transparency around into a cylinder. The latter is thus obtained by identifying two opposite edges of the transparency. The new surface is locally flat (Pythagoras' Theorem stills holds perfectly), but it has a quite different shape. It contains closed geodesics. Next we can identify the two remaining edges and get a flat torus (this is a thought experiment, because the usual torus, or "annulus", visualized as a surface of revolution in 3–dimensional Euclidean space, has positive Gaussian curvature at some places and negative at others). The flat torus has still a locally Euclidean metric although its global shape has drastically changed: it is a *finite* surface without borders.

These elementary operations well illustrate the difference between the curvature (given by the metric) and the topology. Both are needed in order to know the full geometry of a two-dimensional, or more importantly, a three-dimensional space. This applies as well to physical space, namely to cosmological models for describing the Universe as a whole. The shape of space is a fundamental issue in physics. Curiously enough, this Cargese '98 summer school on cosmology is probably the first one to include a couple of lectures on cosmic topology. As we shall see below, both for historical and for practical reasons, cosmic topology has been widely ignored during 80 years of relativistic cosmology, except by some pioneering authors.

After an historical section 2, sections 3-4 are mostly based on the review paper by Lachièze-Rey & Luminet [44] (hereafter LaLu 95). Since the mid 90's, worldwide interest for cosmological topology has blossomed, from a mathematical point of view, mostly related to progress in the understanding of compact hyperbolic manifolds, from a theoretical point of view, as well as from an observational point of view, related to improvements of data on the large scale 3–dimensional distribution of cosmic objects and on the 2–dimensional structure of the cosmic microwave background. Since these lectures are aimed to provide a general view of the subject, only some of the new mathematical results (obtained since the 1995 revival) will be mentioned. An incomplete list of theoretical articles related to the possible physical origin of cosmic topology is provided towards the end of section 2. For the latest developments in all aspects of the subject, see the proceedings of the 1997 and 1998 workshops [75, 66]. Section 5 of this lecture, devoted to the observational aspects of cosmic topology, provides an up-date and complement to the review of [68].

2. A Brief History of Cosmological Topology

Most of this historical introduction is taken from [54], see also [74].

Is space finite or infinite, oriented or not, made of one piece or not, has it holes or handles, what is its global shape? A common misconception is to believe that Einstein's general relativity theory is all one needs to answer such paramount questions. However, general relativity deals only with local geometrical properties of the universe, such as its curvature, not with its global characteristics, namely its topology.

The physical extension of space is one of the oldest cosmological questions, going back to twenty–five centuries of cosmological modelling (e.g. [55]). In the history of cosmology, it is well known that Newtonian physical space, mathematically identified with the infinite Euclidean space $\mathbf{R}^3$, gave rise to paradoxes such as the darkness of the night sky (see e.g. [37]) and to problems of boundary conditions.

Wondering about absolute accelerations, Newton imagined a bucket swung on the end of a rope and compared it with a bucket at rest. If the bucket is at rest, the surface of the water remains flat. If the bucket rotates, the water surface becomes concave. Newton argued that the concave shape of the water could not be due to its relative motion with respect to the bucket, and that it proved the existence of an absolute centrifugal acceleration. Mach considered the same problem and criticised Newton's reasoning [2]: what we observe is that the bucket rotates with respect to the fixed stars, but, who is to say whether the bucket is really rotating with respect to the stars at rest or whether the stars are really rotating with respect to the bucket? According to Newton, we should observe the concavity in the first case and not in the second. This is because Newton assumed an absolute frame related to the global distribution of matter in the universe (the fixed stars). Mach denied the concept of absolute acceleration. According to him, a rotating body in a non-rotating universe or a non-rotating body in a rotating universe should give the same result: centrifugal force. Mach concluded that the inertial mass of a body should result from the contributions of all the masses in the universe. But an obvious divergence difficulty arose, since a homogeneous Newtonian universe with non–zero density has an infinite mass. Hence, Mach supported the idea of a finite universe in order to have a finite local inertia.

By the end of the XIXth century, mathematicians such as Riemann had discovered a variety of finite spaces without boundaries, such as the hypersphere and the projective space. Schwarzschild [71] brought this work to the attention of the astronomical community in 1900, and he considered the possibility that the space could have a non–trivial topology: "*One could imagine that as a result of enormously extended astronomical experience, the entire universe consists of countless identical copies of our Milky Way, that the infinite space can be partitioned into cubes each containing an exactly identical copy of our Milky Way. Would we really cling on to the assumption of infinitely many identical repetitions of the same world? ... We would be much happier with the view that these repetitions are illu-*

sory, that in reality space has peculiar connection properties so that if we leave any one cube through a side, then we immediately reenter it through the opposite side" (English translation [72]).

In modern terms, Schwarzschild told us that if the galaxies were found to lie in a rectangular lattice, with images of the same galaxy repeating at equivalent lattice points, then we could conclude that the Universe is a 3–torus.

Cosmology developed rapidly after Einstein's 1915 discovery of general relativity. General relativity explains gravity as the curvature of spacetime; the latter is determined by the density of matter–energy. The aim of relativistic cosmology is to deduce from the gravitational field equations some physical models of the Universe as a whole. When Einstein (1917) assumed in his static cosmological solution that space was a positively–curved hypersphere, one of his strongest motivations was to provide a model for a finite space, although without a boundary. Following Mach and Riemann, he regarded the closure of space as necessary to solve the problem of inertia and chose the hypersphere as his model of space.

The Einstein space model cleared up most of the paradoxes stemming from Newtonian cosmology in such an elegant way that most cosmologists of the time adopted the new paradigm of a closed space, without examining other geometrical possibilities. Einstein was also convinced that the hypershere provided not only the metric of cosmic space – namely its local geometrical properties – but also its global structure, its topology. However, topology does not seem to have been a major preoccupation of Einstein; his 1917 cosmological article did not mention any topological alternative to the spherical space model.

Some of his colleagues pointed out to Einstein the arbitrariness of his choice. Indeed the global shape of space is not only depending on the metric; it depends primarily on its topology, and requires a complemenraty approach to Riemannian differential geometry. Since Einstein's equations are partial derivative equations, they describe only local geometrical properties of spacetime. The latter are contained in the metric tensor, which enables us to calculate the components of the curvature tensor at any nonsingular point of spacetime. But Einstein's equations do not fix the global structure of spacetime: to a given metric solution of the field equations, correspond several (and in most cases an infinite number of) topologically distinct universe models.

Firstly, de Sitter [14] noticed that the Einstein's solution admitted a different spaceform, the 3-dimensional projective space (also called elliptic space), obtained from the hypersphere by identifying antipodal points. The projective space has the same metric as the spherical one, but a different topology (for instance, for the same curvature radius its volume is half that of the spherical space).

H. Weyl also pointed out the freedom of choice between spherical and elliptical topologies. Einstein's answer [17] was unequivocal: "*Nevertheless I have like an obscure feeling which leads me to prefer the spherical model. I have a feeling that manifolds in which any closed curve can be continuously contracted to a point are the simplest ones. Other persons must share this feeling, otherwise astronomers would have taken into account the case where our space is Euclidean and finite. Then the two-dimensional Euclidean space would have the connectivity properties of a ring surface. It is an Euclidean plane in which any phenomenon is doubly periodic, where points located in the same periodical grid are identical. In finite Euclidean space, three classes of non continuously contractible loops would exist. In a similar way, the elliptical space possesses a class of non continuously contractible loops, contrary to the spherical case; it is the reason why I like it less than the spherical space. Can it be proved that elliptical space is the only variant of spherical space? It seems yes to me*".

Einstein [18] repeated his argumentation in a postcard sent to Felix Klein: "*I would like to give you a reason why the spherical case should be preferred to the elliptical case. In spherical space, any closed curve can be continuously contracted to a point, but not in the elliptical space; in other words the spherical space alone is simply-connected, not the elliptical one [...] Finite spaces of arbitrary volume with the Euclidean metric element undoubtedly exist, which can be obtained from infinite spaces by assuming a triple periodicity, namely identity between sets of points. However such possibilities, which are not taken into account by general relativity, have the wrong property to be multiply-connected*".

From these remarks it seems that the Einstein's preference for simple–connectedness of space was of an aesthetical nature, rather than being based on physical reasoning.

In his answer to Weyl, Einstein was definitely wrong on the last point: in dimension three, an infinite number of topological variants of the spherical space – all closed – do exist, including the so-called lens spaces (whereas in dimension two, only two spherical spaceforms exist, the ordinary sphere and the elliptic plane). However, nobody knew this result in the 1920's: the topological classification of 3–dimensional spaces was still at its beginnings. The study of Euclidean spaceforms started in the context of crystallography. Feodoroff [29] classified the 18 symmetry groups of crystalline structures in $\mathbf{R}^3$, Bieberbach [4] developed a full theory of crystallographic groups, and it took twenty years before Novacki [58] showed how Bieberbach's results could be applied to complete the classification of 3–dimensional Euclidean spaceforms. The case of spherical spaceforms was first treated by Klein [42] and Killing [43]. The problem was fully solved much later [86]. Eventually, the classification of homogeneous hyperbolic spaces shot forward in the 1970's; it is now an open field of intensive mathematical research [78, 79].

Returning to relativistic cosmology, the discovery of non-static solutions by Friedmann [31] and, independently, Lemaître [47], opened a new era for models of the universe as a whole (see, e.g., [53] for an epistemological analysis). Although Friedmann and Lemaître are generally considered as the discoverers of the big bang concept — at least of the notion of a dynamical universe evolving from an initial singularity — one of their most original considerations, devoted to the topology of space, was overlooked. As they stated, the homogeneous isotropic universe models (F–L models) admit spherical, Euclidean or hyperbolic spacelike sections according to the sign of their (constant) curvature (respectively positive, zero or negative). In addition, Friedmann [32] pointed out the topological indeterminacy of the solutions in his popular book on general relativity, and he emphasized how Einstein's theory was unable to deal with the global structure of spacetime. He gave the simple example of the cylinder. Generalizing the argument to higher dimensions, he concluded that several topological spaces could be used to describe the same solution of Einstein's equations.

Topological considerations were fully developed in his second cosmological article, although primarily devoted to the analysis of hyperbolic solutions. Friedmann [33] clearly outlined the fundamental limitations of relativistic cosmology: "*Without additional assumptions, Einstein's world equations do not answer the question of the finiteness of our universe*", he wrote. Then he described how space could be finite (and multi-connected) by suitably identifying points. He also predicted the possible existence of "ghost" images of astronomical sources, since at the same point of a multi-connected space an object and its ghosts would coexist. He added that "*a space with positive curvature is always finite*", but he recognized the fact that the mathematical knowledge of his time did not allow him to "*solve the question of finiteness for a negatively-curved space*".

By contrast with Einstein's reasoning, it is seen that the Russian cosmologist had no prejudice in favour of a simply-connected topology. Certainly, Friedmann believed that only finite volume spaces were physically realistic. Prior to his discovery of hyperbolic solutions, the cosmological solutions derived by Einstein, de Sitter and himself had a positive spatial curvature, thus a finite volume. With negatively-curved spaces, the situation became problematic, because the "natural" topology of hyperbolic space has an infinite volume. It is the reason why Friedmann, in order to justify the physical pertinence of his solutions, emphasized the possibility of compactifying space by suitable identifications of points.

Lemaître fully shared the common belief in the closure of space. He expressed his view that Riemannian geometry "*dissipated the nightmare of infinite space*" [50]. Thus Lemaître [47, 48] assumed positive space curvature, but he thoroughly discussed the possibility of projective space, that he preferred to the spherical one. Later, he [49] also noticed the possibility of hyperbolic and Euclidean spaces with finite volumes for describing the physical universe.

Such fruitful ideas of cosmic topology remained widely ignored by the main stream of big bang cosmology. Perhaps the Einstein–de Sitter [19] model, which assumed Euclidean space and eluded the topological question, had a negative influence on the development of the field. Almost all subsequent textbooks and monographs on relativistic cosmology assumed that the global structure of the universe was either the finite hypersphere, or the infinite Euclidean space, or the infinite hyperbolic space, without mentioning at all the topological indeterminacy. As a consequence, some confusion settled down about the real meaning of the terms "open" and "closed" used to characterize the F–L solutions. Whereas they apply correctly to time evolution (open models stand for ever–expanding universes, closed models stand for expanding–contracting solutions), they do not properly describe the extension of space (open for infinite, closed for finite). Nevertheless it is still frequent to read that the (closed) spherical model has a finite volume whereas the (open) Euclidean and hyperbolic models have infinite volumes. The correspondance is true only in the very special case of a simply–connected topology *and* a zero cosmological constant. According to Friedmann's original remark, in order to know if a space is finite or infinite, it is not sufficient to determine the sign of its spatial curvature, or equivalently in a cosmological context to measure the ratio Ω_0 of the average density to the critical value: additional assumptions are necessary — those arising from topology, precisely.

The idea of a multi–connected universe has never been entirely forgotten. In a very comprehensive article, Ellis [20] detailed the classification of 3–dimensional Riemannian manifolds useful for cosmology and started to explore the observational consequences of a toroidal universe. Others, such as Sokoloff and Schvartsman, Fang and Sato, Gott, Fagundes, investigated flat and non-flat spaces (see complete references in LaLu95). Nevertheless, until 1995, investigations in cosmic topology were rather scarce. From an epistemological point of view, it seems that the prejudice in favour of simply–connected (rather than multi–connected) spaces was of the same kind as the prejudice in favour of static (rather than dynamical) cosmologies during the 1920's. At a first glance, "Occam's razor" (often useful in physical modelling) could be invoked to preferably select the simply–connected topologies. However, on the theoretical side, new approaches to spacetime, such as quantum cosmology, suggest that the smallest closed hyperbolic manifolds are favoured, thus providing a new paradigm for what is the "simplest" manifold. On the observational side, present astronomical data (e.g. [12]) indicate that the average density of the observable universe is less than the critical value ($\Omega_0 = 0.3 - 0.4$), thus suggesting that we live in a negatively-curved F-L universe (unless the cosmological constant is positive and large enough). Putting together theory and observation, cosmologists must face the fact that a negatively-curved space can have a finite volume, and that in that case it must be multi-connected.

In the last few decades, much effort in observational and theoretical cosmology has been directed towards determining the curvature of the universe. The problem of the topology of spacetime was generally ignored within the framework of classical relativistic cosmology. It began to be taken seriously discussed in quantum gravity for various reasons: the spontaneous birth of the Universe from a quantum vacuum requires the Universe to have compact spacelike hypersurfaces, and the closure of space is a necessary condition to render tractable the integrals of quantum gravity [1].

Further work in quantum gravity and particle physics which explores ideas of space-time topology globally and/or as the basis of a quantum gravity theory includes that of [38, 87, 56, 7, 40, 15, 63, 16, 73].

However, the topology of spacetime also enters in a fundamental way in classical general relativity. Many cosmologists were surprisingly unaware of how topology and cosmology could fit together to provide new insights in universe models. Aimed to create a new interest in the field of cosmic topology, the review by LaLu95 stressed what multi-connectedness of the Universe would mean and on its observational consequences. However two different papers ([77] ; [13]) declared that the small universe idea was "no longer an interesting cosmological model". However, the authors drew very general conclusions from considering only a few special 3-manifolds in the Euclidean case, and by adopting assumptions on the primordial perturbation spectrum which are observationally justified on large scales only by assuming simple connectedness. They did not take into account a very interesting class of realistic universe models, namely the compact hyperbolic manifolds, which require a quite different treatment. Ironically, approximately the same amount of papers in cosmic topology have been published within the last 3 years as in the previous 80 years !

3. Mathematical background

3.1. BASICS OF TOPOLOGY

The topological properties of a manifold are those which remain insensitive to continuous transformations. Thus, size and distance are in some sense ignored in topology: stretching, squeezing or "kneading" a manifold change the metric but not the topology; cutting, tearing or making holes and han-

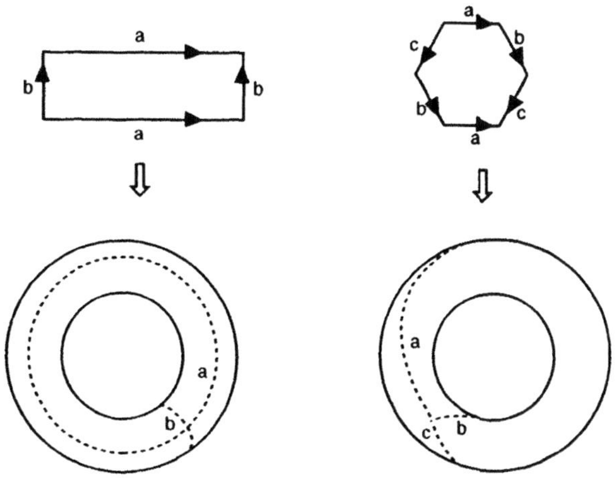

Figure 1. Construction of the flat torus from a rectangle and from an hexagon with suitable identifications

dles change the latter. It is often possible to visualize two-dimensional manifolds by representing them as embedded in three-dimensional Euclidean space (such a mapping does not necessarily exist however, e.g. the Klein bottle and the flat torus). Three-dimensional manifolds require the introduction of more abstract representations, using, for example, the *fundamental domain.*

The different topological surfaces can be represented by polygons whose edges are suitably identified by pairs. As we have seen in the introduction, identifying one pair of opposite edges of a square gives a portion of a cylinder; then, stretching the cylinder and bending it around in such a way that one can glue together the two circular ends generates a simple torus, a *closed* surface (figure 1). The torus is thus topologically equivalent to a rectangle with opposite edges identified. The rectangle is called a *fundamental domain* of the torus. From a topological point of view (namely without reference to size), the fundamental domain can be chosen in different ways: a square, a rectangle, a parallelogram, even a hexagon (since the plane can be tiled by hexagons, the flat torus can be also represented by a hexagon with suitable identification of edges).

The gluing method described above becomes extremely fruitful when the surfaces are more complicated. A two-dimensional g-torus T_g is a torus with g holes. T_g can be constructed as the *connected sum* of g simple tori (figure 2). The g-torus is therefore topologically equivalent to a connected sum of g squares whose opposite edges have been identified. This sum is itself topologically equivalent to a 4g-gon where all the vertices are identical with each other and the sides are suitably identified by pairs.

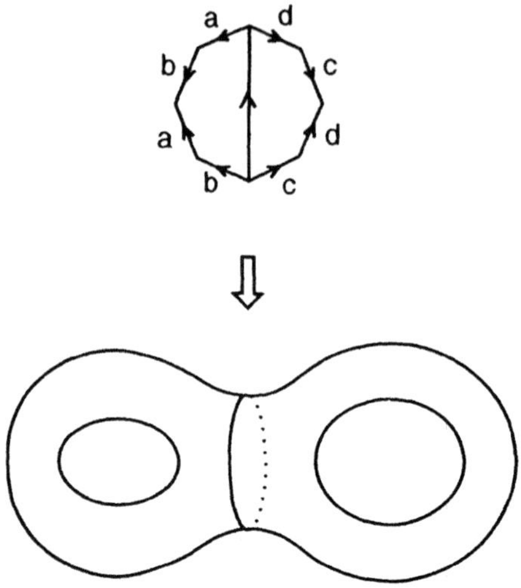

Figure 2. The 2-torus as the connected sum of two simple tori

It would be tempting to visualize the g–torus by gluing together equivalent edges, like for the simple torus. But such an operation is not straightforward when $g \geq 2$. All the vertices of the polygon correspond to the same point of the surface. Since the polygon has at least 8 edges, it is necessary to make the internal angles thinner in order to fit them suitably around a single vertex. This can only be achieved if the polygon is represented in the *hyperbolic plane* $\mathbf{H}^2$ instead of the Euclidean plane $\mathbf{R}^2$: this increases the area and decreases the angles. The more angles to fit together, the thinner they have to be and the greater the surface. The g-torus ($g \geq 2$) is therefore a *compact surface of negative curvature.*

When one deals with more than two dimensions, the gluing method remains the simplest way to visualize spaces. By analogy with the two-dimensional case, the three-dimensional simple torus (also referred to as the *hypertorus*) is obtained by identifying the opposite faces of a parallelepiped. The resulting volume is finite. Let us imagine a light source at our position, immersed in such a structure. Light emitted backwards crosses the face of the parallelepiped behind us and reappears on the opposite face in front of us; therefore, looking forward we can see our back (as in the spherical universe model of Einstein). Similarly, we see in our right our left profile, or upwards the bottom of our feet. In fact, for light emitted isotropically, and for an arbitrarily large time to wait, we could observe ghost images of any object viewed arbitrarily close to any angle. The resulting visual effect would be comparable (although not identical) to what could be seen from inside a parallelepiped of which the internal faces are covered with mirrors. Thus one would have the visual impression of infinite space, although the real space is closed. More generally, any three dimensional compact manifold can be represented as a polyhedron – what we define later more precisely as the fundamental domain (hereafter FD) – whose faces are suitably identified by pairs. But, as soon as the number of faces of the FD exceeds 8, the compact manifold resulting from identifications cannot be developed into the Euclidean space $\mathbf{R}^3$: the FD must be built in hyperbolic space $\mathbf{H}^3$ in order to fit all the angles together at the vertices.

The general method for classifying the topologies of a given manifold $\mathcal{M}$ is:

- to determine its universal covering space $\widetilde{\mathcal{M}}$
- to find the fundamental domain FD
- to calculate the holonomy group acting on the FD.

All these concepts have very formal and abstract definitions that can be found in classical textbooks in topology. We introduce below basic definitions.

The strategy for characterizing spaces is to produce invariants which capture the key features of the topology and uniquely specify each equivalence class. The topological invariants can take many forms. They can be just numbers, such as the dimension of the manifold, the degree of connectedness or the Poincaré – Euler characteristic. They can also be whole mathematical structures, such as the homotopy groups.

A loop at $\mathbf{x} \in \mathcal{M}$ is any path which starts at $\mathbf{x}$ and ends at $\mathbf{x}$. Two loops γ and γ' are homotopic if γ can be continuously deformed into γ'. The manifold $\mathcal{M}$ is *simply-connected* if, for any $\mathbf{x}$, two any loops through $\mathbf{x}$ are homotopic. Equivalently, it is *simply-connected* if every loop is homotopic to a point. If not, the manifold is said to be *multi-connected*. Obviously, the Euclidean spaces $\mathbf{R}^1$, $\mathbf{R}^2, \ldots$, $\mathbf{R}^n$, and the spheres $\mathbf{S}^2$, $\mathbf{S}^3, \ldots$, $\mathbf{S}^n$ are simply-connected, whereas the circle $\mathbf{S}^1$, the cylinder $\mathbf{S}^1 \times \mathbf{R}$ and the torus $\mathbf{S}^1 \times \mathbf{S}^1$ are multi-connected.

The study of homotopic loops in a manifold $\mathcal{M}$ is a way of detecting holes or handles. Moreover the equivalence classes of homotopic loops can be endowed with a group structure, essentially because loops can be added by joining them end to end. The group of loops is called the first homotopy group at $\mathbf{x}$ or, in the terminology originally introduced by Poincaré, the *fundamental group* $\pi_1(\mathcal{M},\mathbf{x})$. The fundamental group is independent of the base point: it is a topological invariant of the manifold.

For surfaces, multi-connectedness means that the fundamental group is non trivial: there is at least one loop that cannot be shrunk to a point. But in higher dimensions the problem is more complex because loops, being only one-dimensional structures, are not sufficient to capture all the

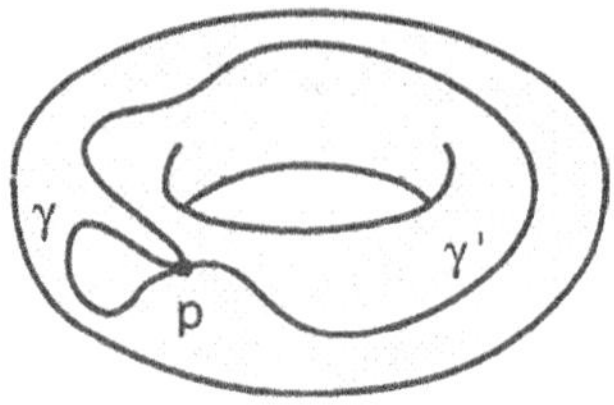

Figure 3. Classes of homotopy of a torus $\mathbf{S}^1 \times \mathbf{S}^1$. Loops can wind m times around the central hole and n times around the body of the torus. Thus the fundamental group consists of pairs (m, n) of integers with addition $(m, n) + (p, q) = (m + p, n + q)$. In other words it is isomorphic to $\mathbf{Z} \oplus \mathbf{Z}$.

topological features of the manifolds. The purpose of algebraic topology, extensively developed during the twentieth century, is to generalise the concept of homotopic loops and to define higher homotopy groups. However the fundamental group (the first homotopy group) remains essential.

To define the universal covering space, start with a manifold $\mathcal{M}$ with metric $\mathbf{g}$. Choose a base point $\mathbf{x}$ in $\mathcal{M}$ and consider the differents paths from $\mathbf{x}$ to another point $\mathbf{y}$. Each path belongs to a homotopy class γ of loops at $\mathbf{x}$. We construct the universal covering space as the new manifold ($\widetilde{\mathcal{M}}$,$\tilde{\mathbf{g}}$) such that each point $\tilde{\mathbf{y}}$ of $\widetilde{\mathcal{M}}$ is obtained as a pair $(\mathbf{y}, \gamma)$, $\mathbf{y}$ varying over the whole of $\mathcal{M}$ while $\mathbf{x}$ remains fixed and γ varies over all homotopy classes. The metric $\tilde{\mathbf{g}}$ is obtained by defining the interval from $\tilde{\mathbf{x}} = (\mathbf{x}, \gamma)$ to a nearby point $\tilde{\mathbf{x}}' = (\mathbf{x}', \gamma)$ in $\widetilde{\mathcal{M}}$ to be equal to the interval from $\mathbf{x}$ to $\mathbf{x}'$ in $\mathcal{M}$. By construction, ($\widetilde{\mathcal{M}}$,$\tilde{\mathbf{g}}$) is locally indistinguishable from ($\mathcal{M}$,$\mathbf{g}$). But its global – namely topological – properties can be quite different. It is clear that, when $\mathcal{M}$ is simply–connected, it is identical to its universal covering space $\widetilde{\mathcal{M}}$. When $\mathcal{M}$ is multi–connected, each point of $\mathcal{M}$ generates an infinite number of points in $\widetilde{\mathcal{M}}$. The universal covering space can thus be thought of as an "unwrapping" of the original manifold.

Consider a point $\mathbf{x}$ and a loop γ at $\mathbf{x}$ in $\mathcal{M}$. If γ lies entirely in a simply-connected domain of $\mathcal{M}$, $(\mathbf{x}, \gamma)$ generates a single point $\tilde{\mathbf{x}}$ in $\widetilde{\mathcal{M}}$. Otherwise, it generates additional points $\tilde{\mathbf{x}}', \tilde{\mathbf{x}}''$, ...which are said to be *homologous* to $\tilde{\mathbf{x}}$. The displacements $\tilde{\mathbf{x}} \mapsto \tilde{\mathbf{x}}'$, $\tilde{\mathbf{x}} \mapsto \tilde{\mathbf{x}}''$, ...are isometries and form the so-called *holonomy group* Γ in $\widetilde{\mathcal{M}}$. This group is discontinuous, i.e., there is a non zero shortest distance between any two homologous points, and the generators of the group (except the identity) have no fixed point. This last property is very restrictive (it excludes for instance the rotations) and allows the classification of all possible holonomy groups.

Equipped with such properties, the holonomy group is said to act freely and discontinuously on $\widetilde{\mathcal{M}}$. The holonomy group is isomorphic to the fundamental group $\pi_1(\widetilde{\mathcal{M}})$.

The geometrical properties of a manifold $\mathcal{M}$ within a simply–connected domain are the same as those of its development in the universal covering $\widetilde{\mathcal{M}}$. The largest simply-connected domain containing a given point $\mathbf{x}$ of $\mathcal{M}$, namely the set $\{\mathbf{y} \in \mathcal{M}, d(\tilde{\mathbf{y}}, \tilde{\mathbf{x}}) \leq d(\tilde{\mathbf{y}}, \gamma(\tilde{\mathbf{x}})), \forall\gamma \in \Gamma\}$, is called the *fundamental domain* (FD).

The FD is always convex and has a finite number of faces (due to the fact that the holonomy group is discrete). These faces are homologous by pairs ; the displacements carrying one face to another are the generators of the holonomy group Γ.

In two dimensions, the FD is a surface whose boundary is constituted by lines, thus a polygon. In three dimensions, it is a volume bounded by faces, thus a polyhedron.

The configuration formed by the fundamental polyhedron FD and its images γFD ($\gamma \in \Gamma$) is called a *tesselation* of $\widetilde{\mathcal{M}}$, each image γFD being a cell of the tesselation.

The FD has two major advantages:

- The fundamental group of a given topological manifold $\mathcal{M}$ is isomorphic to the fundamental group of the FD. Routine methods are available to determine the holonomy group of a polyhedron.
- The FD allows one to represent any curve in $\mathcal{M}$, since any portion of a curve lying outside the FD can be carried inside it by appropriate holonomies.

3.2. TWO-DIMENSIONAL MANIFOLDS

In addition to pedagogical and illustrative interest, the classification of two-dimensional Riemannian surfaces plays an important role in physics for understanding (2+1)-dimensional gravity, a toy model to gain insight into the real world of (3+1)-dimensional quantum gravity.

Any Riemannian surface is homeomorphic to a surface admitting a metric with constant curvature k. Thus, any Riemannian surface can be expressed as the quotient $\mathcal{M} = \widetilde{\mathcal{M}}/\Gamma$, where the universal covering space $\widetilde{\mathcal{M}}$ is either (figure 4):

- the Euclidean plane $\mathbf{R}^2$ if $k = 0$
- the sphere $\mathbf{S}^2$ if $k > 0$
- the hyperbolic plane $\mathbf{H}^2$ if $k < 0$.

and Γ is a discrete subgroup of isometries of $\widetilde{\mathcal{M}}$ without fixed point.

To characterise the quotient spaces we adopt the following abbreviations:

C = closed, O = open, SC = simply-connected, MC = multi-connected, OR = orientable, NOR = non-orientable.

Locally Euclidean surfaces fall into only 5 types: the simply-connected Euclidean plane itself $\mathbf{R}$, the multi-connected cylinder $\mathbf{R} \times \mathbf{S}^1$, the Möbius

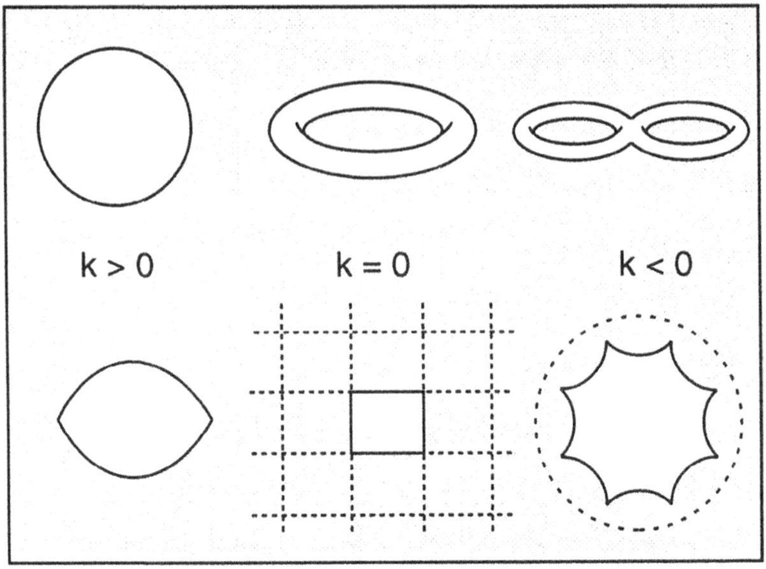

Figure 4. The three kinds of geometries for Riemannian surfaces, as closed surfaces "squeezed" into $\mathbf{R}^3$, and as fundamental domains (FD's). The covering space for $k > 0$ is the sphere $\mathbf{S}^2$, and the covering spaces for the other two curvatures are shown schematically.

band, the torus $\mathbf{S}^1 \times \mathbf{S}^1$ and the Klein bottle. Their characteristics are summarized in figure 5. The Möbius band and the Klein bottle are not orientable. The torus and the Klein bottle are closed spaces.

There are only two spherical surface forms (this result has been generalized to any constant positive curvature manifold of *even* dimension):

- the sphere $\mathbf{S}^2$ itself: C, SC, OR
- the projective plane $\mathbf{P}^2 \equiv \mathbf{S}^2/\mathbf{Z}_2$ (also called the elliptic plane): C, MC, NOR.

Whereas the surface of the unit sphere is 4π, the surface of the unit projective plane is only 2π, and its diameter, i.e., the distance between the most widely separated points, only $\pi/2$.

The hyperbolic plane $\mathbf{H}^2$, historically known as the Lobachevski space, is difficult to visualize because it cannot be isometrically imbedded in $\mathbf{R}^3$. Nevertheless it can be thought of as a surface with a saddle point at every point.

The full isometry group of $\mathbf{H}^2$ is $PSL(2,\mathbf{R}) \equiv SL(2,\mathbf{R})/\mathbf{Z}_2$, where $SL(2,\mathbf{R})$ is the group of real 2×2 matrices with unit determinant.

The best known example of a compact hyperbolic surface is the 2-torus T^2. In this case, the FD is a regular octagon with pairs of sides identified. In the Poincaré representation of $\mathbf{H}^2$, the FD appears curvilinear. The pavement of the unit disk by homologous octagons (which appear dis-

Name	*FD and identifications*	*Shape*	*Closed*	*Orientable*
cylinder	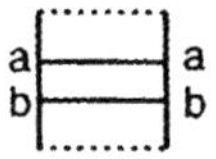	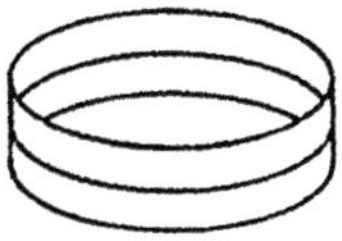	NO	YES
Möbius strip		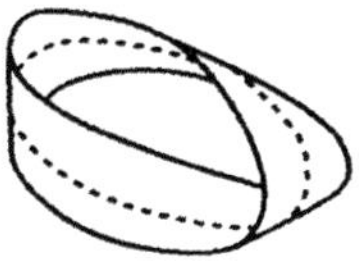	NO	NO
torus	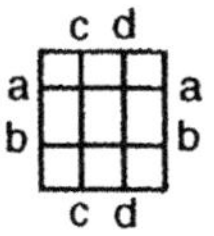	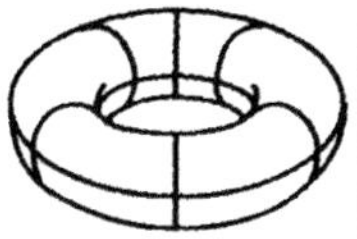	YES	YES
Klein bottle	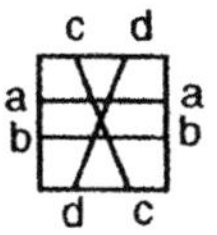	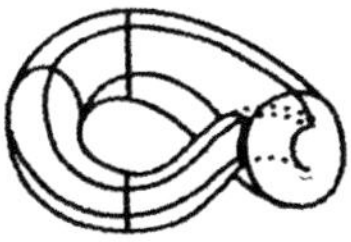	YES	NO

Figure 5. The four types of multi-connected Euclidean surfaces

torted in this representation) corresponds to the tesselation of $\mathbf{H}^2$ by regular octagons (figure 6). The famous Dutch artist M.-C. Escher has designed fascinating drawings and prints using such tilings of the hyperbolic plane.

The topological classification of locally hyperbolic surfaces is complete only for the *compact* $\mathbf{H}^2/\Gamma$, which fall into one of the following categories:

- g-torus T_g, $g \geq 2$ (connected sum of g simple tori): C, MC, OR
- connected sum of n projective planes: C, MC, NOR
- connected sum of a compact orientable surface ($\mathbf{S}^2$ or T_g) and of a projective plane or a Klein bottle: C, MC, NOR

All of these surfaces have a finite area bounded below by 2π, and a diameter greater than $\cosh^{-1}(4) \approx 2.06$.

There are an infinite number of *non-compact* locally hyperbolic surfaces. Clearly, "almost all" Riemannian surfaces are hyperbolic, since:

- Any open surface other than the Euclidean plane, the cylinder and the Möbius band is homeomorphic to a locally hyperbolic surface, for example an hyperbolic plane with or without handles.

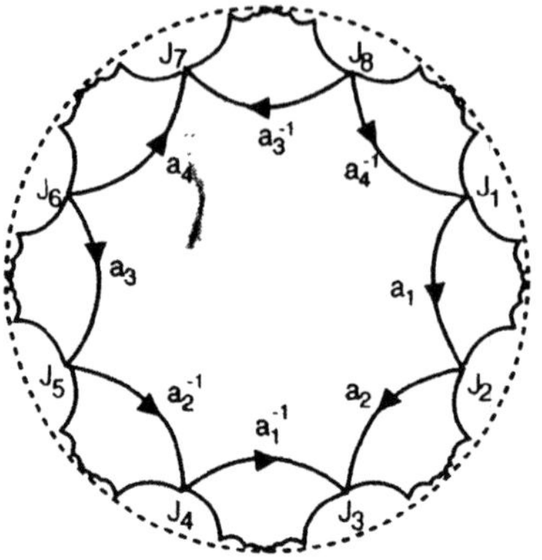

Figure 6. Tesselation of $\mathbf{H}^2$ by octagons in the Poincaré representation. The vertices have the coordinates $z(J_k) = 0.841\ exp\ \frac{(3\ 2k)\pi i}{8}$

- Any closed surface which is not the sphere, the projective plane, the torus or the Klein bottle is homeomorphic to a locally hyperbolic surface.

3.3. THREE-DIMENSIONAL MANIFOLDS OF CONSTANT CURVATURE

Any three-dimensional Riemannian manifold $\mathcal{M}$ admitting at least a 3-dimensional discrete isometry group Γ simply transitive on $\mathcal{M}$ is locally homogeneous and can be written as the quotient $\widetilde{\mathcal{M}}/\Gamma$, where $\widetilde{\mathcal{M}}$ is the universal covering space of $\mathcal{M}$. Let G be the full group of isometries of $\mathcal{M}$ (containing Γ as a discrete subgroup). In the terminology used in the theory of classification of compact three–manifolds, $\mathcal{M}$ is said to admit a *geometric structure* modelled on $(\widetilde{\mathcal{M}}, G)$.

Thurston has classified the homogeneous three–dimensional geometries into eight distinct types, generally used by mathematicians.

On the other hand, the Bianchi types are defined from the classification of all simply-transitive 3–dimensional Lie groups. Since the isometries of a Riemannian manifold form a Lie group, the Bianchi classification is used by workers in relativity and cosmology for the description of spatially homogeneous spacetimes. The correspondance between the locally homogeneous 3–geometries in Thurston's sense and the Bianchi–Kantowski–Sachs classification of homogeneous cosmological models has recently been fully clarified by [62].

Cosmology, however, focuses mainly on locally homogeneous and isotropic spaces, namely those admitting one of the 3 geometries of constant curvature. Any compact 3–manifold $\mathcal{M}$ with constant curvature k can be expressed as the quotient $\mathcal{M} \equiv \widetilde{\mathcal{M}}/\Gamma$, where the universal covering space $\widetilde{\mathcal{M}}$ is either:

- the 3-sphere $\mathbf{S}^3$ if $k > 0$
- the Euclidean space $\mathbf{R}^3$ if $k = 0$
- the hyperbolic 3-space $\mathbf{H}^3$ if $k < 0$.

and Γ is a subgroup of isometries of $\widetilde{\mathcal{M}}$ acting freely and discontinuously.

We give below a schematic description of such spaces.

3.3.1. *Spherical space forms*

Three–manifolds of constant positive curvature have $\mathbf{S}^3$, which is compact, as universal covering space. As a consequence they are all compact.

The metric on $\mathbf{S}^3$ may be written as

$$d\sigma^2 = R^2\{d\chi^2 + sin^2\chi(d\theta^2 + sin^2\theta d\phi^2)\}. \quad (1)$$

The volume is $2\pi^2R^3$.

The full isometry group of $\mathbf{S}^3$ is $SO(4)$. The admissible subgroups Γ of $SO(4)$ without fixed point, acting freely and discontinuosly on $\mathbf{S}^3$, are:

- the cyclic groups of order p, Z_p $(p \geq 2)$.
- the dihedral groups of order $2m$, D_m $(m > 2)$.
- the polyhedral groups, namely:
  - the group T of the tetrahedron (4 vertices, 6 edges, 4 faces), of order 12;
  - the group O of the octahedron (6 vertices, 12 edges, 8 faces), of order 24 ;
  - the group I of the icosahedron (12 vertices, 30 edges, 20 faces), of order 60.

All the homogeneous spaces of constant positive curvature are obtained by quotienting $\mathbf{S}^3$ with the groups described above. They are in infinite number due to parameters p and m.

The volume of $\mathcal{M}= \mathbf{S}^3/\Gamma$ is simply

$$vol(\mathcal{M}) = 2\pi^2\ R^3/\mid\Gamma\mid \quad (2)$$

where $\mid \Gamma \mid$ is the order of the group Γ. For topologically complicated spherical 3–manifolds, $\mid\Gamma\mid$ becomes large and $vol(\mathcal{M})$ is small. There is no lower bound since Γ can have an arbitrarily large number of elements. Hence

$$0 < vol(\mathcal{M}) \leq 2\pi^2\ R^3 \quad (3)$$

In contrast, the *diameter*, i.e., the maximum distance between two points in the space, is bounded below by $\frac{1}{2}arccos((1/\sqrt{3})\ cot(\pi/5)\ R \approx 0.326\ R$, corresponding to a dodecahedral space.

As well–known examples one can cite

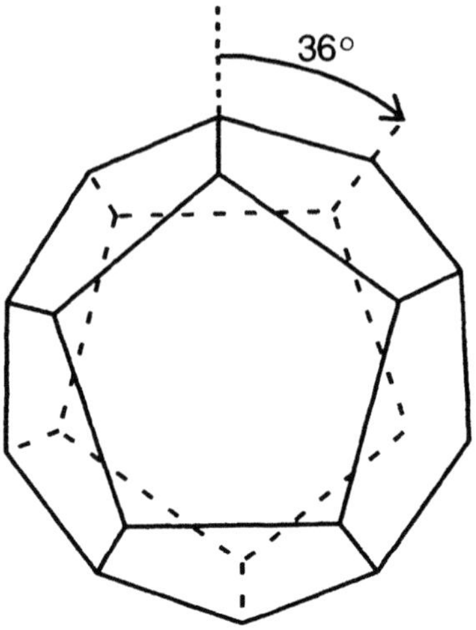

Figure 7. The fundamental domain of the Poincaré spherical space

- The projective space $\mathbf{P}^3 = \mathbf{S}^3/\mathbf{Z}_2$, obtained by identifying diametrically opposite points on $\mathbf{S}^3$. It was used by de Sitter and Lemaître as the space structure of their cosmological models.
- The lens spaces $\mathbf{S}^3/\mathbf{Z}_p$. The simplest one, $\mathbf{S}^3/\mathbf{Z}_3$,divides $\mathbf{S}^3$ into 6 fundamental cells, each having a lens form.
- The Poincaré dodecahedral space is an example of $\mathbf{S}^3/I$. The fundamental polyhedron is a regular dodecahedron of which the faces are pentagons. The compact space is obtained by identifying the opposite faces after rotating by $1/10^{th}$ turn in the clockwise direction around the axis orthogonal to the face (figure 7). This configuration involves 120 successive operations and gives already some idea of the extreme complication of such multi–connected topologies.

3.3.2. *Euclidean space forms*

The line element for the universal covering space $\mathbf{R}^3$ may be written as:

$$d\sigma^2 = R^2\{d\chi^2 + \chi^2(d\theta^2 + sin^2\theta d\phi^2)\} \tag{4}$$

Its full isometry group is $G = ISO(3) \equiv \mathbf{R}^3 \times SO(3)$, and the generators of the possible holonomy groups Γ (i.e., discrete subgroups without fixed point) include the identity, the translations, the glide reflections and the helicoidal motions occurring in various combinations. They generate 18 distinct types of locally Euclidean spaces. The 17 multi–connected space forms are in correspondance with the 17 crystallographic groups discovered more than a century ago by Feodoroff. Eight forms are open (non compact), ten are closed (compact).

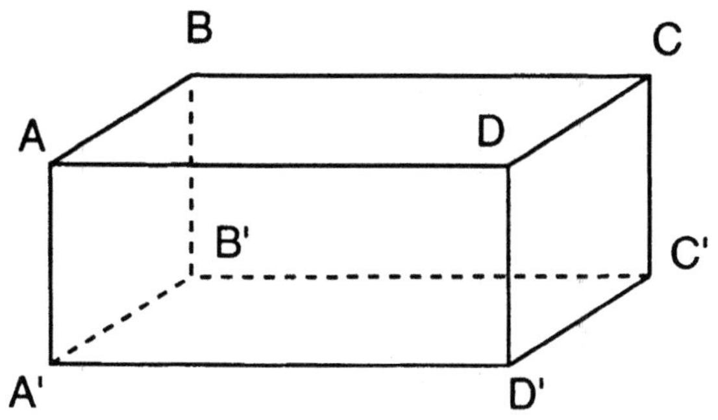

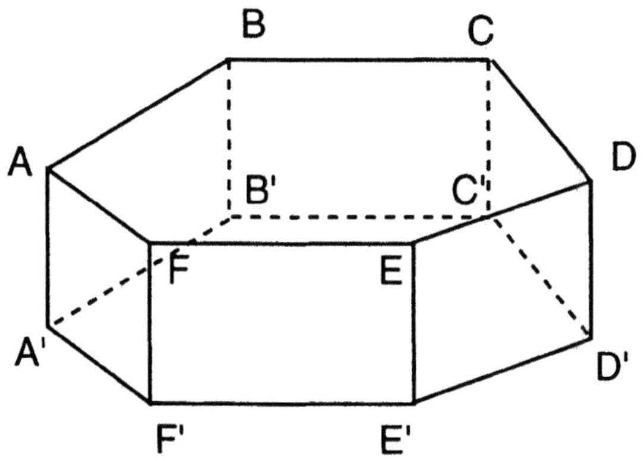

Figure 8. Tesselation of $\mathbf{R}^3$ by parallelepipeds or hexagonal cells

The compact models can be better visualised by identifying appropriate faces of fundamental polyhedra. Six of them are orientable. The fundamental polyhedron can be a parallelepiped or a hexagonal prism, with various possible possible identifications. The 3-torus is the simplest one. Their volume is arbitrary, since there is no rigidity theorem linking the curvature radius to the topological lengths. Until 1985 [22, 23, 25], such spaces were nearly the only ones to be explored for discussion of cosmic topology.

3.3.3. *Hyperbolic space forms*

Locally hyperbolic manifolds are less well understood than the other homogeneous spaces. However, according to the pioneering work of Thurston, "almost all" 3–manifolds can be endowed with a hyperbolic structure.

The induced metric on $\mathbf{H}^3$ may be written as

$$d\sigma^2 = R^2 \left\{ d\chi^2 + sinh^2\chi \, (d\theta^2 + sin^2\theta \, d\phi^2) \right\} \tag{5}$$

The volume of $\mathbf{H}^3$ is infinite. Its group of isometries is isomorphic to

$PSL(2, \mathbf{C})$, namely the group of fractional linear transformations acting on the complex plane. Finite subgroups are discussed in [3].

In hyperbolic geometry there is an essential difference between the 2–dimensional case and higher dimensions. A *surface* of genus $g \geq 2$ supports uncountably many non equivalent hyperbolic metrics. But for $n \geq 3$, a connected oriented n–manifold supports *at most one* hyperbolic metric. More precisely, the *rigidity theorem* proves that (for a fixed value of R) if two hyperbolic manifolds, with dimension $n \geq 3$, have isomorphic fundamental groups, they are necessarily isometric to each other. It follows that, for $n \geq 3$, the volume of a manifold and the lengths of its closed geodesics are topological invariants. This suggested the idea of using the volumes to classify the topologies, which could have seemed, at a first glance, contradictory with the very purpose of topology.

Each type of topology is characterized by some lengths. For compact locally Euclidean spaces, the fundamental domain may possess arbitrary volume, but no more than eight faces. In the spherical case, the volume of $\mathbf{S}^3/\Gamma$ is finite and is that of $\mathbf{S}^3$, the maximum possible value, divided by a whole number. By contrast, it is possible to tesselate $\mathbf{H}^3$ with polyhedra having an arbitrarily large number of faces. In the three-dimensional hyperbolic case, the possible values for the volume of the FD are bounded from below. In other words, there exist hyperbolic 3–manifolds with volumes arbitrarily close to a minimal volume for all hyperbolic 3–manifolds.

Particular interest has been taken by various authors in computing the volumes of compact hyperbolic manifolds. Each topology has a specific volume measured in curvature radius units. The absolute lower bound for the volume of CHMs is given by $V_{min} = 0.16668$ [34]. However, it may have little effect on cosmological applications. The reason is that the true lower bound is expected to be 0.942707, corresponding to the smallest CHM that is presently known [84, 57]. The new V_{min} bound represents an improvement in the techniques of the proof, not an increase in the expected size of the smallest hyperbolic manifold.

In view of cosmological observational effects, the smaller the value of vol_{min}, the more interesting the corresponding manifold for cosmology. Thus the Weeks space has been specially studied, by [24] and more recently by [46]. Its FD is a polyhedron with 26 vertices and 18 faces, of which 12 are pentagons and 6 are quadrilaterals. Its outer structure is represented in figure (9), the Klein coordinates of the vertices and the 18 matrix representations of the generators of the holonomy group are given in [46].

The Weeks manifold leaves room for many topological lens effects, since the volume of the observable universe is about 200 times larger than the volume of Weeks space for $\Omega_0 = 0.3$. Indeed, many CHMs have geodesics

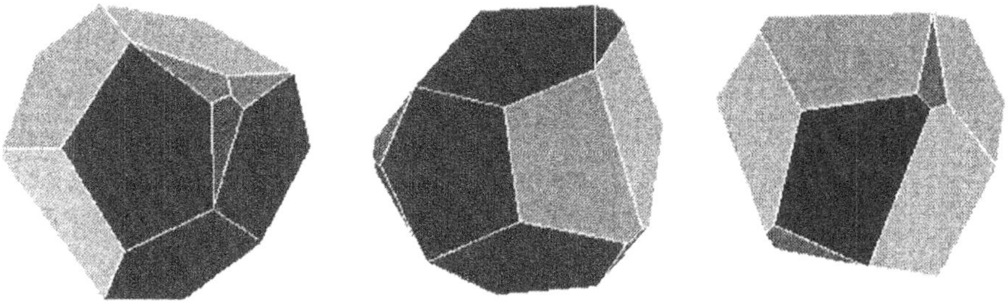

Figure 9. Three views of the fundamental domain of the Weeks manifold

TABLE 1. Small CHMs from SnapPea

Name	Volume	r	r_+	l_{min}
WMF	0.9427	0.5192	0.7525	0.5846
Thurston	0.9814	0.5354	0.7485	0.5780
s556(-1,1)	1.0156	0.5276	0.7518	0.8317
m006(-1,2)	1.2637	0.5502	0.8373	0.5750
m188(-1,1)	1.2845	0.5335	0.9002	0.4804
v2030(1,1)	1.3956	0.5483	1.0361	0.3662
m015(4,1)	1.4124	0.5584	0.8941	0.7942
s718(1,1)	2.2726	0.6837	0.9692	0.3392
m120(-6,1)	3.1411	0.7269	1.2252	0.3140
s654(-3,1)	4.0855	0.7834	1.1918	0.3118
v2833(2,3)	5.0629	0.7967	1.3322	0.4860
v3509(4,3)	6.2392	0.9050	1.3013	0.3458

shorter than the curvature radius, leaving room to fit a great many copies of a fundamental polyhedron within the horizon radius, even for manifolds of volume ~ 10. The publicly available program SnapPea (Weeks) is specially useful to unveil the rich structure of CHMs. Several millions of CHMs with volume less than 10 could be calculated. Table 1 summarizes a sample of the results (r_- is the radius of the largest sphere in the covering space which can be inscribed in the fundamental polyhedron, r_+ is the radius of the smallest sphere in the covering space in which the fundamental polyhedron can be inscribed, $l_{min} \equiv 2r_{\text{inj}}$ is the length of the shortest geodesic; see Fig. 10).

4. Topology and Cosmology

It is presently believed that our Universe is correctly described by a perturbed Friedmann–Lemaître (FL) model. That is, homogeneous and isotropic solutions of Einstein's equations are used, of which the spatial sections have constant curvature. The observational fact that the Universe is, in an exact sense, inhomogeneous and anisotropic, is modelled by perturbing these solutions. Beside the usual "big–bang" solutions, the FL models also include the de Sitter solution, as well as those incorporating a cosmological constant, or a non standard equation of state. From a spatial point of view, the FL models fall into 3 general classes, according to the sign of their spatial curvature $k = -1$, 0, or +1. The spacetime manifold is described by the Robertson–Walker metric

$$ds^2 = c^2\, dt^2 - R^2(t)\, d\sigma^2, \tag{6}$$

where

$$d\sigma^2 = d\chi^2 + S_k^2(\chi)(d\theta^2 + sin^2\theta\, d\phi^2)$$

is the metric of a 3–dimensional homogeneous manifold, flat [$k = 0$, see eq. (4)] or with curvature [$k = \pm 1$, see eqs (1), (5)]. We have defined the function

$$\begin{cases} S_k(\chi) = \sinh(\chi) & \text{if } k = -1 \\ S_k(\chi) = \chi & \text{if } k = 0 \\ S_k(\chi) = \sin(\chi) & \text{if } k = 1 \end{cases}$$

and $R(t)$ is the scale factor, chosen equal to the spatial curvature radius for non flat models, so that k/R^2 is the spatial curvature.

The most usually considered cosmological models are those with simply–connected space. We will refer generically to them as the Simply–Connected Models, hereafter SCM's. However, the assumption of simple–connectedness is arbitrary and can be dropped out. The multi–connected cosmological models (hereafter MCM's) are those having multi–connected spatial sections. To any MCM is associated a unique SCM sharing exactly the same kinematics and dynamics. The universal covering (UC) of the spatial sections of an MCM is the spatial section of the corresponding SCM ($\mathbf{R}^3$, $\mathbf{H}^3$ or $\mathbf{S}^3$ for FL models) at the cosmological time defining the spatial section. In particular, the scale factors $R(t)$ are exactly identical. In fact, most characteristics of the Friedmann–Lemaître models are preserved when we turn to the MCM's, and this is a precious guide for their study.

The values of the metric parameters [the density parameter Ω_0 and the dimensionless cosmological constant, $\lambda_0 \equiv c^2\Lambda/(3H_0^2)$; in the notation of [60] $\lambda_0 \equiv \Omega_\Lambda$] are not known accurately enough to decide the sign of the curvature. In fact, it is difficult to see how dynamical measurements can distinguish between $\Omega_0 + \lambda_0 = 1$ and $\Omega_0 + \lambda_0 = 1 \pm \delta$ where $0 < \delta \ll 1$. On the contrary, significant detection of non–trivial topology would dictate the sign of the spatial curvature, since the structures of the holonomy groups are completely different for the cases $k = 1,\ 0,\ -1$. Moreover, it is worthwhile to note that the multi–connectedness of space would provide precise constraints on the metric parameters independently of the cosmic distance ladder [70].

Thus, beyond its own specific interest, multi–connectedness would offer a very efficient tool of investigation in other aspects of observational cosmology.

In SCM's, the only scale in comoving space which is not fixed is the curvature radius, normally defined as the present value of the scale factor $R(t_0) = R_C$ (there is no scale at all in the flat case). Thus R_C is the natural length unit in comoving space for the UC of an MCM. The Friedmann equations imply the relation

$$\Omega_0 + \lambda_0 - 1 = \frac{k\, c^2}{H_0^2\, R_C^2}. \quad (7)$$

The value of R_C remains a matter of considerable debate. The only cosmological length to which we have a direct observational access is the Hubble length

$$L_{Hubble} = cH_0^{-1} = 3\,000\; h^{-1}\ \text{Mpc}.$$

If we define $f = \sqrt{|\,\Omega + \lambda - 1\,|}$, we have for a non–flat universe (for a flat universe, the value of R_C remains arbitrary):

$$R_C = L_{Hubble}\; f^{-1} = 3\,000\; (f\; h)^{-1}\text{Mpc}. \quad (8)$$

In practice, the Hubble length is of the same order of magnitude as R_H, the particle horizon, which defines the observable universe as a sphere in the UC. This is just slightly larger than the radius to the last scattering surface, which is the present limit to observations.

Another natural cosmological length is associated with the cosmological constant:

$$L_\Lambda = \sqrt{1/\Lambda} = \sqrt{(3H_0^2\lambda_0/c^2)^{-1}} = 1730/\sqrt{\lambda_0}\ \text{h}^{-1}\ \text{Mpc}. \quad (9)$$

The concept of horizon keeps its exact validity in the MCM's, but must be applied to the universal covering space: an image is potentially visible iff its (comoving) distance is smaller than R_H in the universal covering.

In an MCM, additional spatial scales are associated with the topology, those of the fundamental polyhedron. The geometry suggests to compare them with R_C but it is often more convenient, for observations, to compare them to R_H, or to evaluate them in Mpc or h^{-1} Mpc.

Figure 10 and its caption give the basic definitions. Observable effects linked to the multi–connectedness will only occur if these scales are smaller than the size of the observable universe, i.e., the horizon radius.

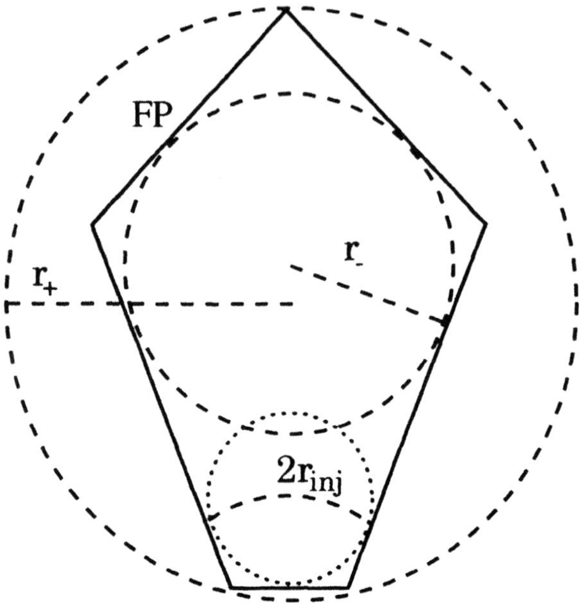

Figure 10. Schematic diagram showing key parameters representing the size of a small universe. The terminology is that of [9]. The in-radius r_- and the out-radius r_+ are characterised by the largest sphere inscribable in the fundamental domain (FD) and the smallest sphere circumscribable around the FD respectively. The injectivity radius $r_{\rm inj}$ is half of the smallest closed (spatial) geodesic from one topological image to another. The particular geometry shown does not necessarily correspond to a projection of a slice through a 3-manifold known to exist, it is for demonstration only.

Dropping the hypothesis that real space is simply–connected has various implications. If at least one of the topological scales is smaller than the horizon, then this will, in principle, be observable: multiple images of the same object or radiation emitting region will exist. The smaller the fundamental domain, the easier it is to observe the multiple topological imaging. It has recently been calculated [46], for a given catalog of observable cosmic sources (discrete or diffuse) with a given depth in redshift, the approximate number of topological images in locally hyperbolic and locally elliptic spaces as a function of the cosmological paramaters Ω_0 and λ_0. How do the present observational data constrain the possible multi–connectedness of the universe? And, more generally, what kinds of tests are conceivable ? The following section deals with these matters.

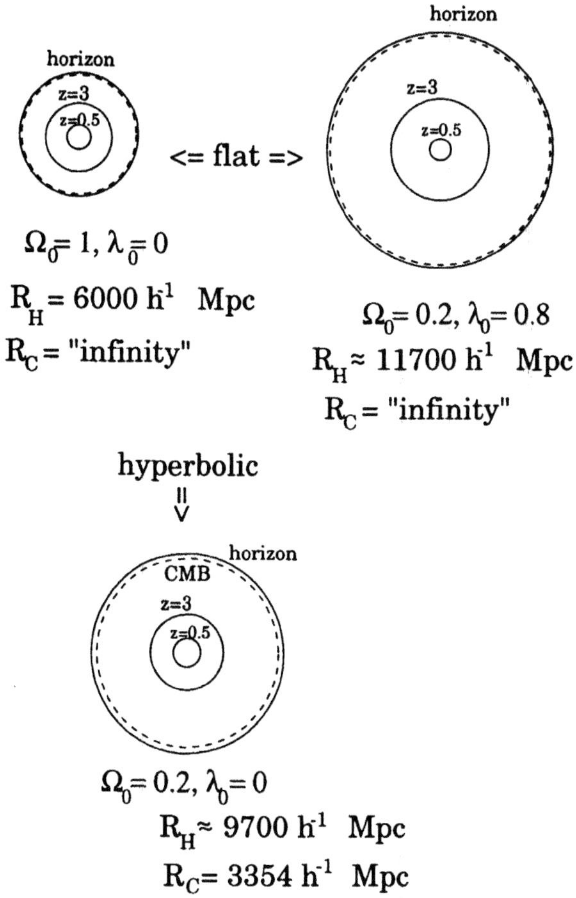

Figure 11. Relative sizes of the observable Universe, depending on different options for the metric parameters Ω_0, λ_0, shown in comoving coordinates on a radially linear (proper distance) scale. The horizon radius R_H is defined by the age of the Universe and the curvature radius is $R_C \equiv c/H_0(|1 - \Omega_0 - \lambda_0|)^{1/2}$. The inner circles for each choice of metric indicate scales to which we can see: X-ray emission from the richest clusters of galaxies ($z \sim 0.5$), large numbers of quasars ($z \sim 3$), the cosmic microwave background ($z \sim 1100$, dashed circle). In the flat models, the tangential distance scale is constant and the same as the radial scales; in the hyperbolic model, the tangential distance scale increases (more Gpc per mm on the page) as a function of increasing radius, i.e. it is $[\sinh(r/R_C)]/(r/R_C)$ times larger than a constant scale, e.g. the horizon circumference is $3.1(2\pi R_H)$.

5. Observational methods, candidates and constraints

5.1. "TOPOLOGY" FOR THE OBSERVER

The simplest observational point of view for approaching cosmological topology is to consider the observable sphere, in comoving coordinates, in the covering space.

For brevity, an abuse of language is often adopted where a "topology" is considered to mean a specific 3–manifold, expressed with some or all of the quantitative parameters in physical units with a definite orientation, in some standard astronomical coordinate system. Formally, this can be expressed as the set

$$\{\widetilde{M} \equiv \kappa_0 \equiv \Omega_0 + \lambda_0 - 1,\ \Omega_0,\ \Gamma,\ \delta\widetilde{M},\ \delta\Omega_0,\ \delta\Gamma\} \tag{10}$$

where $\widetilde{M}$ is the covering space, which both determines and is determined uniquely by the curvature under the assumption of constant curvature; Ω_0 is the present-day value of the density parameter, required (along with λ_0, which satisfies $\lambda_0 = \kappa_0 - \Omega_0 + 1$) in order to relate cosmological redshift z and distance; $\Gamma = \{g_i\}$ is the holonomy group of isometric linear transformations within the covering space which map topological images onto one another, (e.g. combinations of translations, rotations and reflections, if $\widetilde{M} = \mathbf{R}^3$); and δ is used to indicate (in abbreviated form) that the observational uncertainties in all the quantitative parameters need to be determined.

An equivalent, more physical, way of thinking of this set is in terms of the *FD*, or fundamental polyhedron, which we label P, i.e.

$$\{P \equiv \widetilde{M}/\Gamma,\ \Omega_0,\ \Gamma,\ \delta P,\ \delta\Omega_0,\ \delta\Gamma\} \tag{11}$$

where P is just the polyhedron shape, orientation and our location within it (all represented quantitatively in physical units and coordinates), without the generators, and Γ lists the generators (also in physical units and coordinates).

The values of κ_0 of most observational interest are $\kappa_0 \leq 0$. Fig. 11 puts into perspective the length scales corresponding to the range in κ_0 and Ω_0 [expressed as the pair (Ω_0, λ_0)] considered consistent with present observations.

In the cases of zero curvature, no geometric constraint exists on how many copies of P could exist within the observable sphere. In the hyperbolic case, values of $2r_{\mathrm{inj}}/R_C \lesssim 0.5$ are common for manifolds in the census available with SnapPea (e.g. Table 1), i.e. $2r_{\mathrm{inj}} \lesssim 1700\mathrm{h}^{-1}$ Mpc for the hyperbolic covering space $\widetilde{M}$ shown. So, the constraint of the rigidity theorem in the hyperbolic case does not prevent the possibility of multiple topological images at redshifts considerably smaller than $z = 3$.

5.2. EARLY WORK: $\lesssim 100\mathrm{h}^{-1}$ Mpc

Sections 11 and 12 of [44] review many different efforts to detect or constrain the value of r_{inj}.

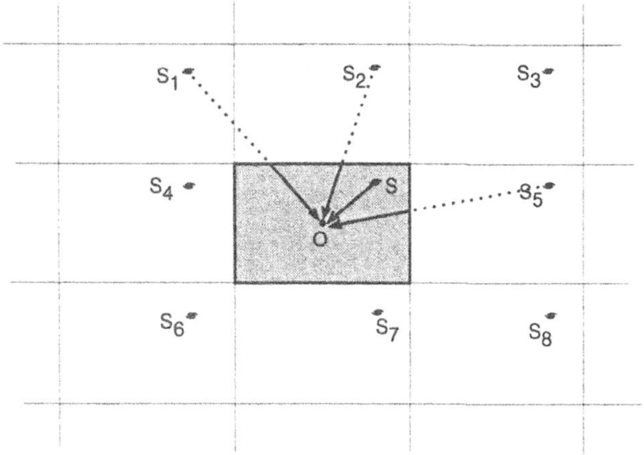

Figure 12. Multiplication of images in the universal covering space of a toroidal universe

Note that the different parameters expressing the size of the Universe, $r_{\rm inj}$, r_- and r_+, in particular, (Fig. 10) have not always been carefully distinguished. For example, the fact that we live in the disk of a spiral galaxy, in which dust obscures most extragalactic observations, implies that many lower limits estimated for the "size" of P are in fact lower limits on r_+, but not on $r_{\rm inj}$.

However, the constraints at $r_{\rm inj} \lesssim 100{\rm h}^{-1}$ Mpc are strong. For example, close images of the Coma cluster do not exist, except if hidden by the galactic plane [36], and large scale structure (walls, filaments, voids) clearly exists and intuitively it is difficult to see how the Universe could be much smaller. So $2r_+ \gtrsim R_H/100$ is a fair estimate.

5.3. SINCE 1993: $\gg 100{\rm h}^{-1}$ Mpc

At larger scales, with increasing amounts of data available from better telescopes, observational work on cosmological topology has increased rapidly in the last few years. Some methods test for specific candidates or families of 3–manifold candidates (e.g. [5]), but several new generic methods have been developed which avoid having to assume a specific manifold to start off with [45, 64, 69, 10].

The principle common to all methods is that multiple topological images of "objects" should exist, if $r_{\rm inj} < R_H$, where in the case of the cosmological microwave background (CMB), regions of space whose black-body temperature can be measured are considered as objects, though they are not astrophysical objects in the ordinary sense. This is illustrated for the flat torus model in Fig. 12.

What varies between the methods depends on many factors. The intrinsic properties of the "objects" (or the lack of knowledge of their properties), their intrinsic distribution throughout the observed time cone and relative to the galactic plane, and their visibility, make finding "spatially well distributed, standard candles" difficult.

Specifically, an ideal "object" (whether a collapsed object, a "configuration" of objects, or a patch of plasma in the CMB) should:

(i) not evolve with lookback time
(ii) have zero *three-dimensional* peculiar velocity
(iii) emit isotropically
(iv) be seen to large redshifts (distances)
(v) be seen over a large total volume (range in redshifts)
(vi) not be obscurable by dust in the galactic plane, nor by high latitude dust (e.g. Ophiuchus, Orion)

Objects which are probably the best standard candles, i.e. clusters of galaxies detected by their hot gas in X-rays, are mostly only seen to very small fractions of R_H (though future observations may improve this, particularly if the curvature κ_0 is negative enough). Objects seen to large fractions of R_H, such as quasars, are probably the worst standard candles.

This is, of course, the basic difficulty in observational cosmology, shared by the attempts to measure the metric parameters Ω_0, λ_0 and H_0.

The different methods, their advantages and disadvantages, the claimed constraints so far, and, moreover, suggested candidates for the 3–manifold [for "the topology" in the sense defined by eq. (10)] are listed in Table 2. At least one of the suggested candidates makes predictions which would be refutable with a modest size observing programme on major telescopes.

5.3.1. *Methods: three-dimensional*

For a full review of three-dimensional observational methods, i.e. those using astrophysical objects distributed throughout the three-dimensional observable sphere of the covering space, see [68]. Only a brief description is provided here.

(i) The principle behind the method of Lehoucq, Lachièze-Rey & Luminet [45], "cosmic crystallography", is that for good standard candles, multiples of the vectors corresponding to the generators, should cause very sharp peaks in a histogram of the separations of pairs of objects (non-normalised two-point correlation function), see Fig. 13. Rich clusters of galaxies, particularly if selected in X-ray surveys, should be fairly good standard candles at redshifts smaller than their formation redshifts, simply because they are dominated by hot gas in kinetic equilibrium which doesn't have time within the age of the Universe to evolve much except to grow by accretion and become hotter and brighter. In addition clusters, again

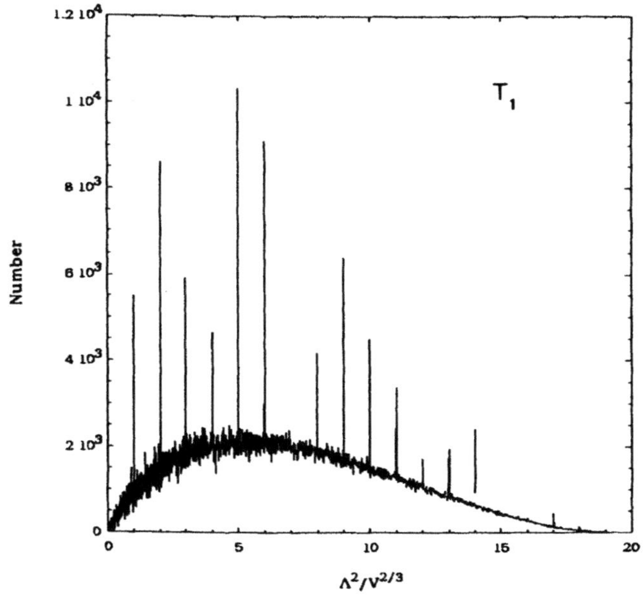

Figure 13. Histogram of pair separations for a computer generated Einstein-de Sitter universe. Fifty galaxies up are randomly distributed in a cubic fundamental domain of size 2500 Mpc. Topological images are calculated up to a redshift 4. The amplitude and relative positions of spikes reveal the holonomies of space and the topological type.

preferably seen by the X-ray gas, should be relatively isotropic emitters (compared to, say, quasars).

Lehoucq & al [45] performed their simulations in Euclidean spaces only and assumed that the other curvature cases gave similar results. The efficiency of the method has also been discussed by [27] when the size of the physical space is comparable to the horizon size.

Lehoucq, Luminet & Uzan [46] have next analysed the applicability of the method to elliptic and hyperbolic spaces. In the first case there may be topological signal in the histogram. However, small elliptical spaces are not experimentally favored by the present estimates of the age of the universe. In the hyperbolic case, no peaks in the pair 3D separations histogram were observed, due to the fact that the number of copies of the fundamental domain in the observable covering space is low and that the points are not moved the same distances by the holonomies of space. The lack of peaks in the hyperbolic case has simultaneously been noticed by [28] and [35].

Both [28] and [82] have devised improvements on the crystallographic method aimed to extract all the topological signal the histograms, either by subtracting out the smooth (non-topological) signal or by considering differential histograms.

(ii) In the absence of deep, wide angle surveys for what are the best known standard candles (galaxy clusters), other approaches are needed.

Another use of standard candles (cf. [36]) with (most likely) monotonic evolutionary properties, i.e. clusters seen in X-rays, is that if the highest redshift cluster in a systematic, complete, all-sky survey is at a high redshift, then no topological images of it can exist at lower redshifts, since they would have to be brighter than the high redshift cluster, which in that case would not be the brightest [69].

(iii) At larger scales, the objects most readily available are quasars. These are not good standard candles, and either exist (as quasars) only for small fractions of the age of the Universe, or recur as bursts several times. In addition, they would be seen (except in special cases) from different angles, and so appear as another form of active galactic nucleus (AGN) much fainter, and not yet catalogued.

The way around this problem [64] is to look for, in a large enough catalogue, the rare cases in which the evolution (and orientation) of the quasars happens to be ideal for an Earth-based observer, and multiple topological images are seen of several quasars in two topological images of a "small region", i.e. of a region of a few $100h^{-1}$ Mpc in size. The method is then to search (in 3-D, using redshifts for distance estimation) for all possible isometries between configurations of quasars in such "small regions". There is not much time for the relative positions of quasars to change much within a configuration, so in a sense, the configurations correspond to pseudo-objects of size $100h^{-1}$ Mpc which, in a small number of cases, effectively do not evolve.

Given contamination by chance isometries of configurations unrelated to topology [64], theoretical work (analytical and/or simulations) is required to find the statistically optimal implementations of this method. For example, although weakening the isometry criteria (e.g. matching quadruplets instead of quintuplets) would increase the number of non-topological (noise) isometries, it might increase the number of topological (signal) matches by a greater factor.

(iv) A method not yet applied directly, though implicit in some sense in the above methods, is to consider units of "large scale structure" (walls, voids, etc. on scales of 50–$150h^{-1}$ Mpc) as objects. Although individual galaxies and quasars will be difficult to identify at different epochs, it is expected that they should trace out thin structures which should not move much over the time scales involved [36]. One approach to analysing the data to be taken by the X-ray satellite XMM is to use the representations of large scale structure by the 2-dimensional topology of contours of constant density as a characteristic to identify potential multiple topological images

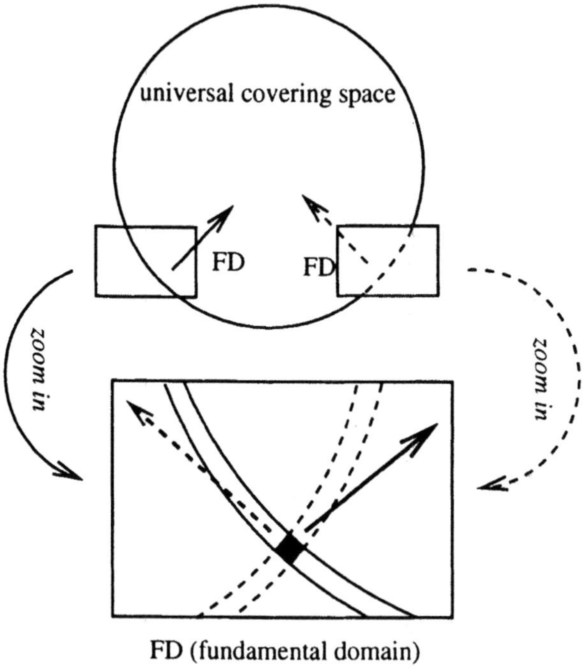

Figure 14. Schematic diagram showing basic principle of CMB methods. A single point of physical space emits radiation isotropically at the recombination epoch, but particularly in two different directions such that after crossing the Universe several times, the photons arrive at the observer. The upper half of the figure is from the point of view of the covering space, with several copies of the FD, while the lower half is a zoomed-in more physical point of view, in a single copy of the FD. The arrows indicate photons' paths. The "object" which is to be seen as multiple topological images is the intersection of the two thin sections of the shell of last scattering, i.e. the intersection shaded in black.

of these structures. In other words, the (2-D) topology of matter would be used to search for the (3-D) topology of space [61].

(v) Other methods to note include [22, 23, 25] specifically for the hyperbolic case and [26] for looking for images of our own galaxy as a quasar. In fact, a challenge to the astronomers and astrophysicists who try understand the dynamical, inter-stellar medium (ISM) and stellar past history of the Milky Way (MW) and Local Group galaxies (LG) is that they should be able to describe this history in a unique enough fashion such that high redshift galaxies can be excluded as possible topological images of the Milky Way just on the basis of intrinsic properties, rather than on checking for further topological images.

Since this school is intended for doctoral (graduate) students, it should be pointed out that this would provide a "safe" thesis project with relevance for cosmic topology, in that the main goal would be to understand the history of the MW and LG, which is already more than enough for a thesis. If it resulted that a "high" redshift galaxy (or a group) were found to have a striking resemblance to how the MW (or LG) should have looked at that epoch, then the student would not necessarily need to conclude that topological images have been found — the resemblance could be ascribed to coincidence and checked for cosmological significance during postdoctoral work (or by other researchers in the field). However, the advances in the understanding of the MW and LG history in order to make such a significant claim of resemblance would be sufficient for several publishable articles.

5.3.2. *Methods: two-dimensional*

The principle behind the use of the CMB, i.e. of a nearly two-dimensional thin shell from the observer's point of view, is that small portions of the shell corresponding to the CMB can be considered in some sense as objects. Depending on the topology of the Universe, some of these "objects" may occupy points of space which have topological images on other parts of the CMB.

A schematic figure of this is shown in Fig. 14. For the "object", i.e. the intersection of two sections of the surface of last scattering (SLS), to be seen as multiple topological images with identical black-body temperatures, it has to have the same average temperature integrated in different directions, and any effects due to the time evolution across the SLS would have to be the same in the two directions (unless the two effects cancelled). In other words, the CMB radiation would have to be *locally isotropic.* This is expected to be the case on large scales for the naïve Sachs-Wolfe effect (gravitational redshift), but on medium scales ($\sim 1°$) the Doppler effect, which is *not* isotropic, is expected to dominate.

Another possible problem with CMB multiple imaging, which has not been explicitly analysed, is that the SLS is much thinner than the resolution of COBE, by a factor of $\sim 10^2$. That is, the black shaded intersection in Fig. 14 which is seen twice, is averaged with $\sim 10^2$ neighbouring patches in each of two directions perpendicular to one line of sight to give a single observed number, let us say in the direction of the solid arrow in the figure. The same black patch is also observed in the direction of the dashed arrow, but this time it is averaged with a set of $\sim 10^2$ patches in two directions perpendicular to the dashed arrow before the single observed value is obtained.

It is not clear whether or not these two averaging processes will lead to the same temperature fluctuation $\delta T/T$, even if the individual patches, un-observable individually, are dominated by the same physical process (gravitational potential). Averaging is a subtle process in physics, and given that the averaging in this case has different relevance depending on whether or not the hypothesis of simple–connectedness is adopted, it is not clear that reasoning in the former case is sufficient for the latter.

(i) The topology-independent method for using the CMB, supposing that these two problems can be overcome, is the "method of circles" [10, 85]. Consider extending the tiling of the covering space by copies of the FD beyond the observable sphere and consider the observable sphere of a second observer in one of the copies of the FD outside the first sphere. The intersection of these two spheres is a circle.

But since each copy of the FD is equivalent, the second observer can in fact perfectly well be physically identical with the first observer. That is, in the covering space, two observers, each of to whom the other is behind the CMB (but not too far), are equivalent to a single observer looking in two different directions towards the CMB, as long as they are at the same position within the FD.

Hence, the effect of multi-connectedness would be that the values of the temperature fluctuations $\delta T/T$ around certain circles on the CMB would map onto one another, since the circles correspond to the same set of points in space and time.

A related two-dimensional method to be applied to future satellites (MAP and Planck Surveyor), is that of searching for patterns of hot and cold spots [52]. This may bypass the problem of the large number of calculations to make for the circles method.

(ii) Other two-dimensional methods require (a) assuming the topology and specific quantitative parameters for individual 3–manifolds of each given topology and (b) modelling the power spectrum of perturbations. The latter is difficult to justify.

The theoretical motivation for Gaussian amplitude distributions, uniform random phases and a $P(k) \propto k^1$ power spectrum are unlikely to be valid at scales approaching r_{inj} and r_+. That is, either for a hyperbolic or for a flat, $\lambda_0 \sim 0.7$ metric to be presently observable, inflation needs some degree of fine-tuning (which can partly be provided by the ergodicity of geodesics in the former case, [8]). Since curvature or $\lambda_0 > 0$ must remain "uninflated" in the sense of being observable at the present epoch in these cases, it is unclear why perturbations on the scale of $R_C \sim R_H$ should necessarily be "inflated" in the sense of exactly satisfying the assumptions on the power spectrum. Moreover, even for other choices of metric, if the Universe is observably an MCM, then scales approaching r_{inj} and r_+ need not necessarily be "inflated" either.

The observational motivation for Gaussian amplitude distributions, uniform random phases and a $P(k) \propto k^1$ power spectrum are equally lacking for tests of MCM's. The only observational justification of these properties on large scales is COBE data, which is analysed *assuming simple–*

connectedness. A self-consistent assumption on the fluctuation spectrum for comparison of an MCM with COBE data would be to calculate this spectrum in three dimensions based on the COBE data shifted into the FD of the MCM assumed. However, use of the resultant spectrum to simulate properties of the COBE observations would give ... exactly the COBE observations (if it has been done correctly).

So a self-consistent alternative to (b) would not enable rejection of an MCM. In other words, this approach tests assumptions on the the fluctuation spectrum (on the scale $\lesssim r_{\text{inj}} < r_{+}$) rather than the MCM itself. (Moreover, some authors find violation of Gaussian amplitude distributions in the COBE data [30, 59].)

Nevertheless, approach (ii) has been applied several times. Generally, a spherical harmonic analysis ("C_l" spectrum) of simulated CMB maps (e.g. [77, 76]) is calculated.

For a hyperbolic spatial section, a standard Fourier analysis is, in a strict sense, invalid, whether or not a trivial topology is assumed. This compounds the problems of assumptions regarding the power spectrum, which cannot be defined in the usual way. Moreover, the classification and listing of the 3–manifolds is an open enough problem, and knowledge of eigenmodes for the equivalent of a Fourier analysis is even less developed. An interesting way to avoid this problem is to use correlation functions rather than power spectra for assumption (b). Bond, Pogosyan & Souradeep [5] have done this for two of the many hyperbolic 3–manifolds known.

A general argument claimed for the methods which assume specific 3–manifolds is that no objects, i.e. no density fluctuations, which are larger than the Universe, i.e. larger than r_{+}, can exist, so there should be a cutoff at small l values (large length scales) in the C_l spectrum.

While this may sound common sense, it is only correct in the FD. In the universal covering space, fluctuations much larger than the Universe can indeed exist (e.g. ~ 30 times larger, see Fig. 1 of [64]). These structures cross the FD many times. While the C_l spectrum may indeed lose power on large scales due to averaging in different directions, it is incorrect to state that fluctuations larger than r_{+} cannot exist. The component eigenmodes in 3-D are of course constrained by some integer multiple of r_{+}, but basis functions of the 2-D sphere cut to $\pm 20°$ from the galactic plane only relate to the 3-D eigenmodes via simplifying assumptions which once again may not be valid at these length scales in MCM's. Recently Inoue [39] computed the low-lying eigenmodes on some compact 3-hyperbolic spaces using the direct boundary element method, and found that the expansion coefficients behave as gaussian pseudo-random numbers.

A summary of the advantages and disadvantages of these different methods is shown in the first part of Table 2.

Another interesting constraint may be derived when we take into account the possibility that phase transitions in the early universe could develop topological defects. Uzan & Peter [83] showed that if space is multi–connected on scales now smaller than the horizon size, strings and domain walls were very unlikely to exist. Uzan [80] generalized these results to textures and studied the cosmological implications of such constraints. He concluded that a large class of multi-connected universes with topological defects accounting for structure formation were ruled out by the cosmic microwave background observations.

5.3.3. *Constraints*

3-D (i,ii) The cosmic crystallography ([45], CC) and brightest cluster ([69], BC) methods indicate lower limits on the out-diameter of around $R_H/20$ and $R_H/10$ respectively, though the former is valid only in Euclidean spaces $\mathbf{R}^3/\Gamma$.

3-D (iii) The isometry search method applied to quasars is specifically designed to work for poor standard candles, so is intended for finding specific candidates for the 3–manifold, not for constraining r_{inj} or r_+.

2-D (i) No application of the identified circles principle to COBE data to constrain r_{inj} or r_+ has yet been published, since the poor signal-to-noise and resolution make the chance of a detection seem unlikely.

2-D (ii) However, many applications of the two-dimensional methods to observations from the COBE satellite [77, 76, 41, 13, 51, 5] have been made. These can be interpreted either as suggesting constraints close to the horizon scale, but only for a (mostly Euclidean) finite number of the infinitely many topologies possible, or as showing that standard assumptions on the statistics of the perturbation spectrum on scales close to r_{inj} and r_+ would have to be wrong for one of these MCM's to be correct. Note that the confidence limits stated for several of these studies are statements about the distribution of random realisations, not about observational inconsistency of multiply imaged observed patches of the CMB.

These various direct constraints and perturbation spectrum based tests are listed in the second part of Table 2.

The poor resolution of COBE ($\sim 10°$) implies that the data from the MAP and Planck satellites will probably be necessary before the circles principle can be applied seriously. This will not be easy, due to the Doppler effect, the integrated Sachs-Wolfe effect, other foregrounds, and the problem of averaging mentioned above.

5.3.4. *Specific candidates*

The most specific candidates suggested are those of [69] and [5], though an implicit candidate may also be implied by the application of the isometry search method to quasars [64].

The former candidate was noticed by chance in a small list of some of the brightest clusters chosen to illustrate the brightest cluster method. Among these seven clusters, three form a right angle to within 2%-3% accuracy (depending on the values of Ω_0 and λ_0) and the two side lengths adjacent to the near-right angle are equal to within 1%. This is better than tolerance considered for special quasar configurations by [26].

An *a posteriori* estimate of the probability of this occurring by chance would be difficult to make objectively, but since the configuration

(i) corresponds to that expected for the most commonly studied candidate in CMB studies (variations on hypertori),
(ii) implies generators which can easily be used to derive observationally refutable predictions, and
(iii) can be used to test in detail the validity of the different statistical methods,

it is useful to consider the implied $\widetilde{M}/\Gamma = T^2 \times \mathbf{R}$ as a candidate 3.

See [69] for object names, positions and values of the generators. A more recent summary of arguments for and against the candidate is provided in the discussion section of [67]; see also [65] for a COBE analysis. If the candidate were correct, then two topological images of the Coma cluster would be already known.

Bond et al. [5] (1998, §4.3) have also noticed a 3–manifold (for the two hyperbolic topologies considered) which, for a certain orientation, is "preferable to standard cold dark matter" when compared to the COBE observations. The volume of this manifold is slightly larger that that of the observable sphere, but r_{inj} is less than R_H, though not enough to imply any multiple topological images at observable sub-SLS distances.

5.4. OTHER OBSERVATIONAL DIFFICULTIES

Another observational complication, which has not generally been considered, particularly in the claims of constraints, i.e. lower limits on r_{inj}, is an important point by [21]: in a strict sense, it is already well established that we do not live in an FL universe. We live in what (locally, at least) is an almost FL universe. Density perturbations (humans, planets, stars, galaxies, large scale structure) exist.

Depending on the way in which space inside the observable sphere and the FD differs from an exact FL universe, the real spatial geodesics will not correspond exactly to multiples (and sums) of the generators derived assuming an exact FL universe. Gravitational lensing is one example, though the effect is probably small.

TABLE 2. Summary of the most recent methods and observational results. See text (§5) for details. Author abbreviations are by authors' surnames' initials and year.

(1) Methods:				
3D:				
clus opt	cosmic crystallog.	LLL 96		
clus Xray	brightest cluster	RE 97		
QSO's	local isom. search		R 96	
2D:				
CMB	ID'd circles			CSS96/98
	C_l — cutoff			many
	correlation fn			BPS98
Ideal object:				
	no Evoln	. monotonic E	strong E	weak E?
	zero pec velocity	prob. small	prob. small	N
	isotropic emitter	Y (nearly)	N	Y/N
	seen to large z	Y ($\kappa_0 < 0$), N (o.w.)	Y	Y
	seen over large vol	N	Y	Y (sph shell)
	seen to $\|b^{II}\| \ll 20$	N	N	N
Assumptions on κ_0, $\{g_i\}$, ideal= none:				
		none	κ_0 (use range)	circles: none C_l: all
(2) Constraints:				
		CC: $2r_+ \gtrsim R_H/20$ BC: $2r_+ \gtrsim R_H/10$		
			N/A	
For the following special cases, but really testing perturbation spectrum assumptions:				
$\kappa_0 = 0$, if $\theta(g_i, g_j) = 90i$ or $60i$, $i \in Z$ then				
				$(2r_{\rm inj} \gtrsim R_H/2)$
$\kappa_0 < 0$, if $\Gamma = m004(-5,1)$ or $\Gamma = v3543(2,3)$ then				
				$(2r_{\rm inj} \gtrsim 2R_H)$
(2) Specific candidates:				
		serendipitous	2σ implicit	"preferable to SCDM"
		$\kappa_0 = 0$	$\kappa_0 < 0?$	$\kappa_0 = -0.2$
		$\widetilde{M}/\Gamma = T^2 \times \mathbf{R}$	[non-orientable]	v3543(2,3)
		$2r_{\rm inj}$ $(\Omega_0) =$		$r_{\rm inj} = 0.95 R_H$
		$965 \pm 5h^{\ 1}$ Mpc (1)		
		$1190 \pm 10h^{\ 1}$ Mpc (0.2)		
		RE97, RB99	R96	BPS98

6. Conclusion

The present surge of research in the mathematics, the physical theory and the observations of cosmological topology promise exciting developments in all three fields. Moreover, some of the spin-offs which would follow from a significant detection have been explicitly calculated: the measurement of *transversal* galaxy velocities [67], precise estimates of the metric parameters [70], position of the topological signature in the Ω_0–λ_0 plane [82].

Mathematics: Continued work by mathematicians to understand and classify hyperbolic 3–manifolds may help theorists develop ideas regarding the physics behind global topology, while further developments in the software (SNAPPEA) will be useful to observational astronomers.

Theory: The development of predictions by quantum gravity theories *before* a 3–manifold candidate is confirmed to high significance would be useful from an epistemological point of view.

Observations: The numerous observational methods offer plenty of scope for improvements in the practical details while waiting for new surveys such as the SDSS (Sloan Digital Sky Survey) and new satellites such as XMM (X-ray Multiple Mission), MAP and Planck Surveyor.

However, it is not necessary to wait: specific candidates for the 3–manifold have been and should continue to be proposed. The benevolence of time allocation committees for the observational refutation (or strengthening) of the candidates should contribute significantly to observational cosmo-topology.

References

1. Atkatz, D., Pagels, H., 1982, Phys. Rv. D **25**, 2065
2. Barbour J. & Pfister H., 1995, (Eds.), Mach's Principle: From Newton's Bucket to Quantum Gravity (Einstein Studies, Vol 6), Birkhauser
3. Beardon, A.F. 1983, The Geometry of Discrete Groups (GTM91), New York:Springer
4. Bieberbach, L., 1911, Mathematische Annalen, **70**,297. 1912; Ibid, **72**, 400.
5. Bond J. R., Pogosyan D., Souradeep T., 1998, Class.Quant.Grav., 15, 2573 (astro-ph/9804041)
6. Briel U. G., Henry J. P., Böhringer H., 1992, A&A, 259, L31
7. Carlip S., 1998, Class.Quant.Grav., 15, 2629 (gr-qc/9710114)
8. Cornish N. J., Spergel D. N., Starkman G. D., 1996, Phys.Rev.Lett., 77, 215
9. Cornish N. J., Spergel D. N., Starkman G. D., 1998a, astro-ph/9708225
10. Cornish N. J., Spergel D. N., Starkman G. D., 1998b, Class.Quant.Grav., 15, 2657 (astro-ph/9801212)
11. da Costa L. N., in Cosmic Velocity Fields, ed. Bouchet F., Lachièze-Rey M., (Gif-sur-Yvette, France: Editions Frontières), p.475
12. Dekel, A., Burnstein, D., White, S., 1996, in Critical Dialogues in Cosmology, ed. N.Turok, World Scientific.
13. de Oliveira Costa A., Smoot G. F., 1995, Ap.J., 448, 477
14. de Sitter W., 1917, MNRAS, 78, 3
15. Dowker H. F., Garcia R. S., 1998, gr-qc/9711042
16. e Costa S. S., Fagundes H. V., 1998, gr-qc/9801066

17. Einstein, A., Postcard to Hermann Weyl, June 1918, from Einstein Archives, Princeton (freely translated from German by J.-P. Luminet).
18. Einstein, A., 1919, Postcard to Kelix Klein, April 16th, ibid.
19. Einstein, A., de Sitter W., 1932, Proc. Nat. Acad. Sci **18**, 213.
20. Ellis, G.F.R. 1971, Gen.Rel.Grav. **2**, 7
21. Ellis G.F.R., Schreiber, G., 1986, PhysLettA, 115, 97
22. Fagundes H. V., 1985, Ap.J., 291, 450
23. Fagundes H. V., 1989, Ap.J., 338, 618
24. H.V. Fagundes, 1993, Phys. Rev. Lett. **70**, 1579
25. Fagundes H. V., 1996, Ap.J., 470, 43
26. Fagundes H. V., Wichoski U. F., 1987, Ap.J., 322, L5
27. Fagundes, H. & Gausmann, E., 1997, *On Closed Einstein-de Sitter Universes*, astro-ph/9704259.
28. Fagundes, H. & Gausmann, E., 1999, astro-ph/9811368
29. Feodoroff, E., 1885, Russian journal for crystallography and mineralogy, St Petersburg, **21**.
30. Ferreira, Pedro G., Magueijo, Joao, Gorski, Krzysztof M., 1998, Ap.J., 503, L1
31. Friedmann, A., 1922, Zeitschrift für Physik, **10**, 37.
32. Friedmann, A., 1923, *Mir kak prostranstvo i vremya* (The Universe as Space and Time), Leningrad, Akademiya (French translation in *Essais de Cosmologie*, Le Seuil/Sources du Savoir, Paris, 1997.)
33. Friedmann A., 1924, Zeitschr.für Phys., 21, 326
34. Gabai, D. Meyerhoff, R., Thurston, N., 1996, *Homotopy Hyperbolic 3-manifolds Are Hyperbolic*, preprint available at http://www.msri.org/MSRI-preprints/online/1996-058.html
35. Gomero, G. I., Teixeira, A. F. F., Reboucas, M. J., Bernui, A., 1999, gr-qc/9811038
36. Gott J. R., 1980, MNRAS, 193, 153
1986, in Inner Space/Outer
37. Harrison, E., 1987, *Darkness at night*, Harvard University Press.
38. Hawking S., 1984, NuclPhysB, 239, 257
39. Inoue, K.T., astro-ph/9810034.
40. Ionicioiu R., 1998, gr-qc/9711069
41. Jing Y.-P., Fang L.-Z., 1994, PhysRevLett, 73, 1882 (astro-ph/9409072)
42. Klein, F., 1890, Mathematisches Annalen, **37**, 544.
43. Killing, W., 1891, Mathematisches Annalen, **39**, 257.
44. Lachièze-Rey M., Luminet J.-P., 1995, PhysRep, 254, 136
45. Lehoucq R., Luminet J.-P., Lachièze-Rey M., 1996, A&A, 313, 339
46. Lehoucq R., Luminet J.-P., Uzan J.-P., 1999, A&A, to be published (astro-ph/9811107)
47. Lemaître, G., 1927, Ann. Soc. Sci. Bruxelles, ser. A, **47**, 29.
48. Lemaître, G., 1931, MNRAS **90**, 490.
49. Lemaître G., 1958, in *La Structure et l'Evolution de l'Univers*, Onzième Conseil de Physique Solvay, ed. Stoops R., (Brussels: Stoops), p.1
50. Lemaître, G., 1978 (posth.) in *L'Univers, problème accessible à la science humaine*, Revue d'Histoire Scientifique, **31**, pp. 345-359.
51. Levin J., Scannapieco E., Silk J., 1998, Class.Quant.Grav., 15, 2689
52. Levin J., Scannapieco E., de Gasperis, G., Silk J., Barrow, J.D., 1998, astro-ph/9807206
53. Luminet, J.-P., 1997 *L'invention du big bang*, suivi de A. Friedmann, G. Lemaître Essais de Cosmologie (Le Seuil/Sources du Savoir, Paris).
54. Luminet, J.-P., 1998, Acta Cosmologica, **24** (gr-qc/9804006)
55. Luminet, J.-P. & Lachièze-Rey, M., 1994, La physique et l'infini (Flammarion/Dominos, Paris).
56. Madore J., Saeger L. A., 1998, Class.Quant.Grav., 15, 811 (gr-qc/9708053)
57. Matveev S.V., Fomenko A.T., 1988, Russian Math. Surveys **43**, 3.

58. Novacki, W., 1934, Commentarii Mathematici Helvetici, **7**, 81.
59. Pando, J., Valls-Gabaud, D., Fang, L.-Zh., 1998, PhysRevLett, accepted, (astro-ph/9810165)
60. Peebles, P.J.E., 1993, Principles of Physical Cosmology, Princeton, U.S.A.: Princeton Univ. Press
61. **Pierre, M., 1999, Proceedings of the workshop *Cosmological Topology in Paris 1998*, Obs. Paris/IAP, Paris 14 December 1998 (http://www.iap.fr/users/roukema/CTP98/resumes.html)**
62. Rainer M., 1996, gr-qc/9602059.
63. Rosales J.-L., 1998, gr-qc/9712059
64. Roukema B. F., 1996, MNRAS, 283, 1147
65. Roukema B. F., 1999, submitted
66. **Roukema, B.F., et al., 1999, Proceedings of the workshop *Cosmological Topology*, Obs. Paris/IAP, Paris 14 December 1998 (http://www.iap.fr/users/roukema/CTP98/resumes.html)**
67. Roukema B. F., Bajtlik, S., 1999, submitted to MNRAS
68. Roukema B. F., Blanloeil V., 1998, Class.Quant.Grav., 15, 2645 (astro-ph/9802083)
69. Roukema B. F., Edge A. C., 1997, MNRAS, 292, 105
70. Roukema B. F., Luminet, J.-P., 1999, submitted
71. Schwarzschild, K. Vierteljahrschrift der Ast. Ges. **35** 67, 1900
72. Schwarzschild, K., 1998, Class.Quant.Grav., 15, 2539 [english translation of Schwarzschild (1900)]
73. Spaans, M., 1999, gr-qc/9901025
74. Starkman G. D., 1998, Class.Quant.Grav., 15, 2529
75. Starkman, G. et al., 1998, Invited papers from the workshop Cosmology and topology (CWRU, Cleveland, 25-27 October 1997), Class. Quant. Grav. **15**, 2529
76. Starobinsky A. A., 1993, JETPLett, 57, 622
77. Stevens D., Scott, Silk J., 1993, PhysRevLett, 71, 20
78. Thurston, W. P., 1979, *The geometry and topology of three manifolds*, (Princeton Lecture Notes)
79. Thurston, W. P., 1997, Three-Dimensional Geometry and Topology, ed. Levy, S., Princeton, U.S.A.: Princeton University Press
80. J-P. Uzan, 1998, Phys. Rev. **D58**, 087301
81. lass. Quant. Grav. **15**, 2711
82. Uzan J.-P., Lehoucq R., Luminet J.-P., 1999, in preparation
83. Uzan, J.-P., Peter P., 1997, Phys.Lett. B **406**, 20
84. Weeks, J., 1985, PhD thesis, Princeton University.
85. Weeks J. R., 1998, Class.Quant.Grav., 15, 2599 (astro-ph/9802012)
86. Wolf, J.A., 1960, C. R. Acad. Si. Paris,**250**, 3443.
87. Zel'dovich Y. B., Grishchuk L. P., 1984, MNRAS, 207, 23P

SYMMETRIES IN GENERAL RELATIVITY AND BEYOND

M. LACHIÈZE-REY
Service d'Astrophysique, C. E. Saclay
91191 Gif sur Yvette cedex, France

1. Introduction

All past, present and prospective physical theories are expressed in (three dimensional) space or (four dimensional) space-time, which appear as the frame to describe physical events. Although fundamental, the concepts of space and space-time strongly differ in different theories, in particular in quantum physics and general relativity. Moreover, the tendencies of contemporary research may involve still different concepts (different number of dimensions, discreteness, ...).

In this lecture, I will examine some general properties of space or space-time, in particular related to their symmetry properties, with geometrical tools like Lie groups, frame bundles, connections, ... In particular, I will analyse the Poincaré group, in relation with the structure of special and general relativities. Then I will indicate how the symmetries of spacetime can be modified or extended to lead to new theories.

2. Spaces and manifolds

Although modern geometry does not require the existence of points, I will consider here that space (I use the word in a general meaning, including space-time) is defined as a collection of points, with certain relations between them. According to some points of view, space is defined by these relations, or by the symmetries that they allow to define. Thus, a space appears as a set, to which a hierarchy of levels of structure can be superposed: a topological structure; then a differential structure; then a metric, or a connection,... I will indicate briefly how these levels are defined, and how they are linked to the symmetries of the space considered, expresses by associated Lie groups or fiber bundles.

M. Lachièze-Rey (ed.), Theoretical and Observational Cosmology, 157–178.

2.1. BASIC DEFINITION

Very generally, a *map* is defined as an application from a space to an other space. A *function* is a map $f : \mathcal{E} \mapsto \mathbb{R}$. A *path* is a map

$$\gamma : \mathbb{R} \mapsto \mathcal{E} : t \mapsto \gamma(t).$$

A basis level of structure is given by a *topology*, which makes a set of points $\mathcal{T}$ a *topological space* : a topology is a collection of sets of $\mathcal{T}$, verifying certain axioms and called *open sets*. This defines the neighborhood relations between the points and, as a consequence, allows to consider the *continuity* of functions over the space.

The derivation properties of functions are insured thanks to an additional level of structure, the *differential structure*. This structure is expressed by *vectors*, from which are constructed *forms* and *tensors*. It may be, for instance, defined by the *differential manifold* structure: a differential manifold $\mathcal{M}$ is piecewise diffeomorphic to $\mathbb{R}^n$: some sets of $\mathcal{M}$ are mapped to the open sets of $\mathbb{R}^n$; the maps conserve, by definition, the derivation properties. Thus, the continuity and derivation properties of functions are defined locally, and derive from those of $\mathbb{R}$. The local correspondence between $\mathcal{M}$ and $\mathbb{R}$ is made through *maps* and *atlases* (the atlas is the ensemble of the maps). Those provide local coordinates labelling the points of $\mathcal{M}$, but most geometrical objects defined on $\mathcal{M}$ do not depend on coordinates.

The differentiable manifold structure allows to derivate the functions over $\mathcal{M}$: a derivation is first defined along a paths $\gamma(t)$, with respect to the parameter t. The functions can be classified with respect to their derivability properties: one defines $C^0, C^1, \ldots, C^n, \ldots, C^\infty$ as the sets of continuous, derivable, ..., n times derivable, ...and smooth (or analytical) functions. Each of these sets has the structure of a commutative algebra (with the ordinary product). The subsets of these algebras (of functions), that take zero value on some subsets of $\mathcal{M}$, are *ideals* of the algebra. For the usual geometry, the structure of these commutative algebras completely define the structure of the space as an ensemble of points. The idea of non commutative geometry is to consider an abstract (non necessarily commutative) algebra as the function algebra of an " artificial space ", so defined (not necessarily composed of points).

2.2. VECTORS AND THE TANGENT SPACE

A vector V at a point $p \in \mathcal{M}$ is a derivation operator (defined for instance from the derivation operation in $\mathbb{R}^n$, via the coordinates). To any function $f(p)$, it associates the value of a derivative $V(f)[p]$ at this point, obeying the *Leibniz rule* $V(f.g) = f.V(g) + V(f).g$. A vector correspond to the derivation along the path tangent to it, but it is also possible to consider vectors in a more abstract way. The derivations with respect to the coordinates $\{x^\mu\}$ are the vectors $\partial/\partial x^\mu = \partial_\mu$.

The set of all vectors at a point p is the vector space $T_p\mathcal{M}$, the tangent space to $\mathcal{M}$ at p. It has the structure of $\mathbb{R}^n$. A *frame* is a basis of $T_p(\mathcal{M})$, i.e., a set $\{e_A\}$ of four independent vectors. Any vector X of $T_p(\mathcal{M})$ can be decomposed as $X = X^A e_A$, the X^A being its components. In particular, the vectors ∂_μ form a *coordinate* or *holonomic* frame, but not any frame is a coordinate basis. In the set of all frames, the transformation of one frame to an other is given by a linear transformation, which is a member of the the linear group $G\ell(n, \mathbb{R})$.

2.3. VECTOR FIELDS

A vector of T_p is a geometrical object defined at a point p. A *vector field* V is defined by the value $V(p)$ of a vector at each point p of $\mathcal{M}$. Vectors and vector fields are different objects. The set $\chi(\mathcal{M})$ of vector fields has the structure of a *module* (a generalization of vector space structure) over the algebra of functions.

A vector field X is an operator acting onto a function f to give another function $X(f)$. Another vector Y can act on the new function $X(f)$. This defines the composition of the vectors $X, Y \to YX$, through $YX(f) = Y[X(f)]$. Care must be taken that YX is not a vector (it does not satisfy the Leibniz rule) ! However, the *commutator* $[X, Y] = XY - YX$ is a vector. This defines a product, and an algebra structure, with the commutator as product. $[X, Y]$ is also called the *Lie derivative* $\mathcal{L}_X Y$ of Y with respect to X.

A *moving frame* is a smooth assignment of a reference frame (an ordered vector basis) above each point of the manifold. From the smoothness requirement, it may not be possible to define globally a moving frame (this is possible only for *parallelisable* manifolds). If $\{e_A\}$ is a moving frame, any vector field X can be decomposed over it; its components X^A are functions : $x = X^A e_A$.

2.4. FORMS AND TENSORS

Considering the tangent space $T_p\mathcal{M}$ at a point p, a map $T_p\mathcal{M} \to \mathbb{R}$ associates a real number to each vector $V \in T_p\mathcal{M}$. Among these operators, the linear ones are called *one-forms* at p: a one-form at p, ω, is a linear operator $T_p \mapsto \mathbb{R}$. Its action over a vector X is denoted $\omega(X) =< \omega, X >$. The set $T_p^*\mathcal{M}$ of all one-forms at p is the *cotangent space* at P, the dual of the tangent space $T_p\mathcal{M}$.

To each vector basis (frame) $\{e_A\}$ of T_p is associated the *dual basis*, or dual frame $\{\omega^A\}$ of T_p^* such that $< \omega^A, e_B >= \delta^A_B$. In particular, to the coordinate basis $\{\partial_\mu\}$ is associated the dual coordinate basis $\{dx^\mu\}$ of T_p^*. A *one-form* is defined by the continuous choice of a one-form at point p, for each $p \in \mathcal{M}$. One-forms act linearly onto vector fields, to give functions. To each function f is associated the one form df defined by

$$< df, X > := X(f), \forall X \in \chi(\mathcal{M}).$$

It is called the *exterior derivative* of the function f.

In general relativity, an example of one-form is given by the *momentum*. The one-forms dt and dx^i are associated to the space and time coordinates t and x^i. The momentum is defined as the one-form $p = -E\ dt + p_i\ dx^i$, $(i = 1, 2, 3)$. The action of the one-form p over the velocity vector u is given by $< p, u >= p_\alpha u^\alpha = -E = -energy$.

Vectors and forms are particular types of *tensors*. Generally speaking, a tensor of type (p, q) is a sum of factors like $v_1 \otimes v_2 \otimes \ldots \otimes v_p \otimes \omega^1 \otimes \omega^2 \otimes \ldots \otimes \omega^q$, where $\otimes$ denotes the Cartesian product. A vector is thus a tensor of type $(1, 0)$, a one-form of type $(0, 1)$. By definition a $q-$form is an antisymmetric tensor of type $(0, q)$. For instance, the antisymmetrised product, or*wedge product* $\alpha \wedge \beta := \alpha \times \beta - \alpha \times \beta$ of two one-forms α and β is a two-form. The associative operation $\wedge$ acts as a product onto all forms. Similarly, the completely antisymmetrized product of q one-forms is a q-form. If $\{\theta_A\}$ is a basis for one-forms, $\{\theta_{A_1} \wedge \ldots \wedge \theta_{A_q}\}$ form a basis for q-forms. A scalar function is called a zero-form. A p-form, with $p > n$, the dimension of the manifold, reduces to zero. There exists a unique n-form (via a multiplicative constant), the volume element $dx^1 \wedge \ldots \wedge dx^n$.

The *exterior derivative* of a one-form $\omega = \omega_\mu\ dx^\mu$ is the 2-form $d\omega$ such that $d\omega(u, v) = u[\omega(v)] - v[\omega(u)] - \omega([u, v])$. It is easy to check that $d\omega = d\omega_\mu \wedge dx^\mu$, where $d\omega_\mu$ is the exterior derivative of the function ω_μ.

To a moving frame $\{e_A\}$ is associated the *dual moving frame* $\{\omega^A\}$ such that $< \omega^A, e_B >= \delta^A_B$. In particular, to the coordinate basis $\{\partial_\mu\}$ is associated the dual coordinate basis $\{dx^\mu\}$.

2.4.1. *Vector valued forms*

A one-form ω associates linearly a scalar $\omega(v)$ to any vector v. A vector valued one-form Ω associates linearly a vector $\Omega(v)$ to any vector v. It can be decomposed on a basis as $\Omega = \Omega^A_B\ e_A \otimes \theta^B$. In other words, it is a tensor of type (1,1).

The simplest example is the " unit form " θ, defined by its action $\theta(v) = v$. It is the unit tensor of type (1,1), decomposed as

$$\theta = \theta^A_B\ e_A \otimes \theta^B = \delta^A_B\ e_A \otimes \theta^B = e_A \otimes \theta^A.$$

It is also sometimes called the " solder form " on $\mathcal{M}$, although this is not the strict definition (see below).

An other example is the covariant derivative ∇v of a vector v (see below). It is a vector valued form, or a (1,1) tensor. Its action on an other vector u gives the covariant derivative $\nabla_u v$ of v with respect to u. We will also meet forms, defined on a manifold $\mathcal{M}$, and taking their value in an other vector space, like for instance a Lie algebra associated to the manifold.

2.4.2. *Components*

Most often (in particular in textbooks), vectors, forms and tensors are expressed by their components in coordinate (or holonomic) frames. It is important to realize that some formulae can take more complicated expressions in non coordinate frames. In the following, I will as far as possible not assume a coordinate frame, and write anything with the less possible reference to frames. I will note, as possible, general components with respect to a (non holonomic) frame with capital latin indices $A, B,$.

The non holonomic nature of a frame $\{e_A\}$ is expressed by the structure coefficients C^C_{AB} defined by $[e_A, e_B] = C^C_{AB}\ e_C$ or equivalently, for the dual frame, $d\theta^A = -\frac{1}{2}C^A_{BC}\ \theta^B \wedge\ \theta^C$. These coefficients vanish for a holonomic frame.

All the previous considerations also hold for the dual frame $\{\theta^A\}$. Given a frame, its basis vector fields (resp. forms) can be decomposed onto an other frame, in particular onto a coordinate frame as $e_A = E^\mu_A\ \partial_\mu$ and $\theta^A = E^A_\mu\ dx^\mu$.

2.5. THE TANGENT AND COTANGENT BUNDLES

2.5.1. *Fiber bundles*

A fiber bundle $(\mathcal{P}, F, \Pi, G)$is a manifold $\mathcal{P}$ with the following structure:

- it has a basis $\mathcal{M}$ and a fiber F, which are both manifolds.
- A projection operator Π projects each point $p \in \mathcal{P}$ onto a point $\Pi(p) \in \mathcal{M}$. All points of $\mathcal{P}$ projected onto the same point $x \in \mathcal{M}$ belong to the same fiber $\Pi^{-1}(x) = F_p$.
- There is a group G acting onto $\mathcal{P}$, called its structure group. A fiber remains invariant under the action of G. The basis is thus $\mathcal{M} = \mathcal{P}/G$.
- Locally, $\mathcal{P}$ has the structure of the Cartesian product $\mathcal{M} \otimes F$. This is insured by the existence of local diffeomorphisms $\phi_i : U_i \otimes F \rightarrow \Pi^{-1}(U_i)$, for open sets U_i which cover the basis manifold $\mathcal{M}$. These maps are " glued together " by transition functions $t_{ij} = \phi_i^{-1}\ \phi_j$, the three functions being defined at a same point p in the intersection of the open sets U_i and U_j.

The simplest example of a fiber bundle is the cylinder. Its basis $\mathcal{M}$ is the line $\mathbb{R}$, its fiber is the circle S^1, the structure group G being the rotation group along the circonference S^1. The cylinder is a *trivial fiber bundle*, since it can be considered, as a whole, as the Cartesian product $\mathbb{R}\otimes S^1$. The Möbius band has the same basis, the same fiber, but a different structure group and is thus a different (non trivial) fiber bundle.

A *section* is a smooth assignment of a point $p \in \Pi^{-1}(x)$ to each point x of the basis $\mathcal{M}$. A fiber bundle is called a *principal bundle* if its fiber is identified with its structure group, $F = G$. To each fiber bundle, is associated its principal fiber bundle. A *vector bundle* is a fiber bundle, with fiber an n-dimensional vector space V^n. Its structure group is the linear group of V^n. To any principal bundle is associated a vector bundle with same group G and same transition functions.

Another example of fiber bundle is encountered in quantum electrodynamics: its basis is Minkowski spacetime, its fiber is the circle S^1, and its structure group is $U(1)$.

2.5.2. *The tangent bundle and the frame bundle*

To a manifold $\mathcal{M}$ are associated different fiber bundles, which express different aspects of its structure. The reunion of all the tangent spaces of a manifold is its *tangent bundle*

$$T(\mathcal{M}) = \bigcup_{p \in \mathcal{M}} T_p\mathcal{M}.$$

Its basis is the manifold $\mathcal{M}$ itself, its fibers are the tangent spaces at the points of $\mathcal{M}$. The *structure group* of $T(\mathcal{M})$ is $GL(n, \mathbb{R})$, the group of non singular linear transformations, or the group of real invertible $n \times n$ matrices. A vector field on the manifold $\mathcal{M}$ (distinct from a vector field on $T\mathcal{M}$), is a section of the tangent bundle $T(\mathcal{M})$. One defines similarly the cotangent bundle $T^*(\mathcal{M})$, as the reunion of all cotangent spaces. A one-form is a section of $T^*(\mathcal{M})$.

The principal bundle associated to $T(\mathcal{M})$ is the *frame bundle* $F(\mathcal{M})$. It has the group $GL(n, \mathbb{R})$ as fiber, and $\mathcal{M}$ as basis. A moving frame is a section of $E(\mathcal{M})$, which can be seen as the collection of all ordered vector bases of $\mathbb{R}^n$.

The orthonormal bundle

When a metric g is defined (see below), this allows to select, among all frames at $x \in \mathcal{M}$, the orthonormal ones. Frames are transformed into each other by an element of $GL(\mathbb{R}, n)$. Orthonormal frames are transformed into each other by an element of the Lorentz group $\mathcal{L}$, a subgroup of $GL(\mathbb{R}, n)$. The *orthonormal bundle* (also called the *tetrad bundle*) has $\mathcal{M}$ as basis. Its fiber at x is the set of orthonormal frames at x. Its structure group is the homogenous Lorentz group. It plays a fundamental role in general relativity.

2.6. METRIC AND TETRADS

A metric over a manifold $\mathcal{M}$ is a $(0,2)$ symmetric tensor g. Its action onto vectors is written $g(X,Y)$. A metric g defines the orthonormal frames $\{e_a\}$, such that $g(e_a, e_b) = \eta_{ab}$, the diagonal Minkowski metric.

If a metric g has been defined, it allows to decide which frames are orthonormal (ON). The transformation between an ON frame to an other ON frame is a *Lorentz rotation*, i.e., a member of the *Lorentz group* $SO(3,1)$, a subgroup of $GL(n,\mathbb{R})$. The metric g, over $\mathcal{M}$, allows to select the orthogonal frames. Conversely, the selection of the orthogonal frames defines the metric, by the requirement that a specific frame is orthonormal.

Expressed in an orthonormal frame $\{e_a\}$, the metric is by definition diagonal: η_{ab}. Its expression in an arbitrary frame is obtained by the change of frame as

$$g_{\mu\nu} = \eta_{ab}\ h^a_\mu\ h^b_\nu.$$

The h^a_μ are called the *tetrad components*. Their definition is equivalent to that of a metric.

To an orthonormal frame is associated a basis of (dual) forms $\{\hat{\theta}^a\}$, the *tetrad, or vierbein, basis of forms*. The metric tensor can be written

$$g = g_{\mu\nu}\ dx^\mu \otimes dx^\nu = \eta_{ab}\ \hat{\theta}^a \otimes \hat{\theta}^b,$$

where $\otimes$ denotes the tensorial product.

It remains to precise the link between the ON frame at one point and the ON frame at an other (neighbor) point. General relativity is specified by the requirement that an ON frame is *parallely transported*. This defines the *Levi-Civita connection*.

3. Connections on a manifold

The differential structure of a manifold is defined by its vector fields (or, equivalently, by its forms). They are tools to derivate the functions. To derive the vectors, a new tool must be introduced, the *connection* or, equivalently, the *covariant derivative*. This adds a new level of structure to the manifold. *Parallel transport* means transport with zero covariant derivative. Once defined, the connection allows to derivate covariantly vectors, tensors and forms.

On a given differentiable manifold, there is a priori an infinity of possible choices of connections. A possible requirement is to demand that a connection has *no torsion* (see below). In general, a connection is independent of any metric. However, if a metric tensor g has been defined, an other possible requirement is to demand that g is *parallely transported*. These two requirements together completely determine the connection, which is then called the *Levi-Civita connection*. It is the one used in general relativity.

3.1. THE LINEAR CONNECTION

A linear connection over a manifold $\mathcal{M}$ is defined as an operator ∇ which maps any vector field X onto a (1,1) type tensor

$$\nabla : \chi(\mathcal{M}) \mapsto \Omega^1(\mathcal{M}) \times \chi(\mathcal{M}) \quad : X \mapsto \nabla X.$$

It obeys $\nabla(X+Y) = \nabla X + \nabla Y$ (linearity), and $\nabla(fX) = f\nabla X + df \otimes X$ where f is a function (Leibniz rule). Thus, the action of ∇ is trivially extended to functions through $\nabla f = df$.

The action of ∇ is extended to tensor fields by requiring commutation with the contracted multiplication, and the Leibniz rule for Cartesian product:

$$\nabla(T \otimes T') = \nabla T \otimes T' + T \otimes \nabla T',$$

for any tensors T and T'. ∇ maps a tensor field into a product of a one-form by a tensor of same rank. With these rules, the action of ∇ onto vector fields determines that onto tensors of any type. In a frame $\{e_A\}$, ∇ is defined by its action onto the basis vectors, developed as

$$\nabla e_A = \omega_A^B \otimes e_B,$$

which defines the connection forms ω_A^B in this frame.

Since ∇X is a (1,1) type tensor, it acts on a vector field u, by contracted multiplication (action on the left), to give a new vector field $\nabla_u X$, the *covariant derivative of X along u*. More generally, if T is a tensor, $\nabla_u T$ is a tensor of the same type as T. Thus the connection acts as a derivative onto its argument, and tensorially onto the indexed vector fields. The covariant derivative with respect to u obeys the covariant derivative axioms:

(i) ∇_u transforms a vector field into vector field and, more generally, a tensor field into a tensor field of same rank;

(ii) ∇_u is linear both in u and in its argument :

$$\nabla_{fX+gZ} T = f\nabla_X T + g\nabla_Z T$$

and

$$\nabla_X[aT + bT'] = a\nabla_X T + b\nabla_X T'.$$

(iii) ∇_X has the usual derivation property (Leibniz rule)

$$\nabla_X(f\,T) = f\nabla_X T + (\nabla_X f)\,T.$$

In addition, its action onto a function is that of $X : \nabla_X f = Xf$. In these formulae, f and g are functions, a and b are constants.

Using the language of forms, exterior derivatives, and wedge products, we have the fundamental relation, where the ω^α are a basis of forms,

$$d\omega^\alpha = \omega^\beta \wedge \omega^\alpha_\beta - T^\alpha, \tag{1}$$

which defines the *torsion two-form* T^α. In the conventional view of general relativity, the torsion is zero.

The connection forms verify an other structure equation

$$d\omega^\alpha_\beta = \omega^\gamma_\beta \wedge \omega^\alpha_\gamma - R^\alpha_\beta, \tag{2}$$

which defines the curvature two-forms R^α_β.

3.1.1. *Covariant derivatives of forms*

The action of ∇ onto vectors determines its action onto any tensor. For instance, let us examine the covariant derivative of a one-form ω. For any vector Y, it is required that

$$\nabla_X \omega(Y) = \omega[\nabla_X Y] + [\nabla_X \omega](Y).$$

Since $\omega(Y)$ is a scalar, its derivative is given by $\nabla_X \omega(Y) = X[\omega(Y)]$, from which results the definition of the covariant derivative of ω, through its action on the arbitrary vector Y :

$$[\nabla_X \omega](Y) = X[\omega(Y)] - \omega[\nabla_X Y].$$

3.1.2. *Components*

The connection ∇ (or the covariant derivative, another expression of the same object) is defined by its action onto tensors and do not depend on any system of coordinates. As soon as such a system has been defined, it can of course be expressed by components. In a basis $\{e_A\}$ (not necessarily a coordinate basis), with dual basis $\{\theta^A\}$, a vector field $X = X^A e_A$ and a form $\xi = \xi_A \theta^A$. We recall the identity $e_A f = df(e_A)$. Noting $\nabla_{e_A} = \nabla_A$, it is easy to check that

$$\nabla_X T = X^A \; \nabla_A T$$

We define the connection coefficients Γ^A_{BC} through

$$\nabla_B e_C = \Gamma^A_{CB} \; e_A.$$

These functions are not components of tensors. They are linked to the connection forms through

$$\omega^A_B = \Gamma^A_{BC} \; \theta^C$$

or, equivalently

$$\Gamma^A_{BC} = \omega^A_B(e_C).$$

Also,

$$\nabla_B e_C = \Gamma^A_{CB} e_A,$$

and

$$\Gamma^A_{BC} = \theta^A(\nabla_C e_B).$$

Care must be taken that the indices are not tensorial indices. For a vector field Y,

$$\nabla_A Y = [e_A(Y^M) + Y^B\, \Gamma^M_{BA}]\, e_M, \tag{3}$$

and, for a form ξ,

$$\nabla_A \xi = [e_A(\xi_M) - \xi_N\, \Gamma^N_{MA}]\, \theta^M. \tag{4}$$

Also,

$$\nabla X = (dX^A + \omega^A_B\, X^B) \otimes e_A.$$

and

$$\nabla \xi = (d\xi_A - \omega^B_A\, \xi_B) \otimes \theta^A.$$

3.2. SYMMETRIES OF THE CONNECTION

A connection ∇ is said *symmetric*, or *with no torsion* iff, for two arbitrary vector fields X and Y,

$$\nabla_X Y - \nabla_Y X = [X, Y].$$

The *torsion* operator τ is defined through its action on two arbitrary vector fields X and Y as

$$\tau(X, Y) = \nabla_X Y - \nabla_Y X - [X, Y].$$

It represents the antisymmetric part of the connection. The $(1, 2)$ torsion tensor is defined as T such that $T(\alpha, U, V) = \alpha[\tau(U, V)]$. In a basis, the torsion form is defined as $T^A(U, V) = T(\theta^A, U, V)$.

In an arbitrary basis, the basis vectors obey the commutation rule $[e_A, e_B] = C^M_{AB}\, e_M$ which define the structure constants C^M_{AB}. It results that the previous relation can be written in components as

$$T^A_{BC} = \gamma^A_{[BC]} - C^A_{BC}.$$

In a coordinate (holonomic) basis only, the structure constants vanish, so that

$$T^\alpha_{\beta\gamma} = \Gamma^\alpha_{[\beta\gamma]}.$$

The curvature ρ of the connection is defined through

$$\rho(U,V) = \nabla_U \nabla_V - \nabla_V \nabla_U - \nabla_{[U,V]}.$$

The curvature tensor R has for components

$$R_i{}^j{}_{kl} = R(e_i, \theta^j, e_k, e_l = e_k(\gamma^j_{kl}) - e_l(\gamma^j_{ki}) - \gamma^i_{km}\ \gamma^m_{li} - \gamma^j_{lm})\ \gamma^m_{ki} - C^m_{kl}\ \gamma^j_{mi}.$$

3.3. THE METRIC CONNECTION

There is a great arbitrariness in the choice of the connection. The symmetry condition, which sets the torsion to zero, is not sufficient to determine it. If a metric tensor g is defined, we may impose the *metricity condition* $\nabla g = 0$, that g is covariantly constant. This implies $\nabla_X g = 0$, for any vector field X. This condition is equivalent to the requirement that the inner product of two vectors remains constant under parallel transport. This defines the *metric connection*.

In an arbitrary basis e_A, the metricity condition is

$$[\nabla_A g]_{MN} = e_A(g_{MN}) - \Gamma^R_{MA}\ g_{RN} - \Gamma^R_{NA}\ g_{MR} = 0.$$

For a given metric manifold, there are many metric connections but only one is also symmetric, the *Levi-Civita connection*. It is the one used in general relativity.

In a coordinate basis, the coefficients of a Levi-Civita connection are linked to the derivatives of the components of g. They reduce to the *Christoffel symbols*

$$\Gamma^\gamma_{\alpha\beta} = \Gamma^\gamma_{(\alpha\beta)} = \{^{\ \gamma}_{\alpha\beta}\} := 1/2\ g^{\gamma\delta}\ (\partial_\alpha g_{\beta\delta}\ + \partial_\beta g_{\alpha\delta}\ - \partial_\delta g_{\alpha\beta}).$$

Then the connection coefficients of a (non symmetric) metric connection are written, very generally, as

$$\Gamma^\gamma_{\alpha\beta} = \{^{\ \gamma}_{\alpha\beta}\} + 1/2\ K^{\ \gamma}_{\alpha\beta},$$

where the *contorsion* is defined as $K^{\ \mu}_{\nu\lambda} = \Gamma^\mu_{[\nu\lambda]} - g^{\lambda\beta}\ g_{\sigma\mu}\ \Gamma^\sigma_{\beta\nu} - g^{\lambda\beta}\ g_{\sigma\nu}\ \Gamma^\sigma_{\beta\mu}.$

3.4. LIE GROUPS

A *topological group* is a Hausdorff topological space, with a group product law $(g,h) \rightarrow gh$, which is continuous, as well as the map $g \rightarrow g^{-1}$. It is a *Lie group* if it possesses a compatible analytical atlas. A *Lie group* G is at the same time a group and an n-dimensional analytic manifold. The left action of the group on itself is denoted L_a and defined as

$$L_a : G \mapsto G; L_a(g) = ag.$$

It is a diffeomorphism of G, seen as a manifold, on itself.

In the neighborhood of each element, it exists a set of n parameters allowing to represent continuously the group elements. For a Lie group, it is required that the applications $(a, b) \to ab$ and $a \to a^{-1}$ must be analytical. The multiplication law ab is consistent with the continuous structure of the manifold: if two more points c and d are " close " to a and b, respectively, the product cd is " close " to the product ab. Lie groups were classified about 100 to 80 years ago, mostly by the Norwegian Sophus Lie, the German Wilhelm Killing and the French Elie Cartan.

Here I will consider Lie groups of spatial transformations, like for instance the groups of rotations, of translations, or of isometries of a given space.

3.4.1. *Lie algebra*

The main properties of a Lie group are expressed by the structure of its associated *Lie algebra* $\underline{g}$ or $\mathcal{G}$. In a Lie group of transformations, any transformation can be obtained as a product of infinitesimal transformations. These infinitesimal transformations can be seen as vector fields on G, which remain invariant under the group action. They form the *Lie algebra* $\mathcal{G}$ of G, the set of these infinitesimal transformations. The structure of $\mathcal{G}$ is that of T_eG, the tangent space to G, at its origin e.

A Lie algebra is in some sense the "logarithm" of a Lie group. Let us consider a Lie group G whose neutral element (unity) is denoted e, and the tangent space at unity $T_e(G)$. The left action L_g induces the map

$$L_{g*} : T_a(G) \mapsto T_{ga}(G)$$

between vectors at a and vectors at $L_g a = ga$.

Given any vector $A \in T_e(G)$, L_{a*} associates the vector $L_{a*}(A)$ at each point $a = L_a(e)$. This associates to $A \in T_e(G)$, a vector field ζ_A over G. By construction, this vector field is left invariant through the action of L_{a*}, i.e., it verifies : $L_{a*}(\zeta \mid_g) = \zeta \mid_{ag}$.

L_{g*} maps a vector field $X \in \chi(G)$ into an other vector field. The Lie algebra $\mathcal{G}$ is the set of left invariant vector fields. It is a (flat) vector space, isomorphic to $T_e(G)$. It is easy to show that if two vectors X and Y are members of the Lie algebra, their commutator $[X, Y]$ is also. This defines a product which gives to the Lie algebra its algebra structure. Its antisymmetry insures the *Jacobi identity*:

$$[X, [YZ]] + [Y, [Z, X]] + [Z, [X, Y]] = 0.$$

A basis of the Lie algebra constitute a system of *generators*. The basis vectors are related by the commutation relations

$$[X_A, X_B] = C^C_{AB}\ X_C,$$

which define the structure of the Lie algebra. The *structure constants* C^C_{AB} are constant functions, which verify $C^C_{AB} = C^C_{BA}$. They vanish if the algebra is *abelian* (i.e., commutative). The corresponding relations

$$d\theta^C = -\frac{1}{2} C^C_{AB}\ \theta^A \wedge \theta^B$$

hold for the dual of the generators, the one-forms $\{\theta^A\}$ verifying $\theta^C(e_A) = \delta^C_A$.

3.4.2. *Lie groups operating on manifolds*
The action of a Lie group G over a manifold $\mathcal{M}$ is defined by associating a permutation $a_{\mathcal{M}}$ of $\mathcal{M}$ to each $a \in G$, such that $(ab)_{\mathcal{M}} = a_{\mathcal{M}}\ b_{\mathcal{M}}$, and that the application

$$a \in G, x \in \mathcal{M} \rightarrow a_{\mathcal{M}}(x)$$

is differentiable.

Typically, we will consider the Lie group G of isometries (or of conformal isometries) acting onto a manifold $\mathcal{M}$, space or spacetime. The dimension of G is generally larger than that of $\mathcal{M}$. Vector fields and forms on G (members of its Lie algebra $\mathcal{G}$) are distinct from vector fields and forms on $\mathcal{M}$. However, we will see that $\mathcal{M}$ can be identified with G/G_0, the transitive subgroup of G. This will allow a correspondence between some vector fields and forms on $\mathcal{G}$ with those on $\mathcal{M}$ (see below). In the following, the symmetries of a spacetime manifold $\mathcal{M}$, will be mainly expressed by vectors (or forms) of the corresponding Lie algebra $\mathcal{G}$.

An example of Lie group is given by the rotations in three dimensions. A rotation in $\mathbb{R}^3$ is defined by a unit vector $\mathbf{u}$ characterizing its (oriented) axis and a rotation angle θ defined modulo 2π. The rotations form a non Abelian three parameter group, the special orthogonal group $SO(3)$. Using the Cartesian coordinates in $\mathbb{R}^3$, rotations are represented by unimodular orthogonal [3*3] matrices Λ, i.e., such that $\Lambda^T\Lambda = \Lambda\Lambda^T = I$, with $det\Lambda = +1$. This group $SO(3)$ is a subgroup of the orthogonal group $O(3)$ defined by $\Lambda^T\Lambda = \Lambda\Lambda^T = I$, with $det\lambda = \pm 1$ (the group $O(3)$ is the group of rotations with inversions). The Lie group structure of $SO(3)$ implies that its Lie algebra has three (its dimension) infinitesimal generators $X_\alpha, \alpha = 1, 2, 3$.

3.4.3. *The Maurer - Cartan form of a Lie group*
Since a Lie group G is also a manifold, it is possible to define vector fields, forms, vector valued forms,...on it. Since G is a group, any point g may

be considered as the action of the left translation L_g onto its unit element e, as $g = L_g(e) = g\ e$. Thus the differential map $(L_g^{-1})'$ maps any vector at g, in $T_g(G)$, to a vector at e, in $T_e(G)$:

$$(L_g^{-1})' : v_g \mapsto \gamma = (L_g^{-1})' v_g \in T_e(G).$$

This allows to define the *canonical form*, or *Maurer - Cartan form*, as a form ω on G, with values in $T_e(G)$, such that

$$\omega(v_g) = \gamma.$$

In fact, γ is a member of the Lie algebra $\mathcal{G}$ of G. It is easy to show that ω can be decomposed as $\omega = \gamma_A \otimes\ \theta^A$, where $\{\gamma_A\}$ is a basis of T_eG and $\{\theta^A\}$ is a dual basis of T_g^*G.

3.5. CONNECTION IN THE FRAME BUNDLE

3.5.1. *The canonical isomorphism in a principal bundle*

Let us consider a principal fiber bundle $\mathcal{P}$, with group G, and the vector fields on it (different from those on its basis $\mathcal{M}$). A vector at $p \in \mathcal{P}$ is called " vertical " iff it is tangent to the fiber. Since the group action is along the fibers (transforms a point into a point of the same fiber), a vertical vector expresses, at the point p, an infinitesimal action of an element of G. The fiber bundle structure determines unequivocally the vertical vector fields. The set of vertical vectors at $p \in \mathcal{P}$ is called the *vertical tangent space* V_p, the subspace of $T_p\mathcal{P}$ which contains the vertical vectors. Any vertical vector field v on $\mathcal{P}$, expressing an action of the group G, is called a *Killing vector field* relative to G. It is associated to an element $\hat{v}$ of the Lie algebra $\mathcal{G}$, by a correspondence called the *canonical isomorphism* between $\mathcal{G}$ and V_p.

3.5.2. *Connection in a principal bundle*

As manifolds, fiber bundles have forms defined on them. However, the connection in a fiber bundle has a special status. In a principal bundle $\mathcal{P}$, with structure group G, we generalize the notion of forms to operators which map the vector fields on $T\mathcal{P}$ into the Lie algebra $\mathcal{G}$. Such forms can be written as sums of factors of the type $A \otimes \omega \in \mathcal{G} \otimes T^*\mathcal{F}$, where ω is a scalar one-form of $\mathcal{P}$ and A an element of $\mathcal{G}$, the Lie algebra of G.

By definition, the *connection form* ω on $\mathcal{P}$ transforms a vertical vector field v into its image $\hat{v}$ through the canonical isomorphism. But this is not sufficient to define ω, since a general vector field on $\mathcal{P}$ is not vertical. The definition of the connection over $\mathcal{P}$ involves the choice of a horizontal subspace $H_p\mathcal{F}$ at each point $p \in \mathcal{P}$, complementary to $V_p\mathcal{F}$ in the tangent space $T_p\mathcal{P}$, so that the latter is decomposed as

$$T_p\mathcal{F} = H_p\mathcal{F} \oplus V_p\mathcal{F}.$$

The horizontal spaces at points of the same fiber are related by the (differential) group action. The choice is continuous.

By definition, $H_p\mathcal{F}$ is the kernel of ω: $\omega(v) = 0$ iff $v \in H_p\mathcal{F}$, so that the choice of $H_p\mathcal{F}$ is equivalent to the definition of the connection form. Any smooth vector field V can be decomposed uniquely as $V = horV + verV$. Although each vertical space is defined uniquely from the fiber bundle structure alone, the vertical component $verV$ of a vector X depends on the connection. More generally, $\omega(X) = (\hat{verX}); X, verX \in T\mathcal{F}; (\hat{verX}) \in \mathcal{G}$.

3.5.3. *The frame bundle*

Given a manifold $\mathcal{M}$, its frame bundle $F(\mathcal{M})$ is defined as a fiber bundle with basis $\mathcal{M}$. Its fiber at $p \in \mathcal{M}$ is the set of all frames at p, with structure group $GL(n)$. The frame bundle expresses most of the structure of the manifold: in fact, the connection, torsion and curvature on the manifold $\mathcal{M}$, although possibly defined in an intrinsic way on $\mathcal{M}$ (as described above), appear as natural objects in $F(\mathcal{M})$.

The solder form

A point p in the frame bundle $F(\mathcal{M})$ may be seen as a couple $p = (x, \{e_A\})$, where $x = \Pi^{-1}(p)$ is a point of the base manifold $\mathcal{M}$, and $\{e_A\}$ a frame (of $\mathcal{M}$) at x. Any vector $v \in T_pF(\mathcal{M})$ is projected to a vector $\Pi_*(v) = V^A\ e_A \in T_x\mathcal{M}$, with components V^A in the frame $\{e_A\}$. By definition, the *solder form*, or *canonical form* is the form $\{\theta^A\}$ with values in $\mathbb{R}^n$, defined by $\theta^A(v) = V^A$.

Projected by Π_*, a vector of $F(\mathcal{M})$ becomes a vector of $\mathcal{M}$. Thus, although the solder form is originally a form on $F(\mathcal{M})$, it can be also seen as a form on $\mathcal{M}$, mapping a vector to its components in the defined frame. In other words, it becomes the identity vector form, or the identity tensor of type (1,1) defined by $\theta = \theta^\alpha_\beta\ e_\alpha\ dx^\beta = \delta^\alpha_\beta\ e_\alpha\ dx^\beta = e_\alpha\ dx^\alpha$ or, equivalently, the set of 4 one-forms $\{\theta^\alpha = dx^\alpha\}$.

From the frame bundle to the manifold

The six scalar connection forms ω^A_B on a manifold $\mathcal{M}$ derive from the connection form ω defined on its frame bundle $F(\mathcal{M})$. Taking their values in $\mathcal{G}$, they express in fact the structure of the latter, i.e., as we will see, the symmetries (isotropies) of $\mathcal{M}$. Similarly, the forms dx^α on $\mathcal{M}$ are associated with the translations. The structure equations (1) above express, in $\mathcal{M}$, the commutation relations of the Lie algebra $\mathcal{G}$ acting on $\mathcal{M}$. Now, we will look at a procedure which allows to construct a space $\mathcal{M}$ which possesses some a priori chosen symmetries, expressed by the generators of a Lie algebra $\mathcal{G}$.

4. Relativity and the Poincaré group

4.1. THE CARTAN - WEYL CONSTRUCTION OF SPACE

The euclidean space $\mathbb{R}^3$ is the three dimensional space admitting maximal symmetries, those of its *isometry group*, the Euclid group

$$ISO(3) = \mathbb{R}^3 \times SO(3).$$

In four dimensions, the Minkowski spacetime is the Riemannian spacetime admitting all the *global* symmetries of the Poincaré group (see below).

It is possible to consider a space, or a spacetime, admitting a predetermined global (Lie) group of symmetries G, according to the following construction. Given G, the first task is to isolate its *isotropy group.* This is a subgroup G_0, containing no normal subgroup of G (a subgroup H of G is normal if $ghg^{-1} \in H$, for any $h \in H$ and $g \in G$). Generally, G_0 is chosen as having one fixed point, but this prescription can be modified, as we will see (a fixed point x of G_0 is such that $xg = x, \forall g \in G$). The isotropy groups of $ISO(3)$ is $SO(3)$. The isotropy groups of the Poincaré group is the *Lorentz group* $\mathcal{L} = SO(3,1)$.

The next step involves the quotient group G/G_0. Then the projection of G on G/G_0 defines a fiber bundle, with base space G/G_0, and fiber G_0. This fiber bundle is called an *elementary geometry* by Klein. Under certain conditions, the base space is identified with a manifold $\mathcal{M}$ (a space or a space-time). The fiber bundle appears as the frame bundle of $\mathcal{M}$. The connection on $\mathcal{M}$ is imposed by the structure of G.

Then it is possible to extend the structure of $\mathcal{M}$, by introducing a curvature associated to its connection : the global symmetries become *local*, in analogy with the gauge theory formalism. $\mathcal{M}$ becomes a Riemannian manifold. Applied to the Poincaré group $\mathcal{P}_0$, this construction leads first to Minkowski spacetime. Introducing a curvature leads to the Einstein's gravity, considered in the spirit of a gauge theory.

4.2. THE POINCARÉ GROUP

The *Poincaré group* is the Lie group of isometries of Minkowski spacetime. The generators of its Lie algebra are the four vectors P_μ for the *translations*, and the six (antisymmetrical) *Lorentz generators* $J_{\mu\nu}$ for the Lorentz rotations. They follow the commutating rules:

$$[P_\mu, P_\nu] = 0 \tag{5}$$

$$[J_{\mu\nu}, P_\rho] = \eta_{\nu\rho}\ P_\mu - \eta_{\mu\rho}\ P_\nu,$$

$$[J_{\mu\nu}, J_{\rho\sigma}] = \eta_{\nu\rho}\ J_{\mu\sigma} + \eta_{\mu\sigma}\ J_{\nu\rho} - \eta_{\mu\rho}\ J_{\nu\sigma} - \eta_{\nu\sigma}\ J_{\mu\rho}.$$

In these relations, $\mu, \nu = 0, 1, 2, 3$ are labels and not components.

It may be more convenient (but equivalent) to consider the *dual algebra*: to the four P_μ and six $J_{\mu\nu}$ are respectively associated the dual one-forms: four p^α and six j^α_β on G. These ten independent one-forms, on the ten dimensional Poincaré group, obey the ten structure equations of the Poincaré group:

$$dp^\alpha = p^\beta \wedge j^\alpha_\beta, \tag{6}$$

$$dj^\alpha_\beta = j^\gamma_\beta \wedge j^\alpha_\gamma. \tag{7}$$

Those follow immediately from the previous one, by duality, and using the general definition of the exterior derivative. Comparison with equ (1) show the absence of torsion.

4.3. FROM THE POINCARÉ GROUP TO MINKOWSKI SPACETIME

The isotropy group of $\mathcal{P}_0$ is the Lorentz group $\mathcal{L}$: it describes space-time rotations which have one fixed point, although space-time translations have no fixed point. The Poincaré group has a fiber bundle structure, with the Minkowski spacetime as a basis, obtained as the quotient of $\mathcal{P}_0/\mathcal{L}$. The fiber is the Lorentz group $\mathcal{L}$. This defines an *elementary geometry*, and $\mathcal{P}_0$ is identified to the bundle of orthonormal frames over Minkowski spacetime. An element of the Lorentz group transforms an orthonormal frame into an other.

To the generators of $\mathcal{P}_0$, translations and Lorentz rotations, are associated the one-forms on Minkowski spacetime: the vierbein (or tetrad) forms θ^α; and the connection forms ω^α_β, respectively. The relations (6,7) involve forms on the Poincaré group. They imply corresponding relations between forms on Minkowski spacetime:

$$d\theta^\alpha = \theta^\beta \wedge \omega^\alpha_\beta,$$

$$d\omega^\alpha_\beta = \omega^\gamma_\beta \wedge \omega^\alpha_\gamma.$$

The Minkowski spacetime is flat, with no curvature. Its structure can be generalized by imposing a curvature to the one forms mentioned, associated to the generators of $\mathcal{P}_0$: to the θ^α, can be associated the torsion two-forms T^α. To the ω^α_β are associated the *Riemann curvature* two-forms R^α_β. The structure equations of the manifold are then generalized as

$$d\theta^\alpha = \theta^\beta \wedge \omega^\alpha_\beta + T^\alpha,$$

$$d\omega^\alpha_\beta = \omega^\gamma_\beta \wedge \omega^\alpha_\gamma + R^\alpha_\beta.$$

This general framework includes Einstein relativity, obtained by putting the curvature associated to the translation generators, i.e., the torsion, to zero: the structure of the theory is expressed by the (Riemann) curvature associated to the generators of the Lorentz group. Its structure appears thus very similar to that of a gauge theory, where the Lorentz group would play the role of a gauge group.

4.4. THE TELEPARALLEL DESCRIPTION OF GENERAL RELATIVITY

It is very encouraging to consider Einstein relativity in a formalism analog to gauge theory of quantum field theory. This opens the door to fruitful comparisons and, hopefully, extensions. But only the Lorentz rotations are gauged here, and not the translations. Aldrovandi & Pereira (1996) have shown that gravitation can also be presented like a gauge theory for translations.

I recall first that, given a manifold $\mathcal{M}$, the choice of a metric g is completely equivalent to the choice of an orthonormal frame, and thus of a tetrad field h^a_μ, such that

$$g_{\mu\nu} = \eta_{ab}\ h^a_\mu\ h^b_\nu.$$

In turn, a metric g defines the *Levi-Civita connection* $\Gamma\ |_{LC}$, from the requirements to preserve the metric g, and to have no torsion. The so defined connection has for curvature the usual Riemann curvature, which gives an account of gravitation as indicated above. Strictly speaking, the curvature is associated to the connection and not to the manifold. However, since *all particles* feel this curvature, it is usually associated to space-time itself.

It is also possible to give an account of gravitation by a non Riemannian structure. Starting from the same manifold $\mathcal{M}$, one considers the same tetrad field h^a_μ, although not associated to a metric, or to orthonormal frame (which are not defined). One can construct a *linear Cartan connection* ∇, with coefficients defined through

$$\Gamma^\sigma_{\mu\nu} = h^\sigma_a\ \partial_\nu h^a_\mu.$$

No metric has been introduced. It is possible to show that the tetrad remains parallel, that there is a non zero torsion, but zero Riemann curvature. In fact,

$$\Gamma^\sigma_{\mu\nu} = \Gamma\ |^\sigma_{LC\ \mu\nu} + K^{\cdot\sigma}_{\mu\nu},$$

where K is the *contorsion* defined from the torsion T. It is possible to construct an action from the torsion T, whose variation leads to the general relativity equations. The latter appears with a structure similar to a gauge theory for the translations. Thus, the same structure (gravitation) is expressed equivalently by a Riemann curvature or by a torsion.

5. Extending

5.1. THE DE SITTER GROUPS

Aldrovandi and Pereira (1996) present arguments, linked to the possibilities of quantization and renormability, against the use of the Poincaré group in physics. They favor instead the use of the *de Sitter groups* $SO(3,2)$ or $SO(4,1)$. Those are the connected components of the groups of transformations of R^5 which respect the quadratic form

$$|x|^2 = +,-,-,-,-$$

or

$$|x|^2 = +,+,-,-,-.$$

They are the only possible isometry groups of maximally symmetric four-dimensional manifolds.

Batakis (1996) abandons the requirement of maximal symmetry. He introduces the generic ten-dimensional Lie group P, including $\mathcal{L}$, with ten generators: four P_a and the six M_{ab} of the Lorentz group. He writes the structure equations for the generators P_a under the most general form

$$[P_a, P_b] = C^c_{ab}\ P_c + C^{cc}_{ab}\ M_{cc}\ ,$$

the structure equations involving the Lorentz generators being unchanged. Then he shows that the structure constants C^c_{ab} must be those of the group $SU(2) \otimes U(1)$, and that P must be isomorphic to the de Sitter group !

The de Sitter groups form in fact a family depending on one parameter $\beta = 1/L$, which has the dimension of the inverse of a length (the case $\beta = 0$ corresponds to the special case of the Poincaré group). Compared to the Poincaré group, the de Sitter groups are characterized by the additional commutation relation concerning the translations

$$[P_a, P_b] = M_{ab},$$

The procedure above then leads to the manifolds $SO(3,2)/\mathcal{L}$ or $SO(4,1)/\mathcal{L}$ respectively, with structure equations:

$$T^a = de^a + \omega^a_b \wedge e^b,$$

$$R^{ab} = d\omega^{ab} + \omega^a_c \wedge \omega^{bc} + \beta e^a \wedge e^b.$$

Gauging the de Sitter groups brings new possibilities: Batakis considers the complexified gauging of P: this leads to a group G isomorphic to de Sitter group, and to the spontaneous broken $SU(2) \otimes U(1)$. However, it seems that the still larger conformal group may still bring even more success.

5.2. THE CONFORMAL GROUP

It is claimed that the geometry of the conformal group may lead to generalizations to quantum physics (Aldrovandi and Pereira, 1996, Wheeler, 1997b and d), to string theory (Wheeler, 1997a), the introduction of electroweak interactions (Batakis, 1996), and to the introduction of Higgs fields (Erdem, 1997).

The *conformal group* $\mathcal{C} = SO(4,2)$ has 15 generators:
- 6 Lorentz rotations,
- 4 translations,
- 4 inverse translations and
- 1 dilation.

It has interesting subgroups, like

– the Poincaré group $\mathcal{P}_0$, with six Lorentz rotations and four translations,
– the *Homogeneous Weyl group* (HW), with 7 generators, six Lorentz rotations + one dilation,
– the *Inhomogeneous Weyl group* IW with 11 generators = Homogeneous Weyl group + 4 translations

The *Weyl theory* (Weyl, 1921) for electromagnetism was constructed from the Weyl group, as

$$[\mathcal{P}_0 + dilation's]/[\mathcal{L} + dilation's].$$

Wheeler proposes a " biconformal model ", obtained from the conformal group over HW. Since HW has 7 dimensions, this leads to an 8-dimensional manifold $\mathcal{C}/HW$ to represent "spacetime".

5.3. THE GENERATORS OF CONFORMAL GROUP

The generators of the conformal group include those of the Poincaré group, which are convenient to use. The larger number of dimensions requires however more indexes. We write the list of generators as ($a, b = 1,2,3,4$ for spacetime).

– four solder forms ω_0^a, associated to translations,
– six connection forms ω_b^a, associated to Lorentz rotations,
– four co-solder forms ω_a^0, associated to inverse translations,
– one Weyl vector ω_0^0, associated to dilation's.

Formally, the structure equations can be written

$$d\omega_B^A = \omega_B^C \wedge \omega_A^C,$$

where the index A =1 15 spans all generators. More explicitly(with summation convention),

$$d\omega_b^a = \omega_b^c \wedge \omega_c^a + \omega_b^0 \wedge \omega_0^a + \eta_{bc}\eta^{ad}\omega_d^0 \wedge \omega_0^c + \Omega_b^a,$$

$$d\omega_0^a = \omega_0^0 \wedge \omega_a^0 + \omega_0^b \wedge \omega_b^a + \Omega_0^a,$$

$$d\omega_a^0 = \omega_0^a \wedge \omega_0^0 + \omega_a^b \wedge \omega_b^0 + \Omega_a^0,$$

$$d\omega_0^0 = \omega_0^a \wedge \omega_a^0 + +\Omega_0^0,$$

where Ω_b^a is the Riemann curvature, Ω_0^a the torsion, Ω_b^0 the co-torsion, and Ω_0^0 the dilation curvature. Jaekel and Reynaud (1997) consider as generators the conformal transformations to uniformly accelerated frames. From these generators, they define quantum observable associated to positions in spacetime. The Riemann spacetime of general relativity is obtained following the prescriptions

$$\Omega_0^a = \Omega_b^0 = \Omega_0^0 = 0.$$

The four-dimensional Weyl geometry is obtained from

$$\Omega_0^a = \Omega_b^0 = 0.$$

5.4. THE WHEELER BICONFORMAL THEORY

Although it has been proposed to consider the 11 dimensional group IW as the isotropy subgroup of $\mathcal{C}$, the procedure described above then leads to the same base manifold as the gauging of the Poincaré group seen before. But the cotranslations play no role and the extended symmetry (which lies in the fiber) leads to no new physics.

Wheeler (1997c, 1998) suggests the alternative choice of the 7 dimensional homogeneous Weyl group HW as isotropy subgroup of $\mathcal{C}$. To justify this choice, he considers the compactified Minkowski spacetime (= Minkowski space-time + 1 infinite point + its light cone + one ideal 2 spheres). The translations and inverse translations share the property to have one fixed point (at infinity and at origin respectively). The Lorentz rotations and the dilation, on the other hand, have two fixed points.

Thus Wheeler considers G/HW. It has eight dimensions, four associated with the translations X, and four associated with the co-translations Y, associated with the translations in momentum space. He thus considers this eight-dimensional space-time as the *phase space* for one particle. He shows that this formalism contains general relativity, Hamiltonian dynamics, and electromagnetism in the form of the Weyl theory, the usual problem of length scale of the latter being automatically resolved. In addition, he claims that this leads naturally to the Hilbert space structure of quantum physics.

The model is constrained to have no torsion and no dilation curvature, this latter requirement suppressing the scale problem of the Weyl theory. The space coordinates are noted x, the momentum coordinates y: they are the coordinates for the four additional dimensions of the generalized " spacetime ".

6. Conclusion

We refer the reader to the mentionned paper to find a complete exposition of the work presented here. Our intention in this lecture was to exhibit the convenience of the formalism based on fiber bundles and connections. As shown here, it expresses perfectly the structure of space-time in special or general relativity. It is also well known that it is the convenient formalism for gauge theories (see for instance O'Raifeartaigh L., 1997). Thus, it allows an unified view on these theories. In addition, the examples have shown its ability to be generalized, to incorporate new structure, new symmetries. Thus it seems to offer a very convenient framework for the search of unified theories.

References

Aldrovandi R. and Pereira J. G. 1996, SISSA preprint gr-qc/9610068

Batakis N. A. 1996, SISSA preprint hep-th/9605217

Erdem R. 1997, SISSA preprint hep-th/9701067

Jaekel M.-T. and Reynaud S. 1997, SISSA preprint quant-ph/9706036

O'Raifeartaigh L., The Dawning of Gauge theory, Princeton University Press 1997, and references therein

Weyl H. 1921, Nature, 106, 781; 1925, Ann. d. Physik. ,65, 541; 1929, Zeit. f. Physik, 56, 330

Wheeler J. T. 1997a, Strings without strings, SISSA preprint hep-th/9706209

Wheeler J. T. 1997b, Why Quantum Mechanics is Complex, SISSA preprint hep-th/9708088

Wheeler J. T. 1997c, Normal Biconformal Spaces, SISSA preprint hep-th/9706215

Wheeler J. T. 1997d, Quanta Without Quantization, Mod. Phys. Lett. A12 (1997) 2175-2182 (hep-th/9705235)

Wheeler J. T. 1998, New Conformal Gauging and the Electromagnetic Theory of Weyl, J. Math. Phys. 39, 299-328 (hep-th/9706214)

GRAVITATIONAL LENSES

FRANCIS BERNARDEAU
Service de Physique Théorique, C.E. de Saclay
F-91191 Gif-sur-Yvette Cedex, France

1. Introduction

Gravitational lenses are becoming a precious mean of probing the matter distribution in the Universe. From the search of matter in the Galactic halo to the study of the large-scale structures of the Universe, the gravitational lens effects offer a unique alternative to light surveys and are now widely used. This evolution is due in particular to the use of new observation devices, such as the wide field CCD cameras.

The aim of this course is to study the effects of gravitational lenses in those different astrophysical contexts. These notes are voluntarily focused on the fundamental mechanisms and the basic concepts that are useful to describe these effects. The observational consequences will be presented in more details in these proceedings by Y. Mellier (see also his review paper 1998). Related textbooks are,

- "Gravitation" by Misner, Thorne and Wheeler (1973) for General Relativity and in particular for the presentation of the geometric optics.
- "Large-scale Structures of the Universe" by Peebles (1993) for the description of the large-scale structures of the Universe.
- "Gravitational lenses" by Schneider, Ehlers & Falco (1992) for a general (but rather mathematical) exhaustive presentation of the lens physics.

The content of these notes is the following. In the first section I describe of the basic mechanisms of gravitational lenses, techniques and approximations that are usually employed. The second section is devoted to the case of a very simple deflector, a point-like mass distribution. This corresponds to microlensing events in which the deflectors are compact objects of a fraction of a solar mass that may populate the halo of our Galaxy. The last two sections are devoted to cosmological applications. After a presentation of

M. Lachièze-Rey (ed.), Theoretical and Observational Cosmology, 179–209.

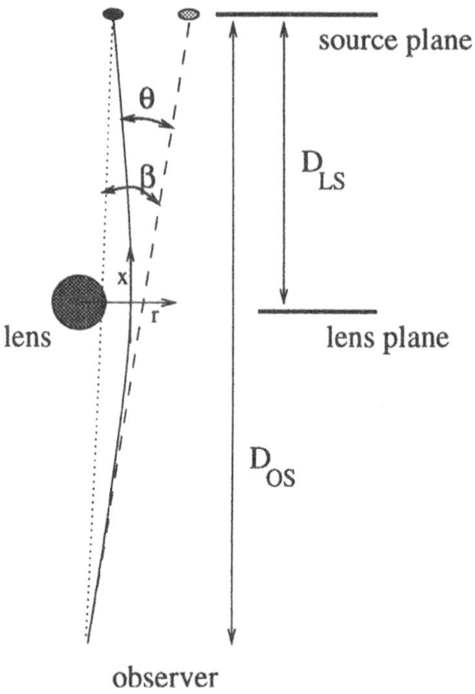

Figure 1. The geometrical relationship between the deflection angle θ and the displacement angle β.

the geometrical quantities that are specific to cosmology, I will present the various phenomena that can be observed in this context. Finally I describe the weak lensing regime. This is a rapidly developing area that should eventually allow us to map the mass distribution in the Universe. I explore how this can be used to constrain the cosmological parameters.

2. Physical mechanisms

The physical mechanisms of gravitational lenses are well known since the foundation of General Relativity. Any mass concentration is going to deflect photons that are passing by with a fraction angle per unit length, $\delta\theta/\delta s$, given by

$$\frac{\delta\vec{\theta}}{\delta s} = -2\,\vec{\nabla}_x \frac{\phi}{c^2} \tag{1}$$

where the spatial derivative is taken in a plane that is orthogonal to the photon trajectory and ϕ is the Newtonian potential[1].

[1] We will see in section 4 what is its meaning in a cosmological context.

2.1. BORN APPROXIMATION AND THIN LENS APPROXIMATION

In practice, the total deflection angle is at most about an arcmin. This is the case for the most massive galaxy clusters. It implies that in the subsequent calculations it is possible to ignore the bending of the trajectories and calculate the lens effects as if the trajectories were straight lines. This is the Born approximation.

Eventually, one can do another approximation by noting that in general the deflection takes place along a very small fraction of the trajectory between the sources and the observer. One can then assume that the lens effect is instantaneous and is produced through the crossing of a plane, the lens plane. This is the thin lens approximation.

2.2. THE INDUCED DISPLACEMENT

The direct consequence of this bending is a displacement of the apparent position of the background objects. This apparent displacement depends on the distance of the source plane, D_{OS}, and on the distance between the lens plane and the source plane D_{LS}. More precisely we have (see Fig. 1),

$$\vec{\beta} = \vec{\alpha} - \frac{2}{c^2} \frac{D_{LS}}{D_{OS}\, D_{OL}} \vec{\nabla}_\alpha \left(\int \mathrm{d}s\ \phi(s, \alpha) \right) \tag{2}$$

where $\vec{\alpha}$ is the position in the image plane, $\vec{\beta}^S$ is the position in the source plane. The gradient is taken here with respect to the angular position (this is why a D_{OL} factor appears). The total deflection is obtained by an integration along the line of sight, assuming the lens is thin. In a cosmological context the exact expressions of the angular distances are not trivial, they depend on the local curvature of the background.

3. The case of a point-like mass distribution

3.1. MULTIPLE IMAGES AND DISPLACEMENT FIELD

The potential of a point-like mass distribution is given by,

$$\phi(r) = \frac{-G\,M}{r}, \tag{3}$$

for an object of mass M. Let me calculate the instantaneous deflection angle at an apparent distance r. We suppose that the impact parameter of the trajectory is r and $\mathbf{x}$ is the abscissa to the point of the trajectory that is the closest to the lens (see Fig. 1). Along the trajectory the potential is given by,

$$\phi(x) = \frac{-G\,M}{\sqrt{r^2 + x^2}}. \tag{4}$$

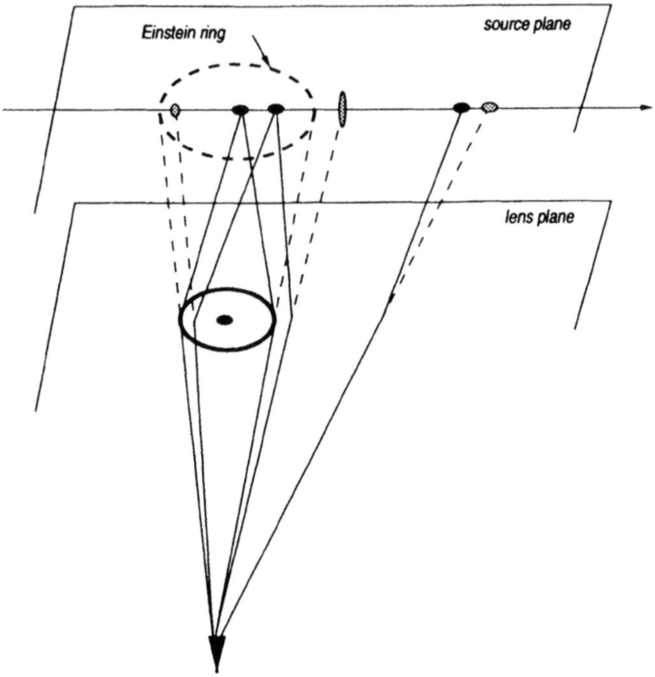

Figure 2. Position of the Einstein ring for a point-like mass distribution.

Then the deflecting angle is given by,

$$\frac{\delta\theta}{\delta x} = -2\,\frac{G\,M\,r}{c^2\,(r^2+x^2)^{3/2}}. \tag{5}$$

The total deflection angle θ is given by the result of the integration of this quantity with respect to x. It gives,

$$\theta = \frac{4\,G\,M}{r\,c^2}. \tag{6}$$

This is a well known expression which served as the first test of General Relativity with the observed displacement of stars around the solar disc.

It implies that the true position of an object on the sky, β, is related to its apparent position, α, with

$$\vec{\beta} = \vec{\alpha} - \frac{R_E^2}{\alpha^2}\,\vec{\alpha} \tag{7}$$

where $\vec{\beta}$ and $\vec{\alpha}$ are 2D angular position vectors and R_E is the Einstein radius,

$$R_E = \sqrt{\frac{4\,G\,M}{c^2}\,\frac{D_{LS}}{D_{OS}\,D_{OL}}}. \tag{8}$$

One can see that when $\alpha^2 = R_E^2$ the lens and the background objects are necessarily aligned. It implies that, since the optical bench is symmetric around its axis, the observed object appears as a perfect ring (see Fig. 2). It is worth noting that for this potential, except for this particular position, all background objects have two images. This is however quite specific to a point like mass distribution which has a singular gravitational potential.

The problem is that none of these features are observable when the lens is a star. Let for example assume that we have a one solar mass star in the halo of our galaxy (therefore at a distance of about 30 kpc). The apparent size of such a star is about 10^{-8} arcsec. Its Einstein ring is about 10^{-4} arcsec[2]. None of these dimensions are accessible to the observations (the angular resolution of telescope is at best a few tens of arcsec). The Einstein radius is therefore much too small to be actually seen!

Note however that these numbers show that the point-like approximation is entirely justified for a star (the Einstein ring is much more bigger that the apparent size of a star). Simple examination of the scaling in those relations shows that this would not be true for massive astrophysical objects such as galaxies or galaxy clusters.

The detection of gravitational effects due to stars should then be done by another mean: the amplification effect.

3.2. THE AMPLIFICATION MATRIX

The case of circular lenses has already given us a clue: when the source is precisely aligned with the lens, the image is no more a point but a circle. One consequence is that the observed total luminosity is much larger than what would have been observed without lenses. The effect is basically due to the variations of the displacement field with respect to the apparent position. These variations induce a change of both the size and shape of the background objects. To quantify this effect one can compute the amplification matrix $\mathcal{A}$ which describes the linear change between the source plane and the image plane,

$$\mathcal{A} = \left(\frac{\partial \alpha_i}{\partial \beta_j}\right). \tag{9}$$

Its inverse, $\mathcal{A}^{-1}$, is actually directly calculable in terms of the gravitational potential. It is given by the derivatives of the displacement with respect to the apparent position,

$$\mathcal{A}^{-1} \equiv \frac{\partial \beta_i}{\partial \alpha_j} = \delta_{ij} - 2\frac{D_{LS}}{D_{OS}\, D_{OL}}\phi_{,ij}. \tag{10}$$

[2]To do this calculation it is useful to know that the horizon of a one solar mass black hole, $r = 2\,G\,M_{\odot}/c^2$, is about 3 km.

In case of a point-like mass distribution it is easy to see that,

$$\mathcal{A}^{-1} = \left(\delta_{ij}\left[1 - \frac{R_E^2}{\alpha^2}\right] + \frac{\alpha_i \alpha_j}{\alpha^2} R_E^2\right). \tag{11}$$

The amplification effect for each image is given by the inverse of the determinant of the amplification matrix computed at the apparent position of the image. The amplification factor is usually noted μ,

$$\mu = 1/\det(\mathcal{A}^{-1}). \tag{12}$$

In case of the point-like distribution we have (the calculation is simple at the position $\alpha_1 = \alpha,\ \alpha_2 = 0$),

$$\mu = \left|\frac{\alpha^4}{\alpha^4 - R_E^4}\right|, \tag{13}$$

for each image. The *total* amplification effect is given by the summation of the two effects for the 2 images,

$$\mu_{\rm tot} = \frac{u^2 + 2}{u(u^2 + 4)^{1/2}} \quad \text{with} \quad u = \frac{b}{R_E}, \tag{14}$$

where b is the impact parameter of the background object in the source plane. The amplification effect is obviously dependent on the impact parameter. If it is changing with time, this effect is detectable.

3.3. THE MICROLENSING EXPERIMENTS

The microlensing experiments are based on this effect. When a compact object of the halo of our galaxy reaches, because of its proper motion, the vicinity of the light path of a background star (from the SMC or the LMC) the impact parameter is changing with time and can be small enough to induce a detectable amplification (when u is about unity, the amplification is about 30%). In practice one observes changes in the magnitude of the remote stars that obey specific properties,

- the time dependence of the amplification is symmetric and has a specific shape;
- the amplification effect is unique;
- the magnitude of the amplification effect is the same in all wavelengths.

The time scale of such an event is about a few days to a few month depending on the mass of the deflectors. Currently a fair number of such events have been recorded (see contribution of J. Rich, these proceedings) and constraints on the content of our halo with low massive compact objects have been put.

4. Gravitational lenses in Cosmology

The extension of the lens equations to a cosmological context raises some technical difficulties because the background in which the objects are embedded is not flat. The aim of this section is to clarify these points. However, readers that are not familiar with cosmology can jump to section 5.

The basic equations, that describe jointly the evolution of the expansion parameter and the mean density, are the following,

$$3\,\ddot{a} = -4\pi G\,\rho\,a + \Lambda\,a; \tag{15}$$

$$\dot{a}^2 = \frac{1}{a}\left[\frac{8\pi\,G}{3}\rho\,a^3 - k\,a + \frac{\Lambda\,a^3}{3}\right]. \tag{16}$$

where Λ is a possible cosmological constant and k a possible curvature term. The Hubble constant reads,

$$H = \frac{\dot{a}}{a} = \sqrt{\frac{8\pi\,G}{3}\rho - \frac{k}{a^2} + \frac{\Lambda\,a^3}{3}}. \tag{17}$$

To simplify the discussions the reduced quantities are introduced,

$$\Omega = \frac{8\pi\,G}{3\,H^2}\rho \quad \text{and} \quad \lambda = \frac{\Lambda}{3\,H^2}. \tag{18}$$

They have an index 0 when they are taken at present time.

4.1. THE ANGULAR DISTANCES

We consider an object of size l (either because of its proper size or because of a peculiar physical process) at redshift z. When this size l is seen under an angle α, then by definition,

$$\alpha = \frac{l}{\mathcal{D}_0}, \tag{19}$$

where $\mathcal{D}_0$ is the angular distance. This is the distance at which this object would be in an Euclidean metric. What is then the relationship between $\mathcal{D}_0$ and z? The Friedmann-Robertson-Walker metric is given by,

$$\mathrm{d}s^2 = -c^2\,\mathrm{d}t^2 + a^2(t)\left(\frac{\mathrm{d}x^2}{1 - k\,x^2} + x^2\mathrm{d}\theta^2 + x^2\,\sin^2\theta\,\mathrm{d}\varphi^2\right). \tag{20}$$

The fact that the size of this object is l means that it takes a time interval $dt = l/c$ for light to travel to one end to the other. The corresponding angle $\alpha = d\theta$ can be obtained by writing $ds = 0$ with $dx = 0, d\varphi = 0$, which gives

$$l = a(t)\, x\, \alpha. \tag{21}$$

The angular *comoving* distance is thus given by $\mathcal{D}_0(z) = x(z)$. The expression of $x(z)$ can then be computed by the relation $ds = 0$ along the line of sight with $d\theta = 0$ and $d\varphi = 0$,

$$\int_{t_0}^{t_1} \frac{c\,dt}{a} = \int_0^{x(z)} \frac{dx}{\sqrt{1-kx^2}} \equiv \chi(z). \tag{22}$$

For an open Universe, $k < 0$, and we have,

$$x(z) = \mathcal{D}_0(z) = \frac{1}{\sqrt{-k}} \sinh[\sqrt{-k}\; \chi(z)]. \tag{23}$$

Obviously when $k = 0$, $x = \chi$.

The relation $\chi(z)$ depends on the function $a(t)$ for a given cosmology, and therefore on the matter content of the Universe, on Λ and k . For instance, for an Einstein-de Sitter Universe, in the matter dominated era, we have

$$a(t) = \left(\frac{t}{t_0}\right)^{2/3}, \tag{24}$$

so that,

$$x(z) = \int_{t_0}^{t_1} t^{-2/3} dt = 3(t_0^{1/3} - t_1^{1/3}) t_0^{2/3}, \quad H_0 = \frac{2}{3\, t_0}, \tag{25}$$

and eventually,

$$\mathcal{D}_0(z) = \chi(z) = \frac{c}{H_0}\left(2 - \frac{2}{\sqrt{1+z}}\right). \tag{26}$$

This is the comoving angular distance for an Einstein-de Sitter Universe. More generally we have an explicit solution if $\Lambda = 0$ only. Finally the lens equation also requires the angular distance between two different redshifts z_1 and z_2. The calculation is actually quite simple. This distance is given formally by $\mathcal{D}_0(z_2)$ when it is calculated at a time when the observer is at redshift z_1. k being time independent we should formally have,

$$\mathcal{D}_0(z_1, z_2) = \frac{1}{\sqrt{-k}} \sinh[\sqrt{-k}\; (\chi(z_2, z_1))], \tag{27}$$

and to compute $\chi(z_2, z_1)$ one only needs to remark,

$$\chi(z_2, z_1) \equiv \int_{t_1}^{t_2} \frac{c\,\mathrm{d}t}{a} = \chi(z_2) - \chi(z_1), \tag{28}$$

which gives the expression of the angular distance we need. Eventually, it is fruitful to notice that,

$$\frac{c\,\mathrm{d}t}{a} = -\frac{c\,\mathrm{d}z}{H}, \tag{29}$$

which gives,

$$\mathcal{D}_0(z_1, z_2) = \frac{c}{H_0\sqrt{1-\Omega_0-\lambda_0}} \sinh\left[H_0\sqrt{1-\Omega_0-\lambda_0} \int_{z_1}^{z_2} \frac{\mathrm{d}z}{H(z)}\right], \tag{30}$$

and

$$\mathcal{D}_0(z) = \frac{c}{H_0\sqrt{1-\Omega_0-\lambda_0}} \sinh\left[H_0\sqrt{1-\Omega_0-\lambda_0} \int_{0}^{z} \frac{\mathrm{d}z'}{H(z')}\right]. \tag{31}$$

The whole geometrical part of the lens equation is thus established.

4.2. GEOMETRIC OPTICS IN A WEAKLY INHOMOGENEOUS UNIVERSE

What is now the source term for the deflection angle? We should first notice that in absence of lenses the light rays follow the geodesics of the Friedmann-Robertson-Walker metric. And in the applications we are interested in, the metric fluctuations are always weak. These fluctuations are given by, $G\,M/(R\,c^2)$. For instance, for

- 1 star: $M = 1\,M_\odot$, $R = 7\,10^5$km, $\delta\phi \approx 10^{-6}$;
- 1 galaxy cluster: $M = 10^{15}\,M_\odot$, $R = 1$Mpc $= 3\,10^{19}$km, $\delta\phi \approx 10^{-5}$.

The metric inhomogeneities are thus always extremely weak, even in the most extreme cosmological situations.

Following[3] Sachs (1961), we consider two nearby geodesics, $\mathcal{L}$ and $\mathcal{L}'$, in a light bundle in an FRW Universe with small metric fluctuations. We denote α_i the bi-dimensional angular distance between $\mathcal{L}$ and $\mathcal{L}'$ as it is seen by the observer. This is the distance in the image plane, that is the difference between the angular coordinates with which the photons arrive. We denote $\xi_i(z)$ the real distance between $\mathcal{L}$ and $\mathcal{L}'$ at redshift z (see Fig. 3). It implies that the geodesics are straight enough so that light always travels towards the observer. We also assume that the deflections are small enough so that it is possible to make the small angle approximation,

$$\xi_i(z) = a\,\mathcal{D}_{ij}(z)\,\alpha_j, \tag{32}$$

[3]See Misner, Thorne and Wheeler for an exhaustive presentation of the geometric optics.

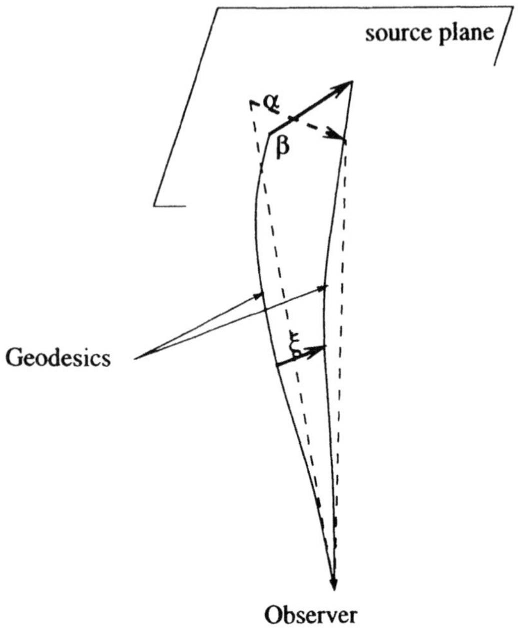

Figure 3. Sketch showing the geometrical quantities that appear in Eq. 32

that is that we assume the the position vector ξ_i can be obtained by a simple linear transform of the angular coordinates. For an homogeneous space $\mathcal{D}_{ij}(z)$ is simply given by $\mathcal{D}_0(z)\,\delta^K_{ij}$ where δ^K_{ij} is the Kronecker symbol. Obviously $\mathcal{D}_{ij}$ changes as a function of redshift along the trajectories. The "virtual" angular position in the source plane is then given by the ratio of the real distance (at time of light emission for instance) by the angular distance of the emitter in an homogeneous space,

$$\vec{\beta} = \frac{\vec{\xi}(z)}{a\mathcal{D}_0(z)}. \tag{33}$$

The amplification matrix, or rather its inverse, $\mathcal{A}^{-1}$, is then given by,

$$\mathcal{A}^{-1}(z) = \frac{\mathcal{D}_{ij}(z)}{a\mathcal{D}_0(z)}. \tag{34}$$

for a source plane at redshift z.

Sachs (1961) gave the master equation which governs the evolution of the distance between the geodesics. The derivation of this equation goes beyond these lecture notes and I give only the final answer (as it has been given by Seitz, Schneider and Ehlers 1994),

$$\frac{\mathrm{d}^2[a\ \mathcal{D}_{ij}(\vec{\beta}, z)]}{\mathrm{d}\eta^2} = a(z)\mathcal{R}_{ik}(\vec{\beta}, z)\ \mathcal{D}_{kj}(\vec{\beta}, z) \tag{35}$$

where the derivatives are taken with respect to η,

$$\mathrm{d}\eta = -\frac{\mathrm{d}a}{H(a)} = -a\,\mathrm{d}t, \quad \eta(z=0) = 0. \tag{36}$$

with the boundary conditions,

$$(\mathcal{D}_{ij})_{z=0} = 0; \quad \left(\frac{\mathrm{d}\mathcal{D}_{ij}}{\mathrm{d}\eta}\right)_{z=0} = \frac{c}{H_0}. \tag{37}$$

The matrix $\mathcal{R}_{ij}$ represents the tidal effects. It can by written in terms of the gravitational potential ϕ given by,

$$\Delta_x\phi = 4\pi\, G\,\overline{\rho}\, a^2\, \delta_{\mathrm{mass}}. \tag{38}$$

The Laplacian is taken with respect of the comoving angular distances. We have

$$\mathcal{R}_{ij} = -\frac{4\pi\, G\,\overline{\rho}}{H_0^2\, a^2}\begin{pmatrix} 1 & 0 \\ 0 & 1 \end{pmatrix} - \frac{2}{H_0^2\, a^2}\begin{pmatrix} \phi_{,11} & \phi_{,12} \\ \phi_{,21} & \phi_{,22} \end{pmatrix}. \tag{39}$$

Since $8\pi\, G\,\overline{\rho}\, a^3 = 3H_0^2\,\Omega_0$, (e.g. Eq. 18) for an homogeneous Universe we have,

$$\mathcal{R}_{ij}^{(0)} = -\frac{3}{2}\,(1+z)^5\,\Omega_0\,\delta_{ij}^K \tag{40}$$

(the superscript (0) means here that it is the value of $\mathcal{R}$ without perturbations). In this case the matrix $\mathcal{D}_{ij}$ is proportional to δ_{ij}^K and we have,

$$\frac{\mathrm{d}^2[a\,\mathcal{D}_0(z)]}{\mathrm{d}\eta^2} = -\frac{3}{2}(1+z)^4\,\Omega_0\,\mathcal{D}_0(z). \tag{41}$$

We recover in fact the comoving angular distance the expression of which we know,

$$\mathcal{D}_0(z) = \frac{c}{H_0\sqrt{1-\Omega_0-\lambda_0}}\,\sinh\left[\sqrt{1-\Omega_0-\lambda_0}\int_0^z \frac{\mathrm{d}z'}{E(z')}\right], \tag{42}$$

with

$$E(z) = \frac{H(z)}{H_0} = \sqrt{\lambda_0 + (1+z)^2(1-\Omega_0-\lambda_0) + (1+z)^3\Omega_0}. \tag{43}$$

This integral has a closed form only when the cosmological constant, Λ, is zero.

4.3. THE LINEARIZED EQUATION OF GEOMETRIC OPTICS

We can remark that the equation (35) is not linear since $\mathcal{D}$ is not simply proportional to $\mathcal{R}$. This expresses the fact that the deformation of the angular distance is made all along the light trajectory by multiple deflections.

The general resolution of Eq. (35) is in general very complicated. It can however handled when it is linearized. Let's assume we can expand $\mathcal{D}_{ij}$ with respect of the local density contrast,

$$\mathcal{D}_{ij}(z) = \mathcal{D}_0 + \mathcal{D}_{ij}^{(1)} + \dots \tag{44}$$

It implies that, at first order,

$$\frac{\mathrm{d}^2[a\ \mathcal{D}_{ij}^{(1)}(\vec{\beta}, z)]}{\mathrm{d}\eta^2} - a(z)\mathcal{R}_{ik}^{(0)}(\vec{\beta}, z)\mathcal{D}_{kj}^{(1)}(\vec{\beta}, z) = -3\Omega_0(1+z)^4\mathcal{D}_0(z)\varphi_{,ij}(\vec{\beta}, z) \tag{45}$$

with,

$$\left(\mathcal{D}_{ij}^{(1)}\right)_{z=0} = 0, \quad \left(\frac{\mathrm{d}\mathcal{D}_{ij}^{(1)}}{\mathrm{d}\eta}\right)_{z=0} = 0, \tag{46}$$

and we define the field φ so that,

$$\Delta_x\varphi = \delta_{\mathrm{mass}}(\vec{\beta}, z) = \frac{\Delta_x\phi}{4\pi G\,\overline{\rho}\,a^2}. \tag{47}$$

To solve this differential equation it is easier to write it with the variable z. It then reads,

$$\begin{aligned}\frac{\mathrm{d}^2[a\ \mathcal{D}_{ij}^{(1)}(\vec{\beta}, z)]}{\mathrm{d}z^2} + \frac{1}{E(z)}\frac{\mathrm{d}E(z)}{\mathrm{d}z}\frac{\mathrm{d}\mathcal{D}_{ij}^{(1)}(\vec{\beta}, z)}{\mathrm{d}z} - \\ -\frac{1}{1+z}\frac{1}{E(z)}\frac{\mathrm{d}E(z)}{\mathrm{d}z}\mathcal{D}_{ij}^{(1)}(\vec{\beta}, z) + \frac{3}{2}\frac{\Omega_0(1+z)}{E^2(z)}\mathcal{D}_{ij}^{(1)}(\vec{\beta}, z) = \\ -3\frac{\Omega_0(1+z)}{E^2(z)}\mathcal{D}_0(\vec{\beta}, z)\varphi_{,ij}(\vec{\beta}, z).\end{aligned} \tag{48}$$

The differential homogeneous equation it is associated with has two known solutions. One describes the angular distance $\mathcal{D}_0$, the other is given by,

$$\mathcal{U}_0(z) = \frac{1}{\sqrt{1-\Omega_0-\lambda_0}}\cosh\left[\sqrt{1-\Omega_0-\lambda_0}\int_0^z\frac{\mathrm{d}z'}{E(z')}\right], \tag{49}$$

The general solution of 48 then reads,

$$\mathcal{D}_{ij}^{(1)}(\vec{\beta}, z) = -3\Omega_0\int_0^z\mathrm{d}z'\frac{(1+z')\mathcal{D}_0(z')\varphi_{,ij}(z')}{E^2(z)}\frac{\mathcal{U}_0(z)\mathcal{D}_0(z') - \mathcal{U}_0(z')\mathcal{D}_0(z)}{\mathcal{U}_0'(z')\mathcal{D}_0(z') - \mathcal{U}_0(z')\mathcal{D}_0'(z')}, \tag{50}$$

which, after elementary mathematical transforms, gives,

$$\mathcal{D}^{(1)}_{ij}(\vec{\beta}, z) = -3\,\Omega_0 \int_0^z \frac{\mathrm{d}z'}{E(z)} \frac{1}{\sqrt{1-\Omega_0-\lambda_0}} \times$$
$$\times \sinh\left[\sqrt{1-\Omega_0-\lambda_0} \int_z^{z'} \frac{\mathrm{d}z''}{E(z'')}\right] (1+z')\mathcal{D}_0(z')\varphi_{,ij}(z'). \qquad (51)$$

It can be rewritten by introducing the physical distance χ along the line of sight. We eventually have,

$$\mathcal{A}^{-1}(z) = \mathrm{Id} - \frac{3\,\Omega_0}{(c/H_0)^2} \int_0^{\chi(z)} \mathrm{d}\chi' \frac{\mathcal{D}_0(z', z)\mathcal{D}_0(z')}{\mathcal{D}_0(z)} (1+z')\varphi_{,ij}(z'), \qquad (52)$$

where the angular distances $\mathcal{D}_0(z)$ and $\mathcal{D}_0(z, z')$ are comoving. This equation actually gives the expression of the amplification matrix for a non-trivial background. We find that the amplification matrix is given by the superposition of lens effects of the different mass layers. We can remark that the lens term is given by the gravitational potential, ϕ, that is by the potential the source term of which is given by the density contrast.

Note finally that this equation is valid in two limit cases, either for a single lens plane with an arbitrary strength or the superposition of any number of weak lenses. This equation naturally extents the previous result, (10), obtained for a single lens in an Euclidean background. The higher orders of Eq. (35) give the intrinsic lens coupling effects (i.e. their nonlinear parts). We will not consider them here.

5. Galaxy clusters as gravitational lenses

The study of galaxy clusters has become a very active field since the discovery of the first gravitational arc by Soucail et al. (1988) in Abell cluster A370. Galaxy clusters give the most dramatic example of gravitational lens effects in a cosmological context. The difficulty is however to describe the shape of their mass distribution.

5.1. THE ISOTHERMAL PROFILE

For an isothermal profile we assume that the local density $\rho(r)$ behaves like,

$$\rho(r) = \rho_0 \left(\frac{r}{r_0}\right)^{-2}. \qquad (53)$$

With such a density profile the total mass is not finite. So this is not a realistic description but it is a good starting point for the central part of clusters. It is actually more convenient to parameterize the depth of a potential well with the velocity dispersion it induces. The velocity dispersion is due to the random velocity that particles acquire when they reach a sort of thermal equilibrium. Such a dispersion is in principle measurable with the observed galaxy velocities along the line of sight. The velocity dispersion is related to the mass $M(<r)$ of the potential well that is included within a radius r,

$$\sigma^2(r) \sim \frac{G\,M(<r)}{r}. \tag{54}$$

In case of a isothermal profile, the velocity dispersion is *independent* of the radius and we have

$$\sigma^2 = 2\pi\,G\,\rho_0\,r_0^2. \tag{55}$$

The integrated potential along the line of sight is given by,

$$\varphi(r) = 2\pi\,\sigma^2\,r. \tag{56}$$

As a consequence the amplitude of the displacement is independent of the distance to the cluster center and

$$\vec{\beta} = \vec{\alpha} - \frac{4\pi}{c^2}\frac{D_{LS}}{D_{OS}}\sigma^2\,\frac{\vec{\alpha}}{\alpha}. \tag{57}$$

The position of the Einstein ring is obviously given by,

$$\alpha = R_E = \frac{4\pi}{c^2}\frac{D_{LS}}{D_{OS}}\sigma^2, \tag{58}$$

which depends both on the velocity dispersion and on the angular distances. The number of images depends in this case on the value of the impact parameter. If it is too large (i.e. larger than R_E) then each background object has only one image. For a galaxy cluster of a typical velocity dispersion of 500 km/s, and for a source plane situated at twice the distance of the lens, the size of the Einstein ring is about 0.5 arcmin. It is interesting to note that the size of the Einstein ring is directly proportional to the square of velocity dispersion (in units of c^2) and to the ratio D_{LS}/D_{OS}.

The amplification matrix reads,

$$\mathcal{A}^{-1} = \begin{pmatrix} 1 & 0 \\ 0 & 1-\frac{1}{x} \end{pmatrix}, \tag{59}$$

where we have,

$$x = \frac{r}{R_E}. \tag{60}$$

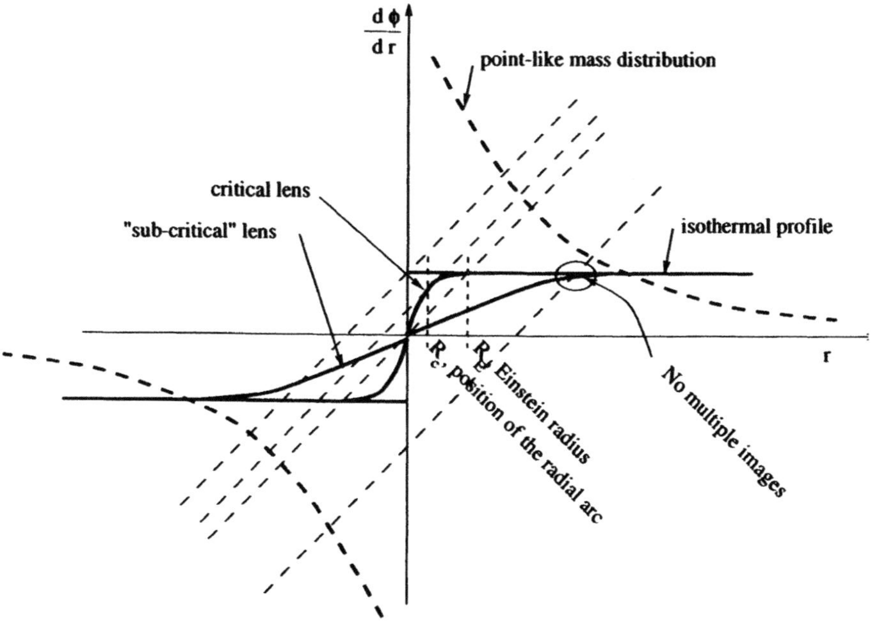

Figure 4. Graphical determination of the position and number of images from the shape of the potential.

As a result the amplification is given by,

$$\mu = \frac{x}{1-x}. \tag{61}$$

Once again the amplification becomes infinite when $x \to 1$ that is, close to the critical line.

5.2. THE CRITICAL LINES FOR A SPHERICALLY SYMMETRIC MASS DISTRIBUTION

The two previous cases correspond to specific profiles. In this part I only assume a spherical symmetric profile. The displacement is then given by the derivative of the potential,

$$\vec{\beta} = \vec{\alpha} - \frac{d\varphi}{dr}\frac{\vec{\alpha}}{\alpha}. \tag{62}$$

It is interesting to visualize this relation with a graphic representation. This is proposed in Fig. 4. The number and position of the images of a given background object are given by the number of intersection points between the curve and a straight line of slope unity. This is a direct consequence of the relation,

$$b - a = -\frac{\mathrm{d}\varphi(a)}{\mathrm{d}a} \tag{63}$$

when the potential is computed along a given axis that crosses the cluster through the center and a and b are the abscissa on this axis of one given object in respectively the source and the image plane.

The interesting quantity is also the amplification matrix that indicates the position of the critical lines. In general this matrix reads,

$$\mathcal{A}^{-1} = \begin{pmatrix} 1 - \frac{\partial^2\varphi}{\partial r^2} & 0 \\ 0 & 1 - \frac{1}{r}\frac{\partial\varphi}{\partial r} \end{pmatrix}, \tag{64}$$

when it is written in the basis $(\vec{e}_r, \vec{e}_\theta)$. Then the amplification is infinite in two cases, when

$$\frac{\partial^2\varphi}{\partial r^2} = 1 \quad \text{or} \quad \frac{1}{r}\frac{\partial\varphi}{\partial r} = 1. \tag{65}$$

The second eigenvalue corresponds to the same case as for a singular isothermal profile. At this particular position the source forms an Einstein ring. The first eigenvalue, however, is associated with an eigenvector that is along the x direction, that is along the radial direction. It means that the "arc" which is thus formed is radial. It graphically corresponds to the case of two merging roots. It is therefore directly associated with the behavior of the potential near the origin.

5.3. THE ISOTHERMAL PROFILE WITH A CORE RADIUS

Let us consider a simple case where the projected potential is made regular near the origin,

$$\varphi(r) = \varphi_0\sqrt{1 + (r/r_c)^2} \tag{66}$$

The constant φ_0 is related to the velocity dispersion with

$$\varphi_0 = \frac{4\pi\sigma^2}{c^2}\frac{D_{LS}\,D_{OL}}{D_{OS}\,r_c} \tag{67}$$

where σ is here the velocity dispersion at a radius much larger that r_c (the velocity dispersion decreases to zero at the origin in this model). This is a more realistic case. It is interesting to note that in this case the potential is not necessarily critical (there may be no region of multiple images, see Fig. 4). When it is critical the discovery of a radial arc is an extremely precious indication for the value of the core radius.

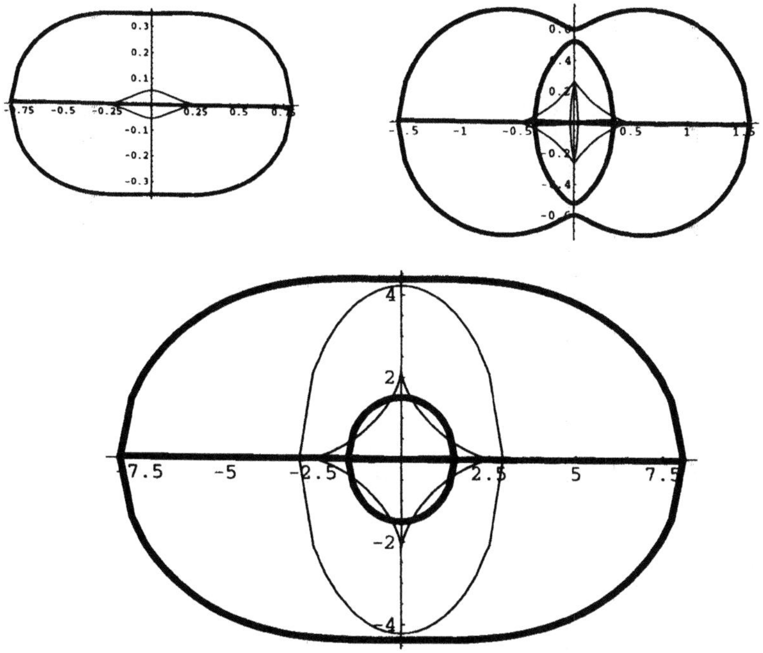

Figure 5. Shape of the caustic lines (thick lines) and critical lines (thin lines) for an elliptic potential and for different values of the central potential.

5.4. CRITICAL LINES AND CAUSTICS IN REALISTIC MASS DISTRIBUTIONS

In realistic reconstructions of lens potential however, it is very rare that the lens is circular. Most of the time the mass distribution of the lens is much more complicated. It induces complex features and series of multiple images.

The simplest assumption beyond the spherically symmetric models is to introduce an ellipticity ϵ in the mass distribution (Kassiola & Kovner 1993),

$$\varphi = \varphi_0 \sqrt{1 + r_{\rm em}^2 / r_c^2} \;\; \text{with} \;\; r_{\rm em}^2 = \frac{x^2}{(1-\epsilon)^2} + \frac{y^2}{(1+\epsilon)^2}. \tag{68}$$

To understand the physics it induces one should introduce the caustics and critical lines. The *critical lines* are the location on the image plane of the points of infinite magnification. The *caustics* are the location of these points on the source plane. These points are determined by the lines on which $\det(\mathcal{A}^{-1}) = 0$. It means that arcs are along the critical lines and that they are produced by background galaxies that happen to be located on the caustics. On Fig. 5 one can see the shapes of the critical lines and caustics for different depth of the potential (68) (or equivalently for different positions of the source plane).

Eventually the reconstruction of galaxy cluster mass maps requires the use of more complicated models and it can be necessary to perform non-parametric mass reconstructions. Recent results have been obtained by AbdelSalam et al. (1997) for few clusters.

6. The weak lensing regime

In this section I consider the possibility of using the lens effects to probe the large-scale structures of the Universe. The difficulty is here that the distortion induced by the lenses can be very small. The projected potential should then be reconstructed with a statistical analysis on the deformation measured on a lot of background objects. Let me define more precisely the different regimes.

6.1. THE MATHEMATICAL DESCRIPTION OF THE WEAK LENSING REGIME

Depending on the magnitude of the lens effect, different regimes are possible:

- The strong regime is such that several light path are possible between the sources and the observer. It thus induces multiple images, and the images of background galaxies are often extremely distorted: this is the regime of the giant arcs. The one that has been investigated so far.
- The regime of arclets correspond to a case where there are no multiple images although a significant distortion of the background objects can be observed.
- The weak lensing regime corresponds to cases of modest distortion (typically a few %). Such effect cannot be detected with a single object and therefore should be measured in a statistical way, by averages over a large number of background galaxies.

In all cases the displacement field is not directly observable. In the weak lensing regime, the deformation only is measurable. For slightly extended objects such as background galaxies the deformation in shape is induced by the variations of the displacement field. It can actually be described by the amplification matrix, the components of its inverse are in general written,

$$\mathcal{A}^{-1} = \begin{pmatrix} 1-\kappa-\gamma_1 & -\gamma_2 \\ -\gamma_2 & 1-\kappa+\gamma_1 \end{pmatrix}, \tag{69}$$

taking advantage of the fact that it is a symmetric matrix. The components of this matrix are expressed in terms of the convergence, κ, (a scalar field) and the shear, γ (a pseudo vector field) with

$$\kappa = \frac{1}{2}\nabla^2\psi; \quad \gamma_1 = \frac{1}{2}(\psi_{,11} - \psi_{,22}) \ ; \quad \gamma_2 = \psi_{,12}, \ \text{with} \ \psi = 2\frac{D_{LS}}{D_{OS}\,D_{OL}}\phi. \tag{70}$$

The convergence describes the linear change of size and the shear describes the deformation. The consequences of such a transform can be decomposed in two aspects:

- The magnification effect. Lenses induce a change of size of the objects. As the surface brightness is not changed by this effect, the change of surface induces a direct magnification effect, μ. This magnification is directly related to the determinant of $\mathcal{A}$ so that,

$$\mu = \det(\mathcal{A}) = 1/\left[(1-\kappa)^2 - \gamma^2\right]. \tag{71}$$

- The distortion effect. Lenses also induce a change of shape of the background objects. The eigenvalues of the matrix $\mathcal{A}^{-1}$ determine the direction and amplitude of such a deformation.

6.2. THE MAGNIFICATION EFFECT

In the weak lensing regime, the observational consequences of the magnification effect is a combination of a change the apparent area of the objects, that makes their detection easier, and their dilution (the total area is enlarged as well). The mean local number density of galaxies is then related to the slope of the galaxy counts, α, through

$$n(\gamma) = n_g\,\mu^{2.5\alpha-1} \ \text{with} \ \alpha = \frac{\mathrm{d}\log N}{\mathrm{d}m}, \tag{72}$$

where m is the apparent magnitude (in a given band) and n_g is the mean number density of galaxies in the absence of lenses. Whether the number density of galaxies increases or decreases in magnified area thus depends on α and consequently on the selected population of objects.

Such an effect has been advocated (Broadhurst 1995 and Broadhurst et al. 1995) as a way to detect the lens effect. In general this method suffers from the fact that the background galaxies have intrinsic number density fluctuations. It is therefore more appropriate for mapping the mass fluctuations in galaxy clusters where the magnification effect is large enough to dominate. In galaxy clusters it is a cheap way to map the mass profiles. More sophisticated analysis allow even to have access to the cosmological constant Λ by probing the extension of the depletion area (Fort et al. 1997, see contribution of Y. Mellier).

6.3. THE GALAXY SHAPE MATRICES TO MEASURE THE DISTORTION FIELD

The distortion effects change the shape of the background objects. The objects appear elongated along the eigenvalues of the amplification matrix. When background objects are only moderately extended (this excludes the case of arcs), their shapes can be described by the matrix,

$$\mathcal{S} \equiv \int \mathrm{d}^2\theta\, \theta_i\, \theta_j\, \mathcal{I}(\theta). \tag{73}$$

It is easy to relate the shape matrix in the source plane to the one in the image plane. This is obtained by a simple change of variable that uses the fact that the surface brightness of the objects is not changed. It implies

$$\mathcal{S}^S = \mathcal{A}^{-1} \cdot \mathcal{S}^I \cdot \mathcal{A}^{-1}. \tag{74}$$

By averaging over the shape matrices in the source plane, assuming the intrinsic shape fluctuations are not correlated, one can eventually get[4] the value of $\mathcal{A}^{-1}/\sqrt{\det \mathcal{A}^{-1}}$. The combination we have access to is totally independent of the amplification factor. As a consequence, the quantity which is measurable is the reduced shear field,

$$\mathbf{g} = \vec{\gamma}/(1-\kappa). \tag{75}$$

This quantity identifies with γ only in the limit of very weak lensing (i.e. when $\kappa \ll 1$).

6.4. THE CONSTRUCTION OF THE PROJECTED MASS DENSITY

The elaboration of methods for reconstructing mass maps from distortion fields is not a trivial issue. In a pioneering paper, Kaiser & Squires (1993) showed that this is indeed possible, at least in the weak lensing regime. It is indeed not too difficult to show that (in the single lens approximation),

$$\nabla\kappa = -\begin{pmatrix} \partial_1 & \partial_2 \\ -\partial_2 & \partial_1 \end{pmatrix} \cdot \begin{pmatrix} \gamma_1 \\ \gamma_2 \end{pmatrix} \tag{76}$$

when $\kappa \ll 1$ and $\gamma_i \ll 1$. By simple Fourier transforms it is then possible to recover κ from a distortion map.

Such a method was further extended in many ways. Bartelmann, Schneider and his collaborators (Bartelmann et al 1996, Schneider 1995, Seitz & Schneider 1995, Seitz & Schneider 1996, Seitz et al. 1998) have worked

[4]See Mellier, these proceedings, for a more detailed presentation of the data analysis techniques.

in detail on the edge effects, the possibility of having adaptive smoothing procedures, the use of maximum entropy method etc... This is particularly important when structures of different sizes and contrasts are present at the same time. Finally Kaiser (Kaiser 1995) exhibited the relation between the local convergence and the distortion field $\mathbf{g}$ which is valid in all regimes. This relation reads

$$\nabla \log(1-\kappa) = \begin{pmatrix} 1-g_1 & g_2 \\ g_2 & 1+g_1 \end{pmatrix}^{-1} \cdot \begin{pmatrix} \partial_1 & \partial_2 \\ -\partial_2 & \partial_1 \end{pmatrix} \cdot \begin{pmatrix} g_1 \\ g_2 \end{pmatrix}. \tag{77}$$

This is a non-linear and non-local relationship.

The first reconstruction of a mass map of a galaxy cluster has been done on MS1224 by Fahlman et al. (1994). Many other reconstructions have now been done or are under preparation (see Mellier 1998 and these proceedings).

7. The weak lensing as a probe of the Large-Scale Structures

7.1. THE LARGE-SCALE STRUCTURES

The idea of probing the large-scale structures with gravitational effects is very attractive. The gravitational survey offers indeed a unique way of probing the mass concentrations in the Universe since, contrary to galaxy survey, it can provide us with mass maps of the Universe that are free of any bias. Its interpretations in terms of cosmological parameters would then be straightforward and independent on hypothesis on galaxy or cluster formation schemes.

The physical mechanisms are the same in the context of large-scale structures and the source for the gravitational effects is the gravitational potential ϕ given by Eq. (38). It is important to remember that the source term of this equation is $\overline{\rho}(t)\,\delta_{\rm mass}(t,\mathbf{x})$. The density contrast $\delta_{\rm mass}(\mathbf{x})$ is usually written in terms of its Fourier transforms,

$$\delta_{\rm mass}(\mathbf{x}) = \int \frac{{\rm d}^3\mathbf{k}}{(2\pi)^{3/2}}\,\delta_{\rm mass}(\mathbf{k})\,D_+(t)\,\exp({\rm i}\mathbf{k}.\mathbf{x}). \tag{78}$$

The density field is then entirely defined by the statistical properties[5] of the random variables $\delta_{\rm mass}(\mathbf{k})$. At large enough scale the field is (almost) Gaussian (at least for Gaussian initial conditions which is the case in inflationary scenarios). The amplitude of the fluctuations grows with time in linear theory in a known way $D_+(t)$. This function is simply proportional to the expansion factor for an Einstein-de Sitter Universe.

[5]For a detailed introduction to large-scale structure formation theory and phenomenology see lecture notes of Bertschinger, 1996, and these proceedings.

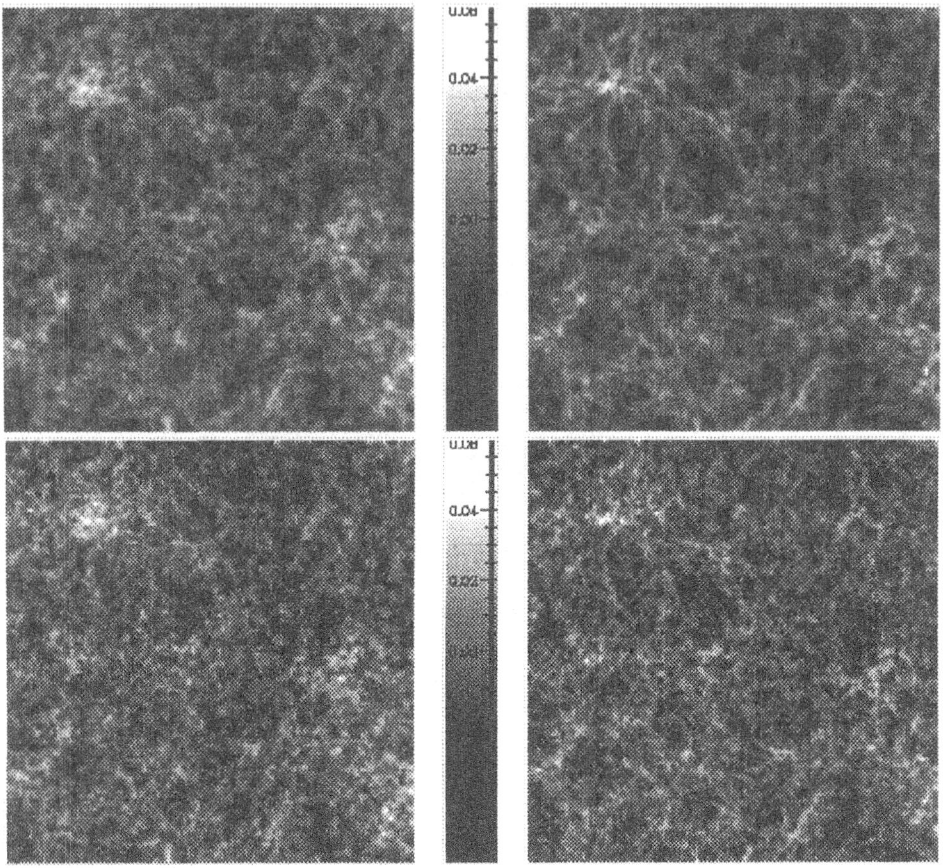

Figure 6. Example of reconstructions of projected mass maps. The top panels show the initial noise-free κ map for either $\Omega = 1$ (left panel) or $\Omega = 0.3$ (right panel) with the same underlying linear random field and the same rms distortion. The bottom panels show the reconstructed κ maps with noise included in the shear maps. The maps cover a total area of 25 degrees^2. Each pixel has an angular size of 2.5 arcmin^2 and averages the shear signal expected from deep CCD exposures (about 30 galaxy/arcmin^2). The sources are assumed to be all at redshift unity and to have a realistic intrinsic ellipticity distribution. Such a survey is easily accessible to MEGACAM at CFHT. The precision with which the images can be reconstructed and the striking differences between the two cosmological models demonstrate the great interest such a survey would have.

The variables are then entirely determined by the power spectrum $P(k)$,

$$\langle \delta_{\rm mass}(\mathbf{k})\delta_{\rm mass}(\mathbf{k}')\rangle = \delta_{\rm Dirac}(\mathbf{k}+\mathbf{k}')\, P(k). \tag{79}$$

The cosmological model is therefore completely determined by the power spectrum, Ω and Λ as long as the the dark matter distribution is concerned.

7.2. THE RELATION BETWEEN THE LOCAL CONVERGENCE AND THE

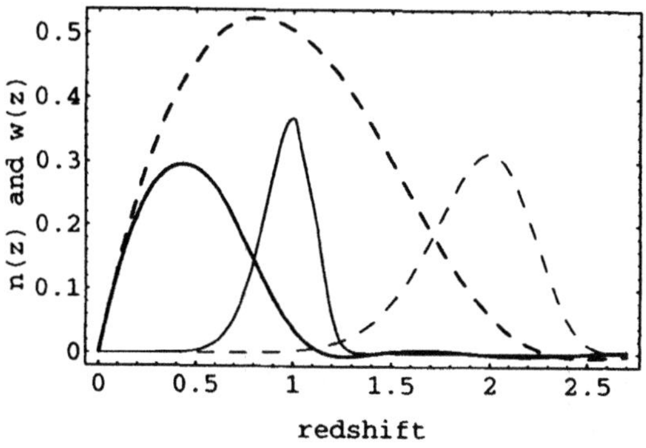

Figure 7. shape of the efficiency function, $w(z)$ (thick lines), for two different hypothesis on the shape of the redshift distribution of the sources (thin lines).

LOCAL DENSITY CONTRAST

The relation between the convergence and the local density contrasts in the local universe can be derived easily from Eqs. (38, 52),

$$\kappa(\gamma) = \frac{3}{2}\Omega_0 \int \mathrm{d}z_s\, n(z_s) \int \mathrm{d}\chi\, \frac{\mathcal{D}(\chi_s, \chi)\, \mathcal{D}(\chi)}{\mathcal{D}(\chi_s)}\, \delta_{\mathrm{mass}}(\chi, \gamma)\, (1+z). \tag{80}$$

In this relation the redshift distribution of the sources in normalized so that,

$$\int \mathrm{d}z_s\, n(z_s) = 1. \tag{81}$$

All the distances are expressed in units of c/H_0. The relation (80) is then totally dimensionless.

7.3. THE EFFICIENCY FUNCTION

It is convenient to define the efficiency function, $w(z)$, with

$$w(z) = \frac{3}{2}\Omega_0 \int \mathrm{d}z_s\, n(z_s)\, \frac{\mathcal{D}(\chi_s, \chi)\, \mathcal{D}(\chi)}{\mathcal{D}(\chi_s)}\, (1+z) \tag{82}$$

On Fig. 7 one can see the shape of the efficiency function for different hypothesis for the source distribution. Obviously the further the sources are the more numerous the lenses that can be detected are, and the larger the effect is.

From this equation it is obvious that the amplitude of κ is directly proportional of the density fluctuation amplitude and that the two point correlation function of the κ field is related to the shape of the density power spectrum. In the following its amplitude is parameterized with σ_8 which is the r.m.s. of the density contrast in a sphere of radius $8\,h^{-1}$Mpc. The relation (80) also shows that κ depends on the cosmological parameters. There is a significant dependence in the expression of the distances but the dominant contribution comes from the overall Ω_0 factor.

The amplitude of the fluctuations of κ depends on the angular scale at which the convergence map is filtered. We can introduce the filtered convergence κ_θ, with

$$\kappa_\theta(\gamma) = \int \mathrm{d}^2\gamma'\,\kappa(\gamma+\gamma')\,W_\theta(\gamma'). \tag{83}$$

It is convenient to introduce the Fourier transform of the window function $W(k)$. This function is,

$$W(k) = 2\frac{J_1(k)}{k}, \tag{84}$$

where J_1 is the Bessel function, in case of a angular top-hat filter. Then the filtered convergence reads,

$$\begin{aligned}\kappa_\theta(\gamma) = &\int \mathrm{d}\chi\, w(z) \int \frac{\mathrm{d}^2\mathbf{k}_\perp}{2\pi}\,\frac{\mathrm{d}k_r}{(2\pi)^{1/2}}\,\delta(\mathbf{k})\,D_+(z)\,\times\\ &\exp\left[\mathrm{i}k_r\chi(z) + \mathrm{i}\mathbf{k}_\perp.\gamma\mathcal{D}(z0)\right]\,W\left[k_\perp\theta\mathcal{D}(z)\right],\end{aligned} \tag{85}$$

where the wave vector $\mathbf{k}$ has been decomposed in two parts k_r and $\mathbf{k}_\perp$ that are respectively along the line of sight and perpendicular. The computation of the r.m.s. of κ_θ is analytic in the small angle approximation only. In such an approximation we have,

$$\theta \ll 1 \;\;\Rightarrow\;\; k_r \ll k_\perp \;\;\Rightarrow\;\; P(k)\,D_+^2(z) \approx P(k_\perp). \tag{86}$$

Eventually the variance reads,

$$\langle\kappa_\theta^2\rangle = \int \mathrm{d}\chi\, w^2(\chi) \int \frac{\mathrm{d}^2\mathbf{k}}{2\pi}\,P(k)\,W^2(k_\perp\theta\mathcal{D}). \tag{87}$$

For realistic models of the power spectrum (e.g. Baugh and Gaztañaga, 1995), the numerical result is (Bernardeau et al. 1997),

$$\langle\kappa_\theta^2\rangle^{1/2} \approx 0.01\,\sigma_8\,\Omega_0^{0.8}\,z_s^{0.75}\left(\frac{\theta}{1\,\mathrm{deg}}\right)^{-(n+2)/2}. \tag{88}$$

To be noticed is the dependence on the redshift of the sources. This was noticed by Villumsen (1996) who pointed out that the Ω_0 dependence is roughly given by the Ω value at the redshift of the sources. These results are slightly affected by the introduction of the non-linear effects in the shape of the power spectrum (Miralda-Escudé 1991, Jain & Seljak 1997).

7.5. THE EXPECTED SIGNAL TO NOISE RATIO

Are the effects from large-scale structures measurable? It depends on the number density of background objects for which the shape matrices are measurable. In current deep galaxy survey the typical mean number density of objects is about 50 arcmin^{-2}. The precision of the measured distortion at the degree scale is then about,

$$\Delta_{\rm noise}\kappa = \frac{0.3}{\sqrt{50\ 60^2}} \approx 10^{-3}, \tag{89}$$

for an intrinsic ellipticity of sources of about 0.3. This number is to be compared with the expected amplitude of the signal coming from the large-scale structures, about 1% according to Eq. (88) (see also earlier computations by Blandford et al. 1991, Miralda-Escudé 1991, Kaiser 1992). This makes such detection a priori possible with a signal to noise ratio around 10 (provided the instrumental noise can be controlled down to such a low level).

7.6. SEPARATE MEASUREMENTS OF Ω AND σ_8

In Eq. (88) one can see that the amplitude of the fluctuations depend both on σ_8 and on Ω_0. A question that then arises is whether it is possible to separate the amplitude of the power spectrum from the cosmological parameters. A simple examination of the equation (80) shows that it should be the case, because, for a given value of σ_κ, the density field is more strongly evolved into the non-linear regime when Ω_0 is low.

The consequences of this are two fold. The nonlinearities change the angular scale at which the non-linear dynamics starts to amplify the growth of structures. This effect was more particularly investigated by Jain & Seljak (1997) who showed that the emergence of the nonlinear regime is apparent in the shape of the angular two-point function. This effect is however quite subtle since it might reveal difficult to separate from peculiarities in the shape of the initial power spectrum.

The other aspect is that nonlinear effects induce non-Gaussian features due to mode couplings. These effects have been studied extensively in Perturbation Theory. Technically one can write the local density contrast as an expansion with respect to the initial density fluctuations,

$$\delta_{\rm mass}(\mathbf{x}) = \delta^{(1)}_{\rm mass}(\mathbf{x}) + \delta^{(2)}_{\rm mass}(\mathbf{x}) + \ldots \tag{90}$$

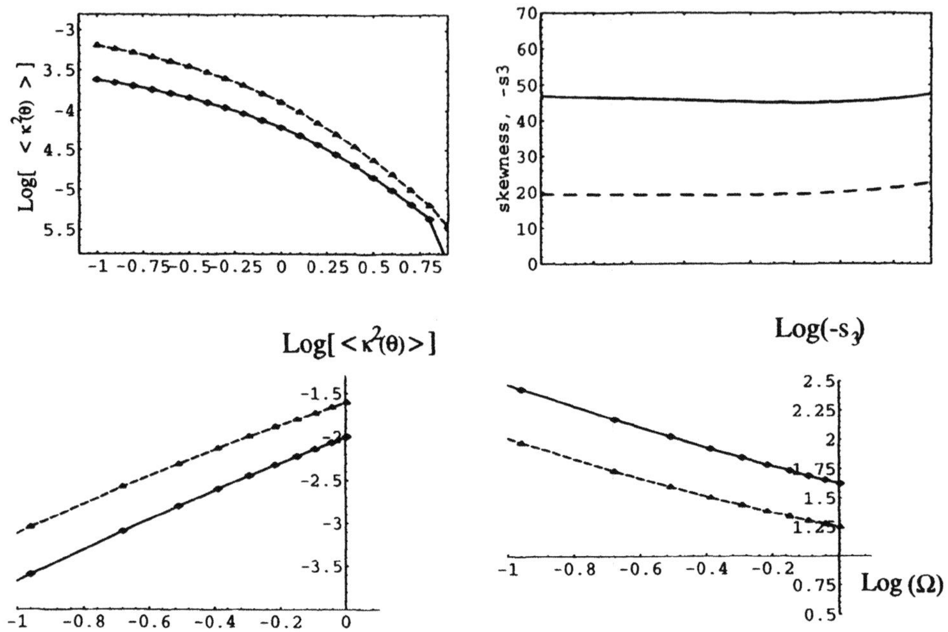

Figure 8. Perturbation Theory results (Bernardeau et al. 1997) obtained for the width (left panels) and skewness (right panels) of the probability distribution function of the local convergence. The results are plotted as a function of the angular scale in the top panels, of Ω_0 in the middle ones. The solid lines correspond to source redshifts of 1, and the dashed lines to redshifts of 2. In the bottom panels the iso-values of the width and the skewness are plotted in the Ω_0-Λ plane (for source redshifts of 1). It shows that the dependence on Ω_0 is slightly degenerate with Λ.

where $\delta^{(1)}_{\rm mass}(\mathbf{x})$ is proportional to the initial density field (this is the term we have considered so far), $\delta^{(2)}_{\rm mass}(\mathbf{x})$ is quadratic, etc. Second order perturbation theory provides us with the expression of $\delta^{(2)}_{\rm mass}(\mathbf{x})$ (there are many references for the perturbation theory calculations, Peebles 1980, Fry 1984, Goroff et al. 1986, Bouchet et al. 1993 for the Ω dependence of this result),

$$\delta^{(2)}_{\rm mass}(t,\mathbf{x}) = \int \frac{d^3\mathbf{k}_1}{(2\pi)^{3/2}}\frac{d^3\mathbf{k}_2}{(2\pi)^{3/2}}\, D_+^2(t)\,\delta_{\rm lin.}(\mathbf{k}_1)\delta_{\rm lin.}(\mathbf{k}_2)\,\exp[{\rm i}(\mathbf{k}_1+\mathbf{k}_2)\cdot\mathbf{x}]\times \left[\frac{5}{7}+\frac{\mathbf{k}_1\cdot\mathbf{k}_2}{k_1^2}+\frac{2}{7}\frac{(\mathbf{k}_1\cdot\mathbf{k}_2)^2}{k_1^2\,k_2^2}\right], \tag{91}$$

where $\delta_{\rm lin.}(\mathbf{k})$ are the Fourier components of the *linear* density field. It behaves essentially as the square of the linear term, with a non-trivial geometric function that contains the non-local effects of gravity.

Equivalently it is possible to expand the local convergence in terms of the initial density field,

$$\kappa(\gamma) = \kappa^{(1)}(\gamma) + \kappa^{(2)}(\gamma) + \dots \tag{92}$$

The apparition of a non-zero $\kappa^{(2)}$ induces non-Gaussian effects that can be revealed for instance by the computation of the skewness, third moment, of κ_θ (Bernardeau et al. 1997),

$$\langle\kappa_\theta^3\rangle = \langle\left(\kappa_\theta^{(1)}\right)^3\rangle + 3\langle\left(\kappa_\theta^{(1)}\right)^2\kappa_\theta^{(2)}\rangle + \dots \tag{93}$$

The actual dominant term of this expansion is $3\langle\left(\kappa^{(1)}\right)^2\kappa^{(2)}\rangle$ since the first term vanishes for Gaussian initial conditions. For the computation of such term one should plug in Eq. (80) the expression of $\delta^{(2)}_{\rm mass}$ in Eq. (91) and do the computations in the small angle approximation (and using specific properties of the angular top-hat window function, Bernardeau 1995).

Eventually perturbation theory gives the following result for a realistic power spectrum (Bernardeau et al. 1997),

$$s_3(\theta) \equiv \frac{\langle\kappa_\theta^3\rangle}{\langle\kappa_\theta^2\rangle^2} = 40\;\Omega_0^{-0.8}\;z_s^{-1.35}. \tag{94}$$

The origin of this skewness is relatively simple to understand: as the density field enters the non-linear regime the large mass concentrations tend to acquire a large density contrast in a small volume. This induces rare occurrences of large negative convergences. The under-dense regions tend on the other hand to occupy a large fraction of the volume, but can induce only moderate positive convergences. This mechanism is clearly visible on the maps of figure (6). When the mean source redshift grows the skewness diminishes since the addition of independent layers of large-scale structures tend to dilute the non-Gaussianity.

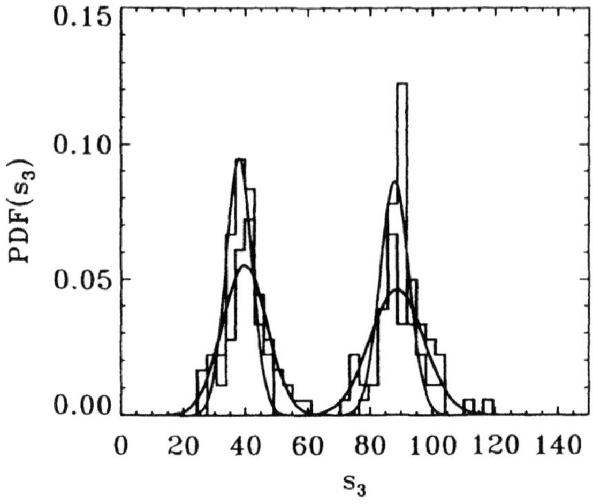

Figure 9. Histograms of the values of s_3, top-hat filter, for $\Omega = 1$ and $\Omega = 0.3$ for a 5×5 degree survey (thick lines) and a 10×10 degree survey (thin lines).

What the Eq. (94) demonstrates is that distortion maps can be used to determine the cosmic density parameter, Ω_0, provided the redshift distribution of the sources is well known. The hierarchy exhibited in this relation is also a direct consequence of the hypothesis of Gaussian initial conditions. Such a hierarchy has been observed for instance in galaxy catalogues (see Bouchet et al. 1993 for results in the IRAS galaxy survey). It can be very effective in excluding models with non-Gaussian initial conditions (see the attempt of Gaztañaga & Mähönen 1996). To be more precise I present the actual histograms of the measured skewness in numerical simulations (Fig. 9) which clearly demonstrate that the two cosmologies are easily separated. One can see that the scatter in s_3 is roughly the same in the two cases and that the difference in the relative precision is due to the differences in the expectation values.

The validity of Eq. (94) has been confirmed numerically by Gaztañaga & Bernardeau (1998), who showed it is valid for scales above a few tens of arcmins. A non-zero skewness has also been observed in the numerical experiment of Jain et al. (private communication, in preparation) and van Waerbeke et al. (1998). Large angular convergence maps can therefore provide new means for constraining fundamental cosmological parameters. Numerical results show that in maps of 5×5 square degrees it is reasonable to expect a precision of a few percent on the normalization and about 5% to 10% on the cosmological density parameter depending on the underlying cosmological scenario (see Fig. 10).

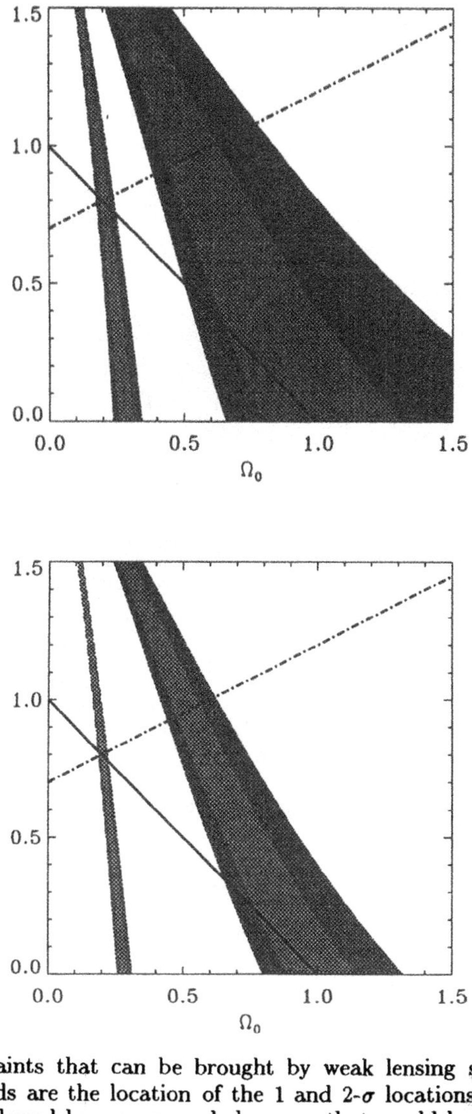

Figure 10. Constraints that can be brought by weak lensing survey in an $\Omega_0 - \Lambda$ plane. The grey bands are the location of the 1 and 2-σ locations (respectively darker and lighter bands) allowed by a measured skewness that would be obtained with either $\Omega_0 = 0.3$ (left bands) or $\Omega_0 = 1$ (right bands). The solid straight lines corresponds to a zero curvature universe, and the dot-dashed lines to a fixed acceleration parameter, q_0. The panels correspond to survey of either 5×5 (top) or 10×10 degrees (bottom).

7.7. PROSPECTS

From an observational point of view, the investigation of the large-scale structures of the Universe with gravitational lenses is in a very preliminary stage. After an early claim by Villumsen (1995), a direct evidence of the detection distortion signal of gravitational origin has been reported recently by Schneider et al (1997).

There are at present many studies, either theoretical or numerical, that aim to examine all possible systematic errors (Bonnet & Mellier 1995, Kaiser et al. 1995), to optimize the data analysis concepts (such as the pixel autocorrelation function by van Waerbeke et al. 1997) and the scientific interpretations of the resulting mass maps (Bernardeau 1998, Bernardeau et al. 1997, Seljak 1997, van Waerbeke et al. 1998). A few observational surveys are now emerging,

- the ESO key program jointly done by MPA and IAP;
- the DESCART project, part of the scientific program of the wide field CCD camera to be installed at the CFHT.

Acknowledgements

The author thanks Y. Mellier for innumerable discussions on the lens physics and IAP for its hospitality.

References

AbdelSalam, H.M., Saha, P. & Williams, L.L.R. astro-ph/9710306
Bartelmann, M., Narayan, R., Seitz, S. & Schneider, P., 1996, *Astrophys. J.* **464**, 115
Baugh, C.M. & Gaztañaga, E. 1996, *Mon. Not. R. astr. Soc.* **280**, L37
Bernardeau, F. astro-ph/9802243
Bernardeau, F. 1995, *Astr. & Astrophys.* **301**, 309
Bernardeau, F., van Waerbeke, L. & Mellier, Y. 1997, *Astr. & Astrophys.* **324**, 15
Bertschinger, E. 1996, in "Cosmology and Large Scale Structure", Les Houches Session LX, August 1993, NATO series, eds. R. Schaeffer, J. Silk, M. Spiro, J. Zinn-Justin, Elsevier Science Press
Blandford, R. D., Saust, A. B., Brainerd, T. G., Villumsen, J. V. 1991, *Mon. Not. R. astr. Soc.* **251**, 600
Bonnet, H. & Mellier, Y. 1995 *Astr. & Astrophys.* **303**, 331
Bouchet, F., Juszkiewicz, R., Colombi, S. & Pellat, R., 1992, *Astrophys. J.* **394**, L5
Bouchet, F., Strauss, M.A., Davis, M., Fisher, K.B., Yahil, A. & Huchra, J.P. 1993, *Astrophys. J.* **417**, 36
Broadhurst, T. astro-ph/9511150.
Broadhurst, T., Taylor, A.N., Peacock, J. 1995 *Astrophys. J.* **438**, 49
Eke, V.R., Cole, S. & Frenk, C.S. 1996 *Mon. Not. R. astr. Soc.* **282**, 263
Fry, J., 1984, *Astrophys. J.* **279**, 499
Fort, B., Mellier, Y. & Dantel-Fort, F. 1997, *Astr. & Astrophys.* **321**, 353
Fahlman, G., Kaiser, N., Squires, G., Woods, D. 1994, *Astrophys. J.* **437**, 56
Gaztañaga, E. & Bernardeau, F. 1998, *Astr. & Astrophys.* **331**, 829
Gaztañaga, E. & Mähönen, 1996, *Astrophys. J.* **462**, L1
Goroff, M.H., Grinstein, B., Rey, S.-J. & Wise, M.B. 1986, *Astrophys. J.* **311**, 6
Jain, B., Seljak, U. 1997, *Astrophys. J.* **484**, 560
Kaiser, N. 1992 *Astrophys. J.* **388**, L72
Kaiser, N. 1995 *Astrophys. J.* **439**, 1
Kaiser, N. & Squires, G. 1993 *Astrophys. J.* **404**, 441
Kaiser, N., Squires, G. & Broadhurst, T. 1995 *Astrophys. J.* **449**, 460

Kassiola, A. & Kovner, I. 1993, *Astrophys. J.* ,, 417,450
Limber, D.N., 1954 *Astrophys. J.* **119**, 655
Mellier, Y. 1998, astro-ph/9812172 to appear in ARAA
Miralda-Escudé, J. 1991 *Astrophys. J.* **380**, 1
Misner, C.W. Thorne, K. & Wheeler, J.A. 1973, Gravitation, San Francisco, Freeman.
Palanque-Delabrouille, et al. astro-ph/9710194
Peebles, P.J.E. 1980; The Large-Scale Structure of the Universe; Princeton University Press, Princeton, N.J., USA;
Sachs, R. K. 1961, *Proc. Roc. Soc. London* A**264**, 309
Schneider, P. astro-ph/9706185
Schneider, P., Ehlers, J., Falco, E. E. 1992, *Gravitational Lenses*, Springer.
Schneider, P., Van Waerbeke, L., Mellier, Y., Jain, B., Seitz, S., Fort, B. 1997, astro-ph/9705122.
Seitz, C., Schneider, P., 1995 *Astr. & Astrophys.* **297**, 247
Seitz, S., Schneider, P., 1996 *Astr. & Astrophys.* **305**, 388
Seitz, S., Schneider, P., Ehlers, J., 1994, Class. Quant. Grav. 11, 2345
Seitz, S., Schneider, P., Bartelmann, M., astro-ph/9803038
Seljak, U. astro-ph/9711124
Soucail, G., Mellier, Y., Fort, B., Mathez, G. & Cailloux, M. 1988 *Astr. & Astrophys.* **191**, L19
Squires, G. & Kaiser, N. 1996, *Astrophys. J.* **473**, 65
Villumsen, J. V., astro-ph/9507007
Villumsen, J. V. 1996, *Mon. Not. R. astr. Soc.* **281**, 369
van Waerbeke, L., Mellier, Schneider, P., Fort, B. & Mathez, G. 1997 *Astr. & Astrophys.* **317**, 303
van Waerbeke, L., Bernardeau, F. & Mellier, Y. astro-ph/9807007

COSMOLOGICAL APPLICATIONS OF GRAVITATIONAL LENSING

Y. MELLIER
Institut d'Astrophysique de Paris
and Observatoire de Paris, DEMIRM
98 bis Boulevard Arago
75014 Paris
France

1. Introduction

Because gravitational lensing provides a unique tool to probe almost directly the dark matter, its use for cosmology is of considerable interest. The discovery of giant arc(let)s in clusters of galaxies (Soucail et al 1987, Lynds & Petrosian 1986, Fort et al 1988) or Einstein rings around galaxies (Hewitt et al 1988), and the spectroscopic proofs that they are produced by gravitational lensing effects (Soucail et al 1988) have revealed that gravitational distortion can probe with a remarkable amount of details the mass distribution of clusters (Tyson et al 1990, Kaiser & Squires 1993) and galaxies (Kochanek 1990).

The applications of gravitational lensing to cosmology are so important that one cannot ignore them in a course dedicated to observational cosmology. The most obvious applications are the determination of the mass of gravitational systems, because the total mass inferred from a simple gravitational lens model is remarkably robust. In this respect microlensing experiments as well as strong and weak lensing studies provide the most powerful techniques to probe the dark matter of the Universe from the Jupiter-like planets up to large-scale structures. That is more than 16 orders of magnitude in size, more than 19 orders of magnitude in mass, and more than 25 orders of magnitude in density contrast! These limits are only due to technical limitations of present-day instrumentation, and in principle gravitational lensing can probe a much broader range. Therefore, measuring the mass distribution with gravitational lensing can put important constraints on the gravitational history of our Universe and the formation of its structures and its virialized gravitational systems.

M. Lachièze-Rey (ed.), Theoretical and Observational Cosmology, 211–243.

Gravitational lensing also can provide valuable constraints on the cosmological parameters. For examples, the fraction of quasars with multiple-image, or the magnification bias in lensing-clusters depend on the curvature of the Universe. Deep surveys devoted to these lenses have already put boundaries on the cosmological constant λ. Furthermore, the measurement of time delays of transient events observed in multiple images of lensed sources can potentially provide useful constraints on the Hubble constant, H_0, on cosmological scales where distortion of the measurements from any local perturbation is negligible. Of equal importance, gravitational lenses can also be used as gravitational telescopes in order to observe the deep Universe. When strongly magnified, detailed structures of extremely high redshift galaxies can be analyzed spectroscopically in order to understand the dynamical stage and the merging history of the young distant galaxies (Soifer et al 1998) . Furthermore, the joint analysis of the dark matter distribution in clusters and the shape of the lensed galaxies can be used to recover their redshift distribution.

In this course, I will focus on some applications to cosmology. However, in order to avoid self-duplications, I will only address some aspects about the mass determination from gravitational lensing studies. The other topics more detailed presentations (namely, distant galaxies, cosmological parameters, lensing on the CMB) or more detailed presentations can be found in Bernardeau (this proceedings), Blandford & Narayan (1992), Fort & Mellier (1994), Mellier (1998, 1999), Refsdal & Surdej (1994) and obviously in the textbook written by Schneider, Ehlers & Falco (1992). The microlensing experiments will be also presented elsewhere (Rich, this proceedings). Since F. Bernardeau has already discussed the theoretical aspects in his lecture, I will only make some addenda on some specific quantities in order to clarify the order of magnitude, but I assume that all the basic concepts are know already.

2. Some important quantities and properties

2.1. IMAGE MULTIPLICITY AND EINSTEIN RADIUS

Let us assume the lens being a point mass of mass M and a one-dimensional configuration. In that case, the solutions for θ_I are given by

$$\theta_S = \theta_I + \frac{D_{LS}}{D_{OL} D_{OS}} \frac{4GM}{c^2 \theta_I} \tag{1}$$

which in general has 2 solutions. This illustrates the fact that gravitational lenses can produce multiple images. In the special case of a point mass, the divergence of the deviation angle at the origin is meaningless. For a more realistic mass density, the divergence vanishes and this produces another

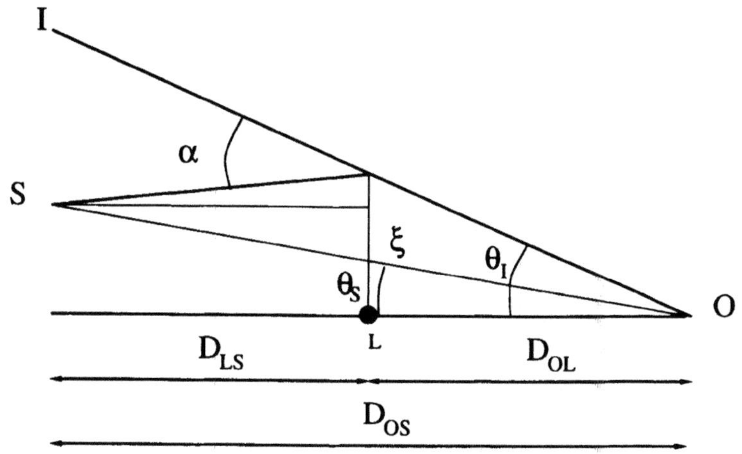

Figure 1. Simple lens configuration which explains the basic notations used in this lecture (see Bernardeau for more details).

solution for the images close to the center. This reflects a more general theorem that the number of lensed images is odd for non-singular mass density (Burke 1981). A critical case occurs when $\boldsymbol{\theta}_S = 0$, that is when there is a perfect alignment between the source, the lens and the observer. The positions of the images are degenerated and form a strongly magnified circle at the *Einstein radius* $\boldsymbol{\theta}_E$:

$$\theta_E = \left[\frac{4GM}{c^2}\frac{D_{LS}}{D_{OL}D_{OS}}\right]^{1/2} . \tag{2}$$

For example,

- for a star of 1 $M_\odot$ at distance D=1 kpc, $\boldsymbol{\theta}_E = 0.001$ arcsec,
- for a galaxy of 10^{12} $M_\odot$ at D=1 Gpc, $\boldsymbol{\theta}_E = 1$ arcsec,
- and for a clusters of 10^{14} $M_\odot$ at $z = 0.3$ and sources at $z_S = 1$, $\boldsymbol{\theta}_E = 30$ arcsec.

When the source if slightly off-centered with respect to the axis defines by the lens and the observed, a circular lens produces two elongated arcs diametrally opposite with respect to the center of the lens.

2.2. CRITICAL MASS

By definition, the magnification matrix is given by $A = d\boldsymbol{\theta}_S/d\boldsymbol{\theta}_I$. This defines the convergence, κ, and the shear,(γ_1, γ_2),

$$\begin{pmatrix} 1-\kappa-\gamma_1 & -\gamma_2 \\ -\gamma_2 & 1-\kappa+\gamma_1 \end{pmatrix} = \begin{pmatrix} 1-\partial_{xx}\varphi & -\partial_{xy}\varphi \\ -\partial_{xy}\varphi & 1-\partial_{yy} \end{pmatrix} , \tag{3}$$

where "∂_{ij}" denotes the partial derivative of the projected potential with respect to the coordinates ij. The *magnification* is given by

$$\mu = \frac{1}{|\mathrm{det}A|} = \frac{1}{|(1-\kappa^2) - \gamma^2|} , \tag{4}$$

where the shear amplitude $\gamma = \sqrt{\gamma_1^2 + \gamma_2^2}$ expresses the amplitude of the anisotropic magnification. On the other hand, κ expresses the isotropic part of the magnification. In is worth noting that $2\kappa = \Delta\varphi = \Sigma/\Sigma_{crit}$, is directly related to the projected mass density. It may happen that the determinant vanishes. It that case, the magnification becomes infinite. From an observational point of view, these cases correspond to the formation of very extended images, like Einstein rings or giant arcs. The points of the image plane where the magnification is infinite are called *critical lines.* To these critical lines correspond points in the source plane which are called *caustic lines.*
The critical density is

$$\Sigma_{crit} = \frac{c^2}{4\pi G} \frac{D_{OS}}{D_{LS} D_{OL}} \tag{5}$$

expresses the capability of a gravitational system to produce strong lensing effect ($\Sigma_{crit} \geq 1$). If we scale to $2H_0/c^2$ in order to define the "normalized" angular distances d_{ij}, then

$$\Sigma_{crit} \approx 0.1 \left(\frac{H_0}{50\mathrm{km/s/Mpc}} \right) \frac{d_{os}}{d_{ls} d_{ol}} \ \mathrm{g.cm}^{-2} \tag{6}$$

For example, for a lensing-cluster at redshift $z_L = 0.3$ and lensed sources at redshift $z_S = 1$, $d_{os}/(d_{ls} d_{ol}) \approx 3$. If the cluster can be modeled by a isothermal sphere with a core radius R_c and with $M(R_c) = 2 \times 10^{14}\ M_\odot$, then

- For R_c=250 kpc, Σ_{crit}=0.05 g.cm^{-2},
- For R_c=50 kpc, Σ_{crit}=1. g.cm^{-2} .

Hence, the existence of giant arcs in clusters implies that clusters should be strongly concentrated.

2.3. RELATION WITH OBSERVABLE QUANTITIES

Since, to first approximation, faint galaxies look like ellipses, their shapes can be expressed as function of their weighted second moments,

$$M_{ij} = \frac{\int S(\ )(\theta_i - \theta_i^C)(\theta_j - \theta_j^C) d^2\theta}{\int S(\ ) d^2\theta} , \tag{7}$$

where the subscripts i j denote the coordinates $\boldsymbol{\theta}$ in the source and the image planes, $S(\boldsymbol{\theta})$ is the surface brightness of the source and $\boldsymbol{\theta}^C$ is the center of the source.
Since gravitational lensing effect does not change the surface brightness of the source (Etherington 1933), then, if one assumes that the magnification matrix is constant across the image, the relation between the shape of the source, M^S and the lensed image, M^I is

$$M^I = A^{-1}\, M^S\, A^{-1} \tag{8}$$

Therefore, the gravitational lensing effect transforms a circular source into an ellipse. Its axis ratio is given by the ratio of the two eigenvalues of the magnification matrix. The shape of the lensed galaxies can then provide information about these quantities. However, though γ_1 and γ_2 describe the anisotropic distortion of the magnification, they are not directly related to observables (except in the weak shear regime). The *reduced complex shear*, $\mathbf{g}$, and the complex polarization (or distortion), $\boldsymbol{\delta}$,

$$\mathbf{g} = \frac{\gamma}{(1-\kappa)} \quad ; \quad \boldsymbol{\delta} = \frac{2g}{1+|\mathbf{g}|^2} = \frac{2\gamma(1-\kappa)}{(1-\kappa)^2+|\gamma|^2} \,, \tag{9}$$

are more relevant quantities because $\boldsymbol{\delta}$ can be expressed in terms of the observed major and minor axes a^I and b^I of the image, I, of a circular source S:

$$\frac{a^2-b^2}{a^2+b^2} = |\boldsymbol{\delta}| \tag{10}$$

In this case, the 2 components of the complex polarization write:

$$\delta_1 = \frac{M_{11}-M_{22}}{Tr(M)} \quad ; \quad \delta_2 = \frac{2M_{12}}{Tr(M)} \,, \tag{11}$$

where $Tr(M)$ is the trace of the magnification matrix. For non-circular sources, it is possible to relate the ellipticity of the image $\boldsymbol{\epsilon}^I$ to the ellipticity of the lensed source, $\boldsymbol{\epsilon}^S$. In the general case, it depends on the sign of $Det(A)$ (that is the position of the source with respect to the caustic lines) which expresses whether images are radially or tangentially elongated:

$$\boldsymbol{\epsilon}^I = \frac{1+b^I/a^I}{1+b^I/a^I}e^{2i\vartheta} = \frac{\boldsymbol{\epsilon}^S+\mathbf{g}}{1-\mathbf{g}^*\boldsymbol{\epsilon}^S} \text{ for } Det(A) > 0 \,, \tag{12}$$

and

$$\boldsymbol{\epsilon}^I = \frac{1+\boldsymbol{\epsilon}^{S*}\mathbf{g}}{\boldsymbol{\epsilon}^{S*}+\mathbf{g}^*} \text{ for } Det(A) < 0 \,. \tag{13}$$

3. Academic examples

3.1. THE SINGULAR ISOTHERMAL SPHERE

The projected potential at radius r, $\varphi(r)$, of a singular isothermal sphere with 3-dimension velocity dispersion σ is

$$\varphi = 4\pi \frac{\sigma^2}{c^2} \frac{D_{LS}}{D_{OS}} r \tag{14}$$

and the images are described by the lensing equation

$$\boldsymbol{\theta}_S = \boldsymbol{\theta}_I - 4\pi \frac{\sigma^2}{c^2} \frac{D_{LS}}{D_{OS}} \frac{\boldsymbol{\theta}_I}{|\boldsymbol{\theta}_I|} \tag{15}$$

and the magnification matrix

$$\begin{pmatrix} 1 & 0 \\ 0 & 1 - 4\pi \frac{\sigma^2}{c^2} \frac{D_{LS}}{D_{OS}} \frac{1}{|\boldsymbol{\theta}_I|} \end{pmatrix} \tag{16}$$

Hence, there is only one critical line which is given by the Einstein radius

$$\theta_{SIS} = 4\pi \frac{\sigma^2}{c^2} \frac{D_{LS}}{D_{OS}} \approx 16" \left(\frac{\sigma}{1000 \text{ km.s}^{-1}} \right)^2 \tag{17}$$

for an Einstein de Sitter universe, with $z_L = 0.3$ and $z_S = 1$. The total mass included within the radius θ is then

$$M(\theta) = 5.7 \times 10^{13} \; M_\odot h_{50}^{-1} \left(\frac{\theta}{16"} \right) \left(\frac{\sigma}{1000 \text{ km.s}^{-1}} \right)^2 \tag{18}$$

From Eq.(28) and (29), it is obvious that the magnification writes

$$\mu(\theta_I) = \frac{\theta_I}{\theta_I - \theta_{SIS}} \tag{19}$$

The singular isothermal sphere permits to keep in mind the various properties of gravitational lenses and order of magnitude estimates of the associated physical quantities. However, more complex lenses can produce somewhat different configurations of strong lensing. Some example are shown in Figure 2, where a series of arcs generated by an elliptical potential well is displayed. We see that various kind of multiple images can be produced, with some strange radial arcs for some configurations. Each panel can be found in the universe so this kind of template of lens features can be helpful for the understanding of the lens modeling. This has been extensively used,

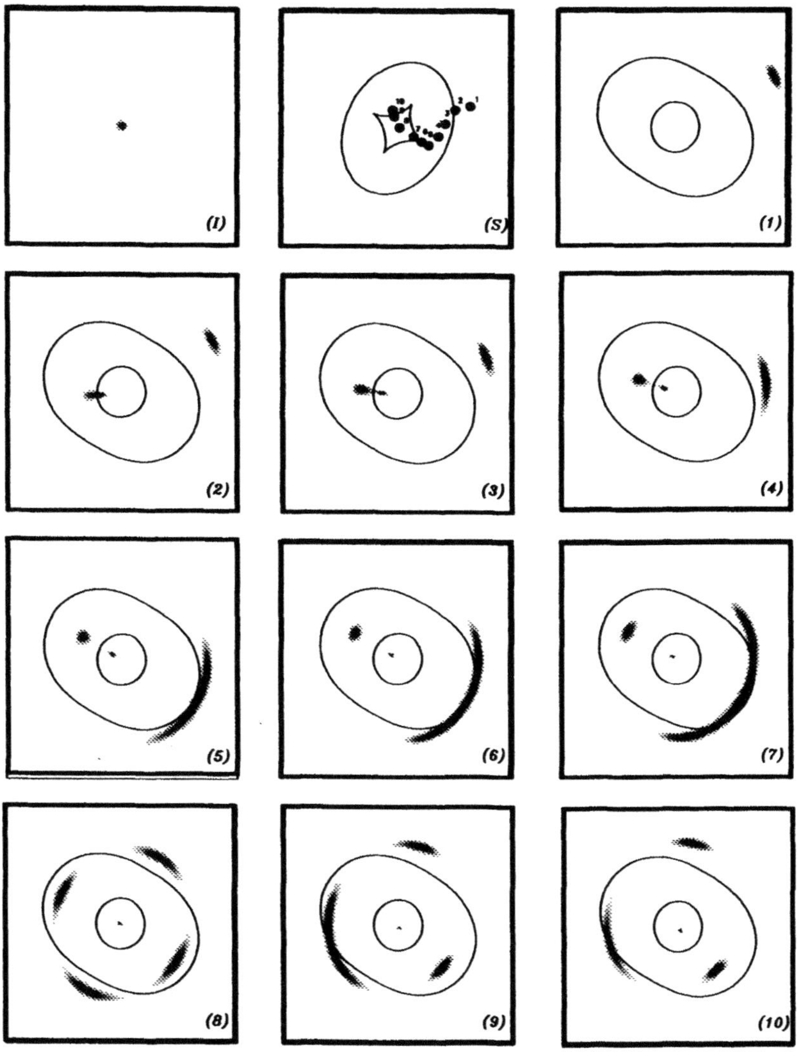

Figure 2. Gravitational distortion produced by an elliptical potential as a function of source position. The top left panel shows the shape of the source in the source plane. The second panel shows 10 positions of the source in its source plane (referenced from 1 to 10) with respect to a simulated cluster lens. The thin lines show the inner and outer caustics. Panels 3 to 12 show the inner and outer critical lines and the shape of the image(s) of the lensed source. Positions 6 and 7 correspond to cusp configurations, and position 9 is typical fold configuration. On the fifth panel we see two inner merging images forming a typical radial arc (from Kneib 1993, PhD thesis).

in particular on HST images (see Fort & Mellier 1994, Kneib et al 1996, Natarajan & Kneib 1997, Mellier 1998 and references therein).

3.2. THE GENERAL CASE OF AN AXIALLY SYMMETRIC LENS

Assuming the rescaled angle writes (see Bernardeau, this proceedings)

$$\boldsymbol{\alpha} = \boldsymbol{\nabla}\varphi \quad , \quad \Delta\varphi = 2\kappa(x) \; , \tag{20}$$

for an axially symmetric potential, we have

$$\alpha = \frac{d\varphi}{dx} \quad , \quad \frac{1}{x}\frac{d}{dx}\left(x\frac{d\varphi}{dx}\right) = 2\kappa(x) \; , \tag{21}$$

and the mass at radius x is $m\,(\boldsymbol{x}) = m\,(|\boldsymbol{x}|) = m(x)$. Hence

$$\frac{dm}{dx} = 2x\kappa(x) \implies \frac{d\varphi}{dx} = \frac{2}{x}\int_0^x \kappa\,(x')\,x'dx' \; . \tag{22}$$

Therefore, in general the magnification matrix writes

$$A = \mathcal{I} - \frac{m\,(x)}{x^4}\begin{pmatrix} x_2^2 - x_1^2 & -2x_1x_2 \\ -2x_1x_2 & x_1^2 - x_2^2 \end{pmatrix} - \frac{1}{x^3}\left(\frac{dm(x)}{dx}\right)\begin{pmatrix} x_1^2 & x_x x_2 \\ x_x x_2 & x_2^2 \end{pmatrix} , \tag{23}$$

where $\mathcal{I}$ is the identity matrix. The expression of the shear terms is then immediate, and the total magnification writes:

$$Det\,(A) = \left(1 - \frac{m}{x^2}\right)\left(1 - \frac{d}{dx}\left(\frac{m(x)}{x}\right)\right) \tag{24}$$

where the first term on the right-hand part provides the position of the tangential arcs, whereas the second term provides the location of the radial arcs. These equations are useful in order to compute the lensing properties of mass distributions, like the universal profile discussed in Section 4.1.2 .

4. Astrophysical examples

4.1. MEASURING THE MASSES OF CLUSTERS OF GALAXIES

4.1.1. *The case of MS2137-23*

Strong lensing refers to lensing configurations where the source is located close to a caustic line and produces arcs or rings. In addition to the accurate estimate of the total mass, strong lensing features can probe some details on the mass density profile.

A nice example is the case of the lensing cluster MS2137-23 which contains a radial and a tangential arc. A simple investigation already permits to have some constraints on the lens configuration. First, since there is no counter-arc associated to the tangential arc on the diametrally opposite side of the cluster center, it is unlikely that the projected mass density is circularly distributed. Second, since the tangential arc and the radial arc are on the same side, it is unlikely that they are images of the same source. An interesting tentative to model the lens is to assume that the dark matter follows the light distribution of the central galaxy (see Figure 3). The isophotes then provide the center, the orientation and the ellipticity of light as constraints to dark matter. Let us assume that the projected potential as the following geometry

$$\varphi(r,\theta) = \varphi_0\sqrt{1+\left(\frac{r}{r_c}\right)^2(1-\epsilon\cos(2\theta))} \tag{25}$$

where r_c is the core radius, ϵ the ellipticity, and φ_0 the depth of the potential, respectively. For an elliptical lens, the magnification matrix can be written

$$\begin{pmatrix} 1-\frac{1}{r}\frac{\partial\varphi}{\partial r}-\frac{1}{r^2}\frac{\partial^2\varphi}{\partial\theta^2} & 0 \\ 0 & 1-\frac{\partial^2\varphi}{\partial r^2} \end{pmatrix} \tag{26}$$

which gives immediately the position or the radial and tangential arcs,

$$\left[1-\frac{\partial^2\varphi}{\partial r^2}\right]_R = 0 \ ; \ \left[1-\frac{1}{r}\frac{\partial\varphi}{\partial r}-\frac{1}{r^2}\frac{\partial^2\varphi}{\partial\theta^2}\right]_T = 0 \tag{27}$$

From these two constraints, we can have a useful information of the core radius itself (Mellier et al 1993):

$$\left(\frac{r_R}{r_c}\right)+1 = \left(\frac{D_{OL_R}}{D_{OS_R}}\right)\left(\frac{D_{OS_T}}{D_{OL_T}}\right)\left[\left(\frac{r_T}{r_c}\right)^2+1\right]^{3/2} \tag{28}$$

where the subscripts R and T denote the radial and tangential arc respectively. Hence, provided the redshifts of the lens and the two sources are known, one can infer the core radius, or more generally the steepness of the dark matter density.

The modeling of the central regions of clusters of galaxies shows that the geometry of the dark matter follows the geometry of the diffuse light associated to the dominant galaxies. This is a remarkable success of the strong lensing analysis. However, there are still some uncertainties about the total mass which does not agree with the mass inferred from X-ray analysis. The mass-to-light ratio ranges in the same domain ($\approx$ 100-300), but one can find frequently a factor of two difference.

4.1.2. *The X-ray/lensing mass discrepancy*

As it is illustrated in Figure 4, it turns out that the peaks of dark matter revealed by giant arcs in lensing-clusters correspond to the the peaks

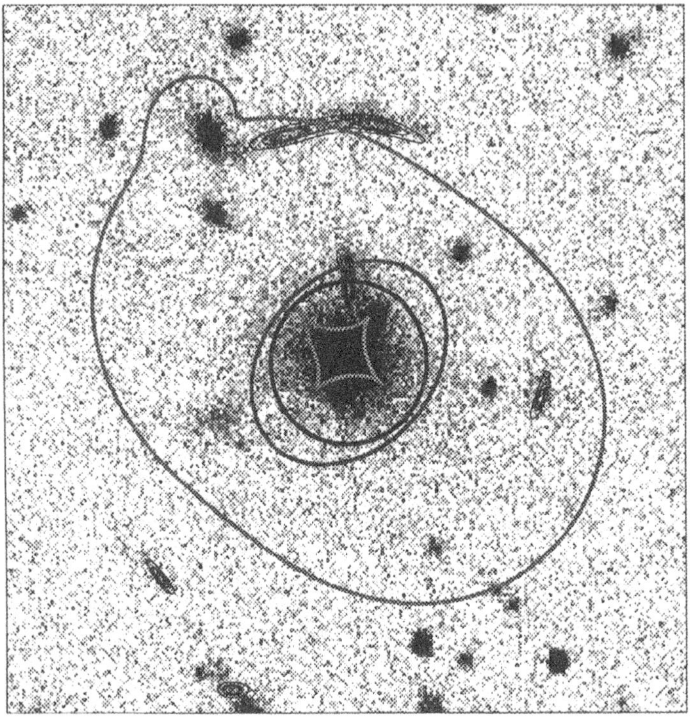

Figure 3. Model of MS2137-23. This cluster is at redshift 0.33 and shows a tangential arc and the first radial arc ever detected. Up to now it is the most constrained cluster and the first one where counter images were predicted before being observed (Mellier et al 1993). The external and the dark internal solid lines are the critical lines. The internal grey ellipse and the diamond are the caustic lines. The thin isocontours shows the positions of the arcs and their counter images.

of X-ray emission as well as the location of the brightest components of optical images. On the other hand, Miralda-Escudé & Babul (1995) have pointed out an apparent contradiction between the mass estimated from X-ray data and the lensing mass ($M_{lensing} \approx 2 - 3M_X$). Further works done by many groups lead to somewhat inconclusive statements about this contradiction. Böhringer et al (1998) find an excellent agreement between X-ray and lensing masses in A2390 which confirms the view claimed by Pierre et al (1996); Gioia et al (1998) show that the disagreement reaches a factor of 2 at least in MS0440+0204; Schindler et al (1997) find a factor of 2-3 discrepancy for the massive cluster RXJ1347.4-1145, but Sahu et al (1998) claim that the disagreement is marginal and may not exist; Ota et al

TABLE 1. Results obtained from strong lensing analyses of clusters. These are only few examples. σ_{obs} is the velocity dispersion obtained from spectroscopy of cluster galaxies. σ_{DM} is the velocity dispersion of the dark matter obtained from the modeling of the lens.

Cluster	z	σ_{obs}	σ_{DM}	M/L	Scale
		(kms^{-1})	(kms^{-1})	(h_{100})	(h_{100}^{-1} Mpc)
MS2137-23	0.33	-	900-1200	680-280	0.5)
A2218	0.18	1370	$\approx$ 1000	200	0.9
A2390	0.23	1090	1260	250	0.4
Cl0024+17	0.39	1250	1300	>200	0.5
A370	0.38	1370	850	>150	0.5
Cl0500-24	0.316	-	<1200	<600	0.5

(1998) and Wu & Fang (1997) agree that there are important discrepancies in A370, Cl0500-24 and Cl2244-02.

There is still no definitive interpretation of these contradictory results. It could be that the modeling of the gravitational mass from the X-ray distribution is not as simple. By comparing the geometry of the X-ray isophotes of A2218 to the mass isodensity contours of the reconstruction, Kneib et al (1995) found significant discrepancies in the innermost parts. The numerous substructures visible in the X-ray image have orientations which do not follow the projected mass density. They interpret these features as shocks produced by the in-falling X-ray gas, which implies that the current description of the dynamical stage of the inner X-ray gas is oversimplified. Recent ASCA observations of three lensing-clusters corroborate the view that substructures are the major source of uncertainties.

In order to study this possibility in more details, Smail et al (1997) and Allen (1998) have performed a detailed comparison between the lensing mass and X-ray mass for a significant number of lensing clusters. Both works conclude that the substructures have a significant impact on the estimate of X-ray mass. More remarkably, the X-ray clusters where cooling flows are present do not show a significant discrepancy with X-ray mass, whereas the others X-ray clusters do (Allen 1998). This confirms that the discrepancy is certainly due to wrong assumptions on the physical state of the gas. The interpretation of this dichotomy in cluster samples may be the

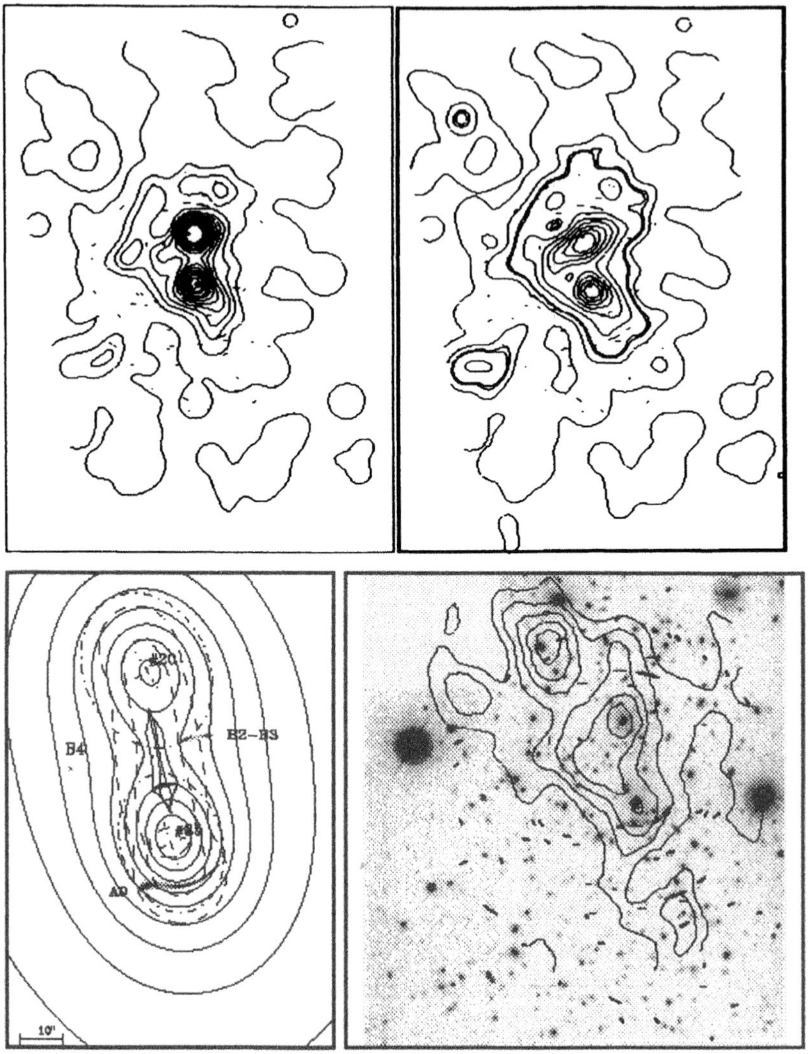

Figure 4. Analysis of the the matter distribution in the cluster A370 using strong lensing features (arcs and arclets). The top right panel shows the R-light distribution from the isoluminosity contours of galaxies. The top left panel shows the number density contours. The bottom right is a B CCD image of the clusters with the X-ray luminosity contours overlayed. The arclets directly show the shape of the projected mass density. In particular, we see that the arclet pattern indicates the presence of an extension toward the eastern region, which is also seen in the isoluminosity contours in the R-band and the X-ray maps. This clearly shows that in the center of this cluster light from galaxies and from the hot gas trace the mass. Note the arc reconstruction and the mass model in the fourth panel (the triplet B2-B3-B4 is discussed by Kneib et al. 1993).

following. Clusters with cooling flows are compact and rich systems which have probably virialised and have a well-defined relaxed core. Therefore, when removing the cooling flow contribution, the assumptions that the gas is in hydrostatic equilibrium is fully satisfied. Conversely, non-cooling flow clusters are generally poor, do have lot of substructures and no very dense core dominates the cluster yet. For these systems, the gas cannot be described simply (simple geometry, hydrostatic equilibrium) and the over-simplification of its dynamical stage produces a wrong mass estimator. This interpretation needs further confirmations. However, from these two studies we now have the feeling that we are now close to understand the origin of the X-ray and lensing discrepancy.

An alternative has been suggested by Navarro, Frenk & White (1997) who proposed that the analytical models currently used for modeling mass distributions may be inappropriate (hereafter NFW). They argue that the universal profile of the mass distribution produced in numerical simulations of hierarchical clustering may reconcile the lensing and X-ray masses. This kind of profile must be considered seriously because the universal profile is a natural outcome from the simulations which does not use external prescriptions.

For the reader who want to look more deeply at the lensing properties of this profile, it is interesting to describe the NFW properties into more details. Let us assume that the cluster 3-dimension mass density has the NFW shape:

$$\rho(x) = \frac{\rho_s}{x\,(1+x)^2} \quad \text{, with} \quad x = \frac{r}{r_s} \; . \tag{29}$$

The projected mass density of the NFW profile is (use equations of Sect. 3.2)

$$\Sigma(x) = \int_{-\infty}^{+\infty} \rho(x) dz = 2\rho_s r_s \frac{f(x)}{x^2 - 1} \tag{30}$$

where

$$f(x) = \begin{cases} 1 - \dfrac{2}{\sqrt{x^2-1}} \text{arctg}\left(\sqrt{\dfrac{x-1}{x+1}}\right) & (x > 1) \\ 0 & (x = 1) \\ 1 - \dfrac{2}{\sqrt{1-x^2}} \text{argth}\left(\sqrt{\dfrac{1-x}{1+1}}\right) & (x < 1) \end{cases} \tag{31}$$

Then the convergence and the mass write as follows:

$$\kappa(x) = \frac{2\rho_s r_s}{\Sigma_{crit}} \frac{f(x)}{x^2 - 1} \quad , \quad m(x) = 2\kappa g(x) \quad , \tag{32}$$

where

$$g(x) = \ln\left(\frac{x}{2}\right) + \begin{cases} \frac{2}{\sqrt{x^2-1}}\text{arctg}\left(\sqrt{\frac{x-1}{x+1}}\right) & (x > 1) \\ 1 & (x = 1) \\ \frac{2}{\sqrt{1-x^2}}\text{argth}\left(\sqrt{\frac{1-x}{1+1}}\right) & (x < 1) \end{cases} . \quad (33)$$

Now, from the analytical shape of $M(x)$, it is clear that $d(M/x)/dx \longrightarrow \infty$ when $x \longrightarrow 0$ and that $(d/dx)(M/x) \longrightarrow 0$ when $x \longrightarrow \infty$. Therefore, there is always a radial line, whatever ρ_s and r_s.
In the fortunate cases of $MS2137-23$ and $A370$ which both have a tangential and a radial arc, one can infer r_s and ρ_s from the analysis of the tangential and radial arcs. As far as the redshifts of the two arcs are known, then the ratio of the critical mass density at these two positions is,

$$\frac{\Sigma_{Crit,r}}{\Sigma_{Crit,t}} = \left(\frac{x_t^2}{g(x_t)}\right)\left[\frac{1}{R_{rt}}\left(\frac{dg(R_{rt}x_t)}{dx}\right) - \left(\frac{1}{R_{rt}x_t}\right)^2 g(R_{rt}x_t)\right] , \quad (34)$$

where $R_{rt} = x_r/x_t$. Once used jointly, the positions and redshifts of the two arcs and the two independent equations, permit to infer r_s and ρ_s.
For example, for $A370$ (see the radial arc in Figure 8), we have

- $z_{lens} = 0.375$, $z_t = 0.724$ (measured), $z_r \approx 1.5$ (assumed),
- $R_{rt} = 0.7$
- $\Sigma_{Crit,t} = 1.4\ h_{100}$ g.cm^{-2}, $\Sigma_s = \rho_s r_s = 0.28\ h_{100}$ g.cm^{-2}
- $r_s \approx 250\ h_{100}^{-1}$ kpc,
- and the overdensity $\delta_c = \rho(0)/\rho_{critic} = 2 \times 10^4$, if $\Omega_0 = 1$.

It is worth noting that the statement that NFW profiles do predict radial arcs contradicts the general view that their existence rules out mass profiles with singularity.
Despite this interesting prediction which makes the NFW profile even more attractive, Bartelmann (1996) has shown that the caustics produced by this profile predict that radial arcs should be thicker than observed in MS2137-23 and in A370, unless the sources are very thin ($\approx$ 0.6 arcsecond for MS2137-23). This is not a strong argument against the universal profile because this is possible in view of the shapes of some faint galaxies observed with HST that some distant galaxies are indeed very thin. But it is surprising that no radial arcs produced by "thick galaxies" have been detected so far. Even a selection bias would probably favor the observation of large sources rather than small thin and hardly visible ones.

4.1.3. *Clusters from weak lensing analysis*

Deep images of lensing-clusters show many weakly lensed galaxies having a correlated distribution of ellipticity/orientation which maps the projected mass density (Fort et al 1988 and Figure 5). The first attempt to use this distribution of arclets as a probe of dark matter has been done by Tyson et al (1990), but the rigorous inversion technique was first proposed by Kaiser & Squires (1993).

The weak lensing analysis starts from the following hypotheses:

- Assume that the orientation of the sources is isotropic.
- Assume that the orientation of the source is not correlated to their ellipticity.
- Assume that the redshift distribution of sources is known.

Then it proceeds as follows:

- Measure the averaged ellipticity and orientation of the galaxies inside all subareas of the field.
- Produce a (ellipticity, orientation) map (see Figure 6).
- Provide a relation between the (ellipticity, orientation) and the components of the shear.
- Provide a relation between the shear and the mass density.
- Provide a relation between the shape of the source and the shape of the image.

The hypotheses can be rather well controled (in principle) from the observations of unlensed areas. Ellipticities of field galaxies provide control fields for the first and second assumptions. Spectroscopic surveys and photometric redshifts allows to model the redshift distribution of galaxies. The procedure itself requires technical analysis of the data (see Mellier 1998 for the technical issues) and theoretical relations provided by gravitational lensing theory. The crucial points are the relations between shear, mass density and geometry of the lensed galaxies. This is done by combining the following equations:

$$\left\{ \begin{array}{l} \gamma = \gamma_1 + i\gamma_2 = \frac{1}{2}\left(\partial_{xx} - \partial_{yy}\right)\varphi + i\partial_{xy}\varphi \\ \kappa = \frac{1}{2}\left(\partial_{xx} + \partial_{yy}\right)\varphi \\ \varphi = \frac{1}{\pi}\int \kappa\left(\boldsymbol{\theta}'\right)\ln\left(\left|\boldsymbol{\theta} - \boldsymbol{\theta}'\right|\right) d\boldsymbol{\theta}' \ , \end{array} \right. \tag{35}$$

from which one can express the complex shear as a function of the convergence, κ (see Seitz & Schneider 1996 and references therein):

$$\gamma(\boldsymbol{\theta}) = \frac{1}{\pi}\int \mathcal{D}(\boldsymbol{\theta} - \boldsymbol{\theta}')\ \kappa(\boldsymbol{\theta}')d^2\theta' \ , \tag{36}$$

where

$$\mathcal{D}(\boldsymbol{\theta}-\boldsymbol{\theta}') = \frac{(\theta_2-\theta_2')^2-(\theta_1-\theta_1')^2-2i(\theta_1-\theta_1')(\theta_2-\theta_2')}{|(\boldsymbol{\theta}-\boldsymbol{\theta}')|^4} \ . \tag{37}$$

This equation can be inverted in order to express the projected mass density, or equivalently κ, as function of the shear:

$$\kappa(\boldsymbol{\theta}) = \frac{1}{\pi}\int \Re[\mathcal{D}^*(\boldsymbol{\theta}-\boldsymbol{\theta}')\gamma(\boldsymbol{\theta}')]\ d^2\theta' \ + \kappa_0 \ , \tag{38}$$

where $\Re$ denotes the real part. Finally, from Eq.(9-12) we can express the shear as a function of the complex ellipticity. Hence, if the background ellipticity distribution is randomly distributed, then $< |\epsilon^S| >= 0$ and

$$< |\epsilon^I| >= |\mathbf{g}| = \frac{|\gamma|}{1-\kappa} \tag{39}$$

(Schramm & Kayser 1995). In the most extreme case, when $\kappa << 1$ (the linear regime), $< |\epsilon^I| >\approx |\gamma|$, and therefore, the projected mass density can be recovered directly from the measurement of the ellipticities of the lensed galaxies.

Alternatively, one can measure the total mass within a circular radius using the *Aperture densitometry* technique (or the "ζ-*statistics*"), which consists in computing the difference between the mean projected mass densities within a radius r_1 and within an annulus $(r_2 - r_1)$ (Fahlman et al 1994, Kaiser 1995) as function of the *tangential shear*, $\gamma_t = \gamma_1\cos(2\vartheta)+\gamma_2\sin(2\vartheta)$, averaged inside the ring. Let us denote $\bar{\kappa}$ the averaged value of κ inside the loop of a circle with radius r and $\langle\kappa\rangle_\theta$ the averaged value of κ over the loop. We have

$$\bar{\kappa} = \frac{1}{\pi r^2}\int_0^{2\pi}\int_0^r \kappa\left(r',\theta'\right) r'dr'd\theta' \tag{40}$$

and

$$\frac{d\bar{\kappa}}{dr} = -\frac{2}{r}\frac{1}{\pi r^2}\int_0^{2\pi}\int_0^r \kappa\left(r',\theta'\right) r'dr'd\theta' + \frac{1}{\pi r^2}\frac{d}{dr}\left(\int_0^{2\pi}\int_0^r \kappa\left(r',\theta'\right) r'dr'd\theta'\right) \tag{41}$$

therefore,

$$\frac{d\bar{\kappa}}{dr} = -\frac{2}{r}\bar{\kappa} + \frac{2}{r}\langle\kappa\rangle_\theta \ . \tag{42}$$

or equivalently

$$\frac{1}{2}\frac{d\kappa}{d\ln r} = \langle\kappa\rangle_\theta - \bar{\kappa} \tag{43}$$

Now, the mean tangential shear writes

$$\langle\gamma_t\rangle = \frac{1}{2\pi}\int_0^{2\pi} \gamma_t d\theta' \tag{44}$$

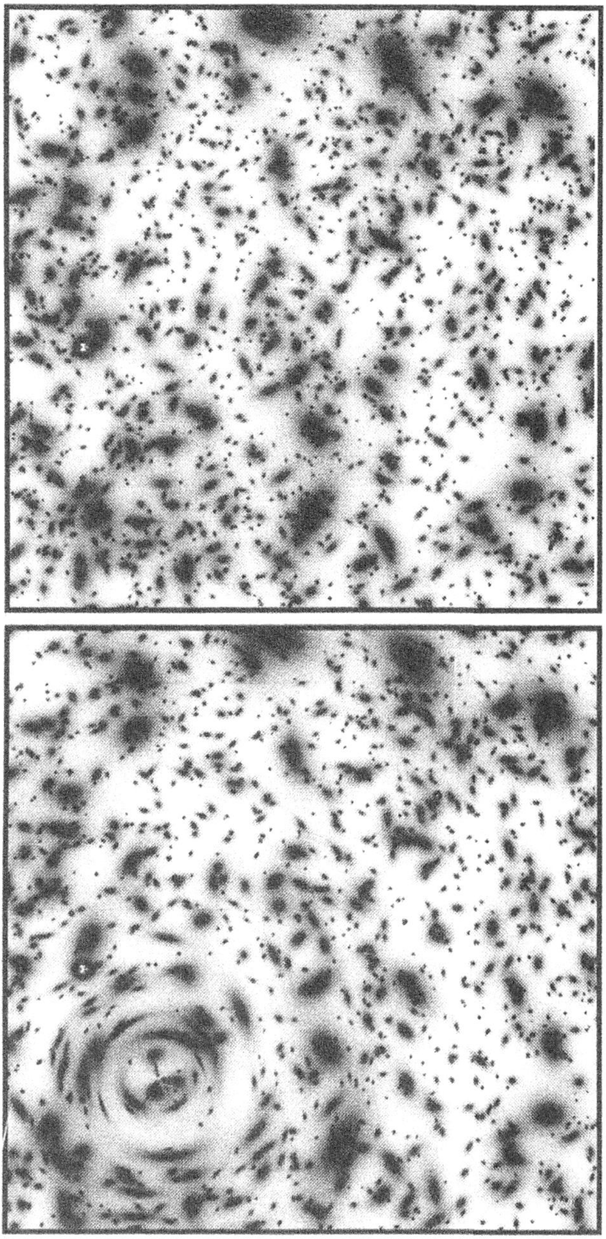

Figure 5. Distortion field generated by a lens. The top panel shows the grid of randomly distributed background sources as it would be seen in the absence of the lens. The projected number density corresponds to very deep exposure, similar to the HDF. The bottom panel shows the same population once they are distorted by a foreground (invisible) circular cluster with a typical velocity dispersion of 1300 kms^{-1}. *The geometrical signature of the cluster is clearly visible. The potential can be recovered by using the formalism defined in part 4. In this simulation, the sources are at* $z = 1.3$, *and the cluster at* $z = 0.15$.

whereas in polar coordinates

$$\gamma_t = \frac{1}{2}\left(\partial_{rr}\varphi - \partial_{ll}\varphi\right) = \partial_{rr} - \kappa \ . \tag{45}$$

Therefore

$$\langle\gamma_t\rangle = \frac{d}{dr}\left(\int_0^{2\pi} \partial_r\varphi \frac{d\theta'}{2\pi}\right) - \langle\kappa\rangle_\theta \ , \tag{46}$$

which implies that

$$\langle\gamma_t\rangle = -\frac{1}{r^2}\left\langle r\frac{d\varphi}{dr}\right\rangle_\theta + \frac{1}{r}\frac{d}{dr}\left(\left\langle r\frac{d\varphi}{dr}\right\rangle_\theta\right) - \langle\kappa\rangle_\theta \ . \tag{47}$$

Therefore, from Eq.(42), $\langle\gamma_t\rangle$ is related to $\bar{\kappa}$ by this simple relation:

$$\langle\gamma_t\rangle = \frac{1}{2}\frac{d\bar{\kappa}}{d\ln r} \tag{48}$$

which provides the ζ estimator:

$$\zeta(r_1, r_2) = <\kappa(r_1)> - <\kappa(r_1, r_2)> = \frac{2}{1 - r_1^2/r_2^2}\int_{r_1}^{r_2} <\gamma_t> d\ln r \ . \tag{49}$$

This quite robust mass estimator minimizes the contamination by foreground and cluster galaxies and permits a simple check that the signal is produced by shear, simply by changing γ_1 in γ_2 and γ_2 in $-\gamma_1$ which should cancel out the true shear signal.

The generalization to the non-linear regime (Kaiser 1995 and Seitz & Schneider 1996) can be done by solving the integral equation obtained from Eq.(38) where γ is replaced by $(1-\kappa)\mathbf{g}$. Alternatively one can use the fact that both κ and γ depend on second derivatives of the projected gravitational potential φ:

$$\left\{\begin{array}{l} \dfrac{\partial\kappa}{\partial x_1} = \dfrac{\partial\gamma_1}{\partial x_1} + \dfrac{\partial\gamma_2}{\partial x_2} \\[2ex] \dfrac{\partial\kappa}{\partial x_2} = \dfrac{\partial\gamma_2}{\partial x_1} - \dfrac{\partial\gamma_1}{\partial x_2} \end{array}\right. \tag{50}$$

which permits to recover the mass density by this relation:

$$\nabla\log(1-\kappa) = \frac{1}{1-|\mathbf{g}|^2}\cdot\begin{pmatrix} 1+g_1 & -g_2 \\ -g_2 & 1-g_1 \end{pmatrix}\begin{pmatrix} \partial_1 g_1 + \partial_2 g_2 \\ \partial_1 g_2 - \partial_2 g_1 \end{pmatrix} \tag{51}$$

Both Eq.(36) and Eq.(51) express the same relation between κ and γ and can be used to reconstruct the projected mass density.

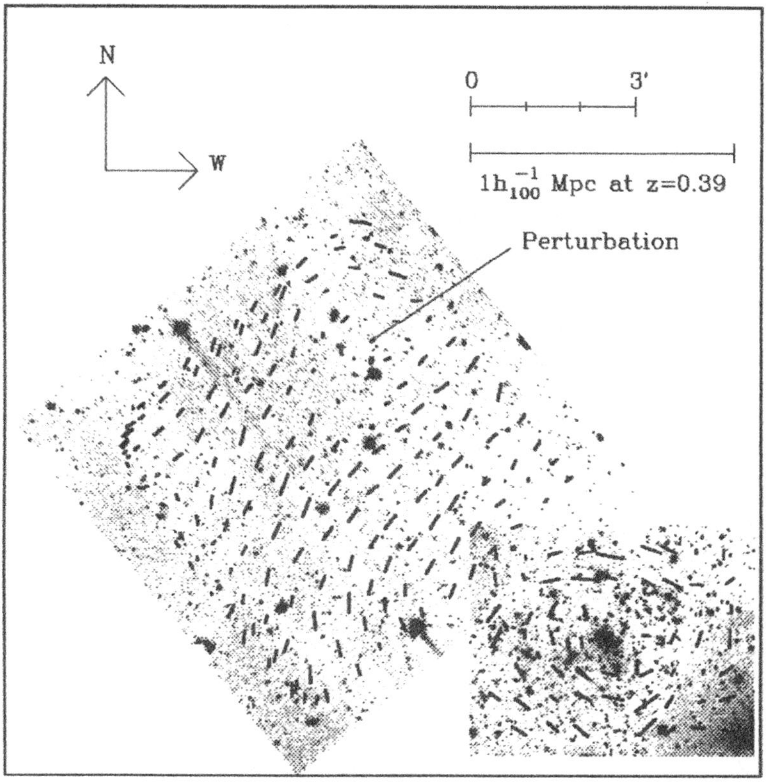

Figure 6. Detection of the shear field around Cl0024+1654. The figure is composed of two deep CCD images obtained at CFHT. The small field on the right is the central region of the cluster. The off-centered field on the left covers a much larger field and has been observed in order to detect the mass distribution at the periphery of the cluster. The thick full lines indicate the local average ellipticity. Each line displays the amplitude and the orientation of the distortion. The pattern is typical of a coherent gravitational shear produced by the mass of the cluster. Note also the perturbation of the shear field in the upper left. This effect is due to a secondary deflector which locally modifies the shear field.

Although mass reconstruction is now as a robust technique (see the comparison of various algorithms in Mellier 1998), the mass distribution recovered is not unique because the addition of a lens plane with constant mass density keeps the distortion of the galaxies unchanged. Furthermore, the inversion only uses the ellipticity of the galaxies regardless of their dimension, so that changing $(1-\kappa)$ in $\lambda(1-\kappa)$ and γ in $\lambda\ \gamma$ keeps $\mathbf{g}$ invariant. This is the so-called *mass sheet degeneracy* (Gorenstein et al 1988).
The degeneracy could in principle be broken if the magnification can be measured independently, since it is not invariant under the linear transformation mentioned above, but instead it is reduced but a factor $1/\lambda^2$. The magnification can be measured directly by using the magnification bias

(Broadhurst et al 1995), which changes the galaxy number-counts.
The magnification bias expresses the effects of the gravitational magnification, which increases the flux received from lensed galaxies and magnifies by the same amount the area of the projected lensed sky and thus decreases the apparent galaxy number density. The total amplitude of the magnification bias depends on the slope of the galaxy counts as a function of magnitude and on the magnification factor of the lens. For a circular lens, the radial galaxy number density of background galaxies writes:

$$N(< m, r) = N_0(< m)\ \mu(r)^{2.5\alpha-1} \approx N_0\ (1+2\kappa)^{2.5\alpha-1} \quad \text{if } \kappa \text{ and } |\gamma| \ll 1\ , \tag{52}$$

where $\mu(r)$ is the magnification, $N_0(< m)$ the intrinsic (unlensed) number density, obtained from galaxy counts in a nearby empty field, and α is the intrinsic count slope:

$$\alpha = \frac{d log N(< m)}{dm}\ . \tag{53}$$

A radial magnification bias $N(< m, r)$ shows up only when the slope $\alpha \neq 0.4$; otherwise, the increasing number of magnified sources is exactly compensated by the apparent field dilatation. For slopes larger than 0.4 the magnification bias increases the galaxy number density, whereas for slopes smaller than 0.4 the radial density will show a depletion. Hence, no change in the galaxy number density can be observed for $B(< 26)$ galaxies, since the slope is almost this critical value (Tyson 1988). But it can be detected in the $B > 26$, $R > 24$ or $I > 24$ bands when the slopes are close to 0.3 (Smail et al 1995). The change of the galaxy number density can be used as a direct measurement of the magnification and can be included in the maximum likelihood inversion as a direct observable in order to break the mass sheet degeneracy .
The *lens parallax method* (Bartelmann & Narayan 1995) which compares the angular sizes of lensed galaxies with an unlensed sample can be also used as an alternative to break the degeneracy. Another approach consists in using wide field cameras with a field of view much larger than clusters of galaxies. In that case κ should vanish at the boundaries of the field, so that the degeneracy could in principle be broken.

Since 1990, many clusters have been investigated using the weak lensing inversion, either using ground-based or HST data. They are summarized in Table 2, but the comparison of these results is not straightforward because of the different observing conditions which produced each set of data and the different mass reconstruction algorithms used by each author. Nevertheless, all these studies show that on scales of about 1 Mpc, the geometry of mass distributions, the X-ray distribution and the galaxy distribution are similar (see Figure 4), though the ratio of each component with respect to the others may vary with radius. The inferred median M/L value

TABLE 2. Results obtained from weak lensing analyses of clusters. The scale is the typical radial distance with respect to the cluster center. The last cluster has two values for the M/L ratio. This corresponds to two extreme redshifts assumed for the lensed population, either $z = 3$ or $z = 1.5$. For this case, the two values given for the velocity dispersion are those inferred when $z = 3$ or $z = 1.5$ are used.

Cluster	z	σ_{obs} (kms^{-1})	σ_{wl} (kms^{-1})	M/L (h_{100})	Scale (h_{100}^{-1} Mpc)
A2218	0.17	1370	-	310	0.1)
A1689	0.18	2400	1200-1500	-	0.5
			-	400	1.0
A2163	0.20	1680	740-1000	300	0.5
A2390	0.23	1090	$\approx$1000	320	0.5
Cl1455+22	0.26	$\approx$ 700	-	1080	0.4
AC118	0.31	1950	-	370	0.15
Cl1358+62	0.33	910	780	180	0.75
MS1224+20	0.33	770	-	$\approx$ 800	1.0
Q0957+56	0.36	715	-	-	0.5
Cl0024+17	0.39	1250	-	150	0.15
			1300	$\approx$900	1.5
Cl0939+47	0.41	1080	-	120	0.2
			-	$\approx$250	0.2
Cl0302+17	0.42	1080		80	0.2
RXJ1347-11	0.45	-	1500	400	1.0
3C295	0.46	1670	1100-1500	-	0.5
			-	330	0.2
Cl0412-65	0.51	-	-	70	0.2
Cl1601+43	0.54	1170	-	190	0.2
Cl0016+16	0.55	1700	-	180	0.2
			740	740	0.6
Cl0054-27	0.56	-	-	400	0.2
MS1137+60	0.78	859	-	270	0.5
RXJ1716+67	0.81	1522	-	190	0.5
MS1054-03	0.83	1360	1100-2200	350-1600	0.5

is about 300, with a trend to increase with radius. Contrary to the strong lensing cases, there is no evidence of discrepancies between the X-ray mass and the weak lensing mass. It is worth noting that the strong lensing mass and the weak lensing mass estimates are consistent in the region where the amplitude of two regimes are very close. This is an indication that the description of the X-ray gas, and its coupling with the dark matter on the scales corresponding to strong lensing studies is oversimplified, whereas on larger scales, described by weak lensing analysis, the detailed description of the gas has no strong impact.
The large range of M/L could partly be a result of one of the issues of the mass reconstruction from weak lensing. In particular, the deviation angle depends on the ratios of the three angular-diameter distances (see Bernardeau, these proceedings), which depends on the redshift we assume for the sources. For low-redshift lenses, the dependence with redshift of the background galaxies is not important, so the calibration of the mass can be provided with a reasonable confidence level. However, distant clusters are highly sensitive to the redshift of the sources, and it becomes very difficult to scale the total mass without this information.

From the investigation of about 20 clusters, it turns out that the median M/L is lower than 400. This implies that weak lensing analyses predict $\Omega < 0.3$ with a high significance level. These constraints on Ω are in good agreement with other observations.
Another strong statement results from the mass reconstruction obtained by Luppino & Kaiser (1997) and Clowe et al (1998) or from the detection of giant arcs in very distant clusters (Deltorn et al 1997): massive clusters do exist at redshift $\approx$1 ! This is a strong but reliable statement, though the total mass and the M/L cannot be given with a high accuracy. Therefore, unless unknown important systematics have been disregarded, we now have the first direct observational evidences that high mass-density peaks have generated massive clusters of galaxies at redshift 1. These promising results are corroborated by weak lensing studies around radio sources and quasars (Mellier 1998).

4.2. MEASURING THE MASSES OF GALAXIES

Gravitational lensing can also provide valuable insight on the halos of galaxies. Since it works on all scales, in principle the halos of galactic dark matter could be probed from their gravitational lensing effects on background galaxies.

TABLE 3. Results on Einstein ring analyses. This is not a complete survey of the rings detected. I only report on those for which enough data have been obtained and a model has been presented. [1] Kochanek 1995, [2] Impey et al 1998, [3] Warren et al 1998.

Lens	z_{lens}	z_{source}	σ_{DM}	M/L
			(kms^{-1})	(h_{100})
MG 1654+134[1]	0.25	1.74	≈ 220	≈ 20.4 (in B)
PG 1115+080[2]	0.310	1.722	≈ 240	≈ 8.2 (in I)
0047-2808[3]	0.485	3.595	≈ 270	-

4.2.1. *Einstein rings*

Rings occur when the alignment of the observer, the lens and the source is almost perfect, and if the source is covering the whole internal caustic, forming the so called "Einstein ring". The first rings were observed around galaxies in radio surveys (see Refsdal & Surdej 1994 for a recent review). They have provided unique targets to measure the mass-to-light ratios and to probe the mass profiles of galaxies (Kochanek 1991). In the case of rings, the mass of the lensing galaxies can be very well constrained (see for instance Kochanek 1995 and Table 3), so the properties of the halos inferred from modeling are reliable. The results are somewhat reasonable, with typical velocity dispersion and mass-to-light ratio in good agreement with other techniques. However, I would like to stress again that, due the simplicity of these lens configurations and the very strong constraints provided by the size of the ring, these results are very robust.

4.2.2. *Perturbations near giant arcs*

Perturbations of caustics by intervening masses can locally change the length and shape of arcs or locally increase the intensity of unresolved arc substructures. Dramatic perturbations could even be responsible for the complete vanishing of an arc segment. The perturbation of caustics by a smaller interloping lens can be understood by considering that the magnification matrix degenerates to a single eigenvector tangent to the critical curve. Indeed the distortion of the images of objects close to the critical line corresponds mainly to a stretching along the direction of merging. Therefore, the angular coordinates of the source can be developed in polynomial form along the direction of merging (Kassiola, Kovner & Fort 1992). Figure

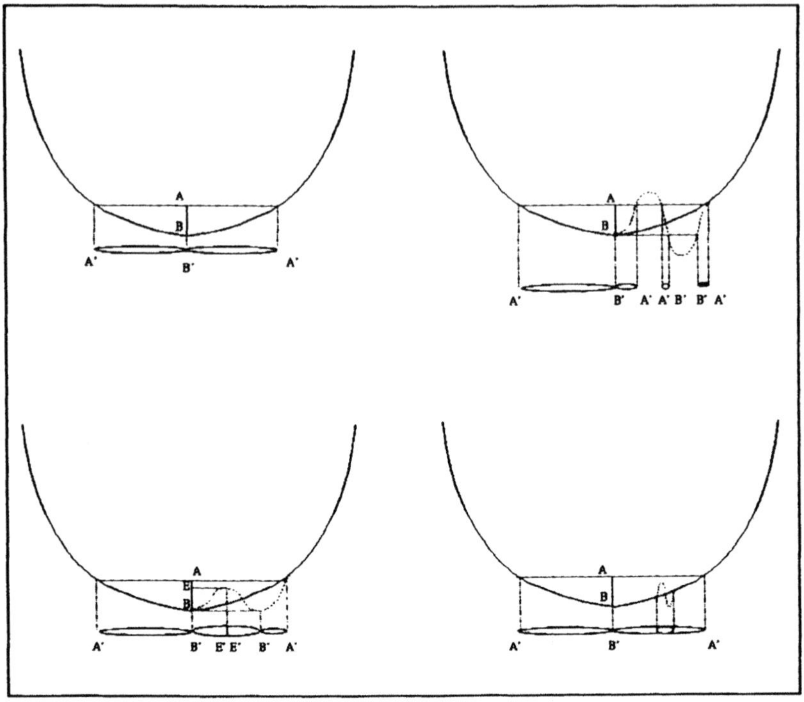

Figure 7. Gravitational distortion induced by a perturbation close to a giant fold arc. The top left panel shows the formation of two elongated images by a fold catastrophe. The vertical segment (A,B) is the length of the source in the source plane. The images are given by the antecedent of the parabola (fold caustic) and therefore two images are formed. In the next panel we introduce a perturbation represented as the dashed line which co-adds to the parabola. This perturbation roughly represents the deviation angle expected from an isothermal sphere with soft core, assuming its influence is zero beyond a given radius. The difference between the three configurations is the intensity of the perturbation. When the intensity is large enough (top right panel) it can break the nearest image and form multiple small images. When the intensity decreases (bottom left) the perturbation can break the images but can also form sub-ellipses of merging sub-images.

7 gives a short description of the effect of perturbations on two merging images of an extended object near a fold. In this case the functional form of the unperturbed fold is approximated by a second-order polynom. If a large perturbation from a nearby galaxy is added, the image can be split into many components (Figures 7 and 8)

Large perturbations of caustics have been used to constrain the galaxies located close to the giants arcs in A370, Cl0024+17, Cl2244 or MS2137-23. In general, the absence of breaks along a well-defined arc provides robust upper limits to the mass of perturbing galaxies. In summary, the results, as those shown in Table 4 are not suprising. The masses found for these cluster galaxies range between 10^{10} $M_{\odot}$ and 2×10^{11} $M_{\odot}$, with typical mass-to-light

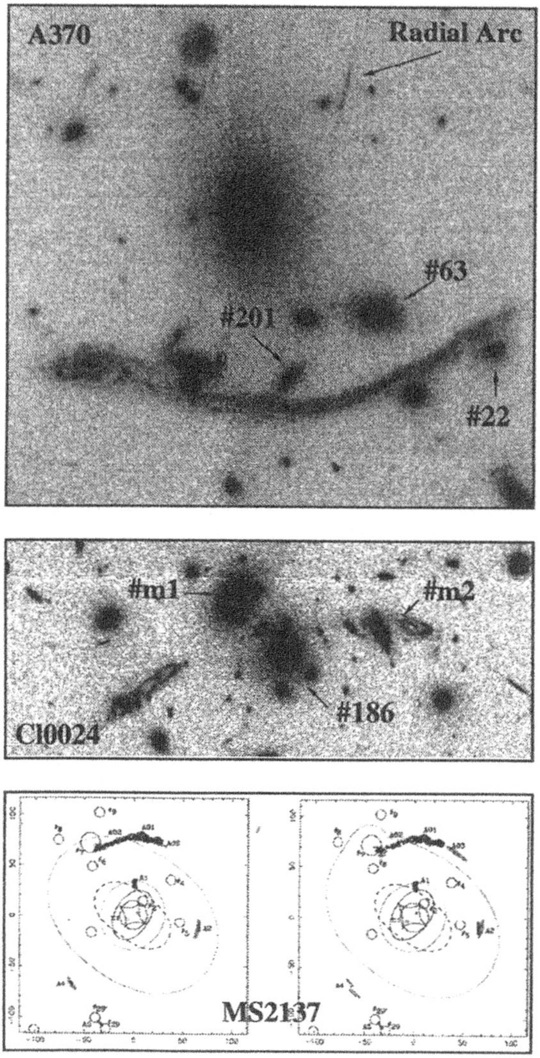

Figure 8. Perturbations close to giant arcs produced by galaxies. The top panel is the giant arc in A370. The galaxies with a number are those reported in Table IV. On this HST images the effects of these galaxies is clear, in particular for the galaxy #22. The middle panel shows similar pertubation in Cl0024+17. In particular, the effect of #158 is important. Though the central arc should be twice as long as the others (prediction of cups arcs), it is clearly smaller. The contraction is produced by the two galaxies which are located at the top and the bottom of the central arc. The bottom panel shows, the best lens model (left) of MS2137-23 and an example of the perturbation of the galaxy #7. When two much mass is put in this galaxy one can see that the giant arc is broken in three sub-arcs. This effect permits to put upper limits on the mass of this galaxy.

TABLE 4. Results cluster galaxies near arcs (perturbations). The table gives the constrains on the masses of the halos of galaxies around giant arcs which are shown in Figure 8.

Lens	z_{lens}	# Galaxy	σ_{DM} (kms^{-1})	M/L (h_{100})
A370	0.38	201	-	$2.4 < M/L < 8$ (in B)
A370	0.38	22	-	$4.4 < M/L < 12.6$ (in B)
A370	0.38	63	-	$9 < M/L < 30$ (in B)
MS2137-23	0.33	7	≤ 190	-
Cl0024+17	0.39	m1	≈ 180	-
Cl0024+17	0.39	186	≈ 250	-
Cl0024+17	0.39	m2	≈ 120	-

ratios between 5 and 30.

4.2.3. *Galaxy-galaxy lensing on field galaxies*

A more promising approach consists in a statistical study of the deformation of distant galaxies by foreground galactic halos. The galaxy-galaxy lensing analysis uses the correlation between the position of foreground galaxies and the orientation of background population. If the correlation is produced by the gravitational shear of the foreground halos, then it is possible to probe their mass, if the redshift distributions of the foregrounds and the backgrounds are known.

Let us define the shape of a galaxy by the vector $\boldsymbol{\epsilon}$ as defined in Eq.(12). In the weak lensing regime, the shear simply translates in the $\boldsymbol{\epsilon}$ plane at the new position $\boldsymbol{\epsilon} = \boldsymbol{\epsilon}_0 + \boldsymbol{\delta}$, where $\boldsymbol{\epsilon}_0$ is the intrinsic shape and $\boldsymbol{\delta}$ the distortion produced by the shear. If we assume that the translation is done in the x-direction, then the shape distribution is simply translated:

$$f_\epsilon\left(\epsilon_x, \epsilon_y\right) = f_\epsilon^0\left(\epsilon_0, \epsilon_y\right) = f_\epsilon^0\left(\epsilon_x - \delta, \epsilon_y\right) \ , \tag{54}$$

where f_ϵ^0 is the intrinsic shape distribution. In that case, the orientations of the galaxies with respect to the x-axis, ϕ, is modified and the final distribution, averaged over the ellipticity of the galaxies, is

$$P_\phi\left(\phi\right) = \int f_\epsilon^0 \ \epsilon \ d\epsilon + \delta \cos\left(2\phi\right) \int \epsilon \ \frac{df_\epsilon^0}{d\epsilon} d\epsilon \tag{55}$$

that is

$$P_\phi(\phi) = \frac{2}{\pi}\left(1 - \langle\delta\rangle \cos(2\phi)\left\langle \epsilon^{-1}\right\rangle\right) \tag{56}$$

Therefore, we expect a deficit of radially-oriented galaxies and conversely an excess of tangentially-oriented images.
A statistical analysis is then possible, if one assumes that all the foreground galaxies have similar halos, which can be scaled from observations. The procedure is the following. Assume an analytical shape for the projected potential φ (or for the projected mass density Σ, or the 3-dimension mass density ρ). Then, assuming the potential of halos are circular, the deflection angle is

$$\alpha(r) = \frac{2}{c^2}\frac{D_{LS}}{D_{OS}}\frac{d\varphi}{dr} \tag{57}$$

and the polarization writes

$$p(r) = D_{OL}\, r\frac{d}{dr}\left(\frac{\alpha(r)}{r}\right) \tag{58}$$

which depends on the typical scale, say r_c and the depth of the potential well (*i.e.* for an ellipical galaxy or a cluster of galaxies the velocity dispersion of the lens, σ_{los}, and for a spiral galaxy its rotation velocity).
In the case of spiral galaxies, the depth of the potential can be related to the circular velocity of the galaxy, which can be obtained either directly or from the apparent magnitude and the redshift of the galaxy. Let us assume for simplicity that the mass-to-light ratio of the galaxy is independent of its luminosity. Then the typical scale

$$r_c \propto \sqrt{M} \quad , \text{ that is, from Tully} - \text{Fisher} : r_c \propto \sqrt{L} \propto V_c^2 \ . \tag{59}$$

Therefore, the photometry of the foreground galaxies can, to first approximation, scale their mass, from the calibrated Tully-Fisher relation, as well as their redshift, from a magnitude-redshift relation. The main objective of the galaxy-galaxy lensing is then to calibrate the Tully-Fisher relation by measuring the physical scales r_c^* and L^* from the averaged polarization produced by the foreground galaxies onto the background lensed sources.
The expected gravitational distortion is very weak: for foregrounds at redshift $< z_l >= 0.1$, backgrounds at $< z_s >= 0.5$, and typical halos with velocity dispersion of 200 kms^{-1} and radius of 100 kpc, $|\gamma| \approx 1\%$ at about 20 kpc from the center. But if the observations go to very faint magnitudes there is a huge number of background lensed galaxies, so that the weakness of the signal is compensated by the large statistics.
The first reliable results came from deep sub-arcsond seeing CCD observations (Brainerd et al 1996). The distortion was compared with simulations,

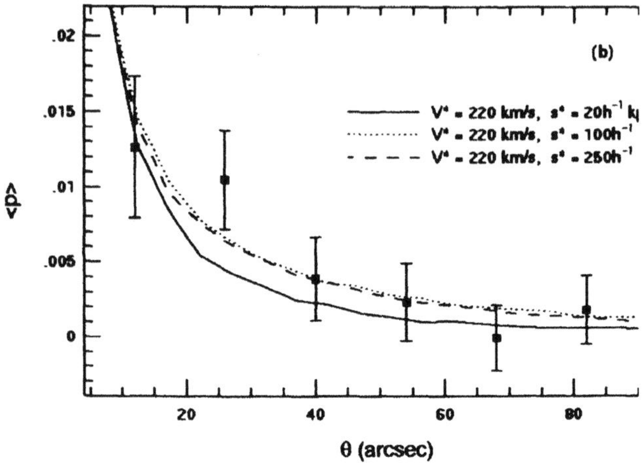

Figure 9. Angular variation of polarization produced by weak lensing of foreground galaxies on the background (lensed) sources in the Brainerd et al (1996) sample. The lines show theoretical expectations for three models of halos having different velocity dispersion and scales.

based on analytical profiles assumed for the dark matter halos as well as the Tully-Fisher relation and a magnitude-redshift relation, in order to relate mass models to observations. They detected a significant polarization of about 1% , averaged over separation between 5" and 34" (see Figure 9). They concluded that halos smaller that $10h^{-1}$ kpc are excluded at a 2σ level, but the data are compatible with halos of size larger that $100h^{-1}$ kpc and circular velocities of 200 kms^{-1}.
The HST data look perfectly suited for this kind of program which demands high image quality and the observation of many field galaxies. Griffiths et al (1996) used the Medium Deep Survey (MDS) and measured the distortion produced by foreground elliptical and spiral galaxies. They found similar results as Brainerd et al (1996) but with a more significant signal for foreground elliptical than spiral galaxies. The comparison with shear signals expected from various analytical models seems to rule out de Vaucouleur's law as mass density profile of ellipticals. Dell Antonio & Tyson (1996) and Hudson et al (1998) analyzed the galaxy-galaxy lensing signal in the HDF. As compared with the ground-based images or the MDS, the field is small but the depth permits to use many background galaxies even on scale smaller than 5 arcseconds. Furthermore, the UBRI data of the HDF permit to infer accurate photometric redshifts for the complete sample of galaxies. By comparing the lensing signal with predictions from an analytical model for the halo, Dell'Antonio & Tyson found a significant distortion of about 7% at 2" from the halo center which corresponds to halos with

typical circular velocities of less than 200 $km.s^{-1}$. All these results seem consistent with those of Brainerd et al.

4.2.4. *Galaxy-galaxy lensing on cluster galaxies*

As it has been shown from the observation of perturbations along giant arcs, the clumpiness of dark matter on small-scale can be probed by anomalies in lensing configurations (see Sect. 4.2.3). With the details visible on the HST images of arclets in A2218, AC114 or A2390, the sample of halos which can be constrained by this method is much larger and can provide more significant results by extending the method to perturbations along arclets. The number of details permits also to use more sophisticated methods of investigation.

The simplest strategy is to start with an analytical potential which reproduces the general features of the shear pattern of HST images, and in a second step, to include in the model analytical halos around the brightest cluster members. In practice, additional mass components are put in the model in order to interpret the arc(let)s which cannot be easily explained by the simple mass distribution. Some guesses are done in order to pair unexplained multiple images. The colors of the arc(let)s as well as their morphology help a lot to make these associations. This approach has been proposed by Natarajan & Kneib (1997), and Natarajan et al (1998). The detailed study done in AC114 by Natarajan et al (see Figure 10) indicates that about 10% of the dark matter is associated with halos of cluster galaxies. These halos have truncation radii smaller than field galaxies ($r_t \approx$15kpc) with a general trend of S0-galaxies to be even more truncated than the other galaxies. If this result is confirmed it would be a direct evidence that truncation by tidal stripping is really efficient in rich clusters of galaxies. This result is somewhat contradictory with the absence a clear decrease of rotation curves of spiral galaxies in nearby clusters (Amram et al 1993) which is interpreted as a proof that massive halos of galaxies still exist in cluster galaxies. However, it could be explained if the spirals which have been analyzed appear to be in the cluster center only by projection effects but are not really located in the very dense region of the clusters where stripping is efficient.

Geiger & Schneider (1997, 1998) used a maximum likelihood analysis which explores simultaneously the distortions induced by the cluster as a whole and by its individual galaxies. They applied this analysis to the HST data of Cl0939+47 and reached similar conclusions as Natarajan et al (1998). Several issues limit the reliability of their analyses and of the other methods as well (Geiger & Schneider 1998). First, depending on the slope of the mass profile of the cluster, the contributions of the cluster mass density and of the cluster galaxies may be difficult to separate. Second, it is

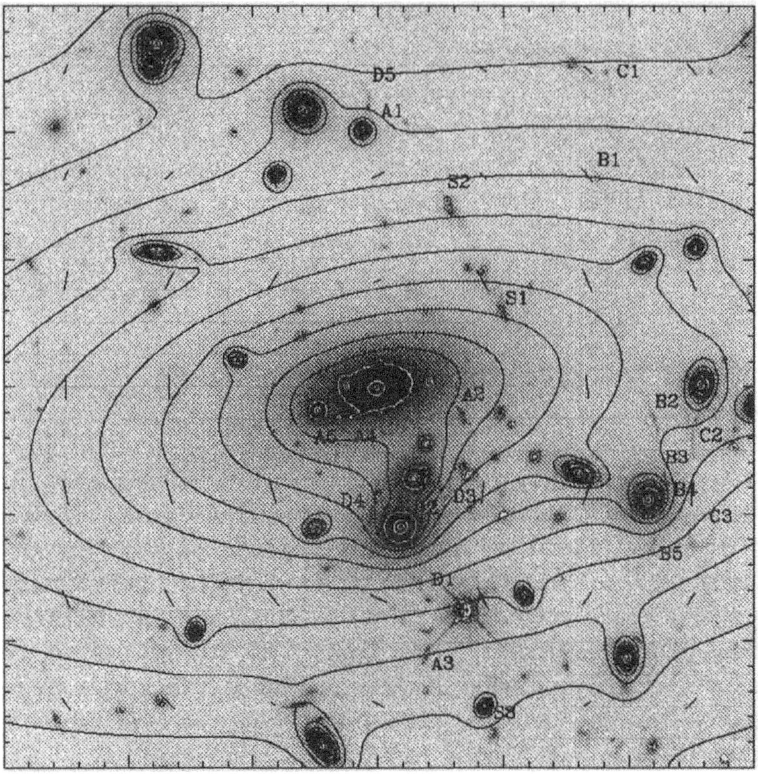

Figure 10. A nice example of a complete lensing analysis using arc(let)s. The figure shows the mass density contours of AC114 overlayed to the HST image and obtained from a fitting of potential wells which include the clumps associated to the clusters and the halos of the galaxies. Therefore, both the massive clumps and the perturbations are included. This permits to estimate the mass and size or dark halos of galaxies (from Natarajan & Kneib 1997).

necessary to have a realistic model for the redshift distribution of the background and foreground galaxies. Finally, the mass sheet degeneracy is also an additional source of uncertainties. Regarding these limitations, Geiger & Schneider discuss the capability of the galaxy-galaxy lensing in clusters to provide valuable constrains on the galactic halos from the data they have in hands. Indeed, some of the issues they raised can be solved, like for instance the redshift distribution of the galaxies. It would be interesting to look into more details how the analysis could be improved with more and better data.

5. Conclusion and future prospects

The measurement of mass of galaxies and cluster of galaxies made a major step after the discovery of gravitational lenses like rings, arcs and arclets.

The results on clusters of galaxies seem quite robust. In summary, the total masses and the mass-to-light ratio recovered from lensing theory are similar to those found by other techniques, which confirms that investigation of dark matter in clusters favors $\Omega < 0.4$. The new result is the evidence that dark matter is strongly concentrated at the cluster center. More interesting, it seems that the geometry of the light distribution of the brightest cluster members traces the dark matter with a good accuracy. Furthermore, despite the discrepancies found between the X-ray mass and the lensing mass on small scales, the agreement is good on Megaparcsec scales. This is still a matter of debates, but the dichotomies between large scale and small scale as well as between non-cooling flows and cooling flow clusters are strong arguments that the discrepancy is produced by over simplifications on the dynamical stage of the gas.
The investigation of galaxies are more debated. Though Einstein rings provide very accurate mass, the galaxy-galaxy lensing approach is still at its infancy and needs much more attention before providing something reliable.
In the future, the investigation of gravitational systems will be extended to nearby clusters. For those systems, the shear amplitude is small but each angular scale probes a much smaller physical scale, so that one can use much more galaxies than in distant systems in order to map the same physical scale. Groups of galaxies will be soon analyzed extensively as well. In parallel, we are still looking for dark clusters that gravitational lensing could reveal easily. It these systems do exist, then their discovery would be a major breakthrough.
Indeed, the wide field CCD cameras and the future NGST observations will permit to increase considerably the scientific return from lensing studies. The two key technical points are the understanding of systematics and the statistics. Both wide field CCD cameras and HST/NGST will provide important insights and intrinsically much better data that present-day observations. In parallel, since the masses are scaled by angular distances, the redshifts of the lenses and the sources are needed. The new giant telescopes equipped with visible and infrared spectrographs, as well as the joint use of visible and near infrared photometric redshifts open a new area. We expect that the number of Einstein rings and giant arcs with redshift will increase significantly in the next decade, putting much better constrains on the mass and mass-to-light ratios of galaxies and clusters.

Acknowledgments

I am grateful to M. Lachièze-Rey for his invitation to give this lecture in Cargèse. I thank F. Bernardeau, D. Elbaz, B. Fort, J.-P. Kneib, J. Rich, P.

Schneider, L. van Waerbeke for the numerous stimulating discussions.

References

Allen, S. W. 1998. *MNRAS* **296**, 392-406.
Amram, P., Sullivan, W.T., Balkowski, C., Marcelin, M., Cayatte, V. 1993. *Ap. J. Lett.* **403**, L59-L62.
Bartelmann, M. 1996. *Astron. Astrophys.* **313**, 697-702.
Bartelmann, M., Narayan, R. 1995. *Ap. J.* **451**, 60-75.
Blandford, R.D., Narayan, R. 1992. *Annu. Rev. Astron. Astrophys.* **30**, 311-58.
Böhringer, H., Tanaka, Y., Mushotzky, R. F., Ikebe, Y., Hattori, M. 1998. *Astron. Astrophys.* **334**, 789-98.
Brainerd, T. G., Blandford, R. D., Smail, I. 1996. *Ap. J.* **466**, 623-37.
Broadhurst, T., Taylor, A. N., Peacock, J. 1995. *Ap. J.* **438**, 49-61.
Burke, B. F. 1981. *Ap. J.* **244**, L1.
Clowe, D., Luppino, G. A., Kaiser, N., Henry, J. P., Gioia, I. M. 1998. *Ap. J. Lett.* **497**, L61-L64.
Dell'Antonio, I. P., Tyson, J. A. 1996. *Ap. J. Lett.* **473**, L17-L20.
Deltorn, J.-M., Le Fèvre, O., Crampton, D., Dickinson, M. 1997. *Ap. J. Lett.* **483**, L21-L24.
Etherington, I. M. M. 1933. *Phil. Mag.* **15**, 761-75.
Fahlman, G., Kaiser, N., Squires, G., Woods, D. 1994. *Ap. J.* **437**, 56-62.
Fort, B., Prieur, J.-L., Mathez, G., Mellier, Y., Soucail, G. 1988. *Astron. Astrophys. Lett.* **200**, L17-L20.
Fort, B., Mellier, Y. 1994. *Astron. Astrophys. Rev.* **5**, 239-92.
Geiger, B., Schneider, P. 1997. astro-ph/9805034.
Geiger, B., Schneider, P. 1998. *MNRAS* **295**, 497-510.
Gioia, I. M., Shaya, E. J., Le Fèvre, O., Falco, E. E., Luppino, G., Hammer, F. 1998. *Ap. J.* **497**, 573-86.
Gorenstein, M.V., Falco, E.E., Shapiro, I.I. 1988. *Ap. J.* **327**, 693-711.
Griffiths, R. E., Casertano, S., Im, M., Ratnatunga, K. U. 1996. *MNRAS* **282**, 1159-64.
Hewitt, J. N., Turner, E. L., Schneider, D. P., Burke, B. F., Langston, G. I., Lawrence, C. R., 1988. *Nature* **333**, 357.
Hudson, M. J., Gwyn, S., Dahle, H., Kaiser, N. 1998. *Ap. J* **503**, 531-42.
Impey, C., Falco, E., Kochanek, C., Lehar, J., McLeod, B., Rix, H.-W., Peng, C., Keeton, C. 1998. astro-ph/9803207.
Kaiser, N., Squires, G. 1993. *Ap. J.* **404**, 441-50.
Kaiser, N. 1995. *Ap. J. Lett.* **439**, L1-L3.
Kassiola, A., Kovner, I., Fort, B. 1992. *Ap. J.* **400**, 41-57.
Kneib, J.-P., Mathez, G. Fort, B., Mellier, Y., Soucail, G., Longaretti, P.-Y. 1994. *Astron. Astrophys.* **286**, 701-17.
Kneib, J.-P., Mellier, Y., Pelló, R., Miralda-Escudé, J., Le Borgne, J.-F., Böhringer, H., Picat, J.-P. 1995. *Astron. Astrophys.* **303**, 27-40.
Kneib, J.-P., Ellis, R. S., Smail, I., Couch, W., Sharples, R. M. 1996. *Ap. J.* **471** 643-56.
Kochanek, C. S. 1990. *MNRAS* 247:135-51.
Kochanek, C. S. 1991. *Ap. J.* **373**, 354-68.
Kochanek, C. S. 1995. *Ap. J.* **445**, 559-77.
Luppino, G., Kaiser, N. 1997. *Ap. J.* **475**, 20-28.
Lynds, R., Petrosian, V. 1986. *BAAS* **18**, 1014.
Mellier, Y., Fort, B., Kneib, J.-P. 1993. *Ap. J.* **407**, 33-45.
Mellier, Y. 1998. astro-ph/9812172.
Mellier, Y. 1999. *Les Houches lecture on Gravitational Lensing.* October 1997. O. Le Fèvre, S. Charlot Eds.
Miralda-Escudé, J., Babul, A. 1995. *Ap. J.* **449**, 18-27.

Natarajan, P., Kneib, J.-P. 1997. *MNRAS* **287**, 833-47.
Natarajan, P., Kneib, J.-P., Smail, I., Ellis, R. S. 1998. *Ap. J.* **499**, 600-07.
Navarro, J. F., Frenk, C. S., White, S. D. M. 1997. *Ap. J.* **490**, 493-508.
Ota, N., Mitsuda, K., Fukazawa, Y. 1998. *Ap. J.* **495**, 170-78.
Pierre, M., Le Borgne, J.-F., Soucail, G., Kneib, J.-P. 1996. *Astron. Astrophys.* **311**, 413-24.
Refsdal, S., Surdej, J. 1994. *Rep. Prog. Phys.* **56**, 117-85.
Sahu, K. C., Shaw, R., A., Kaiser, M. E., Baum, S. A., Fergusson, H. C., Hayes, J. J. E., Gull, T. R., Hill, R. J., Hutchings, J. B., Kimble, R. A., Plait, P., Woodgate, B. E. 1998. *Ap. J. Lett.* **492**, L125-L129.
Schneider, P., Ehlers, J., Falco, E. E. 1992. *Gravitational Lenses.* Springer.
Schindler, S., Hattori, M., Neumann, D., M., Böhringer, H. 1997. *Astron. Astrophys.* **317**, 646-55.
Schramm, T., Kayser, R. 1995. *Astron. Astrophys.* **299**, 1-10.
Seitz, S., Schneider, P. 1996. *Astron. Astrophys.* 305:383-401.
Smail, I., Hogg, D.W., yan, L., Cohen, J.G. 1995. *Ap. J. Lett.* **449**, L105-L108.
Smail I., Ellis, R. S., Dressler, A., Couch, W. J., Oemler, A., Sharples, R., Butcher, H. 1997. *Ap. J.* **479**, 70-81.
Soifer, B.T., Neugebauer, G., Franx, M., Matthews, K., Illingworth, G.D. 1998. astro-ph/9805219.
Soucail, G., Fort, B., Mellier, Y., Picat, J.-P. 1987. *Astron. Astrophys. Lett.* **172**, L14-L17.
Soucail, G., Mellier, Y., Fort, B., Mathez, G., Cailloux, M. 1998. *Astron. Astrophys. Lett.* **191**, L19-L22.
Tyson, J.A. 1988. *Astron. J.* **96**, 1-23.
Tyson, J.A., Valdes, F., Wenk, R. A. 1990. *Ap. J. Lett.* **349**, L1-L4.
Walsh, D., Carswell, R.F., Weymann, R.J. 1979. *Nature* **279**, 381-383.
Warren, S. J., Iovino, A., Hewett, P. C., Shaver, P. A. 1998. astro-ph/9806075.
Wu, X.-P., Fang, L.-Z. 1997. *Ap. J.* **483**, 62-67.

MEASUREMENT OF COSMOLOGICAL PARAMETERS USING TYPE IA SUPERNOVAE

R. PAIN
LPNHE, CNRS-IN2P3, Univ. of Paris VI & VII
4, place Jussieu, T33 RdC, 75252 Paris Cedex 05

In this lecture, I present the method used to measure the cosmological parameters using type Ia supernovae (SNe Ia) in section 1 and discuss SNe Ia as cosmological tools in section 2. The detection and follow-up of high-redshift supernovae is presented in section 3 and in section 4, recent results on the measurement of H_0, Ω_M and Ω_Λ are reported. Prospects for the near future are briefly discussed in section 5.

1. Which cosmological parameters - How ?

1.1. A FEW DEFINITIONS

In the *Standard Model* of cosmology the Universe is isotropic and homogeneous and can be described using the Friedmann Lemaître Robertson Walker (FLRW) model. Energy density from radiation and matter and the cosmological constant, Λ, contribute to the expansion of the Universe. At present, the energy density in the form of radiation ρ_{rad} is compared to the matter density ρ_m, and the conservation of energy can be written in the form of the Friedmann equation:

$$\left(\frac{\dot{a}}{a}\right)^2 - \frac{k}{a^2} = \frac{8\pi G}{3}\rho_m + \frac{\Lambda}{3}, \tag{1}$$

where a is the Cosmic scale factor and k=-1,0 or 1 for open, flat or closed geometry. Thus, the expansion rate of the Universe $H = \frac{\dot{a}}{a}$, also known as the Hubble parameter, depends on the matter density, the cosmological constant and the geometry of the Universe.

It is convenient to define the fractional energy density contributions, Ω_M and Ω_Λ, and the contribution from the curvature, Ω_K, at the present

M. Lachièze-Rey (ed.), Theoretical and Observational Cosmology, 245–260.

epoch ($z = 0$)

$$\left\{ \begin{array}{lll} \Omega_M & \equiv & \left(\frac{8\pi G}{3H_0^2}\right) \rho_m^0 \\ \Omega_\Lambda & \equiv & \frac{\Lambda}{3H_0^2} \\ \Omega_K & \equiv & \frac{k}{a_0^2 H_0^2} \end{array} \right. \tag{2}$$

In this framework, the deceleration parameter, q_0, is computed as $\frac{\Omega_M}{2} - \Omega_\Lambda$ and equation (1) can then be written as:

$$1 - \Omega_K = \Omega_M + \Omega_\Lambda \tag{3}$$

1.2. PRINCIPLE OF THE MEASUREMENT

A source of known constant luminosity, also called a *standard candle*, can be used to measure relative distances to provide information on the cosmological parameters.

The basic idea is that the apparent luminosity, f, of a given astrophysical object is related to its intrinsic luminosity, $\mathcal{L}$, by equation

$$f = \frac{\mathcal{L}}{4\pi d_L^2} \tag{4}$$

where the *luminosity distance*, d_L, is a function of the cosmological parameters, H_0, Ω_M and Ω_Λ and of the redshift, z (see for example (Carroll, Press, & Turner, 1992)):

$$d_L(z; H_0, \Omega_M, \Omega_\Lambda) = \frac{(1+z)}{H_0\sqrt{|\Omega_K|}} \mathcal{S}\left(\sqrt{|\Omega_K|} \int_0^z F(z')dz'\right) \tag{5}$$

with $\mathcal{S}(x) = \sin(x)$ for $\Omega_K = 1 - \Omega_M - \Omega_\Lambda < 0$ (*closed Universe*), $\mathcal{S}(x) = \sinh(x)$ for $\Omega_K > 0$ (*open Universe*) and $\mathcal{S}(x) = x$ for a *flat Universe*; and

$$F(z) = [(1+z)^2(1+\Omega_M z) - z(2+z)\Omega_\Lambda]^{-\frac{1}{2}} \tag{6}$$

Thus, measuring luminosity distances using standard candles as a function of redshift can be used to derive the values of the cosmological parameters. More precisely, relative distances allow the determination of Ω_M and Ω_Λ whereas the determination of H_0 requires *absolute* distances or luminosity to be measured as well. This is due to the fact that the luminosity distance, d_L is inversely proportional to H_0 which allows only the product $\mathcal{L}H_0^2$ to be derived.

Figure 1 shows the luminosity distance vs redshift for different cosmologies and with $H_0 = 50$ km s^{-1}Mpc^{-1}. This Figure shows that in a $\Lambda = 0$

cosmology, a ~30% precision measurement of the luminosity distance would allow a clear separation between an open Universe and a critical flat Universe i.e. with $\Omega_M = 1$. In the general case, however, the situation is more complicated since, for example, an open cosmology with $\Lambda = 0$ is almost perfectly degenerate with a flat cosmology having equal amount of mass density and cosmological constant (see eq. 1). This shows that an independent determination of Ω_M and Ω_Λ will be difficult to achieve unless precise measurements are obtained on a large enough redshift range. Today's measurements do not yet allow independent determination of Ω_M and Ω_Λ. Instead, a linear combination, $\alpha\Omega_M - \Omega_\Lambda$, is derived where α depends on the mean redshift probed ($\alpha = 1$ *at* $\bar{z} = 0.4$). Consequently, values of Ω_M for $\Lambda = 0$ or for a flat universe are usually derived.

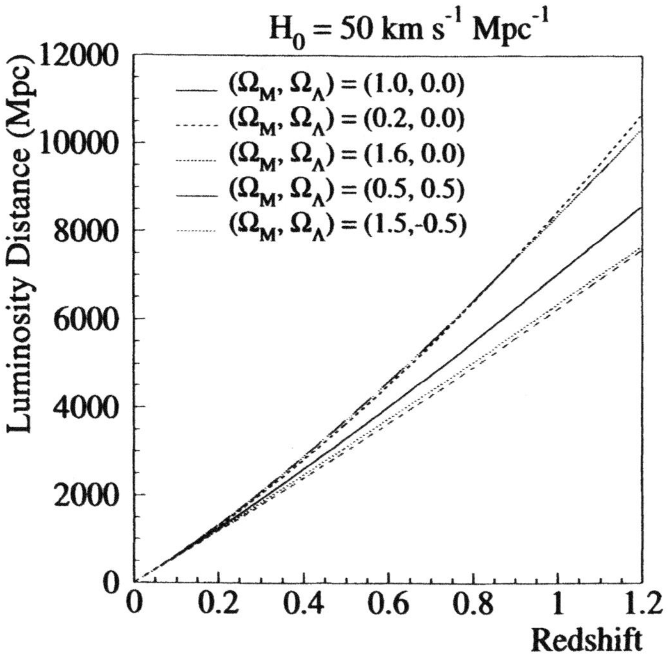

Figure 1. Luminosity distance vs redshift for different cosmological models. Note the degeneracy between an open cosmology with $\Omega_M = 0.2$ and $\Lambda = 0$ and a flat cosmology with equal amount of mass density and cosmological constant.

In practice, optical astronomers measure luminosities in logarithmic units, *magnitudes*[1]. Thus, equation 4 expressed in bolometric magnitudes

[1]In the magnitude system, the brightness of an object through a filter F is defined as

becomes what is known as the Hubble relation:

$$m(z) = M + 5 \log d_L(z; H_0, \Omega_M, \Omega_\Lambda) + 25 \tag{7}$$

with d_L measured in Mpc and where M is the ***absolute magnitude*** (apparent magnitude of the object if it would be at 10 pc distance from the observer) and the Hubble parameter is measured in units of km $s^{-1}Mpc^{-1}$. Thus by measuring the apparent magnitudes of standard candles as a function of redshift one can solve for the unknown cosmological parameters.

2. Type Ia SNe are excellent cosmological tools

Since the earliest studies of supernovae, it has been suggested that these luminous events might be used as standard candles for cosmological measurements (Baade, 1938). At closer distances they could be used to measure the Hubble constant, if an absolute distance scale or magnitude scale could be established, while at higher redshifts they could determine the deceleration parameter (Tammann, 1979),(Colgate, 1979). The Hubble constant measurement became a realistic possibility in the 1980's, when the more homogeneous subclass of Type Ia supernovae (SNe Ia) was identified (see for example (Branch, 1998)). Attempts to measure the other cosmological parameters, however, were stymied for lack of high-redshift supernovae. In a pioneer attempt, Nørgaard-Nielsen et al. (Nørgaard-Nielsen et al., 1989) discovered one SN Ia, at $z = 0.31$. It was only in the mid-nineties that batches of ten or more SNe could be discovered on big telescopes equipped with wide field CCD cameras (see section 3).

In the past years, the case has become quite strong that SNe Ia are a family of very similar events, not all identical. Phillips (1993), Hamuy *et al.* (1995) and Riess, Press, and Kirshner (1995) have presented evidence indicating that this family can be described by a single parameter, essentially representing the shape or width of the light curve, and that this parameter is tightly correlated with the absolute magnitude at maximum. The broad, slow-light-curve supernovae appear brighter, while the narrow, fast-light-curve supernovae are fainter. Hamuy *et al.* used Phillips' (1993) characterizations of this light curve width, Δm_{15}, the magnitude drop in the first 15 days past maximum, and fit their data to template light curves from supernovae representing various Δm_{15} values. Riess *et al.* added and subtracted different amounts of a "correction template" to a Leibundgut "normal" template (Leibundgut et al., 1991) to represent this same lightcurve variation.

$m_F = -2.5 \log \left(\int_0^\infty F(\nu) L(\nu_{app}) d\nu\right) + C_f$, where $F(\nu)$ is the transmission function of the filter used and C_f is a constant specific to the choice of filter. Note that brighter objects have *lower* magnitudes.

A third alternative approach has been used (Perlmutter et al., 1997e). It consists of stretching or compressing the time axis of the Leibundgut template by a "stretch factor" s. This simple parameterization gives variations on the light curve that fit the variety of SNe Ia. On Figure 2, lightcurves of about 10 well measured nearby supernovae discovered and followed-up by the Calan/Tololo group (Hamuy et al., 1996) are shown. No correction for the width-magnitude correlation has been applied on the top Figure whereas the bottom Figure shows the same lightcurves after correction using the "stretch factor" technique.

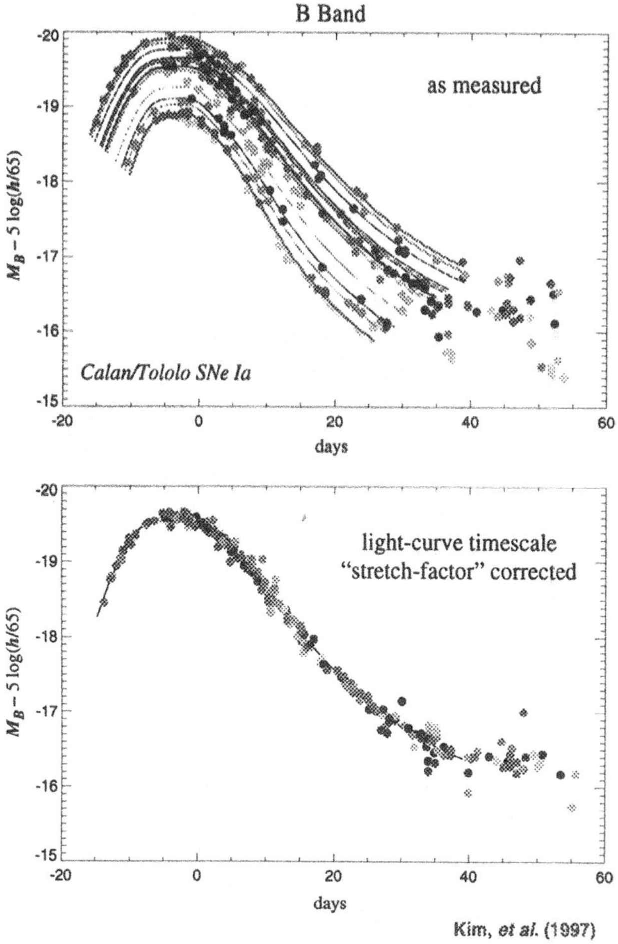

Figure 2. Lightcurves of about 10 well measured nearby supernovae discovered and followed-up by the Calan/Tololo survey: before the width-magnitude correction (top) and after correction using the "stretch factor" technique (bottom).

A number of other supernova observables have been studied as indicators of a given supernova's brightness within the SN Ia family. Branch *et al.* (1998) have discussed the correlation with $B-V$ and $U-B$ color, while Nugent *et al.* (1995) and Riess *et al.* (1997) has pointed out indicative spectral features, suggesting that temperature may provide a single theoretical parameter that characterizes the variations within the SN Ia family. These recent developments suggest that the final reduced "data product" must include not only the K-corrected magnitude at peak, but also any available information on light-curve width, colors, and spectral features that might locate the supernova within the family of fast-and-faint to broad-and-bright SNe Ia.

Applying the width-magnitude correction allows one to "inter-calibrate" Type Ia SNe, reducing the dispersion of the peak magnitude distribution to ~ 0.15 mag. This correction makes SNe Ia particularly good "calibrated candles" and therefore excellent tools for measuring the cosmological parameters.

3. Detection/Follow-up of distant SNe

Supernovae at high redshift are difficult to work with for at least three reasons: they are rare, they are rapid, and they are random. The estimates of SN Ia rates—a few per millennium per galaxy—are daunting, if one wants a statistically useful sample of supernovae. Much of the interesting data must be obtained rapidly, since the supernova rises to maximum light within a few weeks and, at high redshifts, fades below the largest telescopes' limits within a month or two. Furthermore, it is not possible to guarantee photometry and particularly spectroscopy of randomly occurring high-redshift supernovae, since the largest, most over-scheduled telescopes are needed to observe them. Ideally one would like to *schedule* supernova explosions on demand, and then apply for the telescope time to study them, beginning at least a few days before maximum light.

3.1. STRATEGY

To solve these problems, new search techniques were developed. Figure 3 presents a schematic outline of the strategy. Just after a new moon, many tens of high-galactic-latitude fields (including known high-redshift clusters when possible) are observed on a 2.5- to 4-meter telescope. With a wide-field camera, each image contains hundreds of galaxies at high redshifts. Just before the following new moon, the same fields are observed again. The images are compared, thus checking tens of thousands of high redshift galaxies to find the ten or so showing the new light of a supernova that was not there on the previous observation. The supernovae generally do not have time to reach maximum light, with only 2.5 to 3 weeks (or approximately 11 to 14 days in the supernova rest frame) between our after- and before-

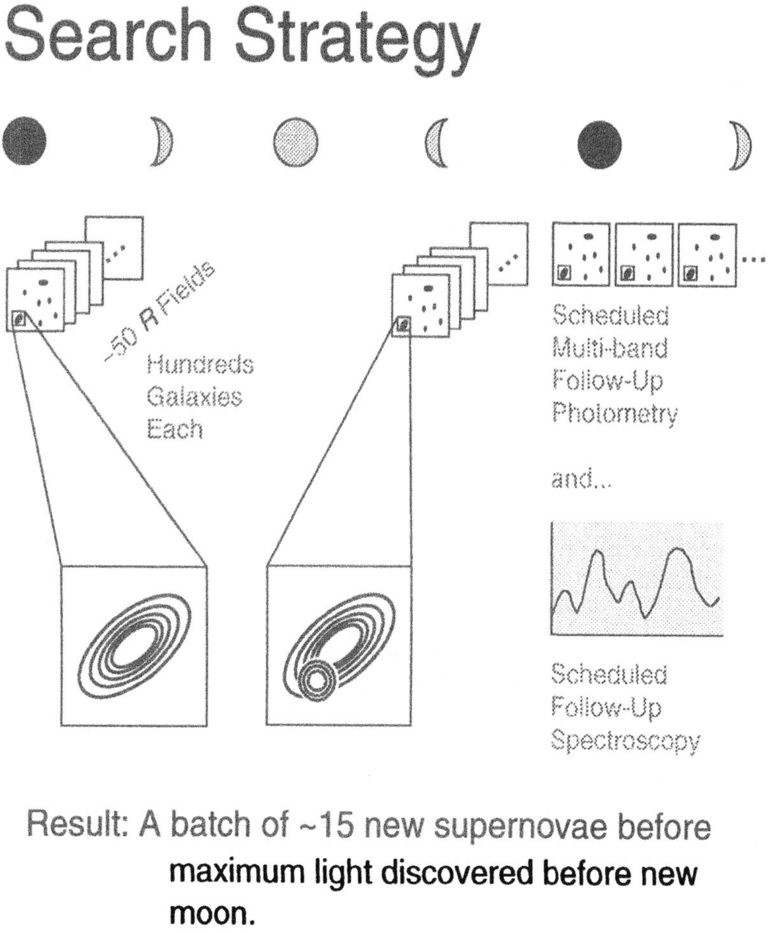

Figure 3. Search strategy designed to discover batches of ∼10 high-redshift supernovae on demand, just before new moon, and while the supernovae are still brightening, i.e. before "maximum light." The follow-up spectroscopy and photometry can therefore be scheduled, and can follow the supernovae over maximum light during dark time.

new-moon comparison images. In order to begin the follow-up photometry and spectroscopy immediately, extensive software has been developed to make it possible to complete the analysis of all the images within hours of the observations.

In short, this search technique allows "batches" of pre-maximum-light supernova discoveries to be *scheduled* just before new moon, the ideal time to begin follow-up spectroscopy and photometry. This follow-up can now be scheduled as well, on the largest telescopes.

3.2. ANALYSIS

The data analysis involves several stages which are summarized on Figure 4.

The supernova light in each image must be measured and the underlying host galaxy light subtracted off. At these redshifts the supernova's seeing disk usually covers a significant amount of the galaxy, so it is important to do this step correctly. Each image has different seeing, and often a different telescope's point-spread-function (PSF) varies both spatially over the field and temporally over the course of an observing night. The amount of galaxy light that must be subtracted therefore varies from image to image.

To compare the high-redshift supernova photometry points to nearby SN Ia light curves, it is necessary to calculate the K correction that accounts for the redshifting of the light observed in a given filter. The standard K corrections give the magnitude difference between the light emitted in a given filter band and the light observed after redshifting in that same band. Since at redshifts of order $z = 0.45$ the light received in the R band corresponds approximately to the light emitted in the B band, a generalization of the K correction, K_{RB}, is calculated that gives the magnitude difference between the rest frame B magnitude and the observed R magnitude(Kim, Goobar, & Perlmutter, 1996).

The K-corrected time-dilated lightcurve template is then fitted to the data points and the peak magnitudes are derived. Finally, the cosmological parameters are extracted from a fit of the FLRW model to the peak magnitude vs redshift data.

4. Recent results on H_0, Ω_M and Ω_Λ from SNe Ia

Very recently, new results have been reported on the measurement of cosmological parameters using SNe Ia.

4.1. RECENT DETERMINATION ON H_0

The Hubble parameter H_0 is derived using SNe Ia at redshifts below $z = 0.1$, where the effect of the mass density and cosmological constant are small. The method relies on fitting data from about 20–40 SNe Ia with a straight line of constant slope (equal to 5), to derive the value of the intercept at $\log(cz) = 0$, $M - 5\log H_0$, where M is the absolute magnitude of SNe Ia. Measuring H_0 requires therefore knowing M. This is done by calibrating SNe Ia that exploded in galaxies or cluster of galaxies whose absolute distance are known from Cepheids calibration. The problem, however is that only a limited number of SNe were found in galaxies (or even cluster of galaxies) of known distances. Moreover some SNe exploded long time ago

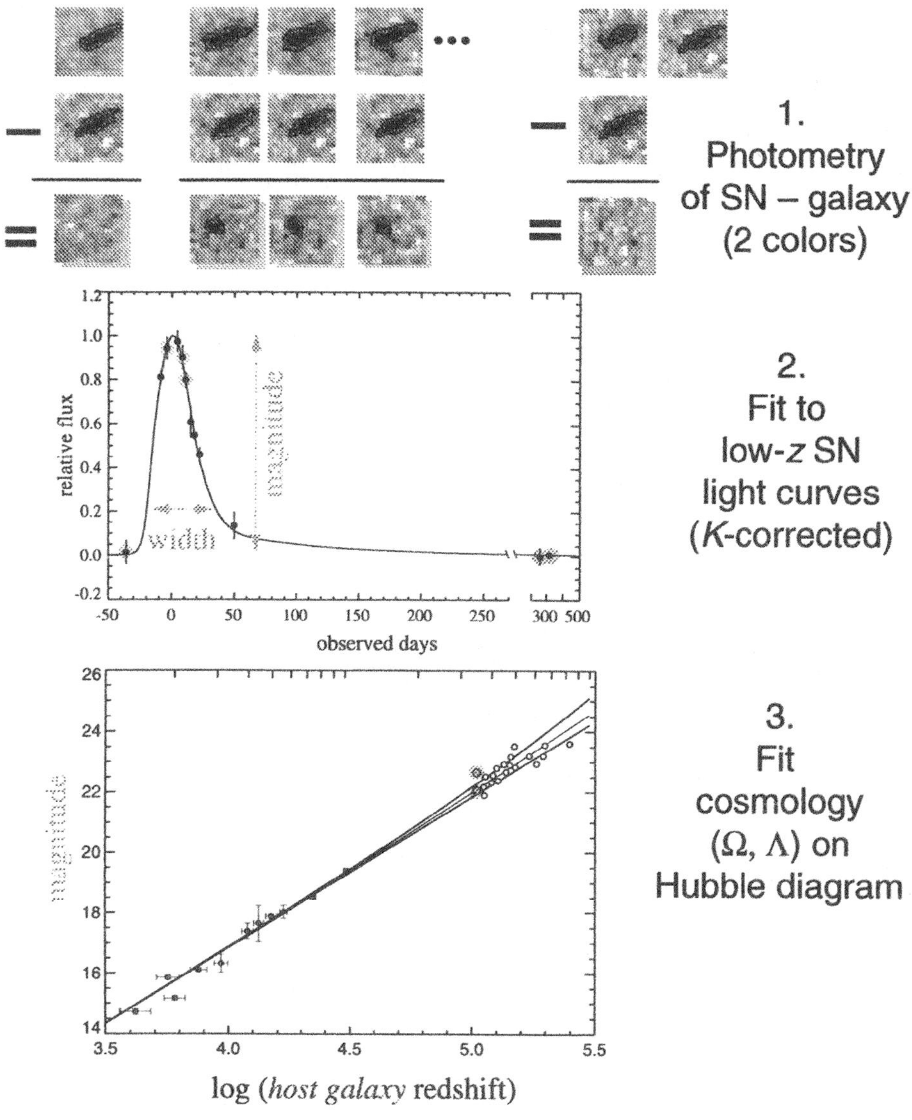

Figure 4. Data analysis stages: (1) reduction of the observed image data to individual photometry points. (2) comparison of these points with nearby SN Ia light curves to determine the luminosity distance and hence the cosmological parameters (3).

and their peak magnitudes are subject to sizeable measurement uncertainties.

Nevertheless, analysis were performed by a few groups. Saha et al. (1997) did a first attempt assuming that type Ia SNe are perfect standard candles and calibrating with supernovae 1895B[2], 1937C, 1960F, 1972E,

[2]Supernovae are named by the International Astronomical Union (IAU) by their year

1981B, 1989B, 1990N. They found $H_0 = 57 \pm 5$ km $s^{-1}Mpc^{-1}$. Riess et al. (Riess, Press, & Kirshner, 1995), (Riess, Press, & Kirshner, 1996) use a larger sample of nearby SNe Ia corrected for the width-luminosity correlation and keeping only the most recent SNe Ia 1972E, 1981B and 1990N for the absolute calibration. They found $H_0 = 65 \pm 7$ km $s^{-1}Mpc^{-1}$. More recently, Freedman et al. (Freedman et al., 1998) did a similar analysis calibrating with 2 more SNe found in the Fornax cluster, 1980N and 1992A, and for which Cepheids calibration had just been done using the Hubble Space Telescope. They found $H_0 = 67 \pm 5(stat) \pm 7(syst)$ km $s^{-1}Mpc^{-1}$.

It should be emphasized again that the measurement of H_0 relies on the knowledge of the absolute distance scale. A change in the Cepheid calibration would directly affect the value of H_0 obtained from SNe Ia. Ultimately, the precision on H_0 will certainly largely be dominated by this uncertainty. On the contrary, measurement of Ω_M and Ω_Λ are free from this calibration uncertainty and should be *in fine* much more precise.

4.2. VERY RECENT DETERMINATION OF Ω_M AND Ω_Λ

Two international collaborations have started the extremely difficult task of measuring Ω_M and Ω_Λ using SNe Ia : the *Supernova Cosmology Project* (SCP) and the *High-Z Supernova Team* (HZT). They use similar techniques both for detecting high-redshift supernovae and to analyse them. Their first results were recently presented (Perlmutter et al., 1998c) & (Perlmutter et al., 1998d),(Schmidt et al., 1998) & (Riess et al., 1998).

The Supernova Cosmology Project (SCP, (1998)), composed of both physicists and astronomers, started in 1988. They developed techniques, including instrumentation, analysis, and observing strategies, that make it possible to systematically study high-redshift supernovae (Perlmutter et al., 1995a). As of summer 1998, about 80 Type Ia supernovae at redshifts $z = 0.18$–0.86 have been discovered and studied (Perlmutter et al., 1995b), (Perlmutter et al., 1996), (Perlmutter et al., 1997a), (Perlmutter et al., 1997b), (Perlmutter et al., 1997c), (Perlmutter et al., 1997d), (Perlmutter et al., 1998a). They presented their analysis of 42 high-z SNe together with 18 nearby SNe previously discovered in Chile by the Calan/Tololo Survey.

The High-Z supernova team (HZT, (1998)) has detected over 50 high-redshift SNe since 1995 (Garnavich et al., 1996a), (Garnavich et al., 1996b), (Garnavich et al., 1998a). They presented their analysis of 16 high-z SNe together with 27 nearby SNe (Schmidt et al., 1998), (Riess et al., 1998).

Analysis of data are done indepently but it should be noted that the results are *not* fully independent since some of the nearby SNe and a couple

of discovery followed by one of more letters of the alphabet reflecting their chronological rank of discovery in the year.

of distant ones are common to both groups and used in their derivation of the cosmological parameters.

Nevertheless, it is quite remarkable that the two collaborations came out with very similar results. Moreover, these results dramatically change our view of the current value of the cosmological parameters Ω_M and Ω_Λ. They are shown on Figure 5 and can be summarized as follows:

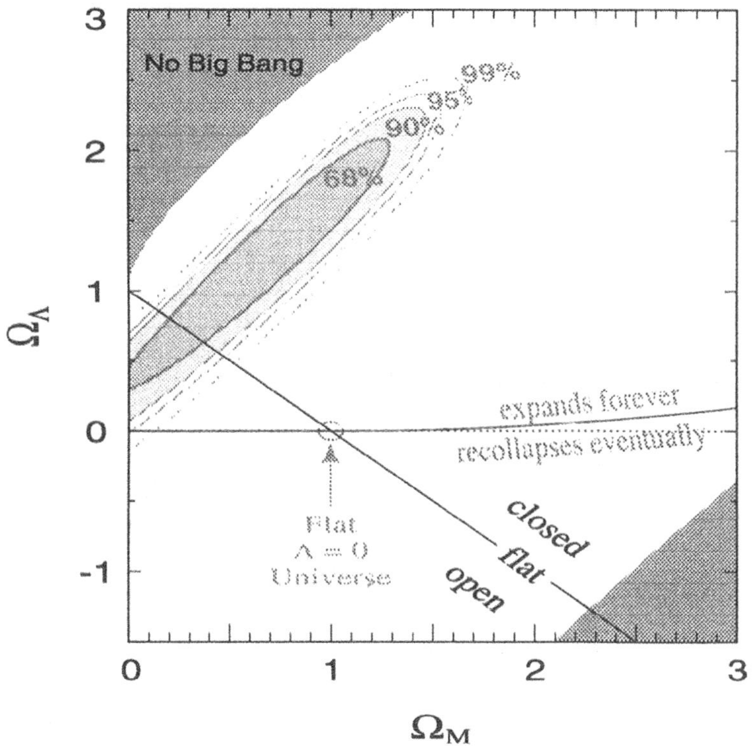

Figure 5. Best-fit confidence regions in the Ω_M-Ω_Λ plane obtained by the SCP. Similar results are obtained by the HZT. The upper-left shaded region represents "bouncing Universe" cosmologies with no big bang and the lower-right shaded region corresponds to a Universe younger than the oldest heavy elements for $H_0 > 50$ km s^{-1}Mpc^{-1}.

- The simplest inflationary solution $\Omega_M = 1$ and $\Lambda = 0$ is strongly excluded ($> 6\sigma$).
- The cosmological constant is non-zero for any value of the mass density: $\Omega_\Lambda > 0$ ($\sim 3\sigma$).
- The Universe will expand forever.
- The deceleration parameter, q_0, is negative i.e. the rate of expansion of the Universe is currently accelerating.

More precisely, the SCP parameterizes the Ω_M, Ω_Λ degeneracy with the equation $1.3\Omega_M - \Omega_\Lambda = -0.4 \pm 0.2$ and finds for a flat Universe ($\Omega_M + \Omega_\Lambda = 1$):

$$\Omega_M = 0.28^{+0.09}_{-0.08}(stat)^{+0.05}_{-0.04}(syst)$$

With the same condition, the HZT gives:

$$\Omega_M = 0.24 \pm 0.1$$

and a 2.8 σ exclusion of the $\Lambda = 0$ cosmologies. The non-zero cosmological constant has also a net effect of increasing the dynamical age of the Universe by a couple of billions years resulting in a value close to 15 billion years for a value of H_0 of 65 km $s^{-1}Mpc^{-1}$.

Given the potentially revolutionary nature of these conclusions, in particular the fact that the cosmological constant is non-zero, it is important to carefully examine all possible sources of systematics. Both groups have gone through this with extreme care, studying possible systematics arising from extinction, sample selection (Malmquist bias) or contamination effect, possible evolution of SNe Ia, effect of a local void and weak lensing effects. So far, none of these sources can reconcile data with the $\Omega_M = 1, \Lambda = 0$ solution and furthermore the $\Lambda = 0$ solution remains very unlikely for any value of Ω_M.

However, work is still going on regarding systematics and more data have to be collected to confirm these very important results.

5. The (near) future

The near future is going to be very interesting for SNe Ia and cosmology. The most important aspect concern the measurement of Ω_M and Ω_Λ. Both the SCP and HZT collaborations are already analyzing more high-redshift data and will continue collecting nearby and distant SNe to increase confidence in the results recently presented.

In the slightly longer term, it is essential that a extensive study of low and high-redshift SNe Ia is done. More precisely to:

- Re-explore the low-redshift region ($z < 0.1$). The current data comes primarily from the Calan/Tololo survey. Discoveries were made using photographic plates with most SNe discovered past their maximum light. The data sample could be Malmquist biased and the detection efficiency is difficult to control. Also it is important to obtain additional high quality photometric and spectroscopic follow-up data at all phases of the supernovae.
- Collect data in the intermediate redshift region, $0.1 < z < 0.4$, where only a handful of SNe have been found so far. It is important to verify

that a Hubble diagram of SNe Ia can reliably be built over the complete redshift range, connecting the $z < 0.1$ nearby region to the $z \sim 0.5$ region where most of the distant SNe were discovered so far.

- Increase statistics in the $z = 0.7$-1.0 region. In that region, the combination of the use of 4 m class telescopes to detect supernovae and 8 m class telescope to follow them spectroscopically and photometrically is needed as well as very precise photometry that can be obtained in space using the HST. Figure 6 is a composite of images taken on the CTIO 4 m telescope before and at discovery time and a follow-up image obtained by pointing the HST in the region of discovery. On this image, the supernova is clearly separated from the host galaxy ensuring a very precise photometry of the supernova. Both groups have data for a dozen or so SNe having HST follow-up images.
- Explore the $z > 1$ region using the largest ground-based telescopes available to search for supernovae. As shown in section 1, an independent determination of Ω_M and Ω_Λ will be possible only if measurements are done on a large enough redshift range.

To conclude, very promising results have recently been obtained on the measurement of cosmological parameters using SNe Ia. SNe Ia already provide one of the most precise measurements of Ω_M and Ω_Λ and the prospects for improvement are very good. These measurements, however, are extremely difficult and every possible systematic effect must be studied in great detail before drawing any firm conclusion on the cosmology. The next few years with the planned observations of more nearby and distant supernovae will certainly provide cosmologists with very exiting new results.

6. References

References

Supernova Cosmology Project - S. Perlmutter et :
G. Aldering, B. J. Boyle, P.G. Castro, W.J. Couch, S. Deustua, R.S. Ellis, S. Fabbro, A. V. Filkippenko, A.S. Fruchter, G. Goldhaber, A. Goobar, D.E. Groom, I.M. Hook, M. Irwin, A.G. Kim, M.Y.Kim, R.A. Knop, J.C. Lee, C. Lidman, R.G. McMahon, T. Matheson, P. Nugent, N.J. Nunes, H.J.M. Newberg, R. Pain, N. Panagia, C.R. Pennypacker, R. Quimby, P. Ruiz-Lapuente, B. Schaefer, N. Walton, cf. http://panisse.lbl.gov:80/public/

High-Z Supernovae Team - B. Schmidt et :
P. Challis, A. Clocchiatti, R. Ciardullo, A. Diercks, A. Dressler, A.V. Filippenko, P. Garnavich, R. Gilliland, M. Hamuy, G. Hogan, R.P. Kirschner, B. Leibundgut, J. Maza, M.M. Phillips, D. Reiss, A.G. Riess, R.A. Schommer, R.C. Smith, J. Spyromilio, C. Stubbs, N.B. Suntzeff, J. Tonry, J. Walsh, cf. http://cfa-www.harvard.edu/cfa/oir/Research/supernova/HighZ.html

Baade 1938, ApJ, 88, 285

Bergström, L. & Goobar, A. 1999, *Cosmology and Particle Astrophysics*. Wiley-Praxis Series in Astronomy & Astrophysics, John Wiley & Sons, Ltd., Chichester, UK.

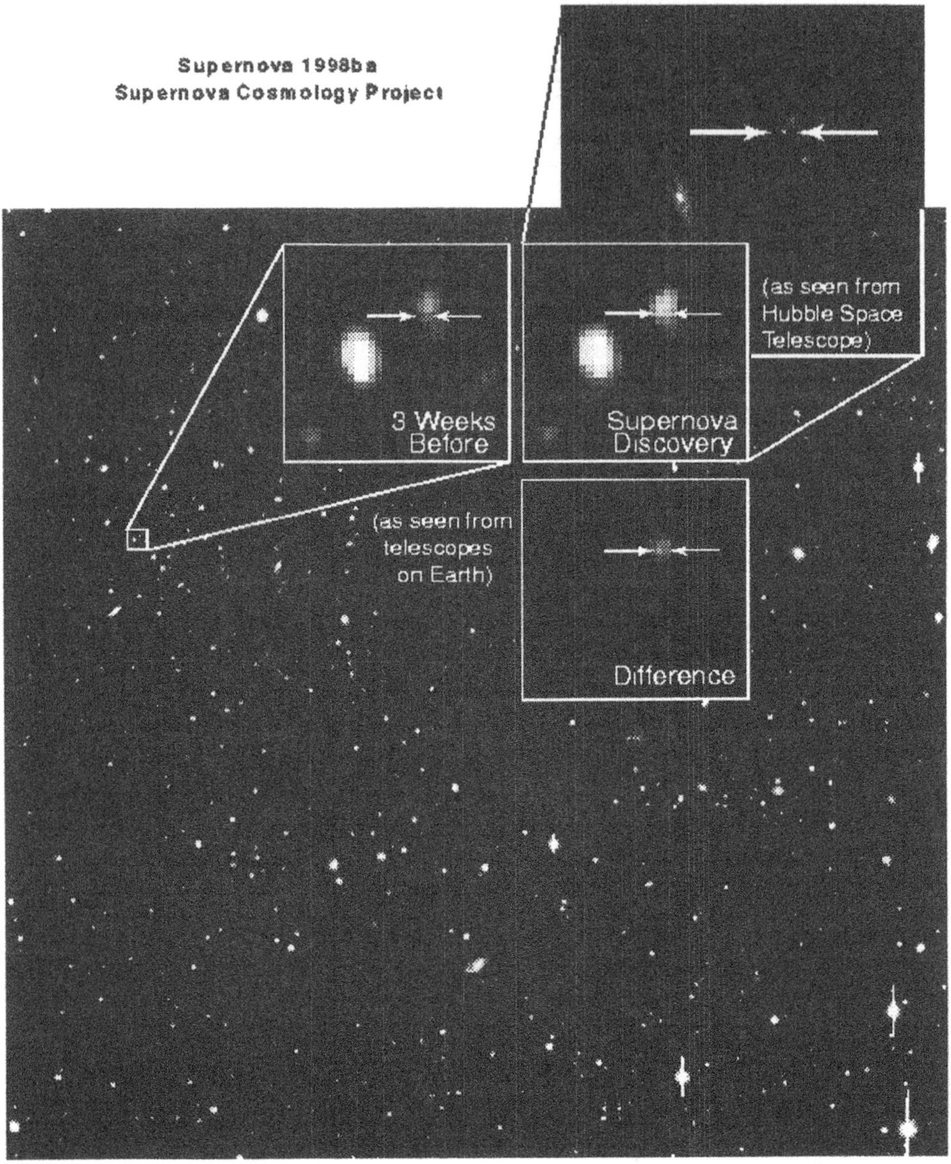

Figure 6. Composite made from images taken on the CTIO 4 m telescope. The supernova has been detected on the difference image resulting from the subtraction of the first image (3 weeks before) to the second one (Supernova Discovery). For comparison, a follow-up image obtained by pointing the HST in the same region of space a few days after discovery is also shown. On this image, the supernova is clearly visible as a point source separated from the host galaxy.

Branch, D. 1998, Ann. Rev. Astro. Astrophys., 36, 17
Branch, D., Lacy, C. H., McCall, M. L., Sutherland, P. G., Uomoto, A., Wheeler, J. C., & Wills, B. J. 1983, ApJ, 270, 123
Carroll, S. M., Press, W. H., & Turner, E. L. 1992, Ann. Rev. Astro. Astrophys., 30, 499
Colgate, S. 1979, ApJ, 232, 404
Filippenko, A. V., et al. 1998, in preparation
Freedman, W. L., Mould, J. R., Kennicut, R. C., & Madore, B. F. 1998. In IAU Symposium 183, pages (astro–ph/9801080)
Garnavich, P., et al. 1996a, IAUC 6332
Garnavich, P., et al. 1996b, IAUC 6358
Garnavich, P., et al. 1998b, IAUC 6633
Garnavich, P., et al. 1998a, ApJ, 493, L53
Garnavich, P., et al. 1998b, ApJ, 509, in press
Goldhaber, G., et al. 1995. In Presentations at the NATO ASI in Aiguablava, Spain, LBL-38400, page III.1; also published in Thermonuclear Supernova, P. Ruiz-Lapuente, R. Canal, and J.Isern, editors, Dordrecht: Kluwer, page 777 (1997)
Goldhaber, G., et al. 1998, Ap.J., in preparation
Goobar, A., & Perlmutter, S. 1995, ApJ, 450, 14
Hamuy, M., Phillips, M. M., Maza, J., Suntzeff, N. B., Schommer, R. A., & Aviles, R. 1995, AJ, 109, 1
Hamuy, M., Phillips, M. M., Maza, J., Suntzeff, N. B., Schommer, R. A., & Aviles, R. 1996, AJ, 112, 2391
Hatano, K., Branch, D., & Deaton, J. 1997, astro-ph/9711311
Höflich, P., & Khokhlov, A. 1996, ApJ, 457, 500
Höflich, P., Wheeler, J. C., & Thielemann, F. K. 1998, ApJ, 495, 61
Holz, D. E., & Wald, R. M. 1998, Phys. Rev. D, 58, 063501
Hook, I. M., Nugent, P., et al. 1998, in preparation
Kim, A., et al. 1998, ApJ, in preparation
Kim, A., Goobar, A., & Perlmutter, S. 1996, PASP, 108, 190
Kirshner, R., et al. 1993, ApJ, 415, 589
Kirshner, R., & Oke, J. 1975, ApJ, 200, 574
Leibundgut, B., Tammann, G.A., Cadonau R. and Cerrito D., AA Suppl. Ser., 89:537, 1991
Leibundgut, B., et al. 1996a, Ap.J.Lett., 466, L21
Leibundgut B. et al. 1996b, Ap.J.Lett., 371, L23
Maddox, S. J., Sutherland, W. J., Loveday, J., & Peterson, B. A. 1990, MNRAS, 247, 1P
Malmquist, K. G. 1924, Medd. Lund Astron. Obs. Ser. II, No. 32, 64
Malmquist, K. G. 1936, Stockholm Obs. Medd., No. 26
Marvin, H., & Perlmutter, S., 1989, IAU Circ. No. 4727
Nørgaard-Nielsen, et al. 1989, Nature, 339, 523
Nugent, P., Branch, D., Baron, E., Fisher, A., Vaughan, T., & Hauschildt, P. 1995, Phys.Rev.Lett., 75, 394, and erratum, 75, 1874
Nugent, P., et al. 1998, PASP, in preparation
Pain, R., et al. 1996, ApJ, 473, 356
Patat, F., et al. 1996, MNRAS, 278, 111
Perlmutter, S., et al. 1995a. In Presentations at the NATO ASI in Aiguablava, Spain, LBL-38400, page I.1; also published in Thermonuclear Supernova, P. Ruiz-Lapuente, R. Canal, and J.Isern, editors, Dordrecht: Kluwer, page 749 (1997)
Perlmutter, S., et al. 1995b, IAU Circ. No. 6270
Perlmutter, S., et al. 1996, IAU Circ. No. 6621
Perlmutter, S., et al. 1997a, IAU Circ. No. 6596
Perlmutter, S., et al. 1997b, IAU Circ. No. 6540
Perlmutter, S., et al. 1997c, IAU Circ. No. 6646
Perlmutter, S., et al. 1997d, IAU Circ. No. 6804
Perlmutter, S., et al. 1997e, ApJ, 483, 565 [P97]

Perlmutter, S., et al. 1998a, IAU Circ. No. 6881
Perlmutter, S., et al. 1998b, Nature, 391, 51 and erratum (on author list), 392, 311
Perlmutter, S., et al. 1998c. In Presentation at the January 1998 Meeting of the American Astronomical Society, Washington, D.C., LBL-42230, available at www-supernova.lbl.gov and astro-ph; referenced in B.A.A.S., volume 29, page 1351 (1997)
Perlmutter, S., et al. 1998b, astro-ph 9812133, to be published in ApJ.
Phillips, M., et al. 1998, ApJ, in preparation
Phillips, M. M. 1993, Ap.J.Lett., 413, L105
Phillips, M. M. 1998, private communication
Phillips, M. M., et al. 1987, PASP, 99, 592
Riess, A., et al. 1997a, AJ, 114, 722
Riess, A., Davis, M., Baker, J., & Kirshner, R. P. 1997b, ApJ, 488, L1
Riess, A. G., et al. 1998, AJ, 116, 1009 (in press)
Riess, A. G., Press, W. H., & Kirshner, R. P. 1995, ApJ, 438, L17
Riess, A. G., Press, W. H., & Kirshner, R. P. 1996, ApJ, 473, 88
Saha, A., et al. 1997, ApJ, 486, 1
Schmidt, B. P., et al. 1998, ApJ, in press
Tammann, G. A. 1979. In ESA/ESO Workshop on Astronomical Uses of the Space Telescope, F. Macchetto, F. Pacini, and M. Tarenghi, editors, Geneva: ESO, page 329
Tripp, R. 1998, A&A, 331, 815
Turner, M. S., & White, M. 1997, Phys. Rev. D, 56, R4439
Vacca, W. D., & Leibundgut, B. 1996, ApJ, 471, L37

PRIMORDIAL NUCLEOSYNTHESIS

K.A. OLIVE
School of Physics and Astronomy, Theoretical Physics Institute
University of Minnesota
Minneapolis MN 55455 USA

1. Introduction

Big Bang Nucleosynthesis (BBN) is the theory explaining the origins of the light elements D,^{3}He, ^{4}He, and ^{7}Li and their primordial abundances. The theoretical framework for BBN is quite simple. It is based on Friedmann-Lemaitre-Robertson-Walker cosmology [1] and a network of nuclear reactions. We can further specify that the standard BBN model refers to homogeneous nucleosynthesis in the context of a FLRW Universe with an electroweak standard model particle content, which for the purposes of BBN really amounts to assuming the existence of three nearly massless and nearly stable neutrinos. The predictions of BBN for the abundances of the light elements are determined by running a code which incorporates the nuclear network in a thermal (and cooling due to the expansion of the Universe) bath. These predictions are then be compared with the observational determinations of the abundances.

In contrast to the theoretical side of BBN, the status of the observational data has changed significantly in the last several years. There is more data on ^{4}He and ^{7}Li, and data on D and ^{3}He that was simply non-existent several years ago. For the most part, the inferred abundances of ^{4}He and ^{7}Li have remained relatively fixed, giving us a higher degree of confidence in the assumed primordial abundances of these isotopes as is reflected in their observational uncertainties. Indeed, the abundances of ^{4}He and ^{7}Li alone are sufficient to probe and test the theory and determine the single remaining parameter in the standard model [2], namely, the baryon-to-photon ratio, η. In contrast, D and ^{3}He are highly dependent on models of chemical evolution (^{3}He is in addition dependent on the uncertain stellar yields of this isotope). New data from quasar absorption systems, on what

M. Lachièze-Rey (ed.), Theoretical and Observational Cosmology, 261–304.

may be primordial D/H is at this time disconcordant, different measurements give different abundances. As a consequence of the uncertainties in D and ^{3}He, one can hope to use the predictions based on ^{4}He and ^{7}Li in order to construct models of galactic chemical evolution. These results also have important implications for the amount of (non)-baryonic dark matter in the galaxy and on the number of allowed relativistic degrees of freedom at the time of BBN, commonly parameterized as N_ν.

1.1. STANDARD MODEL BASICS

Since one of the main inputs of the theoretical side of BBN is the standard hot big bang model, it will be useful to review some of the key concepts as they pertain to BBN. The metric for the FLRW model of the Universe is specified by two quantities, the curvature constant k, and the expansion scale factor $R(t)$. At early times, the curvature is unimportant as can be seen from the Friedmann equation for the Hubble parameter

$$H^2 \equiv \left(\frac{\dot{R}}{R}\right)^2 = \frac{1}{3}8\pi G_N\rho - \frac{k}{R^2} + \frac{1}{3}\Lambda \qquad (1)$$

where Λ is the cosmological constant. Since the density ρ scales as either R^{-3} (for a matter dominated universe) or R^{-4} (for a radiation dominated universe), this term dominates over either the curvature or the cosmological constant. I will ignore both in what follows.

The critical energy density ρ_c is defined such that $\rho = \rho_c$ for $k = 0$

$$\rho_c = 3H^2/8\pi G_N \qquad (2)$$

In terms of the present value of the Hubble parameter this is,

$$\rho_c = 1.88 \times 10^{-29} {h_o}^2 \mathrm{gcm}^{-3} \qquad (3)$$

where

$$h_o = H_o/(100\mathrm{kmMpc}^{-1}\mathrm{s}^{-1}) \qquad (4)$$

The cosmological density parameter is then defined by

$$\Omega \equiv \frac{\rho}{\rho_c} \qquad (5)$$

in terms of which the Friedmann equation, Eq. (1), can be rewritten as (with $\Lambda = 0$)

$$(\Omega - 1)H^2 = \frac{k}{R^2} \qquad (6)$$

so that $k = 0, +1, -1$ corresponds to $\Omega = 1, \Omega > 1$ and $\Omega < 1$. (Very) broad observational limits on h_o and Ω are

$$0.4 \leq h_o \leq 1.0 \qquad 0.1 \leq \Omega \leq 2 \tag{7}$$

The value of Ω, at least on relatively small scales, seems to depend on scale. Indeed, the contribution to Ω from visible matter associated with stars and hot gas is quite small, $\Omega \approx 0.003 - 0.01$. On somewhat larger scales, that of galactic halos or small groups of galaxies, $\Omega \approx 0.02 - 0.1$. On galaxy cluster scales, it appears that Ω may be as large as 0.3. And while there is some evidence, the observations are far from conclusive in indicating a value of Ω as large as 1. It is however possible to obtain a bound on the product, Ωh^2 from

$$H_o t_o = \int_0^1 (1 - \Omega + \Omega/x)^{-1/2} dx \tag{8}$$

(for $\Lambda = 0$). For $t_o > 12$Gyr, and $\Omega \leq 1$, $\Omega h^2 < 0.3$ (This is true even if $\Lambda \neq 0$.)

As indicated above, BBN takes place during the radiation dominated epoch which lasts roughly to the period of recombination (somewhat earlier when dark matter is included) which occurs when electrons and protons form neutral hydrogen through $e^- + p \to \mathrm{H} + \gamma$ at a temperature $T_R \sim$ few $\times 10^3$ K $\sim$1 eV. For $T < T_R$, photons are decoupled while for $T > T_R$, photons are in thermal equilibrium. Today, the content of the microwave background consists of photons with $T_o = 2.728 \pm .002$ K [3]. The energy density of photons in the background can be calculated from

$$\rho_\gamma = \int E_\gamma dn_\gamma \tag{9}$$

where the density of states is given by

$$dn_\gamma = \frac{g_\gamma}{2\pi^2}[exp(E_\gamma/T) - 1]^{-1} q^2 dq \tag{10}$$

and $g_\gamma = 2$ is the number of spin polarizations for the photon, $E_\gamma = q$ is just the photon energy (momentum). (I am using units such that $\hbar = c = k_B = 1$ and will do so through the remainder of these lectures.) Integrating (9) gives

$$\rho_\gamma = \frac{\pi^2}{15} T^4 \tag{11}$$

which is the familiar blackbody result.

In general, at very early times, at very high temperatures, other particle degrees of freedom join the radiation background when $T \sim m_i$ for each particle type i, if that type is brought into thermal equilibrium through interactions. In equilibrium the energy density of a particle type i is given by

$$\rho_i = \int E_i dn_{q_i} \tag{12}$$

and

$$dn_{q_i} = \frac{g_i}{2\pi^2}[exp[(E_{q_i} - \mu_i)/T] \pm 1]^{-1} q^2 dq \tag{13}$$

where again g_i counts the total number of degrees of freedom for type i,

$$E_{q_i} = \left(m_i^2 + q_i^2\right)^{1/2} \tag{14}$$

μ_i is the chemical potential if present and $\pm$ corresponds to either Fermi or Bose statistics.

In the limit that $T \gg m_i$ the total energy density can be conveniently expressed by

$$\rho = \left(\sum_B g_B + \frac{7}{8}\sum_F g_F\right)\frac{\pi^2}{30}T^4 \equiv \frac{\pi^2}{30}N(T)T^4 \tag{15}$$

where $g_{B(F)}$ are the total number of boson (fermion) degrees of freedom and the sum runs over all boson (fermion) states with $m \ll T$. The factor of 7/8 is due to the difference between the Fermi and Bose integrals. Equation (15) defines N(T) by taking into account new particle degrees of freedom as the temperature is raised.

In the radiation dominated epoch, we can obtain a relationship between the age of the Universe and its temperature

$$t = \left(\frac{90}{32\pi^3 G_N N(T)}\right)^{1/2} T^{-2} \tag{16}$$

Put into a more convenient form

$$tT_{MeV}^2 = 2.4[N(T)]^{-1/2} \tag{17}$$

where t is measured in seconds and T_{MeV} in units of MeV.

The value of $N(T)$ at any given temperature depends on the particle physics model. In the standard $SU(3)\times SU(2)\times U(1)$ model, we can specify $N(T)$ up to temperatures of 0(100) GeV. The change in N can be seen in the following table.

At higher temperatures ($T \gg 100$ GeV), $N(T)$ will be model dependent. For example, in the minimal $SU(5)$ model, one needs to add to $N(T)$, 24 states for the X and Y gauge bosons, another 24 from the adjoint Higgs,

TABLE 1. Effective numbers of degrees of freedom in the standard model.

Temperature	New Particles	$4N(T)$
$T < m_e$	γ's + ν's	29
$m_e < T < m_\mu$	$e^\pm$	43
$m_\mu < T < m_\pi$	$\mu^\pm$	57
$m_\pi < T < Tc^*$	π's	69
$T_c < T < m_{strange}$	- π's + $u, \bar{u}, d, \bar{d}$ + gluons	205
$m_s < T < m_{charm}$	$s, \bar{s}$	247
$m_c < T < m_\tau$	$c, \bar{c}$	289
$m_\tau < T < m_{bottom}$	$\tau^\pm$	303
$m_b < T < m_{W,Z}$	$b, \bar{b}$	345
$m_{W,Z} < T < m_{top}$	$W^\pm, Z$	381
$m_t < T < m_{Higgs}$	$t, \bar{t}$	423
$M_H < T$	H^o	427

*T_c corresponds to the confinement-deconfinement transition between quarks and hadrons. $N(T)$ is shown in Figure 1 for $T_c = 150$ and 400 MeV. It has been assumed that $m_{Higgs} > m_{top}$.

and another 6 (in addition to the 4 already counted in $W^\pm, Z$ and H) from the $\bar{5}$ of Higgs. Hence for $T > M_X$ in minimal $SU(5)$, $N(T) = 160.75$. In a supersymmetric model this would at least double, with some changes possibly necessary in the table if the lightest supersymmetric particle has a mass below M_H.

The presence of a particle species in the thermal background assumes thermal equilibrium and hence interaction rates which are sufficiently fast compared with the expansion rate of the Universe. Roughly, this translates to the condition for each particle type i, that some rate Γ_i involving that type be larger than the expansion rate of the Universe or

$$\Gamma_i > H \tag{18}$$

in order to be in thermal equilibrium.

Examples of a processes in equilibrium at early times which drops out of equilibrium or decouples at later times are the processes which involve neutrinos. If we consider the standard neutral or charged-current interactions such as $e^+ + e^- \leftrightarrow \nu + \bar{\nu}$ or $e + \nu \leftrightarrow e + \nu$ etc., the rates for these processes can be approximated by

$$\Gamma = n\langle\sigma v\rangle \tag{19}$$

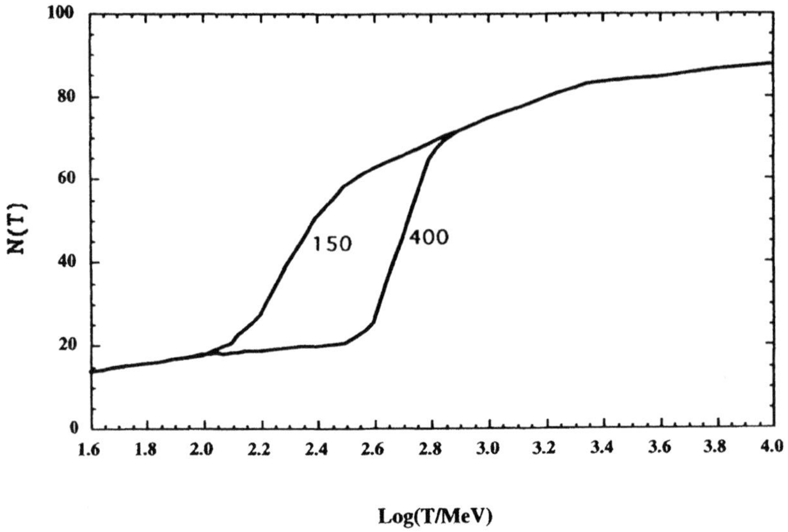

Figure 1. The effective numbers of relativistic degrees of freedom as a function of temperature.

where $\langle\sigma v\rangle$ is the thermally averaged weak interaction cross section

$$\langle\sigma v\rangle \sim 0(10^{-2})T^2/M_W^4 \tag{20}$$

and n is the number density of leptons. Hence the rate for these interactions is

$$\Gamma_{\mathrm{wk}} \sim 0(10^{-2})T^5/M_W^4 \tag{21}$$

The expansion rate, on the other hand, is just

$$H = \left(\frac{8\pi G_N\rho}{3}\right)^{1/2} = \left(\frac{8\pi^3}{90}N(T)\right)^{1/2} T^2/M_P \sim 1.66N(T)^{1/2}T^2/M_P. \tag{22}$$

The Planck mass $M_P = G_N^{-1/2} = 1.22 \times 10^{19}$ GeV.

Neutrinos will be in equilibrium when $\Gamma_{\mathrm{wk}} > H$ or

$$T > (500M_W^4)/M_P)^{1/3} \sim 1MeV. \tag{23}$$

The temperature at which these rates are equal is commonly referred to as the decoupling or freeze-out temperature and is defined by

$$\Gamma(T_d) = H(T_d) \tag{24}$$

For temperatures $T > T_d$, neutrinos will be in equilibrium, while for $T < T_d$ they will not. Basically, in terms of their interactions, the expansion rate is just too fast and they never *"see"* the rest of the matter in the Universe (or themselves). Their momenta will simply redshift and their effective temperature (the shape of their momenta distribution is not changed from that of a blackbody) will simply fall with $T \sim 1/R$.

Soon after decoupling the $e^{\pm}$ pairs in the thermal background begin to annihilate (when $T \lesssim m_e$). Because the neutrinos are decoupled, the energy released heats up the photon background relative to the neutrinos. The change in the photon temperature can be easily computed from entropy conservation. The neutrino entropy must be conserved separately from the entropy of interacting particles. If we denote $T_>$, the temperature of photons, and $e^{\pm}$ before annihilation, we also have $T_\nu = T_>$ as well. The entropy density of the interacting particles at $T = T_>$ is just

$$s_> = \frac{4}{3}\frac{\rho_>}{T_>} = \frac{4}{3}(2 + \frac{7}{2})(\frac{\pi^2}{30})T_>^3 \tag{25}$$

while at $T = T_<$, the temperature of the photons just after $e^{\pm}$ annihilation, the entropy density is

$$s_< = \frac{4}{3}\frac{\rho_<}{T_<} = \frac{4}{3}(2)(\frac{\pi^2}{30})T_<^3 \tag{26}$$

and by conservation of entropy $s_< = s_>$ and

$$(T_</T_>)^3 = 11/4 \tag{27}$$

Thus, the photon background is at higher temperature than the neutrinos because the $e^{\pm}$ annihilation energy could not be shared among the neutrinos, and

$$T_\nu = (4/11)^{1/3}T_\gamma \simeq 1.9K \tag{28}$$

1.2. HISTORICAL PERSPECTIVES

There has always been an intimate connection between BBN and the microwave background as a key test to the standard big bang model. Indeed, it was the formulation of BBN which predicted the existence of the microwave background radiation [4]. The argument is rather simple. BBN requires temperatures greater than 100 keV, which according to eqs. (16) and (17) corresponds to timescales less than about 200 s. The typical cross section for the first link in the nucleosynthetic chain is

$$\sigma v(p + n \to D + \gamma) \simeq 5 \times 10^{-20}\text{cm}^3/\text{s} \tag{29}$$

This implies that it was necessary to achieve a density

$$n \sim \frac{1}{\sigma v t} \sim 10^{17}\text{cm}^{-3} \tag{30}$$

The density in baryons today is known approximately from the density of visible matter to be $n_{B_o} \sim 10^{-7}$ cm^{-3} and since we know that that the density n scales as $R^{-3} \sim T^3$, the temperature today must be

$$T_o = (n_{B_o}/n)^{1/3} T_{\text{BBN}} \sim 10\text{K} \tag{31}$$

A pretty good estimate.

Despite its simplicity, BBN was criticized early on, due to its shortcomings in being able to produce the observed abundances of *all* of the element isotopes due primarily to the gaps in stable nuclei at $A = 5$ and $A = 8$. Attention was therefore turned to stellar nucleosynthesis [5]. However, while the elements from helium on up can be and are produced in stars, no other astrophysical site has ever survived for the production of deuterium [6]. In addition, if one assumes that ^{4}He is entirely of stellar origin, one should be able to find places in the Universe in which the ^{4}He mass fraction is substantially below 25%. The ^{4}He data shown in Figure 2, emphasizes the fact that indeed no such region with low ^{4}He has ever been observed and that (together with the need to produce D) leads one to conclude that BBN nucleosynthesis is a necessary component in any cosmological model. The foundations of modern BBN continued to be laid over time [7] establishing the notions of equilibrium and the nuclear network to obtain the abundances of D through ^{7}Li.

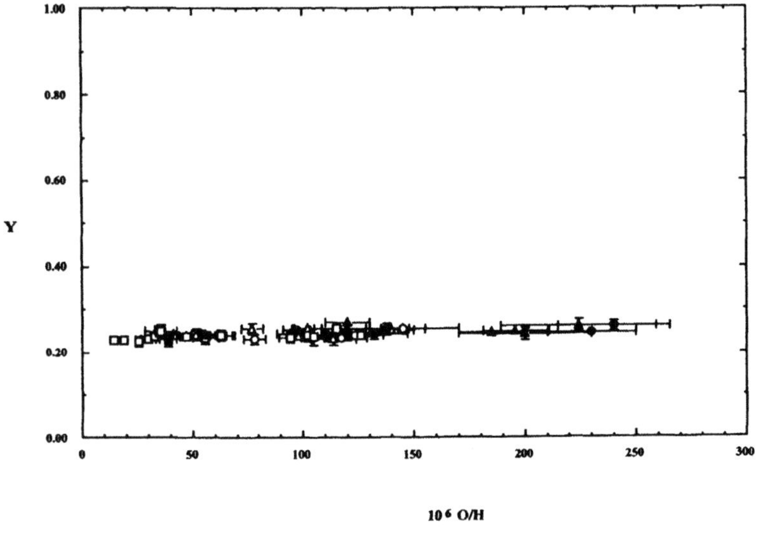

Figure 2. The helium (Y) vs oxygen (O/H) abundances in extragalactic HII regions emphasizing the lack of low ^{4}He regions.

2. Theory

Conditions for the synthesis of the light elements were attained in the early Universe at temperatures $T \lesssim 1$ MeV. At somewhat higher temperatures, weak interaction rates were in equilibrium. In particular, the processes

$$\begin{aligned} n + e^+ &\leftrightarrow p + \bar{\nu}_e \\ n + \nu_e &\leftrightarrow p + e^- \\ n &\leftrightarrow p + e^- + \bar{\nu}_e \end{aligned}$$

fix the ratio of number densities of neutrons to protons. At $T \gg 1$ MeV, $(n/p) \simeq 1$. The energy density (dominated by radiation) in the standard model is

$$\rho = \frac{\pi^2}{30}(2 + \frac{7}{2} + \frac{7}{4}N_\nu)T^4 \tag{32}$$

from the contributions of photons, electrons and positrons, and N_ν neutrino flavors.

As the temperature fell and approached the point where the weak interaction rates were no longer fast enough to maintain equilibrium, the neutron to proton ratio was given approximately by the Boltzmann factor, $(n/p) \simeq e^{-\Delta m/T}$, where Δm is the neutron-proton mass difference. The final abundance of ^{4}He is very sensitive to the (n/p) ratio. As in the case of the neutrino interactions discussed above, freeze out occurs at about an MeV (slightly less than an MeV in this case).

The nucleosynthesis chain begins with the formation of deuterium through the process, $p + n \to \mathrm{D} + \gamma$. However, because the large number of photons relative to nucleons, $\eta^{-1} = n_\gamma/n_B \sim 10^{10}$, deuterium production is delayed past the point where the temperature has fallen below the deuterium binding energy, $E_B = 2.2$ MeV (the average photon energy in a blackbody is $\bar{E}_\gamma \simeq 2.7T$). The point being that there are many photons in the exponential tail of the photon energy distribution with energies $E > E_B$ despite the fact that the temperature or $\bar{E}_\gamma$ are less than E_B. This can be seen by comparing the qualitative expressions for the deuterium production and destruction rates,

$$\begin{aligned} \Gamma_p &\approx n_B \sigma v \\ \Gamma_d &\approx n_\gamma \sigma v e^{-E_B/T} \end{aligned} \tag{33}$$

When the quantity $\eta^{-1}\exp(-E_B/T) \sim 1$ the rate for deuterium destruction ($\mathrm{D} + \gamma \to p + n$) finally falls below the deuterium production rate and the nuclear chain begins at a temperature $T \sim 0.1 MeV$.

In addition to the $p\,(n,\gamma)\,\mathrm{D}$ reaction, the other major reactions leading to the production of the light elements are:

$$\mathrm{D\,(D,}\,p)\,\mathrm{T} \qquad \mathrm{D}\,(n,\gamma)\,\mathrm{T} \qquad {}^3\mathrm{He}\,(n,\,p)\,\mathrm{T}$$
$$\mathrm{D\,(D,}\,n)\,{}^3\mathrm{He} \qquad \mathrm{D}\,(p,\gamma)\,{}^3\mathrm{He}$$

Followed by the reactions producing ^{4}He:

$$\mathrm{D\,(D,}\,\gamma)\,{}^4\mathrm{He} \qquad {}^3\mathrm{He}\,({}^3\mathrm{He},\,2p)\,{}^4\mathrm{He}$$
$$\mathrm{D}\,({}^3\mathrm{He},\,p)\,{}^4\mathrm{He} \qquad \mathrm{T}\,(p,\gamma)\,{}^4\mathrm{He}$$
$$\mathrm{T\,(D,}\,n)\,{}^4\mathrm{He} \qquad {}^3\mathrm{He}\,(n,\gamma)\,{}^4\mathrm{He}$$

The gap at $A = 5$ is overcome and the production of ^{7}Li proceeds through:

$$\begin{array}{l} {}^3\mathrm{He}\,({}^4\mathrm{He},\gamma)\,{}^7\mathrm{Be} \\ \qquad\qquad \to {}^7\mathrm{Li} + e^+ + \nu_e \\ \mathrm{T}\,({}^4\mathrm{He},\gamma)\,{}^7\mathrm{Li} \end{array}$$

The gap at $A = 8$ prevents the production of other isotopes in any significant quantity. The nuclear chain in BBN calculations was extended [8] and is shown in Figure 3.

The dominant product of big bang nucleosynthesis is ^{4}He resulting in an abundance of close to 25% by mass. This quantity is easily estimated by counting the number of neutrons present when nucleosynthesis begins. When the weak interaction rates responsible for $n - p$ equilibrium freeze-out, at $T \approx 0.8$ MeV, the neutron to proton ratio is about 1/6. When free neutron decays prior to deuterium formation are taken into account, the ratio drops to $(n/p) \approx 1/7$. Then simple counting yields a ^{4}He mass fraction

$$Y_p = \frac{2(n/p)}{[1 + (n/p)]} \approx 0.25 \tag{34}$$

In the standard model with $N_\nu = 3$, there is basically one free parameter in BBN, namely the baryon to photon ratio, η. As we have seen above, the value of η controls the onset of nucleosynthesis through the deuterium bottleneck. For larger values of η, the quantity $\eta^{-1}\exp(-E_B/T)$ is smaller, and hence the nuclear chain may begin at a higher temperature. Remember also that a key ingredient in determining the final mass fraction of ^{4}He, is (n/p) [see eq. (34)] and that the final value of (n/p) was determined by the time at which nucleosynthesis begins, thus controlling the time available for free decays after freeze out. If nucleosynthesis begins earlier, this leaves less time for neutrons to decay and the value of (n/p) and hence Y_p is increased. But because the (n/p) ratio is only weakly dependent on η, the ^{4}He mass fraction is relatively flat as a function of η. When we go beyond the standard model, the ^{4}He abundance is very sensitive to changes in the expansion rate which can be related to the effective number of neutrino flavors as

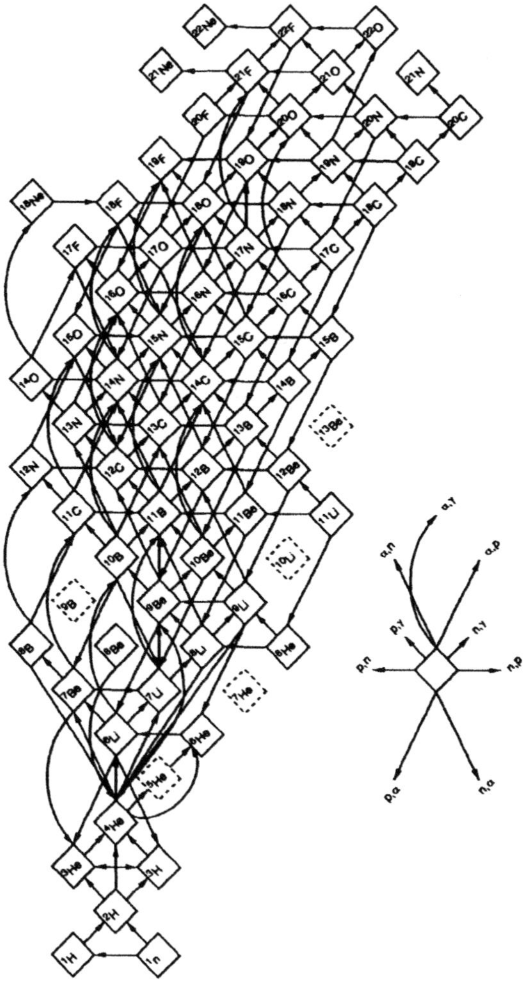

Figure 3. The nuclear network used in BBN calculations.

will be discussed below. Lesser amounts of the other light elements are produced: D and ^{3}He at the level of about 10^{-5} by number, and ^{7}Li at the level of 10^{-10} by number. These abundances (along with ^{6}Li) are shown in Figure 4 [8]. In Figure 5, the produced abundances of the intermediate mass isotopes ^{9}Be, ^{10}B, ^{11}B are also shown. These abundances are far below the observed values and it is believed that these isotopes are formed in cosmic ray nucleosynthesis.

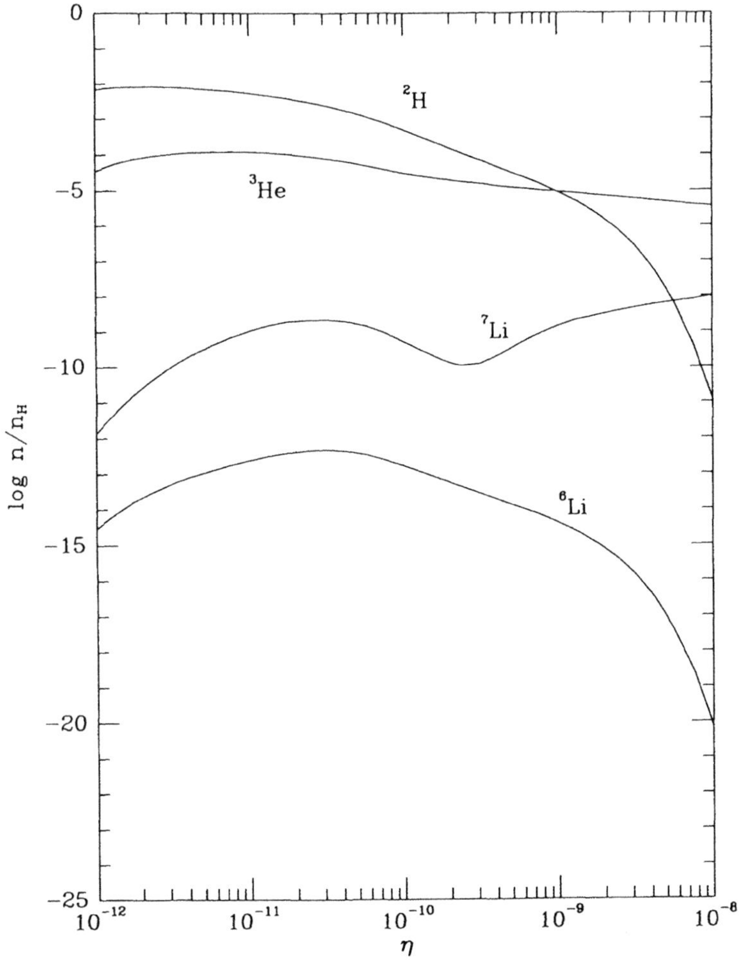

Figure 4. The light element abundances from big bang nucleosynthesis as a function of η, including ^{6}Li.

It is perhaps convenient at this time to note that the value of η is directly related to the fraction of Ω in baryons. Indeed, one can write

$$\Omega_B h^2 = 3.67 \times 10^7 \eta (T_0/2.728\mathrm{K})^3 \tag{35}$$

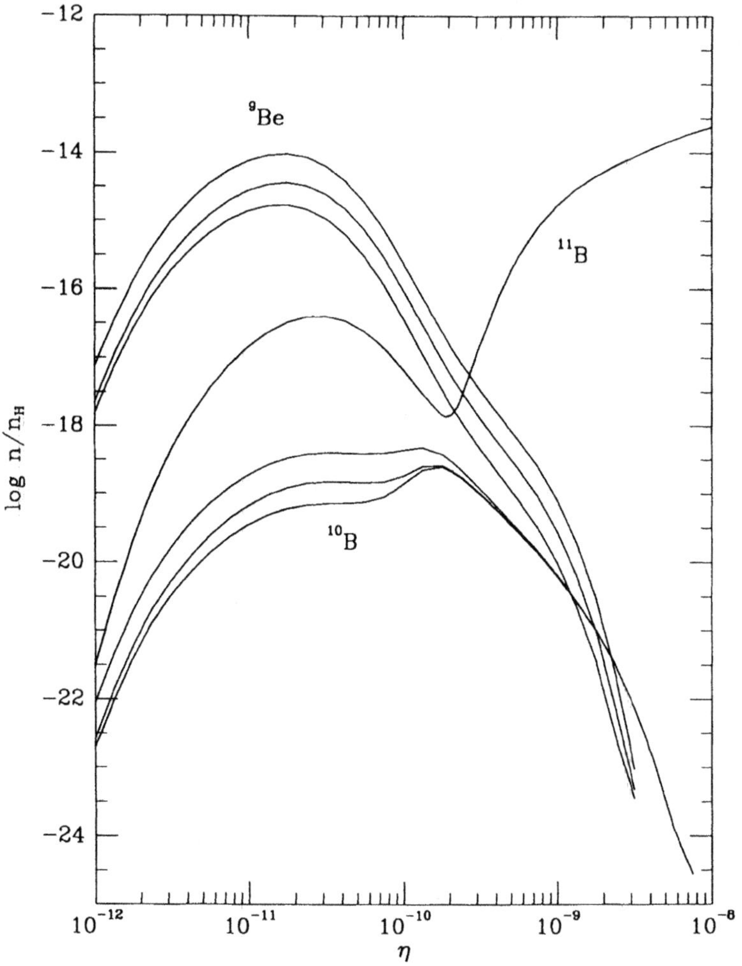

Figure 5. The intermediate mass element abundances from big bang nucleosynthesis as a function of η.

where T_0 is the present temperature of the microwave background.

Historically, it has been common to refer to two other parameter in BBN, the neutron mean life and the number of neutrino flavors. The neutron mean life is now very well determined and its remaining uncertainty can be treated simply as an uncertainty in the calculated abundance of ^{4}He. Although the number of neutrino flavors has also been fixed experimentally,

BBN is sensitive to the number of light degrees of freedom whether or not they interact weakly. It is often convenient to refer to these degrees of freedom as neutrino equivalents. By increasing N_ν in eq. (32), one increases the expansion rate $H \propto \sqrt{\rho}$. As a result, the weak interactions freeze-out at a higher temperature (see eqs. (21) and (22)). This leads once again to a higher value for (n/p) and a higher ^{4}He mass fraction. This effect will be treated in more detail in the last lecture.

For the comparison with the observations, I will use the resulting abundances of the light elements shown in Figure 6, which concentrate on the range in η_{10} between 1 and 10. The curves for the ^{4}He mass fraction, Y, bracket the computed range based on the uncertainty of the neutron mean-life which has been taken as [9] $\tau_n = 887 \pm 2$ s. Uncertainties in the produced ^{7}Li abundances have been adopted from the results in Hata et al. [10]. Uncertainties in D and ^{3}He production are small on the scale of this figure. The boxes correspond to the observed abundances and will be discussed below.

3. Abundances

3.1. ^{4}He

^{4}He is produced in stars, therefore to be able to extract a primordial abundance of ^{4}He, it is advantageous to make abundance measurements in systems of very low metallicity. Low metallicity or low abundances of C, N, and O relative to the solar abundances of these elements would indicate that the level of stellar processing is minimized. The ^{4}He abundance in very low metallicity regions is best determined from observations of HeII $\rightarrow$ HeI recombination lines in extragalactic HII (ionized hydrogen) regions. There is now a good collection of abundance information on the ^{4}He mass fraction, Y, O/H, and N/H in over 70 [11, 12, 13] such regions. In an extensive study based on the data in [11, 12], it was found [14] that the data is well represented by a linear correlation for Y vs. O/H and Y vs. N/H. It is then expected that the primordial abundance of ^{4}He can be determined from the intercept of that relation. A detailed analysis of the data including that in [13] found an intercept corresponding to a primordial abundance $Y_p = 0.234 \pm 0.002 \pm 0.005$ [15]. The stability of this fit was verified by a statistical bootstrap analysis [16] showing that the fits were not overly sensitive to any particular HII region.

To make use of the ^{4}He data, it is crucial to obtain high quality and very low metallicity data. In principle, any one HII region (with non-zero metallicity) should provide an upper limit to Y_p since some stellar processing has taken place augmenting the primordial value. Thus the determination of Y_p by an extrapolation to zero metallicity could be avoided by the obser-

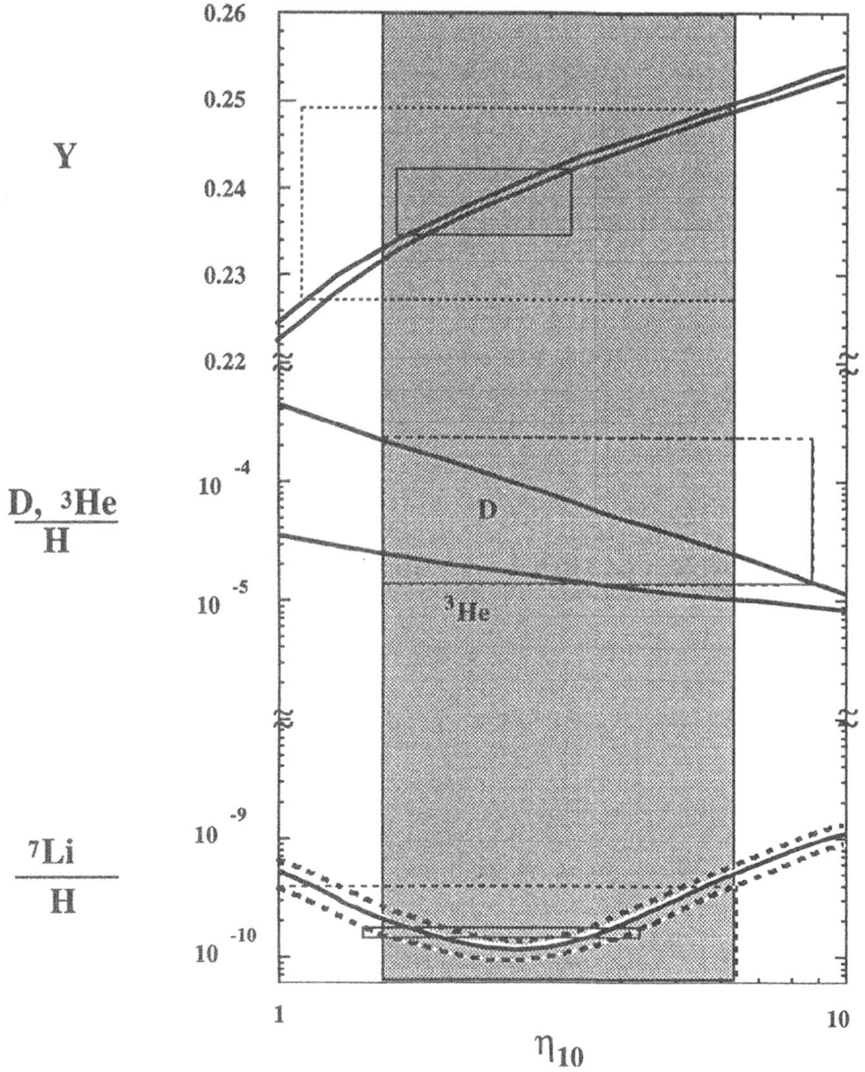

Figure 6. The light element abundances from big bang nucleosynthesis as a function of $\eta_{10} = 10^{10}\eta$.

vations of either low metallicity or low helium HII regions. For a very low metallicity HII region such an upper limit may even provide a reasonable estimate of Y_p.

Another way to avoid an extrapolation to zero metallicity (though such an extrapolation is in fact quite minimal given the low metallicity data available), one can perform a Bayesian analysis [17] in which one makes no other assumption other than the observed ^{4}He abundance is greater

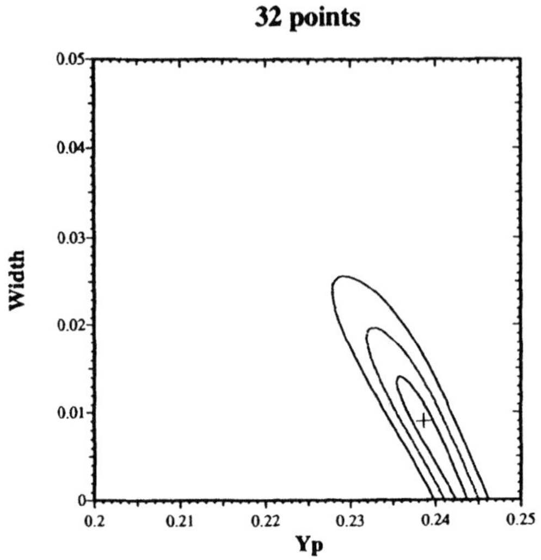

Figure 7. Equal likelihood 1, 2, and 3 σ contours in the $w - Y_p$ plane. The cross shows the position of the peak of the likelihood function.

than or equal to the primordial abundance. That is, one can consider three quantities: Y_T, the true ^{4}He abundance in an H II region about which the observed abundance Y_O is distributed. Both of these differ from the primordial abundance Y_p and the only prior assumed is that for each object, $Y_T \geq Y_p$. If we assume that the true abundance differs from the primordial abundance by no more than w, we can derive a total likelihood function by integrating out the unknown Y_T. We can then plot the equal likelihood contours as a function of Y_p and w. This is shown in Figure 7, where the 32 points of lowest metallicity have been used to calculate the likelihood function. The peak occurs at $Y_p = 0.238$ and the most likely width is $w = 0.009$. The 95% CL upper limit to Y_p in this case is 0.245. For further details on this approach see [17]. The data used in Figure 7 represents an update of that work and includes the data of ref. [18].

Although the above estimates on Y_p are consistent with those based on a linear extrapolation of the data, it has been claimed that the new data in refs. [13] and [18] leads to a significantly higher value for Y_p (in excess of 24%). The higher values of Y_p, quoted by Izotov and Thuan are based on their ^{4}He abundances derived by their method of determining all

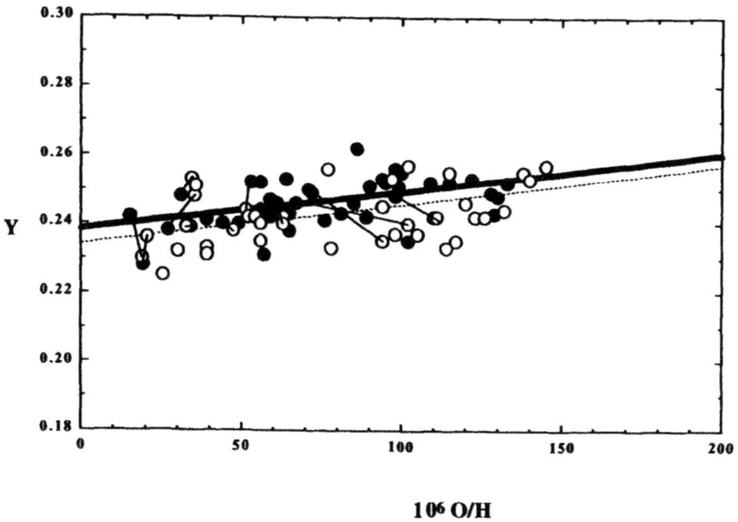

Figure 8. The helium (Y) and oxygen (O/H) abundances in extragalactic HII regions, from refs. [11, 12] (open circles), and from ref. [18] (filled circles). Lines connect the same regions observed by different groups.

of the parameters from a set of 5 helium recombination lines. This gives $Y_p = 0.2444 \pm 0.0015 + (44 \pm 19)\mathrm{O/H}$. However, as argued in [15] there are inherent uncertainties in this method which are not reflected in the error budget. For this reason and because we can more easily compare their data with previous data, we use their results which are based on S II densities. These results are entirely consistent with the data in refs. [11, 12] as can be seen in Figure 8 where the Y versus O/H data from refs. [11, 12] (open circles) is shown along with the newer data (filled circles, from ref. [18]). The fit to the open circles is shown by the dashed line with intercept 0.234, the fit to the filled circles is shown by the this solid line (barely visible) with intercept 0.239. Combining all the data one finds [19] a ^{4}He mass fraction based on 73 distinct HII regions

$$Y_p = 0.238 \pm 0.002 \pm 0.005 \tag{36}$$

The first uncertainty is purely statistical and the second uncertainty is an estimate of the systematic uncertainty in the primordial abundance determination [15]. The fit to all the data is shown by the thick line with intercept given by Eq. (36) above. The small errors quoted in [18] account for the total fit being skewed to the higher value of Y_p. The solid box for ^{4}He in Figure 6 represents the range (at $2\sigma_{\rm stat}$) from (36). The dashed box extends this by including the systematic uncertainty. A somewhat lower primordial abundance of $Y_p = 0.235 \pm .003$ is found by restricting to the 36 most metal poor regions [19].

The primordial ^{4}He abundance can also be determined by examining the correlation between Y and N/H. Indeed in all but one of the H II regions, N/H data is also available. However, unless N/H is directly proportional to O/H, it is not clear that a linear Y vs. N/H fit should give the same result. Indeed, the proportionality of N/H to O/H (or in other terms the primary vs secondary nature of nitrogen) has been studied [14, 15, 19]. Unfortunately from a theoretical point of view this question lies in the realm of very uncertain yields for nitrogen in AGB stars. The data indicate that N is mostly primary. Though the secondary contribution may be responsible for yielding systematically higher intercepts for Y vs N/H relative to Y vs. O/H, however the difference is small $\lesssim 0.003$.

Finally, it also of interest to test our understanding of the slope in the Y vs. O/H data. The data overall show a relatively steep slope $\Delta Y/\Delta O \simeq 110 \pm 25$. Models of chemical evolution typically give a much smaller value of about 20 and even in models with outflow (material ejected from the galaxy) the slopes only go up to about 60. This question, like the N vs. O question is highly sensitive to very uncertain theoretical yields [19].

3.2. ^{7}Li

The ^{7}Li abundance is also reasonably well known. In old, hot, population-II stars, ^{7}Li is found to have a very nearly uniform abundance [20]. For stars with a surface temperature $T > 5500$ K and a metallicity less than about 1/20th solar (so that effects such as stellar convection may not be important), the abundances show little or no dispersion beyond that which is consistent with the errors of individual measurements. Indeed, as detailed in ref. [21], much of the work concerning ^{7}Li has to do with the presence or absence of dispersion and whether or not there is in fact some tiny slope to a [Li] = log ^{7}Li/H + 12 vs. T or [Li] vs. [Fe/H] relationship ([Fe/H] is the log of the Fe/H ratio relative to the solar value).

There is ^{7}Li data from nearly 100 halo stars, from a variety of sources. When the Li data from stars with [Fe/H] < -1.3 is plotted as a function of surface temperature, one sees a plateau emerging for $T > 5500$ K as shown in Figure 9 for the data taken from ref. [21]. As one can see from the figure, at high temperatures, where the convection zone does not go deep below the surface, the Li abundance is uniform. At lower temperatures, the surface abundance of Li is depleted as Li passes through the hotter interior of the star and is destroyed. The lack of dispersion in the plateau region is evidence that this abundance is indeed primordial (or at least very close to it). Another way to see the plateau is to plot the Li abundance data as a

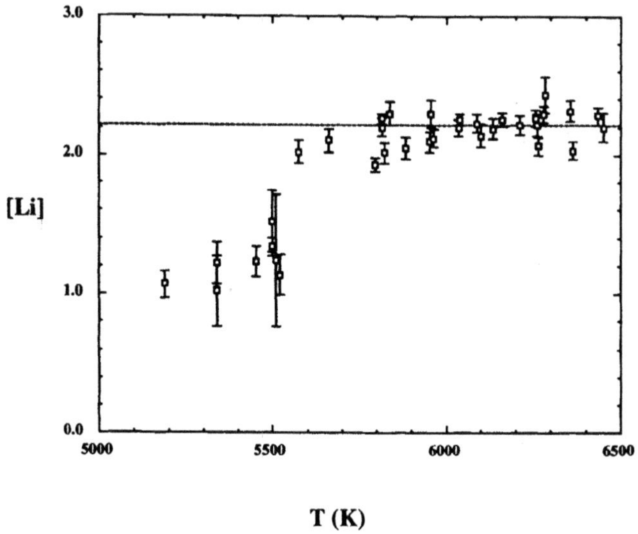

Figure 9. The Li abundance in halo stars with [Fe/H] < -1.3, as a function of surface temperature. The dashed line shows the value of the weighted mean of the plateau data.

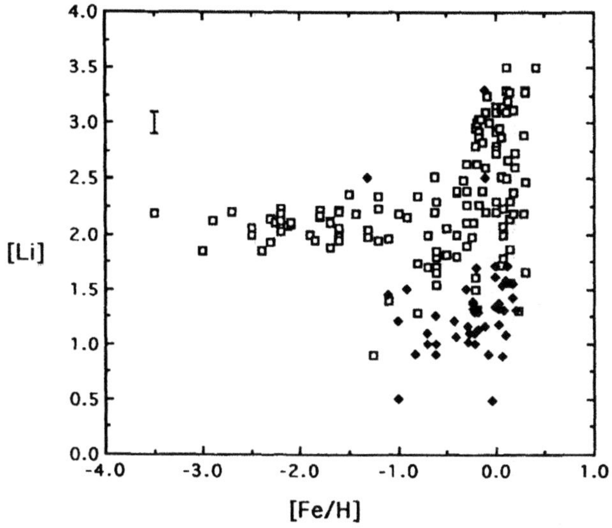

Figure 10. The Li abundance in halo stars with $T > 5500$ K, as a function of metallicity. Filled diamonds represent upper limits.

function of metallicity, this time with the restriction that $T > 5500$ K as seen in Figure 10. In this case, the plateau emerges at low [Fe/H] as would be expected. At higher [Fe/H], the convection zone remains near the surface only for much hotter stars. Thus, for [Fe/H] > -1.3, the effects of depletion are seen. Also apparent in this figure is that at higher metallicities there is evidence for the production of Li which rises by over an order of magnitude at solar metallicity.

I will use the value given in ref. [21] as the best estimate for the mean ^{7}Li abundance and its statistical uncertainty in halo stars

$$\mathrm{Li/H} = (1.6 \pm 0.1) \times 10^{-10} \tag{37}$$

The Li abundance determination is sensitive to stellar parameters such as the assumed surface temperature, the metallicity and the surface gravity. The greatest model dependence is on the conversion of the observed colors (B-V) to temperature. For example, in a sample of 55 stars taken from the papers of ref. [20], one finds [Li] $= 2.08 \pm 0.02$. From Thorburn's[22] sample of 74 stars one finds [Li] $= 2.26 \pm 0.01$. I believe that much of this systematic uncertainty has now been relieved by newer methods for determining the stellar parameters and the Li abundance. The two papers in ref. [21], are based on different methods. The first uses temperatures which are determined by observations of Balmer lines [23] and the second employs the infra-red flux method [24]. The data sets in these papers which contain 24 and 41 stars respectively, both give [Li] $= 2.21 \pm 0.01$. The Li abundance used in (37) corresponds to this value. In fact when correlated with respect to either temperature or iron, the data in the latter paper of [21] shows no statistical trend. With respect to temperature, the data give [Li] $= 1.28 \pm 0.43 + (0.015 + 0.007)T/100\mathrm{K}$, indicating a change in [Li] well within the data over the observed temperature range. With respect to iron, no slope is found [Li] $= 2.17 \pm 0.07 + (-0.018 \pm 0.031)$[Fe/H]. The variance in the plateau data is less than 0.01. The solid box for ^{7}Li in Figure 6 represents the $2\sigma_{\mathrm{stat}}$ range from (37).

There is however an important source of systematic error due to the possibility that Li has been depleted in these stars from their initial abundance. These uncertainties are however limited. As was noted above, the lack of dispersion in the Li data limits the amount of depletion. In addition, standard stellar models[25] predict that any depletion of ^{7}Li would be accompanied by a very severe depletion of ^{6}Li. Until recently, ^{6}Li had never been observed in hot pop II stars. The observation[26] of ^{6}Li (which turns out to be consistent with its origin in cosmic-ray nucleosynthesis and with a small amount of depletion as expected from standard stellar models) is another good indication that ^{7}Li has not been destroyed in these stars[27, 28, 29].

Aside from the big bang, Li is produced together with Be and B in cosmic ray spallation of C,N,O by protons and α-particles. Li is also produced by $\alpha - \alpha$ fusion. Be and B have been observed in these same pop II stars and in particular there are a dozen or so stars in which both Be and

^{7}Li have been observed. Thus Be (and B though there is still a paucity of data) can be used as a consistency check on primordial Li [30]. Based on the Be abundance found in these stars, one can conclude that no more than 10-20% of the ^{7}Li is due to cosmic ray nucleosynthesis leaving the remainder (an abundance near 10^{-10}) as primordial. The dashed box in Figure 6, accounts for the possibility that as much as half of the primordial ^{7}Li has been destroyed in stars, and that as much as 20% of the observed ^{7}Li may have been produced in cosmic ray collisions rather than in the Big Bang. For ^{7}Li, the uncertainties are clearly dominated by systematic effects.

3.3. D

Turning to D/H, we have three basic types of abundance information: 1) ISM data, 2) solar system information, and perhaps 3) a primordial abundance from quasar absorption systems. The best measurement for ISM D/H is [31]

$$(\mathrm{D/H})_{\mathrm{ISM}} = 1.60 \pm 0.09^{+0.05}_{-0.10} \times 10^{-5} \tag{38}$$

Because there are no known astrophysical sites for the production of deuterium, all observed D must be primordial. As a result, a firm lower bound from deuterium establishes an upper bound on η which is robust and is shown by the lower right of the solid box in Figure 6. The solar abundance of D/H is inferred from two distinct measurements of ^{3}He. The solar wind measurements of ^{3}He as well as the low temperature components of stepwise heating measurements of ^{3}He in meteorites yield the presolar (D + ^{3}He)/H ratio, as D was efficiently burned to ^{3}He in the Sun's pre-main-sequence phase. These measurements indicate that [32, 33]

$$\left(\frac{\mathrm{D} + {}^3\mathrm{He}}{\mathrm{H}}\right)_{\odot} = (4.1 \pm 0.6 \pm 1.4) \times 10^{-5} \tag{39}$$

The high temperature components in meteorites are believed to yield the true solar ^{3}He/H ratio of [32, 33]

$$\left(\frac{{}^3\mathrm{He}}{\mathrm{H}}\right)_{\odot} = (1.5 \pm 0.2 \pm 0.3) \times 10^{-5} \tag{40}$$

The difference between these two abundances reveals the presolar D/H ratio, giving,

$$(\mathrm{D/H})_{\odot} \approx (2.6 \pm 0.6 \pm 1.4) \times 10^{-5} \tag{41}$$

It should be noted that measurements of surface abundances of HD on Jupiter show a somewhat higher value for D/H, $\mathrm{D/H} = 5 \pm 2 \times 10^{-5}$ [34]. If this value is confirmed and if fractionation does not significantly alter the D/H ratio (as it was suspected to for previous measurements involving CH_3D), it may have an important impact on galactic chemical evolution models. This value is marginally consistent with the inferred meteoritic values.

Finally, there have been several reported measurements of D/H in high redshift quasar absorption systems. Such measurements are in principle capable of determining the primordial value for D/H and hence η, because of the strong and monotonic dependence of D/H on η. However, at present, detections of D/H using quasar absorption systems do not yield a conclusive value for D/H. As such, it should be cautioned that these values may not turn out to represent the true primordial value and it is very unlikely that both are primordial and indicate an inhomogeneity [35] (a large scale inhomogeneity of the magnitude required to placate all observations is excluded by the isotropy of the microwave background radiation). The first of these measurements [36] indicated a rather high D/H ratio, D/H $\approx$ 1.9 – 2.5 $\times 10^{-4}$. Other high D/H ratios were reported in [37]. More recently, a similarly high value of D/H = 2.0 $\pm$0.5 $\times$ 10^{-4} was reported in a relatively low redshift system (making it less suspect to interloper problems) [38]. However, there are reported low values of D/H in other such systems [39] with values of D/H originally reported as low as $\simeq 2.5 \times 10^{-5}$, significantly lower than the ones quoted above. The abundance in these systems has been revised upwards to about 3.4 $\pm$0.3 $\times$ 10^{-5} [40]. I will not enter into the debate as to which if any of these observations may be a better representation of the true primordial D/H ratio. I only note that it remains a highly contested issue [40, 41] The range of quasar absorber D/H is shown by the dashed box in Figure 6.

There are also several types of ^{3}He measurements. As noted above, meteoritic extractions yield a presolar value for ^{3}He/H as given in Eq. (40). In addition, there are several ISM measurements of ^{3}He in galactic HII regions [42] which show a wide dispersion which may be indicative of pollution or a bias [43]

$$\left(\frac{^3\mathrm{He}}{\mathrm{H}}\right)_{\mathrm{HII}} \simeq 1 - 5 \times 10^{-5} \tag{42}$$

There is also a recent ISM measurement of ^{3}He [44] with

$$\left(\frac{^3\mathrm{He}}{\mathrm{H}}\right)_{\mathrm{ISM}} = 2.1^{+.9}_{-.8} \times 10^{-5} \tag{43}$$

Finally there are observations of ^{3}He in planetary nebulae [45] which show a very high ^{3}He abundance of ^{3}He/H $\sim 10^{-3}$.

Each of the light element isotopes can be made consistent with theory for a specific range in η. Overall consistency of course requires that the range in η agree among all four light elements. However, as will be argued below D and ^{3}He are far more sensitive to chemical evolution than ^{4}He or ^{7}Li and as such the direct comparison between the theoretical predictions of the primordial abundances of D and ^{3}He with the observational determination of their abundances is far more difficult. Therefore in what follows I will for the most part restrict the comparison between theory and observation to the two isotopes who suffer the least from the effects of chemical evolution.

4. Chemical Evolution

Because we can not directly measure the primordial abundances of any of the light element isotopes, we are required to make some assumptions concerning the evolution of these isotopes. As has been discussed above, ^{4}He is produced in stars along with oxygen and nitrogen. ^{7}Li can be destroyed in stars and produced in several (though still uncertain) environments. D is totally destroyed in the star formation process and ^{3}He is both produced and destroyed in stars with fairly uncertain yields. It is therefore preferable, if possible to observe the light element isotopes in a low metallicity environment. Such is the case with ^{4}He and ^{7}Li, and we can be fairly assured that the abundance determinations of these isotopes are close to primordial. If the quasar absorption system measurements of D/H stabilize, then this too may be very close to a primordial measurement. Otherwise, to match the solar and present abundances of D and ^{3}He to their primordial values requires a model of galactic chemical evolution.

The main inputs to chemical evolution models are: 1) The initial mass function, $\phi(m)$, indicating the distribution of stellar masses. Typically, a simple power law form for the IMF is chosen, $\phi(m) \sim m^{-x}$, with $x \simeq -2.7$. This is a fairly good representation of the observed distribution, particularly at larger masses. 2) The star formation rate, ψ. Typical choices for a SFR are $\psi(t) \propto \sigma$ or σ^2 or even a straight exponential $e^{-t/\tau}$. σ is the fraction of mass in gas, $M_{\rm gas}/M_{\rm tot}$. 3) The presence of infalling or outflowing gas; and of course 4) the stellar yields. It is the latter, particularly in the case of ^{3}He, that is the cause for so much uncertainty. Chemical evolution models simply set up a series of evolution equations which trace desired quantities. For example, the mass in gas and the SFR evolve through a relation such as

$$\frac{dM_{\rm gas}}{dt} = -\psi(t) + e(t) + i(t) - o(t) \tag{44}$$

where e represents the amount of gas ejected from stars, i is the gas infall rate, and o is the gas outflow rate. The ejection rate is in turn given by

$$e(t) = \int (m - m_R)\phi(m)\psi(t - \tau(m))dm \tag{45}$$

where m_R is the remnant mass (a function of the stellar mass m as well) and $\tau(m)$ is the stellar lifetime. If we ignore $\tau(m)$, then the ejection rate is simply proportional to the star formation rate ψ, $e(t) = R\psi$. R is referred to as the return fraction and this approximation is known as the instantaneous recycling approximation (IRA). Similar equations can be developed which trace the abundances of the element isotopes [46]. Neglecting both infall and outflow, these take the form

$$\frac{d(XM_{\rm gas})}{dt} = -\psi(t)X + e_X(t) \tag{46}$$

where X is the mass fraction of a particular element of interest and e_X is the mass fraction of the element ejected in the death of a star. In the case of deuterium, $e_D = 0$.

As one can see from (46) deuterium is always a monotonically decreasing function of time in chemical evolution models. The degree to which D is destroyed, is however a model dependent question which depends sensitively on the IMF and SFR. The evolution of ^{3}He is however considerably more complicated. Stellar models predict that substantial amounts of ^{3}He are produced in stars between 1 and 3 $M_\odot$. For M $<$ 8$M_\odot$, Iben and Truran [47] calculate

$$(^3{\rm He/H})_f = 1.8 \times 10^{-4} \left(\frac{M_\odot}{M}\right)^2 + 0.7\left[({\rm D} + {}^3{\rm He})/{\rm H}\right]_i \tag{47}$$

so that for example, when $\eta_{10} = 3$, $((\mathrm{D} + {}^3\mathrm{He})/\mathrm{H})_i = 9 \times 10^{-5}$, and the ratio of the final abundance of ^{3}He/H to the initial (D + ^{3}He)/H abundance denoted by g_3 is $g_3(1M_\odot) = 2.7$. The ^{3}He abundance is nearly tripled. It should be emphasized that this prediction is in fact consistent with the observation of high ^{3}He/H in planetary nebulae [45].

Generally, implementation of the ^{3}He yield in Eq. (47) in chemical evolution models leads to an overproduction of ^{3}He/H particularly at the solar epoch [43, 48]. For example, in Figure 11, the evolution of D and ^{3}He is shown for a model in which only a modest amount of deuterium is destroyed. Namely, by a factor of 5, from D/H $= 7.5 \times 10^{-5}$ to a present value of $\sim 1.5 \times 10^{-5}$. However, due to the production of ^{3}He in low mass stars, ^{3}He is greatly overproduced relative to the solar value. This problem is compounded in models with an intense period of D destruction. In Scully et al. [49], a dynamically generated supernovae wind model was coupled to

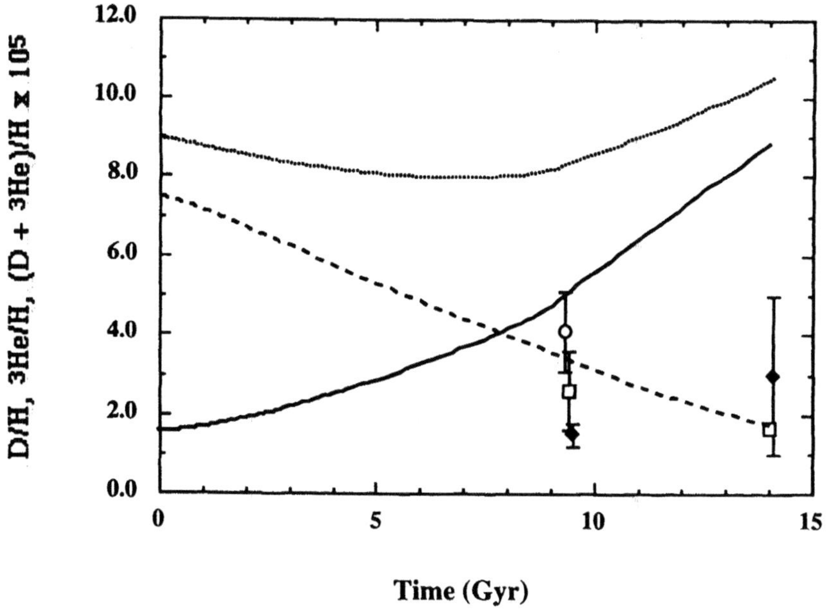

Figure 11. The evolution of D/H and ^{3}He/H and (D+^{3}He)/H with time in units of 10^{-5}. The assumed primordial abundance of D/H is 7.5 $\times 10^{-5}$. The solid curve shows the evolution of ^{3}He/H, the dashed curve for D/H and the dotted curve for the sum (D+^{3}He)/H. The diamonds show the data for ^{3}He, the open squares for deuterium and the open circle for the sum.

models of galactic chemical evolution with the aim of reducing a primordial D/H abundance of 2 $\times 10^{-4}$ to the present ISM value without overproducing heavy elements and remaining consistent with the other observational constraints typically imposed on such models. In Figure 12, the evolution of D/H and ^{3}He/H is shown as a function of time in several representative models with significant deuterium destruction factors (see ref [49] for details). However, as one can plainly see, ^{3}He is grossly overproduced (the deuterium data is represented by squares and ^{3}He by circles).

The overproduction of ^{3}He relative to the solar meteoritic value seems to be a generic feature of chemical evolution models when ^{3}He production in low mass stars is included. This result appears to be independent of the chemical evolution model and is directly related to the assumed stellar yields of ^{3}He. It has recently been suggested that at least some low mass stars may indeed be net destroyers of ^{3}He if one includes the effects of extra mixing below the conventional convection zone in low mass stars on the red giant branch [50, 51]. The extra mixing does not take place for stars

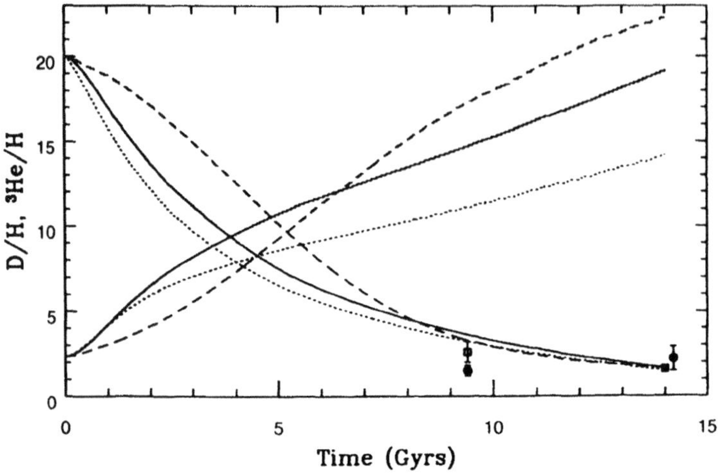

Figure 12. As in Figure 11. The primordial abundance of D/H in this case was chosen to be 2 $\times 10^{-4}$.

which do not undergo a helium core flash (i.e. stars > 1.7 - 2 $M_\odot$). Thus stars with masses *less than* 1.7 $M_\odot$ are responsible for the ^{3}He destruction. Using the yields of Boothroyd and Malaney [51], it was shown [52] that these reduced ^{3}He yields in low mass stars can account for the relatively low solar and present day ^{3}He/H abundances observed. In fact, in some cases, ^{3}He was underproduced. To account for the ^{3}He evolution and the fact that some low mass stars must be producers of ^{3}He as indicated by the planetary nebulae data, it was suggested that the new yields apply only to a fraction (albeit large) of low mass stars [52, 53]. The corresponding evolution [52] of D/H and ^{3}He/H is shown in Figure 13.

The models of chemical evolution discussed above indicate that it is possible to destroy significant amounts of deuterium and remain consistent with chemical evolutionary constraints. To do so however, comes with a price. Large deuterium destruction factors require substantial amounts of stellar processing, which at the same time produce heavy elements. To keep the heavy element abundances in the Galaxy in check, significant Galactic winds enriched in heavy elements must be incorporated. In fact there is some evidence that enriched winds were operative in the early Galaxy. In the X-ray cluster satellites observed by Mushotzky et al. [54] and Loewenstein and Mushotzky [55] the mean oxygen abundance was found to be roughly half solar. This corresponds to a near solar abundance of heavy

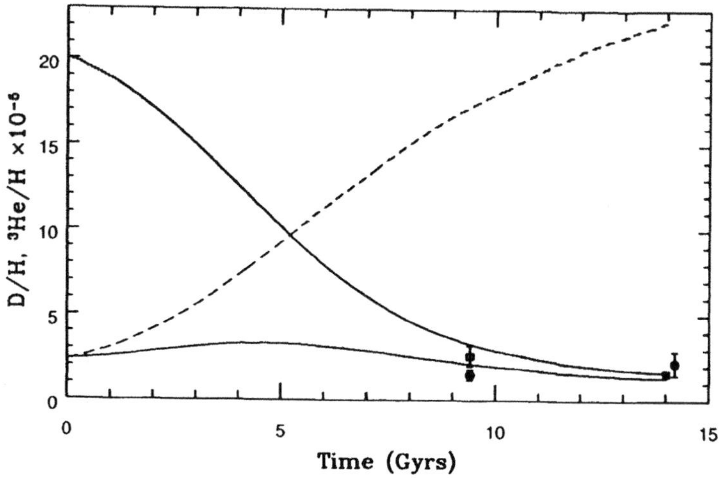

Figure 13. The evolution of D/H and ^{3}He/H with time using the reduced ^{3}He yields of ref. [51] The dashed curve is the same as in Figure 12, using standard ^{3}He yields.

elements in the inter-Galactic medium, where apparently little or no star formation has taken place.

If our Galaxy is typical in the Universe, then the models of the type discussed above would indicate that the luminosity density of the Universe at high redshift should also be substantial augmented relative to the present. Recent observations of the luminosity density at high redshift [56] are making it possible for the first time to test models of cosmic chemical evolution. The high redshift observations, are very discriminatory with respect to a given SFR [57]. Models in which the star formation rate is proportional to the gas mass fraction (these are common place in Galactic chemical evolution) have difficulties to fit the multi-color data from $z = 0$ to 1. This includes many of the successful Galactic infall models. In contrast, models with a steeply decreasing SFR are favored. In Figure 14, the predicted luminosity density based on the model with evolution shown in Figure 13 from [49], as compared with the observations (see ref. [57] for details).

While it would be premature to conclude that all models with large deuterium destruction factors are favored, it does seem that models which do fit the high redshift data destroy significant amounts of D/H. On the other hand, we can not exclude models which destroy only a small amount of D/H as Galactic models of chemical evolution. In this case, however the evolution of our Galaxy is anomalous with respect to the cosmic av-

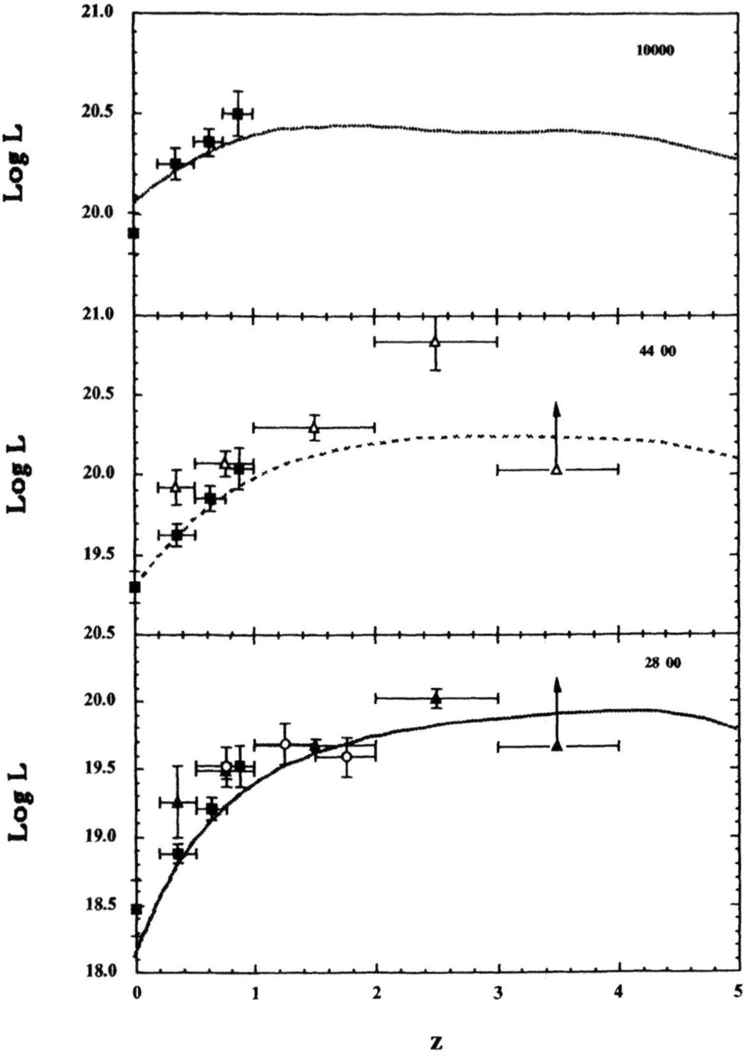

Figure 14. The tricolor luminosity densities (UV, B and IR) at $\lambda = 0.28, 0.44$ and 1.0 μm, in units of (h/.5) $\mathrm{WHz}^{-1}\mathrm{Mpc}^{-3}$ as a function of redshift for a model shown in 13 which destroys significant amounts of D/H. The data are taken from [56].

erage. If the low D/H measurements [39, 40] hold up, then it would seem that our Galaxy also has an anomalously high D/H abundance. That is we would predict in this case that the present cosmic abundance of D/H is significantly lower than the observed ISM value. If the high D/H observations [36, 37, 38] hold up, we could conclude that our Galaxy is indeed representative of the cosmic star formation history.

5. Likelihood Analyses

Monte Carlo techniques have proven to be a useful form of analysis for big bang nucleosynthesis [58, 59, 10]. An analysis of this sort was performed [2] using only ^{4}He and ^{7}Li. Two elements are sufficient for not only constraining the one parameter (η) theory of BBN, but also for testing for consistency. The procedure begins by establishing likelihood functions for the theory and observations. For example, for ^{4}He, the theoretical likelihood function takes the form

$$L_{\rm BBN}(Y, Y_{\rm BBN}) = e^{-(Y-Y_{\rm BBN}(\eta))^2/2\sigma_1^2} \tag{48}$$

where $Y_{\rm BBN}(\eta)$ is the central value for the ^{4}He mass fraction produced in the big bang as predicted by the theory at a given value of η. σ_1 is the uncertainty in that value derived from the Monte Carlo calculations [10] and is a measure of the theoretical uncertainty in the big bang calculation. Similarly one can write down an expression for the observational likelihood function. Assuming Gaussian errors, the likelihood function for the observations would take a form similar to that in (48).

A total likelihood function for each value of η is derived by convolving the theoretical and observational distributions, which for ^{4}He is given by

$$L^{^4\rm He}{}_{\rm total}(\eta) = \int dY \, L_{\rm BBN}\left(Y, Y_{\rm BBN}(\eta)\right) L_{\rm O}(Y, Y_{\rm O}) \tag{49}$$

An analogous calculation is performed [2] for ^{7}Li. The resulting likelihood functions from the observed abundances given in Eqs. (36) and (37) is shown in Figure 15. As one can see there is very good agreement between ^{4}He and ^{7}Li in the range of $\eta_{10} \simeq 1.5 - 5.0$. The double peaked nature of the ^{7}Li likelihood function is due to the presence of a minimum in the predicted lithium abundance. For a given observed value of ^{7}Li, there are two likely values of η.

The combined likelihood, for fitting both elements simultaneously, is given by the product of the two functions in Figure 15 and is shown in Figure 16. The 95% CL region covers the range $1.55 < \eta_{10} < 4.45$, with the two peaks occurring at $\eta_{10} = 1.9$ and 3.5. This range corresponds to values Ω_B between

$$0.006 < \Omega_B h^2 < .016 \tag{50}$$

For the lower value of $Y_p = 0.234 \pm 0.002 \pm 0.005$ as quoted in [15], the ^{4}He peak is shifted to slightly lower values of η and sits on top of the low-η peak as shown in Figure 17. (The difference in the ^{7}Li likelihood distribution is due to the assumed uncertainty in the ^{7}Li abundance which

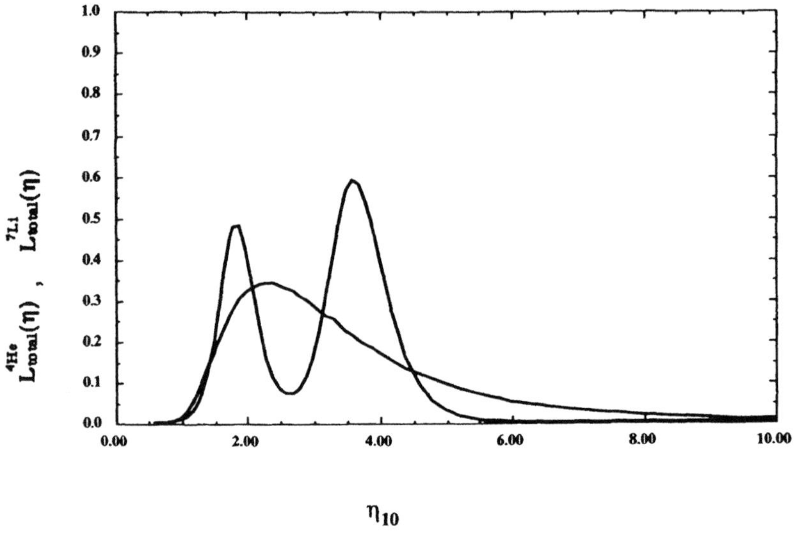

Figure 15. Likelihood distribution for each of ^{4}He and ^{7}Li, shown as a function of η. The one-peak structure of the ^{4}He curve corresponds to the monotonic increase of Y_p with η, while the two peaks for ^{7}Li arise from the minimum in the ^{7}Li abundance prediction.

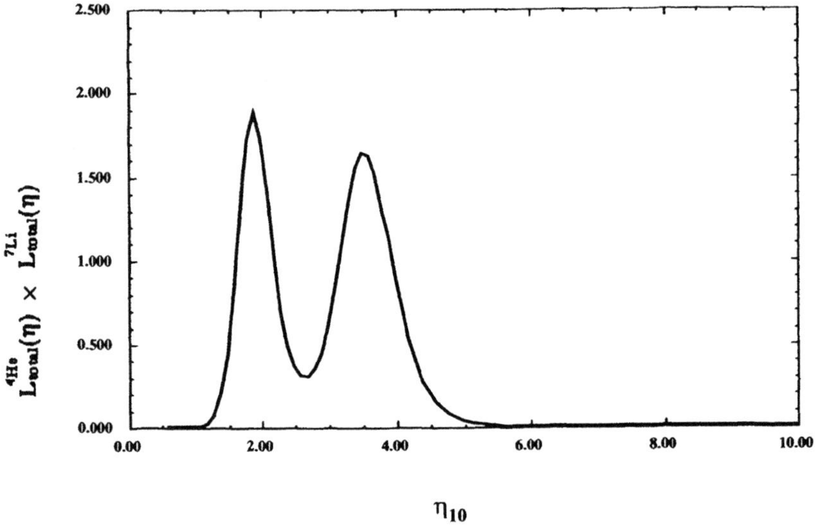

Figure 16. Combined likelihood for simultaneously fitting ^{4}He and ^{7}Li, as a function of η.

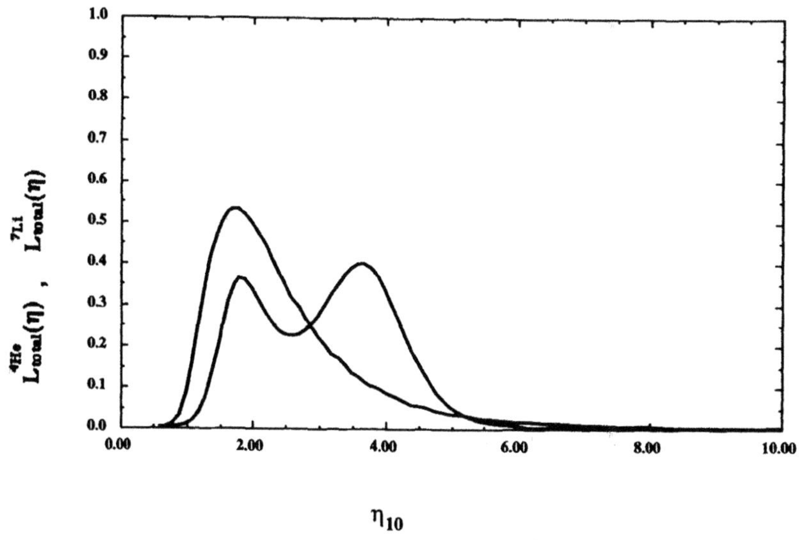

Figure 17. As in Figure 15 with a lower value of $Y_p = 0.234 \pm 0.002 \pm 0.005$.

is slightly higher than that in Figure 15.) The combined likelihood in this case is shown in Figure 18. From Figure 18 it is clear that ^{4}He overlaps the lower (in η) ^{7}Li peak, and so one expects that there will be concordance in an allowed range of η given by the overlap region. This is what one finds in Figure 18, which does show concordance and gives a preferred value for η, $\eta_{10} = 1.8^{+2.4}_{-.4}$ corresponding to (at 95% CL) $\Omega_B h^2 = .007^{+.008}_{-.002}$.

Thus, we can conclude that the abundances of ^{4}He and ^{7}Li are consistent, and select an η_{10} range which overlaps with (at the 95% CL) the longstanding favorite range around $\eta_{10} = 3$. Furthermore, by finding concordance using only ^{4}He and ^{7}Li, we deduce that if there is problem with BBN, it must arise from D and ^{3}He and is thus tied to chemical evolution or the stellar evolution of ^{3}He. The most model-independent conclusion is that standard BBN with $N_\nu = 3$ is not in jeopardy.

It is interesting to compare the results from the likelihood function of ^{4}He and ^{7}Li with that of D/H. Since D and ^{3}He are monotonic functions of η, a prediction for η, based on ^{4}He and ^{7}Li, can be turned into a prediction for D and ^{3}He. The corresponding 95% CL ranges are D/H $= (4.3 - 25) \times 10^{-5}$ and ^{3}He/H $= (1.2 - 2.6) \times 10^{-5}$. If we did have full confidence in the measured value of D/H in quasar absorption systems, then we could perform the same statistical analysis using ^{4}He, ^{7}Li, and D. To include D/H, one would proceed in much the same way as with the other two light

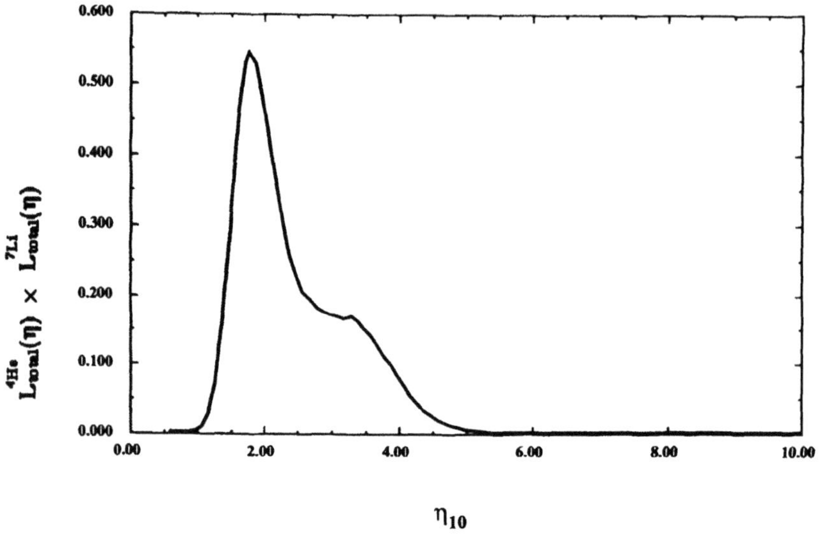

Figure 18. Combined likelihood for simultaneously fitting ^{4}He and ^{7}Li, as a function of η from from Figure 17.

elements. We compute likelihood functions for the BBN predictions as in Eq. (48) and the likelihood function for the observations. These are then convolved as in Eq. (49). Using D/H $= (2.0 \pm 0.5) \times 10^{-4}$ as indicated in the high D/H systems, we can plot the three likelihood functions including $L^{\rm D}_{\rm total}(\eta)$ in Figure 19. It is indeed startling how the three peaks, for D, ^{4}He and ^{7}Li are in excellent agreement with each other. In Figure 20, the combined distribution is shown. We now have a very clean distribution and prediction for η, $\eta_{10} = 1.8^{+1.6}_{-.3}$ corresponding to $\Omega_B h^2 = .007^{+.005}_{-.001}$. The absence of any overlap with the high-η peak of the ^{7}Li distribution has considerably lowered the upper limit to η. Overall, the concordance limits in this case are dominated by the deuterium likelihood function.

If instead, we assume that the low value [40] of D/H $= (3.4 \pm 0.3) \times 10^{-5}$ is the primordial abundance, then we can again compare the likelihood distributions as in Figure 19, now substituting the low D/H value. As one can see from Figure 21, there is now hardly any overlap between the D and the ^{7}Li and ^{4}He distributions. The combined distribution shown in Figure 22 is compared with that in Figure 20. Though one can not use this likelihood analysis to prove the correctness of the high D/H measurements or the incorrectness of the low D/H measurements, the analysis clearly shows the difference in compatibility between the two values of D/H and

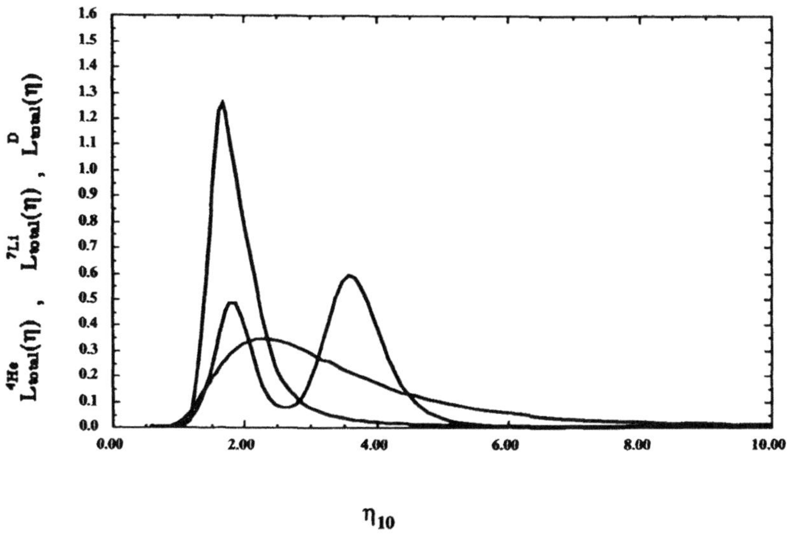

Figure 19. As in Figure 15, with the addition of the likelihood distribution for D/H assuming "high" D/H.

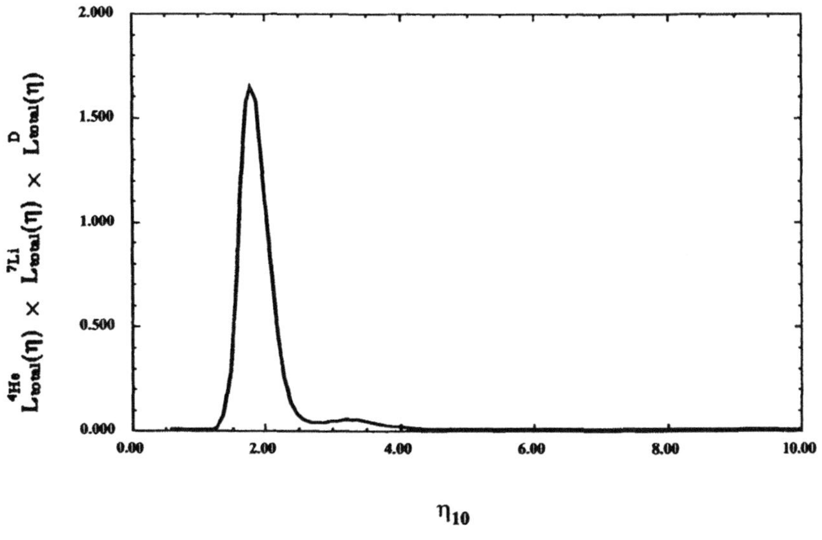

Figure 20. Combined likelihood for simultaneously fitting ^{4}He and ^{7}Li, and D as a function of η from Figure 19.

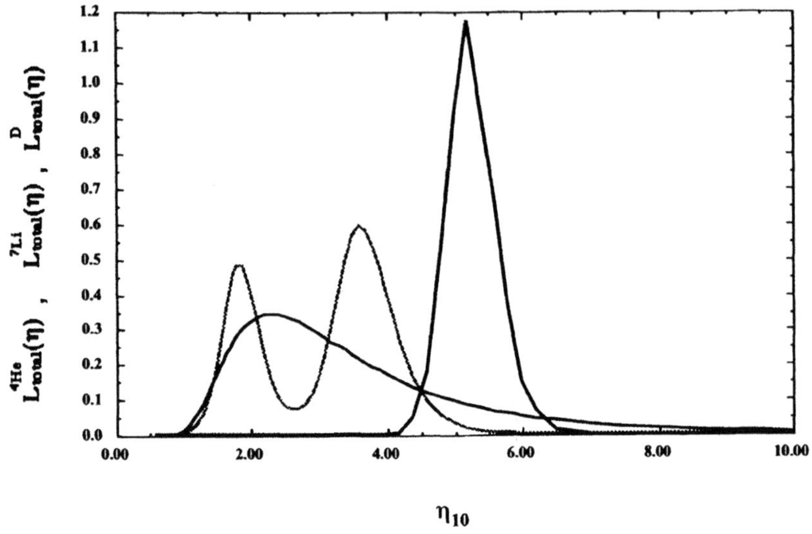

Figure 21. As in Figure 19, with the likelihood distribution for low D/H.

the observational determinations of ^{4}He and ^{7}Li. To *make* the low D/H measurement compatible, one would have to argue for a shift upwards in ^{4}He to a primordial value of 0.247 (a shift by 0.009) which is not warranted at this time by the data, and a ^{7}Li depletion factor of about 2, which is close to recent upper limits to the amount of depletion [60, 28].

The implications of the resulting predictions from big bang nucleosynthesis on dark matter are clear. First, if $\Omega = 1$ (as predicted by inflation), and $\Omega_B \lesssim 0.1$ which is certainly a robust conclusion based on D/H, then non-baryonic dark matter is a necessity. Second, on the scale of small groups of galaxies, which are expected to sample the dark matter in galactic halos, $\Omega \gtrsim 0.05$. This value can be compared with the best estimate for Ω_B from equation (50). The low η peak in Figure 16 corresponds to $\Omega_B h^2 = 0.007$ which for $h = 1/2$ gives $\Omega_B = 0.028$. In this event, some non-baryonic dark matter in galactic halos is required. On the other hand, the high η peak in 16 corresponds to $\Omega_B h^2 = 0.013$ which for $h = 1/2$ gives $\Omega_B = 0.051$ and may be consistent with halo densities (but only just). If we include the data from the high D/H measurements in QSO absorbers this conclusion is unchanged, only the low η peak survives, and some non-baryonic dark matter is needed in galactic halos. In contrast [61], the low D/H measurements would imply that $\Omega_B h^2 = 0.019$ allowing for the possibility that $\Omega_B \simeq 0.08$. In this case, no non-baryonic dark matter is required in galactic

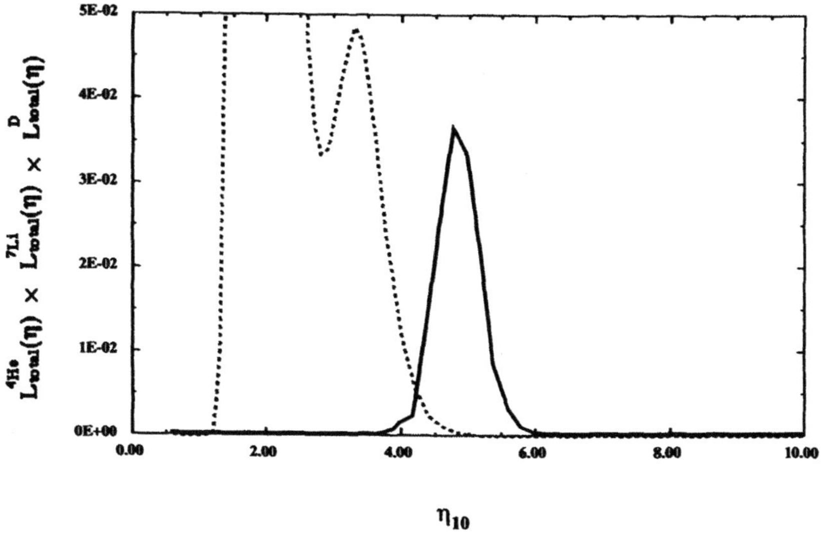

Figure 22. Combined likelihood for simultaneously fitting ^{4}He and ^{7}Li, and low D/H as a function of η. The dashed curve represents the combined distribution shown in Figure 20.

halos. However, I remind the reader that the low D/H value of 3.4 $\times 10^{-5}$ is at present only barely consistent with either the observations of ^{4}He or ^{7}Li and their interpretations as being primordial abundances.

6. Constraints from BBN

Limits on particle physics beyond the standard model are mostly sensitive to the bounds imposed on the ^{4}He abundance. As is well known, the ^{4}He abundance is predominantly determined by the neutron-to-proton ratio just prior to nucleosynthesis and is easily estimated assuming that all neutrons are incorporated into ^{4}He (see Eq. (34)). As discussed earlier, the neutron-to-proton ratio is fixed by its equilibrium value at the freeze-out of the weak interaction rates at a temperature $T_f \sim 1$ MeV modulo the occasional free neutron decay. Furthermore, freeze-out is determined by the competition between the weak interaction rates and the expansion rate of the Universe

$$G_F{}^2 T_f{}^5 \sim \Gamma_{\rm wk}(T_f) = H(T_f) \sim \sqrt{G_N N} T_f{}^2 \qquad (51)$$

where N counts the total (equivalent) number of relativistic particle species. The presence of additional neutrino flavors (or any other relativistic species) at the time of nucleosynthesis increases the overall energy density of the Universe and hence the expansion rate leading to a larger value of T_f, (n/p), and ultimately Y_p. Because of the form of Eq. (51) it is clear that just as one can place limits [62] on N, any changes in the weak or gravitational coupling constants can be similarly constrained (for a recent discussion see ref. [63]). In concluding this lecture, I will discuss the current constraint on N_ν the number of particle species (in neutrino units) and the limit on the strength of new interactions, if 3 right-handed (nearly massless) neutrinos are assumed to exist.

In the past, ^{3}He (together with D) has stood out in its importance for BBN, because it provided a (relatively large) lower limit for the baryon-to-photon ratio [64], $\eta_{10} > 2.8$. This limit for a long time was seen to be essential because it provided the only means for bounding η from below and in effect allows one to set an upper limit on the number of neutrino flavors [62], N_ν, as well as other constraints on particle physics properties. That is, the upper bound to N_ν is strongly dependent on the lower bound to η. This is easy to see: given an observed value for Y_p, for lower η, the ^{4}He abundance drops, allowing for a larger N_ν, which would raise the ^{4}He abundance back to the observed value. However, for $\eta < 4 \times 10^{-11}$, corresponding to $\Omega_B h^2 \lesssim .001 - .002$, which is not too different from galactic mass densities, there is no bound whatsoever on N_ν [65]. Of course, with the improved data on ^{7}Li, we do have lower bounds on η which exceed 10^{-10}. In fact, despite the uncertainty in the D/H abundance in quasar absorption systems, the high D/H values can certainly be regarded as an upper limit to primordial D/H, which also yield a lower limit to η.

Because, of new observations of D and ^{3}He, and the new theoretical work on chemical evolution sparked by these observations, the bound on N_ν which is tied directly to these isotopes, should be called into question. As described earlier, the limits due to ^{3}He are ultimately tied to the assumed yields of low mass stars. Using the reduced yields as depicted in Figure 13, consistent values of $\eta < 2.8$ are certainly possible. Ultimately, as I have said repeatedly, D/H measurements in quasar absorption systems may soon resolve this issue. However, the lower values of η, relax the bounds on the number of neutrino flavors.

As discussed above, the limit on N_ν comes about via the change in the expansion rate given by the Hubble parameter,

$$H^2 = \frac{8\pi G}{3}\rho = \frac{8\pi^3 G}{90}[N_{\rm SM} + \frac{7}{8}\Delta N_\nu]T^4 \qquad (52)$$

when compared to the weak interaction rates. Here $N_{\rm SM}$ refers to the standard model value for N. At $T \sim 1$ MeV, $N_{\rm SM} = 43/4$. Additional degrees of freedom will lead to an increase in the freeze-out temperature eventually leading to a higher ^{4}He abundance. In fact, one can parameterize the dependence of Y on N_ν by

$$Y = 0.2262 + 0.0131(N_\nu - 3) + 0.0135 \ln \eta_{10} \tag{53}$$

in the vicinity of $\eta_{10} \sim 2$. Eq. (53) also shows the weak (log) dependence on η. However, rather than use (53) to obtain a limit, it is preferable to use the likelihood method.

Just as ^{4}He and ^{7}Li were sufficient to determine a value for η, a limit on N_ν can be obtained as well [2, 66]. The likelihood approach utilized above can be extended to include N_ν as a free parameter. Since the light element abundances can be computed as functions of both η and N_ν, the likelihood function can be defined by [66]

$$L_{\rm BBN}(Y, Y_{\rm BBN}) = e^{-(Y - Y_{\rm BBN}(\eta, N_\nu))^2/2\sigma_1^2} \tag{54}$$

and

$$L^{^4{\rm He}}{}_{\rm total}(\eta, N_\nu) = \int dY\, L_{\rm BBN}\,(Y, Y_{\rm BBN}\,(\eta, N_\nu))\, L_{\rm O}(Y, Y_{\rm O}) \tag{55}$$

Again, similar expressions are needed for ^{7}Li and D. A three-dimensional view of the combined likelihood functions [66] is shown in Figure 23 which is based on the slightly lower value of Y_p as used in Figure 17. For updated (but similar) results see [67]. In this case the high and low η maxima of Figure 18, show up as peaks in the $L - \eta - N_\nu$ space (L_{47} when D/H is neglected and L_{247} when high D/H is included). The peaks of the distribution as well as the allowed ranges of η and N_ν are more easily discerned in the contour plot of Figure 24 which shows the 50%, 68% and 95% confidence level contours in the two likelihood functions. The crosses show the location of the peaks of the likelihood functions. L_{47} peaks at $N_\nu = 3.0$, $\eta_{10} = 1.8$ (in agreement with our previous results [2]) and at $N_\nu = 2.3$, $\eta_{10} = 3.6$. The 95% confidence level allows the following ranges in η and N_ν

$$1.6 \le N_\nu \le 4.0 \qquad\qquad 1.3 \le \eta_{10} \le 5.0 \tag{56}$$

Note however that the ranges in η and N_ν are strongly correlated as is evident in Figure 24. Since the deuterium likelihood function picks out a small range of values of η, largely independent of N_ν, its effect on L_{247} is to eliminate one of the two peaks in L_{47}. L_{247} also peaks at $N_\nu = 3.0$, $\eta_{10} = 1.8$. In this case the 95% contour gives the ranges

$$2.0 \le N_\nu \le 4.1 \qquad\qquad 1.4 \le \eta_{10} \le 2.6 \tag{57}$$

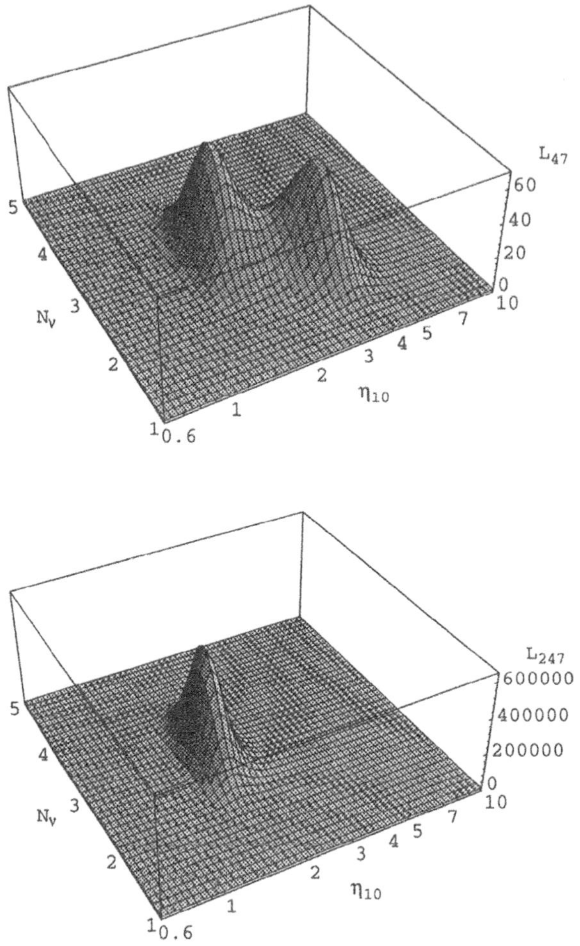

Figure 23. The combined two-dimensional likelihood functions for simultaneously fitting ^{4}He and ^{7}Li in the top panel, and including D in the lower one as functions of both η and N_ν.

Finally, in Figure 25, the resulting 95 % CL upper limit to N_ν is shown as a function of Y_P for several different choices for the primordial value of ^{7}Li/H [67].

One should recall that the limit derived above is not meant for neutrinos

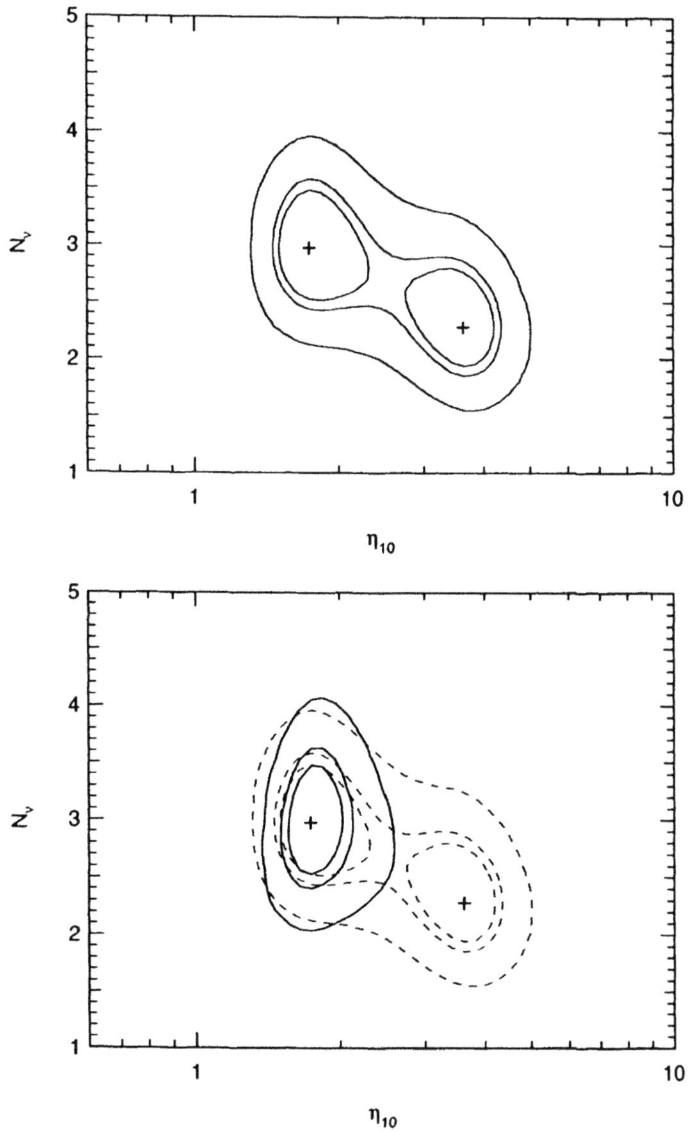

Figure 24. The top panel shows contours in the combined likelihood function for ^{4}He and ^{7}Li. The contours represent 50% (innermost), 68% and 95% (outermost) confidence levels. The crosses mark the points of maximum likelihood. Also shown is the equivalent result when D is included.

in the strictest sense. That is, the limit is only useful when applied to additional particle degrees of freedom which necessarily do not couple to the Z^o. For very weakly interacting particles, one must take into account the reduced abundance of these particles at the time of nucleosynthesis[68].

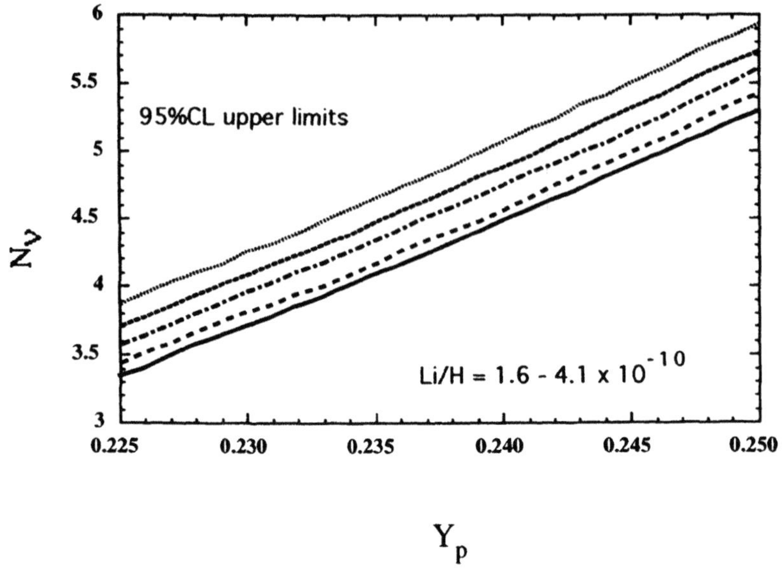

Figure 25. The 95 % CL upper limit to N_ν as a function of Y_p for $(\mathrm{Li/H})_p$ = 1.6, 2.0, 2.6, 3.2, and 4.1 $\times 10^{-10}$.

As discussed in the first lecture, the number of neutrinos today is reduced relative to the number of photons by $(T_\nu/T_\gamma)^3 = 4/11$. For some new particle, χ, which decoupled at $T_d > 1$ MeV, the same argument based on the conservation of entropy tells us that

$$(\frac{T_\chi}{T_\gamma})^3 = \frac{43}{4N(T_d)} \tag{58}$$

Thus we can translate the bound on N_ν, which is really a bound on the additional energy density at nucleosynthesis

$$\Delta\rho = \frac{\pi^2}{30}\left[\sum g_B(\frac{T_B}{T})^4 + \frac{7}{8}\sum g_F(\frac{T_F}{T})^4\right]T^4 \tag{59}$$

for additional boson states with g_B degrees of freedom and fermion states with g_F degrees of freedom. At nucleosynthesis $T = T_\nu = T_\gamma$ and the limit $N_\nu < 4$ becomes

$$\frac{8}{7}\sum\frac{g_B}{2}(\frac{T_B}{T_\nu})^4 + \sum\frac{g_F}{2}(\frac{T_F}{T})^4 < 1 \tag{60}$$

Such a limit would allow a single additional scalar degree of freedom (which counts as $\frac{4}{7}$) such as a majoron. On the other hand, in models with right-handed interactions, and three right-handed neutrinos, the constraint is severe. The right-handed states must have decoupled early enough to ensure

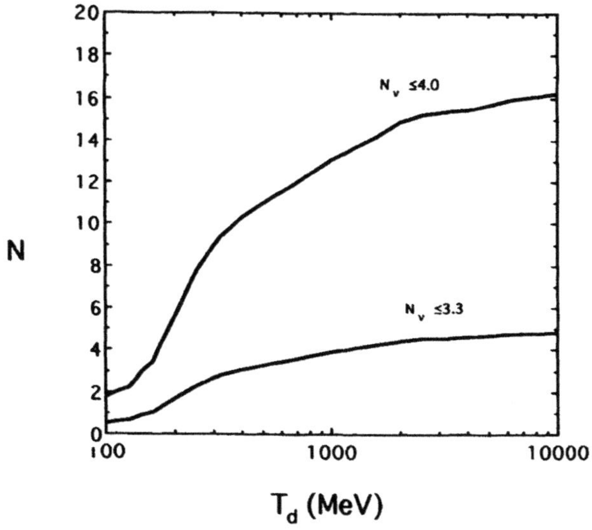

Figure 26. Limits on neutrino-like degrees of freedom.

$(T_{\nu_R}/T_{\nu_L})^4 < 1/3$. The temperature of a decoupled state is easily determined from (58). Three right-handed neutrinos would require $N(T_d) \gtrsim 25$, which from Figure 1 implies that $T_d > 140$ MeV, conservatively assuming a QCD transition temperature of 150 MeV. If right-handed interactions are mediated by additional gauge interactions, associated with some scale $M_{Z'}$, and if we assume that the right handed interactions scale as $M_{Z'}^4$, with a standard model-like coupling, then the decoupling temperature of the right handed interactions is related to $M_{Z'}$ by

$$(\frac{T_{dR}}{T_{dL}})^3 \sim (\frac{M_{Z'}}{M_Z})^4 \tag{61}$$

which for $T_{dL} \sim 3$ MeV (a more accurate value that the 1 MeV estimate) and $T_{dL} \gtrsim 140$ MeV, we find that the associated mass scale becomes $M_{Z'} \gtrsim 1.6$ TeV! In general this constraint is very sensitive to the BBN limit on N_ν. In Figure 26, the allowed number of neutrino degrees of freedom are shown as a function of their decoupling temperature for the case of $N_\nu < 4$ and $N_\nu < 3.3$, shown for comparison.

7. Summary

To summarize on the subject of big bang nucleosynthesis, I would assert that one can conclude that the present data on the abundances of the light element isotopes are consistent with the standard model of big bang nucleosynthesis. Using the the isotopes with the best data, ^{4}He and ^{7}Li, it is possible to constrain the theory and obtain a best set of values for the baryon-to-photon ratio of η_{10} and the corresponding value for $\Omega_B h^2$

$$\begin{array}{ccccccc} 1.55 & < & \eta_{10} & < & 4.45 & 95\%\mathrm{CL} \\ .006 & < & \Omega_B h^2 & < & .016 & 95\%\mathrm{CL} \end{array} \tag{62}$$

For $0.4 < h < 1$, we have a range $.006 < \Omega_B < .10$. This is a rather low value for the baryon density and would suggest that much of the galactic dark matter is non-baryonic [69]. These predictions are in addition consistent with recent observations of D/H in quasar absorption systems which show a high value. Difficulty remains however, in matching the solar ^{3}He abundance, suggesting a problem with our current understanding of galactic chemical evolution or the stellar evolution of low mass stars as they pertain to ^{3}He.

References

1. G.F.R. Ellis, these proceedings.
2. B.D. Fields and K.A. Olive, *Phys. Lett.* **B368** (1996) 103; B.D. Fields, K. Kainulainen, D. Thomas, and K.A. Olive, *New Astronomy* **1** (1996) 77.
3. G.F. Smoot and D. Scott, *Eur. Phys. J.* **C3** (1998) 127.
4. G. Gamow, *Phys. Rev.* **74** (1948) 505; G. Gamow, *Nature* **162** (1948) 680; R.A. Alpher and R.C. Herman, *Nature* **162** (1948) 774; R.A. Alpher and R.C. Herman, *Phys. Rev.* **75** (1949) 1089.
5. E.M. Burbidge, G.R. Burbidge, W.A. Fowler, and F. Hoyle, *Rev. Mod. Phys.* **29** (1957) 547.
6. H. Reeves, J. Audouze, W.A. Fowler, and D.N. Schramm, *Ap.J.* **179** (1973) 909.
7. C. Hayashi, *Prog. Theor. Phys.* **5** (1950) 224; R.A. Alpher, J.W. Follin Jr., R.C. Herman, *Phys. Rev.* **92** (1953) 1347; Ya. B. Zel'dovich, *Sov. Phys. Usp.* **6** (1963) 475; F. Hoyle and R.J. Taylor, *Nature* **203** (1964) 1108; P.J.E. Peebles, *Ap.J.* **146** (1966) 542; R.A. Wagoner, W.A. Fowler, and F. Hoyle, *Ap.J.* **148** (1967) 3; R.A. Wagoner, *Ap.J.* **179** (1973) 343.
8. D. Thomas, D. Schramm, K.A. Olive, and B. Fields, *Ap.J.* **406** (1993) 569.
9. Review of Particle Properties, *Phys. Rev.* **54** (1996) 1.
10. N. Hata, R.J. Scherrer, G. Steigman, D. Thomas, and T.P. Walker, *Ap.J.* **458** (1996) 637.
11. B.E.J. Pagel, E.A. Simonson, R.J. Terlevich and M. Edmunds, *MNRAS* **255** (1992) 325.
12. E. Skillman and R.C. Kennicutt, *Ap.J.* **411** (1993) 655; E. Skillman, R.J. Terlevich, R.C. Kennicutt, D.R. Garnett, and E. Terlevich, *Ap.J.* **431** (1994) 172.
13. Y.I. Izotov, T.X. Thuan, and V.A. Lipovetsky, *Ap.J.* **435** (1994) 647; *Ap.J.S.* **108** (1997) 1.
14. K.A. Olive and G. Steigman, *Ap.J. Supp.* **97** (1995) 49.
15. K.A. Olive, E. Skillman, and G. Steigman, *Ap.J.* **483** (1997) 788.
16. K.A. Olive and S.T. Scully, *IJMPA* **11** (1995) 409.
17. C.J. Hogan, K.A. Olive, and S.T. Scully, *Ap.J.* **489** (1997) L119.
18. Y.I. Izotov, and T.X. Thuan, *Ap.J.* **500** (1998) 188.

19. B.D. Fields and K.A. Olive, *Ap.J.* **506** (1998) 177
20. F. Spite, and M. Spite, *A.A.* **115** (1982) 357; M. Spite, J.P. Maillard, and F. Spite, *A.A.* **141** (1984) 56; F. Spite, and M. Spite, *A.A.* **163** (1986) 140; L.M. Hobbs, and D.K. Duncan, *Ap.J.* **317** (1987) 796; R. Rebolo, P. Molaro, J.E. and Beckman, *A.A.* **192** (1988) 192; M. Spite, F. Spite, R.C. Peterson, and F.H. Chaffee Jr., *A.A.* **172** (1987) L9; R. Rebolo, J.E. Beckman, and P. Molaro, *A.A.* **172** (1987) L17; L.M. Hobbs, and C. Pilachowski, *Ap.J.* **326** (1988) L23; L.M. Hobbs, and J.A. Thorburn, *Ap.J.* **375** (1991) 116; J.A. Thorburn, *Ap.J.* **399** (1992) L83; C.A. Pilachowski, C. Sneden,and J. Booth, *Ap.J.* **407** (1993) 699; L. Hobbs, and J. Thorburn, *Ap.J.* **428** (1994) L25; J.A. Thorburn, and T.C. Beers, *Ap.J.* **404** (1993) L13; F. Spite, and M. Spite, *A.A.* **279** (1993) L9. J.E. Norris, S.G. Ryan, and G.S. Stringfellow, *Ap.J.* **423** (1994) 386.
21. P. Molaro, F. Primas, and P. Bonifacio, *A.A.* **295** (1995) L47; P. Bonifacio and P. Molaro, *MNRAS* **285** (1997) 847.
22. J. Thorburn, *Ap.J.* **421** (1994) 318.
23. K. Fuhrmann, M. Axer, and T. Gehren, *A. A.* **285** (1994) 585.
24. A. Alonso, S. Arribas, and C. Martinez-Roger, C. *A. A. Supp.* **117** (1996) 227.
25. C.P. Deliyannis, P. Demarque, and S.D. Kawaler, *Ap.J.Supp.* **73** (1990) 21.
26. V.V. Smith, D.L. Lambert, and P.E. Nissen, *Ap.J.* **408** (1992) 262; *Ap.J.* (1998) in press; L. Hobbs, and J. Thorburn, *Ap.J.* **428** (1994) L25; M. Spite et al. *A.A.* (1998) submitted.
27. G. Steigman, B. Fields, K.A. Olive, D.N. Schramm, and T.P. Walker, *Ap.J.* **415** (1993) L35; M. Lemoine, D.N. Schramm, J.W. Truran, and C.J. Copi, *Ap.J.* **478** (1997) 554.
28. M.H. Pinsonneault, T.P. Walker, G. Steigman, and V.K. Naranyanan, *Ap.J.* (1998) submitted.
29. B.D.Fields and K.A. Olive, *New Astronomy*, astro-ph/9811183, in press; E. Vangioni-Flam, et al. *New Astronomy*, in press.
30. T.P. Walker, G. Steigman, D.N. Schramm, K.A. Olive and B. Fields, *Ap.J.* **413** (1993) 562; K.A. Olive, and D.N. Schramm, *Nature* **360** (1993) 439.
31. J.L. Linsky, et al., *Ap.J.* **402** (1993) 694; J.L. Linsky, et al.,*Ap.J.* **451** (1995) 335.
32. S.T. Scully, M. Cassé, K.A. Olive, D.N. Schramm, J.W. Truran, and E. Vangioni-Flam, *Ap.J.* **462** (1996) 960.
33. J. Geiss, in *Origin and Evolution of the Elements*, eds. N. Prantzos, E. Vangioni-Flam, and M. Cassé (Cambridge: Cambridge University Press, 1993), p. 89.
34. H.B. Niemann, et al. *Science* **272** (1996) 846.
35. C. Copi, K.A. Olive, and D.N. Schramm, *Proc. Nat. Ac. Sci.* **95** (1998) 2758, astro-ph/9606156.
36. R.F. Carswell, M. Rauch, R.J. Weymann, A.J. Cooke, and J.K. Webb, *MNRAS* **268** (1994) L1; A. Songaila, L.L. Cowie, C. Hogan, and M. Rugers, *Nature* **368** (1994) 599.
37. M. Rugers and C.J. Hogan, *A.J.* **111** (1996) 2135; R.F. Carswell, et al. *MNRAS* **278** (1996) 518; E.J. Wampler, et al., *A.A. 316* (1996) 33.
38. J.K. Webb, R.F. Carswell, K.M. Lanzetta, R. Ferlet, M. Lemoine, A. Vidal-Madjar, and D.V. Bowen, *Nature* **388** (1997) 250; D. Tytler et al., astro-ph/9810217 (1998).
39. D. Tytler, X.-M. Fan, and S. Burles, *Nature* **381** (1996) 207; S. Burles and D. Tytler, *Ap.J.* **460** (1996) 584.
40. S. Burles and D. Tytler, *Ap.J.* **499** (1998) 699; and astro-ph/9712109.
41. A. Songaila, E.J. Wampler, and L.L. Cowie, *Nature* **385** (1997) 137.
42. D.S. Balser, T.M. Bania, C.J. Brockway, R.T. Rood, and T.L. Wilson, *Ap.J.* **430** (1994) 667; T.M. Bania, D.S. Balser, R.T. Rood, T.L. Wilson, and T.J. Wilson, *Ap.J.S.* **113** (1997) 353.
43. K.A. Olive, R.T. Rood, D.N. Schramm, J.W. Truran, and E. Vangioni-Flam, *Ap.J.* **444** (1995) 680.
44. G. Gloeckler, and J. Geiss, *Nature* **381** (1996) 210.
45. R.T. Rood, T.M. Bania, and T.L. Wilson, Nature **355** (1992) 618; R.T. Rood,

T.M. Bania, T.L. Wilson, and D.S. Balser, 1995, in *the Light Element Abundances, Proceedings of the ESO/EIPC Workshop*, ed. P. Crane, (Berlin:Springer), p. 201; D.S. Balser, T.M. Bania, R.T. Rood, T.L. Wilson, *Ap.J.* **483** (1997) 320.
46. B.M. Tinsley, *Fund. Cosmic Phys.*, **5** (1980) 287.
47. I. Iben, and J.W. Truran, *Ap.J.* **220** (1978) 980.
48. D. Galli, F. Palla, F. Ferrini, and U. Penco, Ap.J. **443** (1995) 536; D. Dearborn, G. Steigman, and M. Tosi, *Ap.J.* **465** (1996) 887.
49. S. Scully, M. Cassé, K.A. Olive, and E. Vangioni-Flam, *Ap.J.* **476** (1997) 521.
50. C. Charbonnel, *A. A.* **282** (1994) 811; C. Charbonnel, *Ap.J.* **453** (1995) L41; C.J. Hogan, *Ap.J.* **441** (1995) L17; G.J. Wasserburg, A.I. Boothroyd, andI.-J. Sackmann, *Ap.J.* **447** (1995) L37; A. Weiss, J. Wagenhuber, and P. Denissenkov, *A.A.* **313** (1996) 581.
51. A.I. Boothroyd, A.I. and R.A. Malaney, astro-ph/9512133.
52. K.A. Olive, D.N. Schramm, S. Scully, and J.W. Truran, *Ap.J.* **479** (1997) 752.
53. D. Galli, L. Stanghellini, M. Tosi, and F. Palla *Ap.J.* **477** (1997) 218.
54. R. Mushotzky, et al., *Ap.J.* **466** (1996) 686
55. M. Loewenstein and R. Mushotzky, *Ap.J.* **466** (1996) 695.
56. S.J. Lilly, O. Le Fevre, F. Hammer, and D. Crampton, *ApJ* **460** (1996) L1; P. Madau, H.C. Ferguson, M.E. Dickenson, M. Giavalisco, C.C. Steidel, and A. Fruchter, *MNRAS* **283** (1996) 1388; A.J. Connolly, A.S. Szalay, M. Dickenson, M.U. SubbaRao, and R.J. Brunner, *ApJ* **486** (1997) L11; M.J. Sawicki, H. Lin, and H.K.C. Yee, *A.J.* **113** (1997) 1.
57. M. Cassé, K.A. Olive, E. Vangioni-Flam, and J. Audouze, *New Astronomy* **3** (1998) 259.
58. L.M. Krauss and P. Romanelli, *Ap.J.* **358** (1990) 47; L.M. Krauss and P.J. Kernan, *Phys. Lett.* **B347** (1995) 347; M. Smith, L. Kawano, and R.A. Malaney, *Ap.J. Supp.* **85** (1993) 219.
59. P.J. Kernan and L.M. Krauss, *Phys. Rev. Lett.* **72** (1994) 3309.
60. S. Vauclair and C. Charbonnel, *A.A.* **295** (1995) 715.
61. C.Y. Cardall and G.M. Fuller, *Ap.J.* **472** (1996) 435; N. Hata, G. Steigman, S. Bludman, and P. Langacker, *Phys.Rev.* **D55** (1997) 540.
62. G. Steigman, D.N. Schramm, and J. Gunn, *Phys. Lett.* **B66** (1977) 202.
63. B.A. Campbell and K.A. Olive, *Phys. Lett.* **B345** (1995) 429.
64. J. Yang, M.S. Turner, G. Steigman, D.N. Schramm, and K.A. Olive, *Ap.J.* **281** (1984) 493.
65. K.A. Olive, D.N. Schramm, G. Steigman, M.S. Turner, and J. Yang, *Ap.J.* **246** (1981) 557.
66. K.A. Olive and D. Thomas, *Astro. Part. Phys.* **7** (1997) 27.
67. K.A. Olive and D. Thomas, hep-ph/9811444.
68. G. Steigman, K.A. Olive, and D.N. Schramm, *Phys. Rev. Lett.* **43** (1979) 239; K.A. Olive, D.N. Schramm, and G. Steigman, *Nucl. Phys.* **B180** (1981) 497.
69. E. Vangioni-Flam and M. Cassé, Ap.J. **441** (1995) 471.

SEARCHES FOR DARK MATTER

J. RICH
SPP-DAPNIA
CEA-Saclay
91191 Gif-sur-Yvette
France

1. Introduction

There appears to be an emerging consensus (e.g. (Wang et al., 1999)) that the Universe consists primarily of non-baryonic dark matter (~ 0.3 critical density) and an effective vacuum energy density (~ 0.7 critical density). Evidence for this mixture is highly indirect. It comes from combining data on the clustering properties of galaxies, on the anisotropies of the cosmic background radiation, and on the redshift-magnitude diagram of type Ia supernovae (Perlmutter et al., 1998; Schmidt et al., 1998; Pain, 1998).

We cannot directly measure the vacuum energy density though this may some day be possible through the time dependence of galactic redshifts (Loeb, 1998). On the other hand, many popular non-baryonic dark matter candidates are detectable. Here we review efforts to detect the usual nonbaryonic suspects: wimps, axions and light neutrinos. We also review the microlensing observations of dark compact objects and discuss the possibility that these objects could account for the Galactic dark matter.

2. Wimps

For over a decade, wimps (weakly interacting massive particles) have been the standard cold dark matter candidate. Wimps must have masses greater than $\sim$ MeV to be cold at the epoch of matter-radiation equality and must have lifetimes greater than the Hubble time in order to dominate the dark matter now. Supersymmetric extensions of the standard model of particle physics (invented to solve problems unrelated to dark matter) predict the existence of such particles with masses $m_\chi > 10\text{GeV}$. The particles would have been thermally produced in the early Universe, yielding a cosmological

M. Lachièze-Rey (ed.), Theoretical and Observational Cosmology, 305–316.

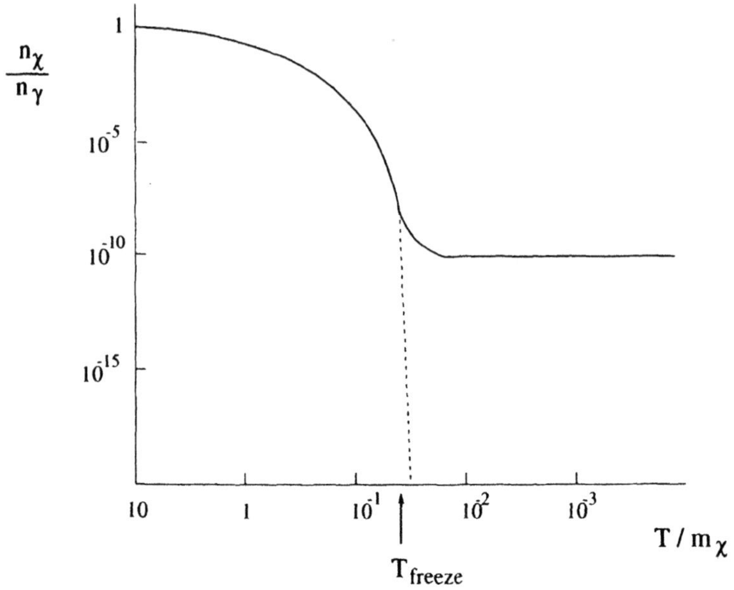

Figure 1. The wimp/photon ratio as a function of temperature in the early Universe. At high temperature, the assumption of thermal equilibrium gives a ratio of order unity. As the temperature dropped below the wimp mass, the number density of wimps became Boltzmann suppressed as the wimps annihilated to lighter particles. This process continued until the annihilation rate dropped below the expansion rate, "freezing" the ratio. For most candidates this happens at $T_{freeze} \sim m_\chi/20$. Setting the annihilation rate $n_\chi \langle \sigma v \rangle$ equal to the expansion rate $(8\pi G\rho/3)^{1/2}$ with $\rho \sim T^4$ yields a wimp/photon ratio proportional to the inverse of the annihilation cross section.

abundance inversely proportional to their annihilation cross section (figure 1). Supersymmetric models contain many free parameters yielding relic densities within a few orders of magnitude on either side of the critical density.

Today, wimps would be expected to inhabit the halos of spiral galaxies like our own. From the Galactic rotation velocity, one can estimate the local density to be about 0.3 GeV.cm^{-3} (Primack et al., 1988). The orbital velocities of objects trapped in the Galaxy are of order 250 km s^{-1} so the local wimp flux is of order $10^7(1\,\mathrm{GeV}/m_\chi)\,\mathrm{cm}^{-2}\mathrm{s}^{-1}$.

Goodman and Witten (1985) suggested that these wimps could be detected via the observation of nuclei recoiling from wimp-nucleus elastic scatters. Galactic wimps with masses in the GeV range have kinetic energies in the keV range so we can also expect nuclear recoils in the keV range. The rate is proportional to the elastic wimp-nucleus scattering cross section which depends on the parameters of the particle physics model. Typical values of the wimp-nucleon cross section are of order $10^{-44}\mathrm{cm}^2$, corresponding to a very weak interaction.

Wimp scatterings can be observed with "calorimetric" techniques (figure 2). Unfortunately, it is difficult to distinguish wimp events from events due to beta- or gamma- radioactivity (also shown in the figure). Statistically, a signal from wimps can be isolated through the expected $\sim 5\%$ seasonal modulation of the event rate (Smith and Lewin, 1990; Freeze et al, 1988). This modulation is due to the fact that while the solar system moves through the (isotropic) wimp gas, the Earth's motion around the Sun alternately adds or subtracts from the wimp-detector velocity.

The present generation of experiments using NaI scintillators (Bernabei et al., 1997), germanium diodes (Baudis et al., 1998), or germanium cryogenic calorimeters (Akerib et al., 1998; Drain et al., 1998) have backgrounds that make them insensitive to supersymmetric dark matter. The models with the highest elastic scattering cross sections give rates about an order of magnitude below the current background levels.

Current efforts in this field are directed towards lowering the background in conventional detectors (NaI scintillators and germanium diodes) and in developing new detectors that can distiguish nuclear recoils from the Compton-electron background (Akerib et al., 1998; Drain et al., 1998). This is possible in certain cryogenic detectors that detect two types of excitations, phonons and ionization or phonons and scintillation. The ratio of the two signals is different for nuclear recoils and Compton electrons, allowing background rejection.

Besides direct detection, it is possible to detect wimps "indirectly" through the detection of particles produced in present-day wimp-antiwimp annihilation. While wimp annihilation ceased in the early Universe because of the universal expansion, it started up again once the wimps became gravitationally bound in galactic halos. The annihilation rate is further enhanced inside material objects like the Sun or Earth. This is because wimps will be trapped in such objects if, while traversing the object, the wimp suffers an elastic collision with a nucleus and has its velocity reduced below the object's escape velocity (figure 3). After the collision, the wimp is in an orbit that passes through the object and will, after repeated collisions, thermalize in its center. In the case of supersymmetric dark matter, the trapping rate in the Sun is sufficiently high that the concentration of wimps reaches a steady state where trapping is balanced by either annihilation (for high-mass wimps) or by evaporation (for low mass wimps). For the Sun, the dividing line between low mass and high mass is about 3 GeV.

The only annihilation products that can be seen emerging from the Sun or Earth are, of course, neutrinos. The flux of such neutrinos can be calculated for a given wimp candidate and the flux compared with that observed in underground detectors. The observed flux is entirely understood as being due to the decay of cosmic-ray pions and kaons in the Earth's

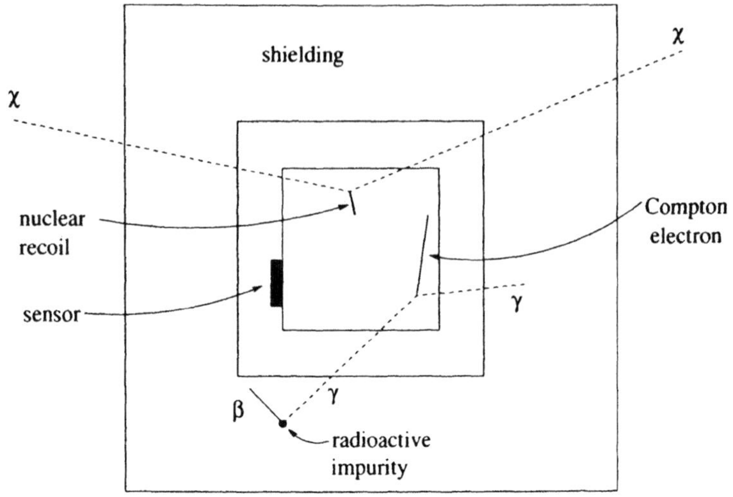

Figure 2. A generic "calorimetric" wimp detector surrounded by its shielding. The Galactic wimp χ enters the detector , scatters off a nucleus , and leaves the detector. The recoiling nucleus creates secondary excitations (e.g. scintillation light, ionization, phonons) that can be detected by the sensor. Also shown is a background event due to the ambiant radioactivity yielding a Compton electron in the detector.

atmosphere (figure 3) . Certain supersymmetric wimps would give a higher flux and are thus excluded (Mori et al., 1991; Mori et al., 1992).

Because the observed flux of neutrinos is due to an unavoidable background, the only improvements in the limits from these techniques would come the observation of a small excess of neutrinos coming from the direction of the Sun or center of the Earth. The most reasonable possibility is to search for upward-going muons coming from ν_μ interactions in the rock below a detector. Calculations (Halzen et al, 1992) indicate that a 1 km^2 detector with a muon energy threshold of $\sim$ 10 GeV would be needed to observe a statistically significant solar signal for typical supersymmetric dark matter. Efforts in this direction are underway by instrumenting the Mediterranean (Antares, 1999) or the Antartic Continental Glacier (Amanda, 1999) to observe Cherenkov light produced by muons.

3. Axions

Axions (Cheng, 1988) are hypothetical light scalar particles invented to prevent CP violation in the strong interaction. They would have been produced in the early Universe via both thermal and non-thermal mechanisms and might produce near-critical relic densities if they have masses in the range $m_a \sim 10^{-5}$eV to $\sim 10^{-3}$eV. Axions act as cold dark matter and should be present in the Galactic Halo.

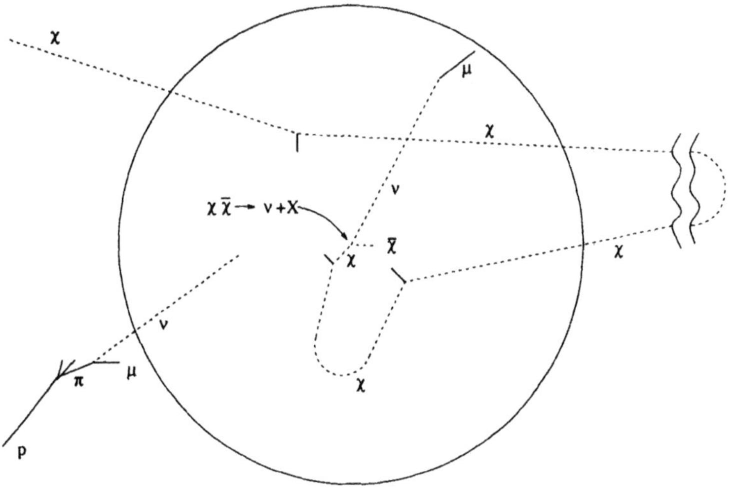

Figure 3. The capture of a wimp χ in the Earth. If the wimp loses sufficient energy in a collision with a nucleus, the wimp's velocity will drop below the the escape velocity placing it in an orbit intersecting the Earth. The subsequent collisions will eventually thermalize the wimp in the center of the Earth. An annihilation with a thermalised anti-wimp may lead to the production of neutrinos that can be detected at the surface. As shown, the background for such events comes from neutrinos produced by the decays of pions and kaons produced by cosmic rays in the Earth's atmosphere.

The most popular detection scheme for Galactic axions is based on the expected axion-2 photon coupling (Sikivie, 1983) which allows the axion to "convert" to a photon of frequency $\nu = m_a c^2/h$ in the presence of a magnetic field. If a microwave cavity is tuned to this frequency, the axions will cause an excess power to be absorbed (compared to neighboring frequencies). If the Halo is dominated by axions, the predicted power is small, about 10^{-21}watts for a cavity of volume 3m^3 and a magnetic field of 10 T. Since the axion mass is not known, it is necessary to scan over the range of interesting frequencies. Pilot experiments (Wuensch et al., 1989; Hagmann et al, 1990) have produced limits on the local axion density about a factor 30 above the expected density. Experiments are now in progress to search for axions at the required level of sensitivity (Hagmann et al, 1998).

4. Neutrinos

The last class of dark matter candidates that we will consider consists of particles that are actually known to exist: ν_e , ν_μ and ν_τ. Unlike wimps, these neutrinos froze when they were still relativistic so they are expected to have a cosmological number density just slightly less than that of photons, the difference being due to the "reheating" of the photons during the epoch of e^+e^- annihilation. If one of these neutrinos has a mass much greater than the present photon temperature, 0.0002 eV, it will now be nonrelativistic and have a cosmological density proportional to its mass:

$$\Omega_\nu h^2 = \frac{m_\nu}{91.5\mathrm{eV}} \tag{1}$$

Thus a neutrino with a mass in the range of tens of eV can close the Universe. It is also expected that fluctuations in density of neutrinos could serve as seeds for clusters of galaxies (hot dark matter).

For the time being there is no practical method of actually observing the cosmological neutrinos. Speculations about how this might one day be done (Smith and Lewin, 1990) provide an interesting case study of the ultimate limits of human knowledge.

There are however techniques that should eventually allow us to know if one of the three neutrinos has a mass in the interesting range. Since the neutrino cosmology is so simple, the measurement of a neutrino mass would indicate clearly its cosmological density.

Mass limits on the three species of neutrinos come from the kinematics of weak decays involving neutrinos. The present limits (Particle Data Group , 1998) are $\nu_e < 10\,\mathrm{eV}$, $\nu_\mu < 170\,\mathrm{keV}$, and $\nu_\tau < 18.2\,\mathrm{MeV}$. Only the ν_e is near being ruled out as a cosmologically important particle, and there is little hope that measurements of the masses of the other two species will soon reach the required precision.

The best known method for measuring directly a neutrino mass in the eV range would be to observe the time distribution of neutrinos arriving at Earth from a distant type II supernova. Such supernovae emit neutrinos of all flavors and with energies in the MeV range over a period of a few seconds. The most easily detected are $\bar{\nu}_e$ in water Cherenkov detectors via the reaction $\bar{\nu}_e p \rightarrow e^+ n$. Assuming the ν_e has a negligible mass, a heavy ν_μ or ν_τ would arrive after the ν_e, about 1 sec for a 50 eV neutrino from the Galactic center (a distance of 8 kpc).

The most practical scheme for detecting ν_μ or ν_τ would use the neutral current ejection of neutrons from a nucleus $\nu(A, Z) \rightarrow \nu(A-1, Z)n$ (Smith, 1997). The rock of a cavern could serve as the target and the ejected neutrons can be easily detected after thermalization. The amount of neutron detectors necessary to observe a Galactic supernova is in the 100 ton range, a reasonable amount for a particle physicist. The primary difficulty with this important experiment is the time scale, of order one century to observe one supernova.

Information on neutrino masses also comes from searches for "neutrino oscillations" a phenomenon which comes about if the neutrinos ν_e , ν_μ and ν_τ, are quantum mechanical mixtures of neutrinos of definite mass ν_1 , ν_2 ,

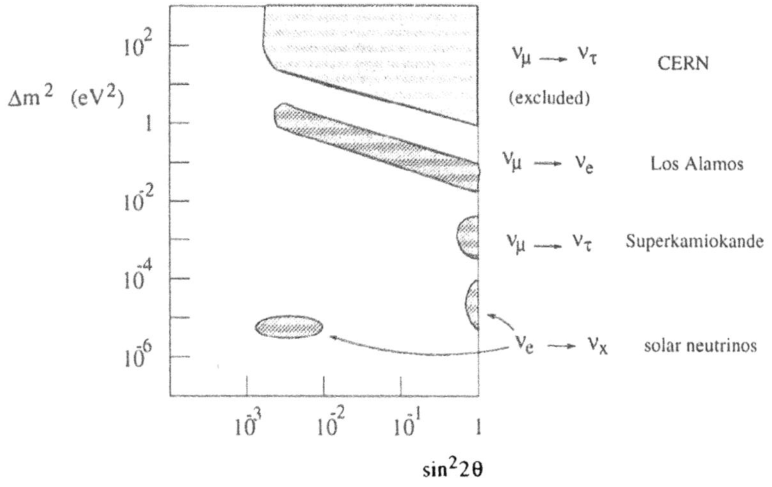

Figure 4. The mixing angle-mass difference parameter space for neutrino oscillations. The excluded region labeled $\nu_\mu \rightarrow \nu_\tau$ (CERN) (Altegoer et al. 1998, Eskut et al. 1998) corresponds to masses yielding near critical densities of neutrinos under the assumption that the mass difference is of order the mass of the heaviest neutrino. The region is excluded by searches for $\nu_\mu \rightarrow \nu_\tau$ oscillations. The allowed regions at lower mass differences correspond to oscillations of $\bar{\nu}_\mu \rightarrow \bar{\nu}_e$ at Los Alamos (Athanassopoulos et al. 1996), to oscillations of atmospheric $\nu_\mu \rightarrow \nu_\tau$ at Superkamiokande (Fukuda et al. 1998) and to oscillations of ν_e (to any other neutrino) that would explain the observed deficit of solar neutrinos (Bahcall et al. 1998).

and ν_3 . For instance, if $\nu_\mu = \nu_1 cos\theta + \nu_2 sin\theta$ and $\nu_\tau = -\nu_1 sin\theta + \nu_2 cos\theta$, a neutrino of energy E_ν produced as a ν_μ will have an oscillating probability $P(\nu_\mu \rightarrow \nu_\tau)$ to be observed as a ν_τ after a distance L:

$$P(\nu_\mu \rightarrow \nu_\tau) \;=\; sin^2 2\theta \; sin^2 \left(\frac{1.27\,\Delta m^2(\mathrm{eV}^2)\; L\,(\mathrm{km})}{E_\nu\,(\mathrm{GeV})} \right) \tag{2}$$

where $\Delta m^2 = m_2^2 - m_1^2$. The probability to observe a ν_μ is just $1 - P(\nu_\mu \rightarrow \nu_\tau)$. Neutrino oscillations would therefore result in a position dependent deficit of neutrinos originally present and an appearance of neutrinos originally absent. The effects are proportional to $sin^2 2\theta$ so we can get information about neutrino masses only if the mixing angle is sufficently large.

Since the charged leptons show a hierarchical mass pattern $m_\tau \gg m_\mu \gg m_e$ it is often assumed that the neutrino masses are well separated implying $\Delta m^2 = m_2^2 - m_1^2 \sim m_2^2$ where by convention ν_2 is the neutrino of heavier mass. If this is indeed the case, searches for neutrino oscillation with $\Delta m^2 \sim 100\mathrm{eV}^2$ are sensitive to cosmologically important neutrinos. Unfortunately, searches (Altegoer et al., 1998; Eskut et al., 1998) have yielded negative results indicating that the mixing angles are small $sin^2 2\theta < 10^{-3}$ if $\Delta m^2 \sim 100\,\mathrm{eV}^2$ (figure 4).

Searches for neutrino oscillations have yielded evidence for neutrino mass differences below that expected for cosmologically important neutrinos (figure 4). Perhaps the firmest evidence comes from the SuperKamiokande collaboration who observed a deficit of upward-going atmospheric neutrinos. They interpret this as evidence for $\nu_\mu \rightarrow \nu_\tau$ oscillations with a mass difference of $\sqrt{\Delta m^2} \sim 0.04$ eV. If we interpret this as the mass of the highest mass neutrino, it would correspond to a neutrino density of about 0.001 critical density. If these results are confirmed, the neutrinos involved in the oscillation cannot be a major component of the nonbaryonic dark matter unless there is a near degeneracy $m_2 \sim m_1$ implying $m_2^2 \gg \Delta m^2$.

5. Machos

The theory of primordial nucleosynthesis predicts that the baryon density is about 0.025 critical density which suggests that baryons cannot account for all of the dark matter. Nevertheless baryons could account for *galactic* dark matter if they are in a form that neither absorbs nor emits light in significant quantities (Carr, 1990; Carr, 1998). The simplest way to hide baryons is to place them in compact objects that either do not burn (e.g. brown dwarfs) or have ceased to burn (e.g. white dwarfs, neutron stars, black holes). If they reside in galactic haloes, they are called "machos" for "MAssive Compact Halo Objects".

Paczyński (Paczyński, 1986) suggested that machos could be discovered through gravitational lensing of individual stars in the Large Magellanic Cloud (LMC) (figure 5). This small galaxy orbits our Galaxy at a distance of 50 kpc from Earth.

Since the angle separating the two images is small (< 1 milliarcsec) the only observable effect is a transient amplification of the total light as the macho moves towards and away from the line of sight. The amplification is

$$A = \frac{u^2+2}{u\sqrt{u^2+4}} \tag{3}$$

where u is the distance of closest approach of the (undeflected) line-of-sight to the deflector in units of the "Einstein Radius" $R_0 = \sqrt{4GMLx(1-x)/c^2}$ where L is the observer-source distance and Lx is the observer-deflector distance and M is the macho mass.

The amplification is greater than 1.34 when the distance to the light-of-sight is less than R_0. This amplification corresponds to a reasonable observational threshold since photometry can "easily" be done to better than 10 percent accuracy. At a given moment, the probability, P, of a given star being amplified by more than a factor 1.34 is just the probability that

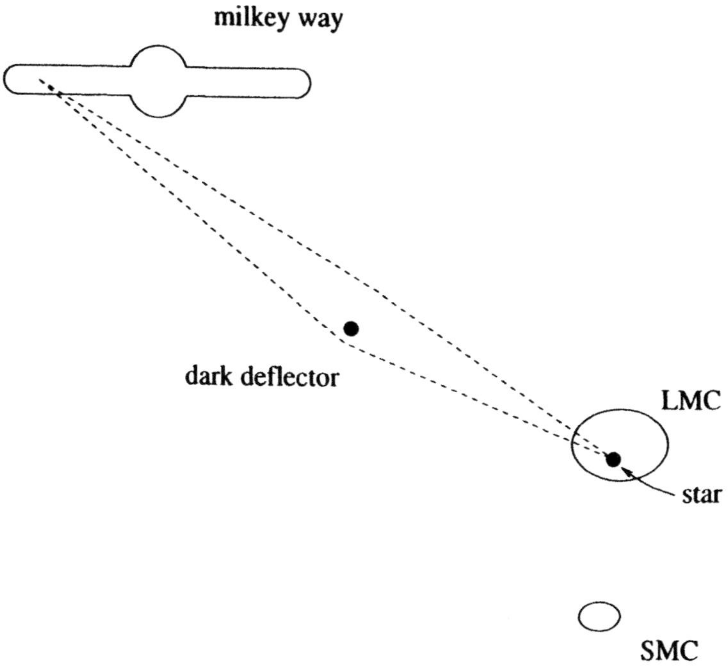

Figure 5. A schematic of the lensing of a star in the Large Magellanic Cloud (LMC) by an unseen object in the Galactic Halo. While the two images cannot be easily resolved, the combined light from the two images gives a transient amplification of the light from the star as the unseen object passes near the line-of-sight. The light curve for a point source is shown in figure 6.

its undeflected light passes within one Einstein radius of a macho:

$$P \sim n_{macho} \, L \, \pi R_0^2 \tag{4}$$

where n_{macho} is the mean number density of machos between us and the LMC and L is the distance to the LMC. The density of machos is roughly $n_{macho} \sim M_{halo}/ML^3$ where M_{halo} is the total halo mass out to the position of the LMC. Using the expression for the Einstein radius, we find that P is independent of M and determined only by the velocity of the LMC:

$$P \sim \frac{GM_{halo}}{Lc^2} \sim \frac{v_{LMC}^2}{c^2} \tag{5}$$

The LMC is believed to orbit the galaxy with $v_{LMC} \sim 200\,\mathrm{kms}^{-1}$ (corresponding to a flat rotation curve out to the position of the LMC). In this case, P is of order 10^{-6}. More detailed calculations give $P = 0.5 \times 10^{-6}$ (Griest, 1991).

Since the observer, star and deflector are in relative motion a sizeable amplification lasts only as long as the undeflected light beam remains within

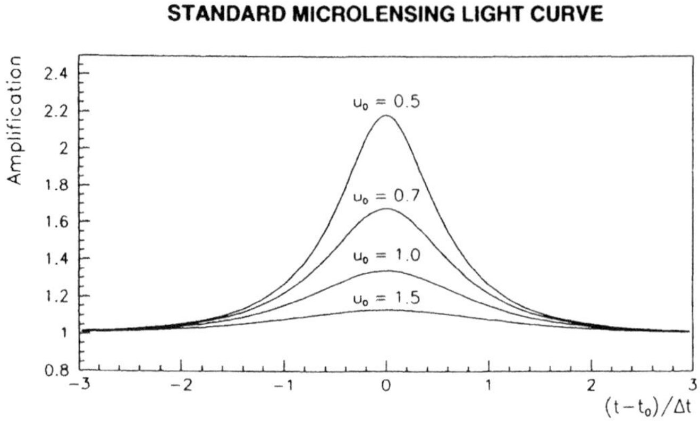

Figure 6. The microlensing light curve for a point source. The curve is shown for four values of the minimum distance of closest approach (0.5, 0.7, 1.0 and 1.5 Einstein radii). The time is normalized to the mass dependent scale Δt given by equation 6.

the Einstein radius. The resulting achromatic, symmetric light curve is shown for various impact parameters in figure 6. The time scale of the amplification is the time Δt for the deflecting object to cross one Einstein radius with respect to the observer and source. For the lensing of stars in the LMC by objects in our Halo, the relative speeds are of order 200 km s^{-1} and the position of the deflectors is roughly midway between the observer and the source ($x \sim 0.5$). The mean Δt is then

$$\Delta t \sim \frac{R_0}{200 \text{km s}^{-1}} \sim 75 \text{days} \sqrt{\frac{M}{M_\odot}}. \quad (6)$$

The observed time scale distribution can therefore be used to estimate the mass of the machos if one assumes that they are in the Halo.

Two groups, the MACHO collaboration and the EROS collaboration, have published results of searches for events in the directions of the LMC and the SMC. The results are sumarized in figure 7 where we show limits on the fraction of the halo comprised of machos as a function of the macho mass. Based on the observation of 7 events the MACHO collaboration (Alcock et al., 1997) reports a LMC rate and time scale distribution corresponding to a Halo fraction $f = 0.5$ and macho mass $\sim 0.5 M_\odot$. EROS has published only upper limits on the fraction of the halo comprised in machos (Renault et al., 1997; Palanque-Delabrouille et al., 1998). For $M = 0.5 M_\odot$ their upper limit, based on the observation of one SMC event, is near the central MACHO value.

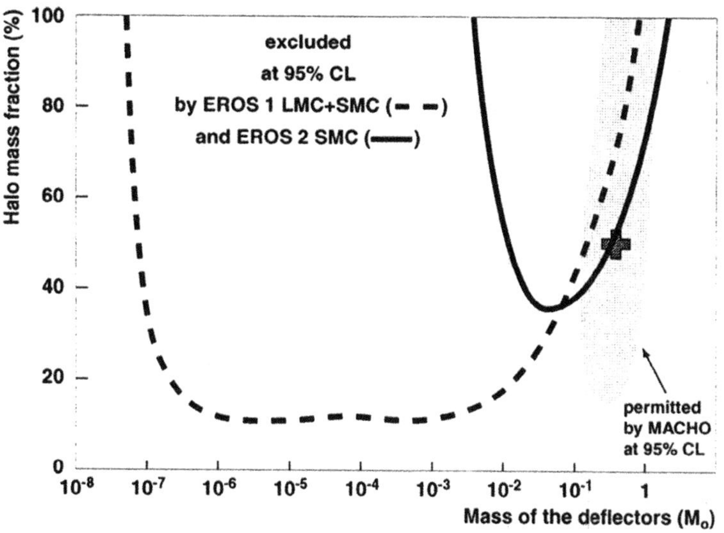

Figure 7. Constraints on the fraction of an isothermal Halo comprised of macho's as a function of the assumed macho mass (in units of the solar mass). The dashed line shows the EROS-I upper limit using the LMC and SMC while the solid line shows the EROS-II upper limit using the SMC. The shaded region show the MACHO favored region centered on a mass fraction of 0.5 and a mass of $\sim 0.5M_{\odot}$.

There are three points that argue against the interpretation of the events as being due primarily to objects in our Halo. First the mass favored by the Macho collaboration ($0.5M_{\odot}$) corresponds to objects that are ruled out by other observations (Carr, 1998). Main-sequence stars of this mass would have been seen by the the HST and stellar remnants (white dwarfs) would have lead, through mass loss, to an unobserved pollution of the halo with heavy elements. Primordial black holes avoid these constraints but no accepted scenario for their production exists.

Second, the time scale of the two observed SMC events are longer than all the published LMC events. This suggest a different origin for events in the two clouds. Lensing by low-mass stars in the clouds themselves is a possibility though the observed event rate is somewhat higher than that expected for self-lensing (Gould, 1995).

Finally, the approximate location of the deflectors for three events has been determined and all of those deflectors are in the Clouds. Constraints on the position of the lenses came from the absence of parallax in an SMC event (Palanque-Delabrouille et al., 1998) and from the breakdown of the point-source approximation in two binary lens events with caustic crossings (Bennett et al., 1996; Afonso et al., 1998).

Only further data will allow the determination of the nature and location of the deflectors and their relevance for the problem of dark matter.

Acknowledgments

It is a pleasure to thank many collegues for discussion on dark matter , especially A. Gould, M. Lachièze-Rey, Y. Mellier, A. Milsztajn R. Pain, B. Sadoulet, P. Smith, M. Spiro, and C. Tao.

References

Afonso, C. et al. 1998, *A.& A.* **337** L17.
Alcock, C. et al. 1998, *Ap. J.* **486** 697.
Altegoer, J. et al. 1998, *Phys. Lett. B* **431** 219-236.
Akerib, D.S. et al. 1998, *Nucl. Phys. B* (Proc. Suppl)**70** 64.
Amanda Collaboration, http://amanda.berkeley.edu .
Antares Collaboration, http://antares.in2p3.fr/ .
Athanassopoulos, C. et al. 1996, *Phys. Rev. Lett.* **77** 3082.
Bahcall, J.N. et al. 1998, *Phys. Rev* **D58** 096016-1.
Baudis, L. 1998, hep-ex/9811045, to appear in Physical Review.
Bennett, D. et al. 1996, *Nucl. Phys. Proc. Suppl.* **51b** 152.
Bernabei, R. et al. 1997, *Phys. Lett.* **389** 757.
Carr, B. J. 1990, *Comments Astrophys.* **14** 257.
Carr, B. J. 1998, in these proceedings.
Cheng, H. Y. 1988. *Phys. Rep.* **158** 1.
Drain, D. et al. 1998. *Phys. Rep.* **307** 297-300.
Eskut, E. et al. 1998. *Phys. Lett. B* **424** 202.
Goodman M.W. and E. Witten 1985 *Phys. Rev.* **D31** 3059.
Griest, K. 1991 *Ap. J.* **366** 412.
Hagmann, C., et al. 1990 *Phys. Rev.* **D42**, 1297.
Hagmann, C., et al. 1998 *Phys. Rev.Lett.* **80**, 2043.
Halzen, F., T. Stelzer and M. Kamionkowski 1992 *Phys. Rev.* **D45** 4439.
Freese,K., J. Freedman and A. Gould 1988 *Phys.Rev.* **D37** 3388.
Fukuda, Y., et al. 1998 *Phys.Rev.Lett.* **81** 1562.
Gould A. 1995 *Ap. J.* **441** 77.
Loeb, A. 1998 astro-ph/9802122.
Mori M. et al. 1991 *Phys. Lett. B* **270** 89.
Mori, M. et al. 1992 *Phys. Lett. B* **289** 463.
Paczyński, B. 1986 *Ap. J.* **304**, 1-5.
Pain, R 1998, in these proceedings.
Palanque-Delabrouille et al. 1998 *A.& A.* **332**, 1.
Particle Data Group 1998 *Eur. Phys. J.* **C3**, 1-794.
Perlmutter, S. et al. 1998 astro-ph/9812133, to appear in Astrophysical Journal .
Primack, J. R., D. Seckel and B. Sadoulet 1988 *Ann. Rev. Nucl. Sci.*
Renault, C. 1997 *A.& A.* **324** L69.
Schmidt, B. P. et al. 1998 astro-ph/9805200, to appear in Astrophysical Journal .
Sikivie, P. 1983. *Phys. Rev. Lett.* **51** 1415.
Smith P.F. 1997, *Astroparticle Physics* **8** 27-42.
Smith P.F. and J. D. Lewin 1990, *Physics Reports* **187** 203.
Wang L., R. R. Caldwell, J. P. Ostriker and Paul J. Steinhardt, 1999 astro-ph/9901388.
Wuensch, W. et al. 1989. *Phys. Rev.* **D40** 3153.

TEMPERATURE ANISOTROPIES OF THE COSMIC MICROWAVE BACKGROUND

N. AGHANIM
Institut d'Astrophysique Spatiale
Bâtiment 121, Université Paris Sud
91405 Orsay Cedex
France

1. Introduction

The Cosmic Microwave Background (CMB), discovered in 1965 by Penzias and Wilson (Penzias & Wilson 1965), is one of the main observational tools for cosmology. It is characterised by:

- the isotropy and homogeneity at large scales: the same intensity is measured in all the directions, suggesting that the physical process at the origin of this emission has concerned the Universe as a whole.
- the emission spectrum of the CMB is very close to the black body radiation.

These characteristics suggest that the whole Universe experienced a period during which the matter and radiation were in thermodynamic equilibrium, and indicate that the Universe knew a hot phase as predicted by the big-bang models. The discovery of the CMB is in fact one of the most powerful arguments in favour of these models.

The big-bang models are based on the theory of general relativity and the so-called cosmological principle. The latter represents two fundamental hypotheses. The first one is the homogeneity hypothesis which stipulates that the properties of the Universe are the same everywhere (at least on scales larger than that of the greatest identified structures: the superclusters). The second hypothesis –isotropy– states that there is no preferred direction in the Universe, once again at large scales.

The content of the Universe rules, through gravity, several dynamical and evolutionary properties. Indeed, gravity slows down the expansion of the Universe and it can even cause the contraction of the Universe, if the

M. Lachièze-Rey (ed.), Theoretical and Observational Cosmology, 317–351.

latter is dense enough. The density of the Universe fixes also its geometry. The density ρ_c, known as the critical density of the Universe, is defined as the density for which the Universe is flat (with a zero space curvature). If the actual density of the Universe, ρ, is larger than the critical one, then the Universe is closed with a positive space curvature (surface of a sphere), if $\rho < \rho_c$ the Universe is open with negative curvature (surface of a saddle). Moreover, if the content of the Universe can be completely described by its density ρ and its pressure, then the cosmological models are known as Friedmann-Lemaitre models. The Friedmann-Lemaitre models are characterised by three main parameters: the Hubble constant H which measures the expansion speed of the Universe, the density parameter $\Omega_0 = \rho/\rho_c$ which measures the ratio of the current density ρ to the critical density ρ_c, and finally the cosmological constant λ, or the normalised cosmological constant Ω_λ. The latter was introduced by Einstein to achieve a static Universe.

Within the context of the Friedmann-Lemaitre models, several cosmogonic scenarios have been invoked to account for the formation and evolution of the cosmic structures observed today, such as galaxies, galaxy clusters, super-clusters... They are based on two hypotheses:

- The presence of perturbations with respect to the average distribution of matter, the primordial density fluctuations, at all scales.
- Due to gravity, the initial density fluctuations grow. It is this gravitational instability which has made the fluctuations evolve to their actual state.

The cosmogonic scenarios are defined, in addition to the three parameters listed previously, by the characteristics of the primordial density fluctuations (statistics, nature...), and those of the dark matter (nature and abundance) which plays an important dynamical role in the Universe. Non-baryonic dark matter has been introduced to explain the galactic rotation curves. In addition, it allows to reach the closure value predicted by the theories $\Omega_0 = 1$, without violating the observational limits of the nucleosynthesis giving a baryon contribution smaller than $\Omega_0 = 0.1$ (Walker et al. 1991). The dark matter interacts with baryons only through gravity. It is called hot when it is constituted of relativistic particles and cold if not.

In the big-bang scenarios, the early Universe was very hot with all particles in thermal equilibrium. The history of the Universe is completely described by its cooling through expansion, and the consequent breaking of the equilibrium conditions. In addition, the Universe was totally ionised, therefore, the matter and radiation were tightly coupled and the Universe was optically thick to radiation. When the temperature had reached $\simeq 4000$ K, the free electrons and protons combined forming neutral hydrogen. The "recombination" decreased the photons probability of scattering and hence increased their mean free path. When the photons mean free path became greater or equal to the horizon size, the photons decoupled from the matter. This process which is commonly called the decoupling or recombination marks a transition between the ionised optically thick, and neutral optically thin states. The recombination is not an instantaneous process, it has a "thickness". In the redshift space, the latter is given by an interval $\delta z \simeq 80$ which corresponds to an angular scale of about $10\,\Omega_0^{1/2}h$ arcminutes (Padmanabhan 1993), where h is the Hubble constant normalised to 100 km/s/Mpc ($h = H_0/100$). The recombination corresponds to the last scattering surface of the photons on the electrons.

The density fluctuations were present before recombination and are called the primary anisotropies. As the CMB represents an image of the photon distribution at the decoupling, the brightness (or temperature) anisotropies of the CMB relates directly to the mass fluctuations because matter and radiation were tightly coupled. The CMB is thus a unique tool to directly observe the density fluctuations at the origin of the cosmic structures (galaxies, galaxy clusters, ...).

In practice the situation is not so ideal. Indeed after decoupling, the CMB photons interact with the matter in the ionised, and/or hot regions, or with the forming structures. These interactions produce the so-called secondary temperature fluctuations, which superimpose to the primary fluctuations, and hence complicate any interpretation of the CMB observations.

In the following, we first start by discussing the principal sources of the primary temperature fluctuations and then the main secondary effects which affect the CMB. Detailed reviews can be found in Padmanabhan (1993), Peebles (1993) and Lachièze-Rey & Gunzig (1999). A second step is to describe the angular power spectrum of the anisotropies and the influence of the different cosmological parameters on its shape and its amplitude. For more details, we refer the reader to Hu & Sugiyama (1995b) and Bunn (1996) and the references therein.

2. Temperature fluctuations of the CMB

2.1. PRIMARY FLUCTUATIONS

The primary fluctuations are the temperature fluctuations (or anisotropies) which existed prior recombination. We know two other sources of primary anisotropies besides the intrinsic temperature fluctuations.

- Intrinsic fluctuations: They represent the initial fluctuations in the energy distribution. Within the framework of the inflationary models, they are due to quantum fluctuations of scalar fields. In this case, the fluctuations are Gaussian distributed and are called "adiabatic".

Any temperature fluctuation is associated with a density fluctuation ($\delta T/T = \frac{1}{3}\delta\rho/\rho$). Fluctuations can be due to topological defects (cosmic strings, textures, domain walls...). In this case their distribution in non-Gaussian and they are known as the "isocurvature" perturbations because the density and radiation (or temperature) fluctuations are adjusted so that the total density, and thus the local curvature, remains unperturbed.

- Sachs-Wolfe effect: Before recombination and because of the tight coupling, the matter–radiation overdensity forms a potential well $\delta\phi$. After decoupling, the photons propagate freely. To come out of the gravitational potential wells, they lose energy and undergo a gravitational redshift (Sachs & Wolfe 1967). This effect produces temperature fluctuations $\delta T/T \simeq \delta\phi/c^2$ (for a flat Universe) which dominate at large scales ($\delta T/T \propto L^2$, where L is the scale of the fluctuation).
- Doppler effect: The photons scattered by a moving source undergo a Doppler shift. As the matter falls in the potential wells, generated by the overdensities, with a velocity v it generates temperature fluctuations $\delta T/T \propto v$. A same overdensity produces, through the potential wells, fluctuations of opposite signs and the average effect is null at the first order. This is the case for perturbations smaller than the size (or thickness) of the last scattering surface for which photons "see" both wings of the well. However for larger perturbations, the photons go out of the potential well when they are already decoupled from the matter. They do not undergo a second Doppler shift and the net effect is different from zero (Peebles 1993). For the scales equal to the horizon size, the balance between the pressure and the gravity is broken and the fluctuations undergo acoustic oscillations.

2.2. SECONDARY FLUCTUATIONS

The secondary fluctuations represent the ensemble of distortions generated after recombination which are superimposed and added to the primary fluctuations.

After recombination, the photons and the matter decouple. If the Universe is neutral (or with a small ionisation fraction), then the photons propagate freely in the Universe. In this case, the local, or integrated, variations of the gravitational potential along the line of sight produce secondary brightness anisotropies. If the Universe is ionised, homogeneously or inhomogeneously, the photons interact by Compton effect with the matter generating secondary anisotropies. Among the most significant secondary effects, there are:

2.2.1. *Gravitational effects*

- Rees-Sciama effect: The Rees-Sciama effect (Rees & Sciama 1968) takes place when the gravitational potential well crossed by the photons varies. These variations are due essentially to the non-linear evolution of the structures. If the photons fall into a gravitational potential well different from the one they go out (because of the evolution) this induces a gravitational redshift (of the Sachs-Wolfe type). The photons gain energy while going towards the center of a structure (blue shift) and lose more energy while going out (red shift), as the evolution deepens the potential well. This results in the generation of an additional temperature fluctuation.
  The Rees-Sciama effect generates temperature fluctuations whose amplitude ($\delta T/T$) are about a few 10^{-7}. This corresponds, at the degree scale, to a contribution of the order of 0.01 to 0.1% of the power. Recent studies (Seljak 1996b) have shown that the contribution of the Rees-Sciama effect is maximum for angular scales between 10 and 40 arcminutes.
- Integrated Sachs-Wolfe effect: It corresponds to the integral along the line of sight of the gravitational redshift effect. It takes into account the time variations of the potentials, in particular during the linear phase (Hu & Sugiyama 1995a Hu & Sugiyama 1995b, Bunn 1996).
- Second order non-linear effects: Many studies among which Martinez-Gonzàles et al. (1990) and Sanz et al. (1996) have evaluated the second order perturbations due to gravitational potential wells. They showed that the quasi-linear and non-linear evolution of these perturbations generate, at low redshifts, secondary anisotropies with a maximum amplitude of about $\delta T/T \simeq 10^{-6}$ at the degree scale.
- Gravitational lensing effects: The integrated Sachs-Wolfe effect can be viewed as an impulse (positive or negative) given to the photons when they cross the density fluctuations along the line of sight. Gravitational lensing is responsible for variations of the photons trajectory in the transverse direction (Blanchard & Schneider 1987, Cayòn et al. 1993, Seljak 1996a, Seljak 1996c). It results in an image distortion of the last scattering surface. The gravitational lensing effect acts like a redistribution of the power between the different angular scales. This effect takes place at the degree and becomes more significant on smaller scales. It can even be responsible for the erasure of the fluctuations at very small scales.

2.2.2. *Effects of the re-ionisation*

The recombination at a redshift of about 1000 is a transition between ionised and neutral states of the Universe. Nevertheless, the Universe has underwent a re-ionisation after recombination. This is indicated by the absence of neutral hydrogen in the direction of the most distant quasars $z \simeq 5$. This test, called the Gunn-Peterson test (Gunn & Peterson 1965), indeed shows the absence of the characteristic Lymann-alpha absorptions for neutral hydrogen. This re-ionisation generates secondary temperature anisotropies on the CMB through two main processes:

- Sunyaev-Zeldovich effect: The inverse Compton scattering of the CMB photons on the free electrons of hot intra-cluster gas produces secondary brightness, and thus temperature, fluctuations on the CMB (Zeldovich & Sunyaev 1969, Sunyaev & Zeldovich 1972, 1980a, 1980b). It is the so-called thermal Sunyaev-Zeldovich (SZ) effect. In addition, if the galaxy cluster moves with a radial peculiar velocity along the line of sight, the motion induces additional anisotropies due to Doppler effect.
- Ostriker-Vishniac effect: The Doppler effect can intervene when the Universe becomes completely re-ionised. Along a line of sight, the density fluctuations and velocities are correlated. The Ostriker-Vishniac effect (Ostriker & Vishniac 1986, Vishniac 1987) represents the second order anisotropies generated from the correlations of the density and velocity perturbations. It takes place when the re-ionisation is complete and produces anisotropies at the few arcminutes scale (Dodelson & Jubas 1995, Hu 1995).

3. Angular power spectrum of the primary anisotropies

At recombination, the CMB exhibits an angular distribution of brightness, or temperature, directly related to the initial density fluctuations. The characteristics of this distribution as a function of the amplitude, the angular scale and the direction on the sky, are predicted by theoretical models. To compare the theoretical predictions and the CMB observations we must carry out a statistical analysis of the temperature fluctuations distribution.

On way to do so, is to perform a spherical harmonics expansion of the temperature fluctuations on the sky:

$$\frac{\delta T(\theta,\phi)}{T} = \sum_{l,m} a_{lm} Y_{lm}(\theta,\phi),$$

where the coefficients a_{lm} represent the intensity (amplitude) of the eigen modes defined by the spherical harmonics. They are given by:

$$a_{lm} = \int_{4\pi} Y_{lm} \frac{\delta T(\theta,\phi)}{T}\, d\Omega.$$

One defines the coefficients C_l as the squared amplitudes for each mode in the spherical harmonics decomposition. They are given by the following relation $C_l =< |a_{lm}|^2 >$, where the "brackets" represent an ensemble average. These coefficients express the power associated with the modes. The distribution of the coefficients, C_l, as a function of the multipoles l is called the angular power spectrum. The multipoles are directly related to the angular scale θ subtended by a fluctuation on the sky by $\theta \simeq l^{-1}$.

The angular power spectrum is related to the initial power spectrum predicted by the cosmological models $P(k)$. The latter corresponds to the second order moment of the correlation function of the initial fluctuations, in the Fourier space. If the initial power spectrum is a power law $P(k) = A\,k^n$ (k is the Fourier mode), then C_l are computed, for each mode, using the spectral index n and the normalisation constant A.
The inflationary scenarios suggest that the initial spectrum of the density fluctuations has an index n equal to unity. The initial spectrum is then known as the Harison-Zeldovich, or scale invariant, spectrum. Indeed, for such a spectrum there is no characteristic scale and the fluctuations of gravitational potential $\delta\phi$ exhibit the same amplitude at all scales.

In an inflationary model with "adiabatic" initial perturbations with Cold Dark Matter (CDM) and defined by the following cosmological parameters: $\Omega_b = 0.05$, $\Omega_{CDM} = 0.95$, $\Omega_\nu = 0$, $\Omega_\Lambda = 0$, $h = 0.5$, $T_{CMB} = 2.728$ K, $n = 1$, and 24% of helium, the angular power spectrum (Fig. 1) is characterised by three main features: a plateau at small l (large angular scales), a cut off at large l (small scales) and a series of peaks and valleys for l between one and a few hundreds (a few arcminutes to one degree).

These features show the main physical processes which affected the cosmic fluid and the Universe.
At large angular scales, the Sachs-Wolfe effect dominates. Its contribution to the power spectrum is proportional to $1/l(l+1)$. Therefore, the product $l(l+1)\,C_l$ is constant (Bond & Efstathiou 1987, Bunn 1996). For the scale invariant spectra ($n = 1$), this results in a plateau. These scales have been observed in particular by the COBE satellite and are used for the normalisation of the initial spectrum of the fluctuations (the constant A).
The oscillations, for $l > 100$ called acoustic, Sakharov or Doppler peaks (Bond & Efstathiou 1987, Hu & Sugiyama 1995a, 1995b) are due to acoustic oscillations of the photon-baryon fluid. A gravitational instability affects the balance between the gravity, due to matter, and the pressure, due to the radiation. The subsequent regime of acoustic oscillations results in a series of peaks and valleys on the power spectrum.
Finally, damping (or the cut off) at $l > 1000$ is related to the scattering of the photons on the free electrons at the recombination as well as the

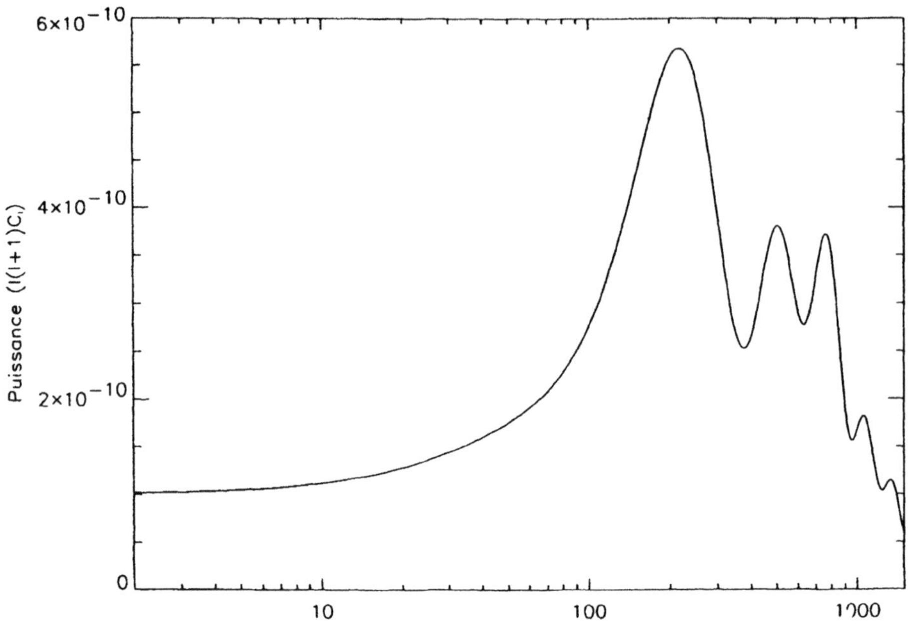

Figure 1. Angular power spectrum normalised to COBE/DMR data. The initial density perturbations are adiabatic and we assume no re-ionisation. The parameters of the model are the following: $\Omega_b = 0.05$, $\Omega_{cdm} = 0.95$, $\Omega_\nu = 0$, $\Omega_\Lambda = 0$, $H_0 = 50$ *km/s/Mpc,* $T_{CMB} = 2.728$ *K,* $n = 1$, *24% He. This plot has been obtained using the CMBFAST code.*

so-called Silk damping. There results a smearing out of the temperature fluctuations of angular scales smaller than, or equal to, the thickness of the last scattering surface.

The CMB angular power spectrum depends on the cosmological model describing the Universe. For this reason, we expect different spectra for models other than CDM. We will review some of the main dependences with respect to the cosmological parameters and their effects on the angular power spectrum (shape and amplitude).

Once the CMB temperature determined, the radiation pressure is also determined. The gravitational potential wells are directly related to the matter contents (baryonic and dark matter). In particular baryons do not contribute to the pressure, but they are sensitive to gravity. The increase of the baryon content, or rather of the product $\Omega_b h^2$, modifies the balance between the pressure and the gravity. It results in an amplification of the temperature fluctuations which in turn increases the amplitude of the acoustic peaks (essentially the first one).

The cut off resulting from the scattering of the photons on the free electrons depends on the product $\Omega_0 h^2$. When the latter increases, the scale associated to the horizon, as well as the scattering length, decreases. At a given Ω_0, an increase in the Hubble constant leads to a shift of the cut to smaller angular scales. The decrease of the product $\Omega_0 h^2$ modifies also the amplitude of the acoustic peaks. When $\Omega_0 h^2$ decreases, the matter dominated era lasts longer, and the gravitational potential decreases. The modes oscillate under a significantly higher pressure. This results in an increase of the amplitude of the acoustic oscillations. When the Hubble constant decreases with a constant Ω_0, the power spectrum exhibits an amplification of its acoustic peak. However, variation of the acoustic peak's amplitude with the Hubble constant, which also depends on the value of Ω_b (Hu & Sugiyama 1995a), complicates the interpretations. Indeed, for low values of Ω_b, increasing h decreases the peak whereas for high values of Ω_b the same variation of h increases the peak. However, Hu & Sugiyama (1995a) have shown that, within the framework of a CDM model such as $\Omega_b = 0.05$, the dependence of the angular power spectrum with the Hubble constant is weak.

In the standard model, the initial power spectrum is scale invariant ($n = 1$). A variation of the spectral index ($n \neq 1$), in tilted models, results in a variation of the slope of the plateau associated with the Sachs-Wolfe contribution at large angular scales.

When the parameter Ω_0, related to the space curvature, varies it induces a modification of the relationship between the angular scale on the sky and the associated physical scale on the last scattering surface. In the case of an open Universe (negative curvature $\Omega_0 < 1$) the geodesics are hyperbolic and a given angular scale corresponds to a physical scale larger than what it would be in a flat Universe with zero curvature. This projection effect induces an overall shift of the power spectrum features (acoustic peaks and cut off) towards small angular scales (large l).

A non zero cosmological constant, $\Omega_\Lambda \neq 0$, increases the distance to the last scattering surface. In this context, an angular scale on the sky will be associated with a physical scale larger than the corresponding scale on the surface of last scattering in a $\Omega_\Lambda = 0$ model. For the angular power spectrum, this effect again shifts the overall features towards small angular scales.

A gravitational wave background, produced during inflation, would have a specific power spectrum (a plateau) at large angular scales. The total angular power spectrum is thus increased in the plateau region and consequently the ratio of the acoustic peak to the plateau is reduced.

Because the anisotropies of the CMB, and thus their power spectrum, depend significantly on the cosmological parameters, an experiment which would carry out a complete mapping of the Universe would allow to measure these parameters and probe the cosmological model. The American space agency NASA has launched a satellite dedicated to cosmological observations in a spectral interval between 30 and 90 GHz: the *COsmic Background Explorer* (COBE). In addition to the instrument DMR (*Diffuse Microwave Radiometer*) conceived to seek for the CMB anisotropies, two other experiments were on board COBE satellite: DIRBE (*Diffuse InfraRed Background Explorer*) for the study of background radiation, and FIRAS (*Far InfraRed Absolute Spectrophotometer*) to study the thermal nature of the cosmic background and to measure of its temperature. It took FIRAS only a few minutes to measure with a precision of 0.1% a nearly perfect black body spectrum (Mather et al. 1994). This result finally established the thermal origin of the cosmological emission and confirmed the prediction of the big-bang models. Currently, the CMB temperature measured by FIRAS is: $T_{CMB} = 2.728 \pm 0.002$ K (Fixsen et al. 1996). DMR instrument has measured temperature fluctuations of amplitudes $\delta T/T \simeq 10^{-5}$ at scales larger than 7° (Smoot et al. 1992) with a signal to noise ratio close to unity. These measurements allowed to evaluate the normalisation constant of the initial power spectrum and to constrain the spectral index n. The latter was found compatible with a scale invariant spectrum. However, the main differences between the cosmological models result in a variation of the amplitude and/or of the shape of the angular power spectrum at the one degree scale, and below, and the cut off at the few arcminutes scale. These differences were not observed by DMR due to its instrumental design (resolution and sensitivity).

4. The Sunyaev-Zel'dovich effect

The galaxy clusters are known to be the largest gravitationally bound structures in the Universe. They have not achieved their evolution but they have, at least in their central part, reached the virial equilibrium. A cluster is thus a multi-component system which contains several hundreds of bright galaxies, an intra-cluster gaseous medium hot and metal rich, observed through its X-ray emission, and a dark matter component.
Given their multi-component structure (not dynamically relaxed in the external regions and strongly non-linear in the centre), the observation of galaxy clusters is particularly interesting for the study of the structure formation and evolution: collapse, merging, clustering. The cluster virial-

isation is obviously still incomplete as 30% of the rich galaxy clusters exhibit substructures (West 1994). These observations are in agreement with numerical simulations (Katz & White 1993) indicating that substructures play a fundamental role in the cluster formation which seems based on the merging of substructures along filaments of matter.

4.0.1. *Cluster mass*

The first estimates of galaxy cluster masses (Zwicky 1933) were based on the measurement of the velocity dispersion of the galaxies inside the cluster. For a spherically symmetric dynamically relaxed cluster with kinetic energy T and potential energy W, the virial theorem ($T+2W=0$) allows to estimate the mass using the following relation: $M \propto R_e\sigma^2$, where σ is the velocity dispersion along the line of sight, and R_e the effective radius of the cluster. The use of the virial theorem for clusters dynamically not relaxed introduces uncertainties in their evaluated masses. Other sources of errors are related to the measurement of the velocity dispersion. "Statistical" uncertainties can also intervene because of the limited number of galaxies, in a cluster, on which we can measure the physical properties.

Another method to estimate the masses uses observations of the X-ray emission of the intra-cluster gas. Estimates of the electrons and protons mean free path as well as time scales of the main physical processes in the intra-cluster gas (Sarazin 1988, 1992) show that the latter can be treated as a single fluid, characterised by a temperature T_{gas}. The dynamic equation of the intra-cluster gas of density ρ and pressure P in a gravitational potential well ϕ is then:

$$\rho\frac{Dv}{Dt} + \nabla P + \rho\nabla\phi = 0, \tag{1}$$

where D/Dt stands for the Lagrangian derivatives. In the spherical symmetry, the gas in hydrostatic equilibrium, in the potential well, is described by the equation 1 which can be written as follow:

$$\frac{1}{\rho}\frac{dP}{dR} = -\frac{GM(R)}{R^2}, \tag{2}$$

where G is the gravitation constant and $M(R)$ is the total mass of the cluster contained in a sphere of radius R. The mass is derived using the equation 2 and the equation of state for an ideal gas. We find:

$$M(R) = -\frac{kT_{gas}(R)\,R}{G\mu m_H}\left(\frac{d\log\rho}{d\log R} + \frac{d\log T_{gas}}{d\log R}\right), \tag{3}$$

with μ the mean molecular mass, m_H the mass of the hydrogen atom and k the Boltzmann constant.

The hot gas seen in X-ray is a tracer of the potential well. The mass distribution of a cluster is, in theory, directly given (Eq. 3) when the temperature and the gas density distributions are known. In practice, X-ray observations do not give the detailed temperature distribution, except for very rare clusters. The mass is thus estimated under some assumptions. The simplest and most commonly used states that the gas is isothermal ($T_{gas}(R) = const$). Recent X-ray observations with the Japanese satellite ASCA (*Advanced Satellite for Astronomy and Astrophysics*), show that the isothermal hypothesis is valid up to about 1 Mpc from the cluster centre (Mushotzky 1994). An alternative description of the temperature distribution would consist on assuming that the central part of the galaxy cluster is isothermal ($T_{gas}(R) = T_{iso} = const$) up to a distance R_{iso} from the centre, and that the external part ($R > R_{iso}$) is described by a polytropic gas with an index Γ ($T_{gas}(R) \propto R^{\Gamma}$).

In order to evaluate the cluster mass, one needs also the density distribution ρ. If the galaxies and gas are in equilibrium in the same gravitational potential well (Sarazin 1988, 1992), the hydrostatic equilibrium equation (1) for the galaxies can be written as follow:

$$\frac{1}{\rho_{gal}} \frac{d(\rho_{gal}\sigma^2)}{dR} = -\frac{GM(R)}{R^2}, \tag{4}$$

ρ_{gal} is the galaxy distribution and σ their velocity dispersion. Under the isothermal assumption, and for a constant dispersion, the equations 4 and 2 give $\rho \propto \rho_{gal}^{\beta}$ with β a parameter written $\beta = \mu m_H \sigma^2 / kT_{gas}$.

The cluster mass estimates using X-ray emission represent an improvement compared to the methods based on the virial theorem:

- The imaging techniques used for X-ray observations reduce the confusion problems due to the superposition of structures along the line of sight.
- The X-ray observations also reduce the errors due to bad sampling of the galaxies.

However, this method is not yet ideal and has some sources of errors which result in an uncertainty on the cluster mass. For example, the gas distribution can be inhomogeneous distributed, the potential well can be non-spherical, or the hydrostatic equilibrium can be affected by turbulent and/or magnetic pressure. Some observational limits, related in particular to the extragalactic X-ray background, also introduce errors.

4.0.2. *Gas profile in clusters*

The gas distribution $\rho(r)$ is completely defined by the hydrostatic equilibrium equation ($\rho \propto \rho_{gal}^{\beta}$). The parameter β represents the ratio of the energy per unit of mass in the galaxies to the energy in the gas. It is given by the following relation:

$$\beta = \frac{\mu m_H \sigma^2}{k_b T_{gas}}. \tag{5}$$

The King approximation (1966) for the galaxy distribution is written $\rho_{gal} \propto (1 + (R/R_c)^2)^{-3/2}$. The gas distribution, called the β profile (Cavaliere & Fusco-Femiano 1978), is thus given by:

$$\rho(R) = \rho_0 \left[1 + \left(\frac{R}{R_c}\right)^2\right]^{-\frac{3\beta}{2}},$$

where ρ_0 is the central density and R_c is the radius for which the density is equal to half of the central density. It is called the core radius.

The parameter β is evaluated in two different ways. The first one is empirical. It consists in fitting the observed surface brightness profile by a β profile: it gives β_{fit}. The second method gives a value β_{def} by using the definition (Eq. 5) of the parameter. It is based on the measurement of the velocity dispersion, σ, of the galaxies and the temperature, T_{gas}, of the intra-cluster gas. There is a disagreement between the results of both methods: $\beta_{def} > \beta_{fit}$ (Henriksen & Mushotzky 1985) which would have several causes, for example:

- if the sample of galaxies used to estimate σ, is taken in the central region of the cluster, then the velocity dispersion could be overestimated.
- the presence of a dominant central galaxy in the galaxy cluster can modify the velocity distribution and influence the motion of the nearby galaxies.
- If the gas is not isothermal, the β value will be representative of only a fraction of the cluster profile.

The King approximation (1966), for which $\beta = 1$, was historically introduced to describe an isothermal self-gravitating sphere at the hydrostatic equilibrium. This approximation is compatible with the observations in the central part of the clusters. In the external regions, the number of galaxies and the density profile of the cluster thus diverge as ln R. Contrary to the King approximation, the observations give $\beta < 1$. These values lie between 0.4 and 0.8, with $0.6 < \beta < 0.8$ for the rich clusters and $0.4 < \beta < 0.5$ for the smallest clusters and groups (David et al. 1990). An average value is $\beta \simeq 2/3$ (Jones & Forman 1992).

These values indicate an excess of energy in the gas as compared to the energy in the galaxies. This could be due to a mechanism of reheating of the gas which would also account for the metal enrichment of the intra-cluster medium. One interpretation is related to the enrichment by the galaxies. Observations of heavy elements (S, If, Ca, N...) formed in stars, or supernovae, suggest that the galaxies of the cluster eject a fraction of their material in the intra-cluster medium. This mechanism (David et al. 1990, Sarazin 1992) would explain the significant reheating of the gas which leads also to $\beta < 1$.

4.1. SUNYAEV-ZEL'DOVICH EFFECT

Hereafter we focus on the effects of the Compton interactions of the CMB photons with the ionised matter: the Sunyaev-Zeldovich (SZ) effect. In the ionised regions of the Universe such as the intra-cluster gas of the galaxy clusters, the CMB photons with average temperature 2.728 K are scattered by the free electrons of the ionised and hot gas. This results in the generation of anisotropies at angular scales of a few arcminutes to the degree with amplitudes up to 10^{-4}. The photon-electron interactions do not change the number of photons. As the temperature of the intra-cluster gas is very high ($T_e \simeq 10^7$ K), the photons undergo an average heating. It results in a redistribution of the photons at the high frequency part of the spectrum.

This effect has been first studied in the case of the galaxy clusters by Zeldovich & Sunyaev (1969) and Sunyaev & Zeldovich (1972, 1980). For extensive reviews on the subject, we refer the reader to two recent review papers (Rephaeli 1995, Birkinshaw 1998) and to the references cited therein.

In order to describe the photon-electron interactions generating the SZ effect, we choose a reference frame in which the CMB is isotropic and at rest. In this frame, the photons scatter on the free electrons of the intra-cluster gas. The motion of the latter induces a frequency variation due to Doppler shift (first order effect) which can be written:

$$\frac{\Delta\nu}{\nu} = \frac{v_e}{c} \times \text{angular factor}, \tag{6}$$

where v_e is the electron velocity.

When the galaxy cluster is at rest in the CMB frame, the average frequency variation $\overline{\Delta\nu/\nu} \propto \overline{v_e}$ due to the Doppler effect is null. Indeed, there are as many electrons with positive velocities than there are with negatives ones (isotropic distribution) and thus $\overline{v_e} = 0$. In this case, only remains an effect of the second order. It is called the **thermal SZ effect** and it corresponds to the heating of a photon gas by an electron gas of higher temperature. It is proportional to $(T_e - T_{CMB})$, where T_e and T_{CMB} are respectively the temperatures of the electron gas and of the CMB. If $T_e = T_{CMB}$ (no heating effect) and the cluster moves with a peculiar radial velocity v_r, the average frequency variation is non-zero. It is proportional to the overall electron velocity v_r. This is called the **kinetic SZ effect.**

In fact, the SZ effect consists in the superposition of both the thermal and kinetic effects.

4.1.1. *Thermal SZ effect*

In this section, we assume that the cluster is at rest in the CMB frame with no bulk motion ($v_r = 0$). Moreover, we assume that the electrons are not relativistic and that they have an electronic temperature $T_e >> T_{CMB}$. For illustrative purpose, we study the photon-electron interactions along one direction only.

In the photon frame (CMB), an interaction between a photon and an electron occurs with the electron moving with $v_e = \beta c$ in the same or the opposite direction of the photon (c is the speed of light). In the first case (electron and photon going same direction) and after interaction, the energy of the photon, in its frame, is given by $E' = h\nu$ if the photon goes in the same direction as the initial one, and $E' = h\nu((1-\beta)/(1+\beta))$ if it goes in the opposite direction. In the second case (electron and photon going opposite directions), the energy after interaction is $E' = h\nu((1+\beta)/(1-\beta))$ for an interaction which reverses the direction of the photon, and $E' = h\nu$ if not. After scattering of the photon on the free electron, the frequency averaged on the four configurations is:

$$\nu' = \frac{\nu}{4}\left(2 + \frac{1-\beta}{1+\beta} + \frac{1+\beta}{1-\beta}\right) = \frac{\nu}{1-\beta^2}.$$

In the non-relativistic hypothesis ($\beta << 1$), we find $\Delta\nu/\nu \simeq \beta^2$. The thermal SZ effect is thus a second order effect in velocity and a first order in temperature. It expresses the frequency, or energy, variation of the CMB photons crossing a hot intra-cluster gas.

The relative frequency variation, per time interval δt averaged on the photon distribution in all directions is given by:

$$\frac{\overline{\Delta\nu}}{\nu} \simeq \frac{1}{3}(\sigma_T c n_e)\,\overline{\beta^2}\,\Delta T, \tag{7}$$

where σ_T is the Thomson cross section, n_e the electron density of the intra-cluster gas. We can introduce the so-called *Compton parameter* noted y by putting $dy = \overline{d\nu}/\nu$. It represents the relative frequency variation due to thermal SZ effect and gives its amplitude. The Compton parameter depends only on the physical parameters of the cluster. It is computed by integrating the equation (7) along the line of sight across the cluster:

$$y = \int \sigma_T n_e \frac{kT_e}{m_e c^2}\,dl, \tag{8}$$

l being a length element along the line of sight and $m_e c^2$ the electron rest mass.

The y parameter can be thought of as the pressure integral of the intracluster gas along the line of sight. In the assumption of an isothermal gas ($T_e = const$), the parameter y is written as a function of the Thomson optical depth $\tau = \int n_e \sigma_T \, dl$ as follows:

$$y = \frac{kT_e}{m_e c^2} \tau. \quad (9)$$

In the empirical approach presented here and also given by Salati (1992, 1996), we assume that the photon distribution is isotropic in the electron frame. We do not take into account neither the energy variation of the photons in the electron frame, nor their stimulated emission. A more accurate description of photon-electron interactions can be made through the Fokker-Planck kinetic equation which takes into account the variations of the photon distribution functions. It is within this framework that Zeldovich & Sunyaev have used the Kompaneets equation (1957). The latter is an approximation of the kinetic equation to the non-relativistic interactions. The more accurate analysis using the Kompaneets equation and an initial black body spectrum for the CMB radiation, gives the spectral dependence of the thermal SZ effect.

The frequency variation due to thermal SZ effect results in a variation of the intensity of the CMB radiation along the line of sight at a frequency ν:

$$\Delta I_\nu = I_\nu \, y \, f(x),$$

where I_ν is the CMB black body intensity, $x = h\nu / kT_{CMB}$ the adimensional frequency and $f(x)$ is the spectral shape function of the thermal SZ effect:

$$f(x) = \frac{x e^x}{e^x - 1} \left[x \left(\frac{e^x + 1}{e^x - 1} \right) - 4 \right].$$

The relative variation of intensity can also be expressed in terms of a relative thermodynamical temperature variation. We find:

$$\frac{\delta T_{CMB}}{T_{CMB}} = \frac{e^x - 1}{x e^x} \frac{\Delta I_\nu}{I_\nu}$$

The distortion induced by thermal SZ effect is characterised by a peculiar spectral signature with a positive extremum at short wavelength and a negative one at longer wavelength as exhibited figure 2. As a result in the direction of a galaxy clusterexhibiting SZ effect, we observe a decrease of the CMB brightness in the Rayleigh-Jeans part of the spectrum (low frequencies) and an increase in the Wien part (high frequencies).

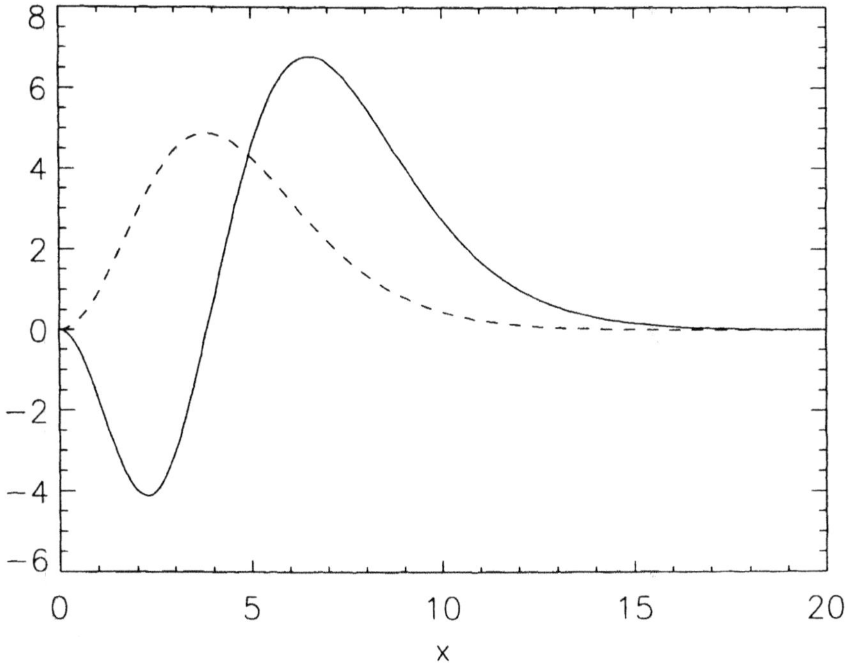

Figure 2. *Spectral shape function of the intensity variation due to the thermal Sunyaev-Zeldovich effect (solid line) and kinetic effect (dashed line) as a function of the adimensional frequency* $x = h\nu/kT_{CMB}$.

4.1.2. *Kinetic SZ effect*

When the galaxy cluster moves with a radial peculiar velocity v_r, the electron bulk velocity is thus equal to v_r and the relative frequency variation associated with the Doppler effect must be taken into account. This induces a distortion, and a relative variation of the black body intensity at the frequency ν, given by:

$$\frac{\Delta I_\nu}{I_\nu} = -\frac{v_r}{c}\tau\, a(x),$$

the spectral shape function $a(x)$ is written:

$$a(x) = \frac{x\, e^x}{e^x - 1},$$

An equivalent expression, in terms of the temperature relative variation $(\delta T/T)_{SZ}$, is:

$$\left(\frac{\delta T}{T}\right)_{SZ} = \frac{e^x - 1}{xe^x}\frac{\Delta I_\nu}{I_\nu} = -\frac{v_r}{c}\tau. \tag{10}$$

The temperature fluctuations $(\delta T/T)_{SZ}$, induced by kinetic SZ effect are spectrally indistinguishable from the primary CMB anisotropies. The kinetic effect is positive for clusters moving towards the observer, that is for negative radial peculiar velocities.

In practice, for the same galaxy cluster, we observe a relative variation of the intensity which is the sum of the thermal and kinetic effects. Therefore, we write:

$$\frac{\Delta I_\nu}{I_\nu} = \tau\frac{xe^x}{e^x - 1}\left[\frac{kT_e}{m_e c^2}\left(x\frac{e^x + 1}{e^x - 1} - 4\right) - \frac{v_r}{c}\right].$$

The figure 2 indicates that the contribution of the kinetic effect to the total distortion has its maximum in the negative part of the thermal effect. A measured amplitude of the SZ effect in this part of the spectrum is thus likely to be significantly reduced. On the other hand, the measured signal in the positive part (short wavelengths) remains dominated by the thermal effect. Similarly, confusion with the CMB, at the scales of the order of the cluster scales, affects and reduces the SZ signal in the negative part (long wavelengths).

4.1.3. *Relativistic corrections*

The calculations of the SZ effect (thermal and kinetic) have been performed using the Kompaneets equation, which corresponds to the non-relativistic approximation for the scattering of the CMB photons on the free electrons. This approximation is not totally justified for two main reasons:

- the cluster optical depth is weak (smaller than a few 10^{-2}). Some of the CMB photons do not interact with the electrons and the scattering approximation is thus inadequate.
- The intra-cluster temperature lies generally between 10^7 and 10^8 K. In such a medium, the electron velocities are quasi-relativistic ($\beta = v_e/c = 10^{-2}$ to a few 10^{-2}). The relative frequency variation, given as a function of β, is not small enough to use the expansion to the second order for the resolution of the Kompaneets equation.

The relativistic corrections to these computations have been studied by many authors (Wright 1979; Taylor & Wright 1989; Loeb et al. 1991; Rephaeli 1995).

Rephaeli (1995) exhibited the main modifications introduced by an exact relativistic calculation. He gave the modifications as a function of the intra-cluster temperature T_e. For increasing T_e, we observe on the one hand, the decrease of the maximum intensity variation of the radiation and the increase in its minimum, and on the other hand a shift of the zero point frequency of the thermal SZ effect towards high frequencies.

4.2. OBSERVATIONS OF THE SZ EFFECT

Soon after the theoretical work of Zeldovich & Sunyaev (1969) and Sunyaev & Zeldovich (1972), Pariskii (1973) claimed the detection of the SZ effect towards the Coma cluster at $\lambda = 4$ cm. But these observations were highly contaminated by the presence of radio sources along the line of sight. The first significant measurements of the SZ effect in the Rayleigh-Jeans part (negative effect) have started in 1984, more than ten years after the first tentative detection. For a detailed review of the observational methods we refer the reader to Lasenby (1992), Rephaeli (1995) and Birkinshaw (1998). Table 1, adapted from Rephaeli (1995), recapitulates some of the most significant measurements of the SZ effect.

TABLE 1. *Main significant measurements of the SZ effect with the observation frequencies as well as the maximum value of their temperature variation and Compton parameter with the* 1σ *errors.*

Cluster	ν (GHz)	ΔT (mK)	y (10^{-4})	Reference
A401	19.5	-0.64 ± 0.18	1.18 ± 0.30	Uson (1986)
A478	32	-0.379 ± 0.029	0.71 ± 0.05	Myers et al.(1995)
A665	20.3	-0.434 ± 0.052	0.8 ± 0.09	Birkinshaw et al. (1991)
A773	32	-0.79 ± 0.16	1.48 ± 0.30	Grainge et al. (1993)
A773	28.7	-0.617 ± 0.042	1.14 ± 0.07	Carlstrom et al. (1995)
A1656	15	-0.495 ± 0.087	0.91 ± 0.16	Herbig et al. (1994)
A2142	32	-0.414 ± 0.026	0.78 ± 0.05	Myers et al. (1995)
A2163	136	$-1.16^{+0.20}_{-0.22}$	$3.78^{+0.74}_{-0.65}$	Wilbanks et al. (1994)
A2163	144	-1.05 ± 0.16	$3.73^{+0.47}_{-0.61}$	Holzapfel et al. (1996)
A2218	20.3	-0.60 ± 0.20	1.11 ± 0.37	Klein et al. (1991)
A2218	20.3	-0.445 ± 0.033	0.82 ± 0.06	Birkinshaw & Hughes (1994)
A2218	15	-1.1 ± 0.2	2.02 ± 0.37	Jones et al. (1993)
A2256	32	-0.240 ± 0.027	0.45 ± 0.05	Myers et al. (1995)
0016+16	20.3	-1.584 ± 0.256	2.93 ± 0.47	Birkinshaw et al. (1984)
0016+16	19.5	-0.48 ± 0.12	0.8 ± 0.22	Uson (1986)
0016+16	28.7	-0.772 ± 0.047	1.43 ± 0.08	Carlstrom et al. (1995)

Three techniques are used for thermal SZ effect observations. Two of them are based on coherent detection techniques (radio) and single dish telescopes, or interferometers, the third consists in total power measurements with bolometers.

The first significant measurements of the thermal SZ effect was made with single dish radio-telescopes (Birkinshaw et al. 1984, Birkinshaw et al. 1991) (40 m Antenna of Caltech at Owens Valley). These observations use beam switching between the direction of the galaxy cluster and a reference field in a nearby direction. They provide a measurement of the temperature (brightness) decrease in the cluster direction in the Rayleigh-Jean part of the spectrum.
This method suffers from various sources of uncertainties, the principal one being the confusion with radio sources, close to or in the cluster direction. Another problem comes from the error associated with the reference level of the unperturbed CMB (reference field). An error on this quantity introduces an uncertainty on the temperature variation related to the SZ effect. Finally, the fluctuations of the atmospheric emission also introduce uncertainties in the measurement of the SZ effect.

Another measurement technique of the SZ effect uses radio-interferometers (Jones et al. 1993). The first advantage of this method, is that it gives an image of the SZ effect rather than a profile. This two-dimension information allows the detection of point sources which could superimpose on the cluster, and eases their subtraction. This can be achieved with a judicious choice of the antenna separations which measures simultaneously the SZ effect and the radio sources emission. It can also constrain the cluster shape and size. Contrary to single dish measurements, the interferometer measurements do not need an evaluation of the reference background. Therefore, they are less sensitive to the variations and perturbations of the CMB, and to the fluctuations of the atmospheric emission.

An even more promising measurement technique uses bolometers. Indeed, they have a very good sensitivity and allow multi-frequency measurements. Contrary to the previous methods, which observe the Rayleigh-Jeans part of the spectrum, this method is also sensitive to the Wien part. Moreover, it allows the use of filters centred on some specific frequency. In particular, a filter centred on the zero point of the thermal effect is promising for the measurement of the kinetic SZ effect. Bolometric observations have been carried out since 1983 (Meyer et al. 1983, Radford et al. 1986, Chase et al. 1987), but the first significant detection was recently obtained by Wilbanks et al. (1994) in direction of the richest cluster A2163 ($T \simeq 15$keV) with the SuZIE instrument (*Sunyaev Zeldovich Infrared Explorer*) developed at Caltech (U.S.A.). Other detections of the SZ effect with bolometers have been made by French groups (collaboration

Grenoble, Orsay, Toulouse). The Diabolo instrument (Désert et al. 1998) has observed the clusters A665, A2163 and 0016+16. The second flight in September 1996 of the balloon borne experiment PRONAOS carrying instrument SPM (*Spectro Photomètre subMillimétrique*) has measured for the first time both the positive and negative part of the SZ effect for the A2163 cluster (Lamarre et al. 1998).

4.3. COSMOLOGICAL APPLICATIONS OF THE SZ EFFECT

The SZ effect is a very powerful tool for cosmology in particular when it is compared and completed by other observations (X-ray emission, gravitational lensing). We mentioned earlier that the amplitude of the thermal SZ effect is characterised by the so-called Compton parameter y (Eq. 9). The latter depends only on the physical properties of the cluster (temperature and electronic density). When we do not take into account of the evolution effects, the SZ effect is thus independent of the redshift. Measurements of the SZ effect of a cluster seem thus more appropriate to probe the early Universe than the X-ray emission. This property is one of the main advantages of the SZ effect, as it can be used to trace the matter distribution through galaxy cluster detection. The clusters being the largest gravitationally bound structures, they are thought to trace the matter at large scales better than galaxies.

4.3.1. *Measurement of the Hubble constant*

Because the SZ effect is independent of the redshift, it can be used to measure the Hubble constant H_0 at distances much larger than those probed by the traditional methods, and, most of all, independently of the distance estimators. It is therefore less affected by the errors associated with such techniques.

The measurement of the Hubble constant H_0 is based on the combination of the SZ effect and the X-ray emission measurements of the same galaxy cluster. The SZ effect is proportional to the electron density, n_e, whereas the X-ray emission is proportional to n_e^2. The basic principle of the measurement is to combine these quantities and derive the size of the cluster which in turn gives the Hubble constant.

The guiding principle of the method is to use the expression of the X-ray surface brightness given by:

$$B_X = \frac{1}{4\pi(1+z)^3} \int n_e^2 \Lambda_e \, dl, \tag{11}$$

and the expression of the temperature variation of the CMB (due to the

SZ effect), in the Rayleigh-Jean part of the spectrum, given by:

$$\Delta T_{RJ} = -2\frac{kT_{RJ}}{m_e c^2}\sigma_T \int n_e T_e \, dl. \tag{12}$$

n_e and T_e are the respectively of electron density and the temperature distributions of the cluster, Λ_e is the X-ray spectral emissivity, and the integral is computed along the line of sight across the cluster. Each quantity n_e, T_e and Λ_e, can be written as the product of a normalisation constant and a shape function describing the spatial distribution of the considered quantity. Moreover, if we assume an isothermal gas, then the equations (11) and (12) give $B_X = N_X \Theta_X$ and $\Delta T_{RJ} = N_{RJ}\Theta_{RJ}$, where $\Theta_{X,RJ}$ represent the integrals along the line of sight of the shape functions. The normalisation constants:

$$N_X = \frac{1}{4\pi(1+z)^3}\Lambda_{e0}\, n_{e0}^2\, D_A$$

and

$$N_{RJ} = -2T_{CMB}\frac{kT_{e0}}{m_e c^2}\sigma_T\, n_{e0}\, D_A$$

are derived from the observation of the X-ray emission and the SZ effect. If we know the cluster temperature and its redshift, we can determine its angular diameter distance D_A. An expression of D_A is indeed obtained when n_{e0} is eliminated between the expressions of N_{RJ} and N_X. It is written:

$$D_A = \frac{N_{RJ}^2}{4T_{CMB}N_X}\,\frac{\Lambda_{e0}}{4\pi(1+z)^3\sigma_T^2}\left(\frac{m_e c^2}{kT_{e0}}\right)^2,$$

In addition, the theoretical expression of D_A is:

$$D_A = \frac{czq_0 + (q_0 - 1)[(1+2q_0 z)^{1/2} - 1]}{H_0 q_0 (1+z)^2}. \tag{13}$$

Within the limit of small redshifts ($z \ll 0.2$), the q_0 dependence is negligible and the equation (13) is equivalent to: $D_A \approx cz/H_0$. The value of the Hubble constant H_0 is derived when we compare the observational estimate of D_A and its value computed in the approximation of small redshifts.

The evaluation of the Hubble constant using X-ray and SZ observations is independent of empirical distance estimator relations. Nevertheless, it has its own limitations and sources of uncertainties. First, it needs a very precise and accurate measurement of the temperature variation ΔT_{RJ}, which is not an easy task. We indeed saw that measurements of the SZ effect using single dish radio-telescopes suffer from errors for the reference background. This translates into errors in ΔT_{RJ}. The interferometric techniques, as well as bolometric measurements, seem more advantageous, but they have still to give evidences of their reliability.

In addition, this method needs a detailed model of the galaxy cluster, in particular, the temperature (isothermal, polytropic) and gas (homogeneous, inhomogeneous) distributions, as well as the cluster geometry (spherical, elliptic) and the projection effects. Any error in one of the physical properties mentioned above results in an error in the value of H_0. Birkinshaw et al. (1991) have listed and estimated the various sources uncertainties. Consequently, the measurement of the Hubble constant using the SZ effect is restricted to a few dynamically relaxed and quasi-spherical clusters.

In order to illustrate the dispersion and errors in the value of the Hubble constant, we quote the table 2 some estimates of H_0 given by Rephaeli (1995). The values obtained for remain comprised between 50 and 100 km/s/Mpc. We also note that this method gave very low H_0 values. For the same cluster we also see that the estimates do not agree which is very likely due to the sources of errors and the differences between the experiments.

TABLE 2. *The Hubble constant derived using SZ effect and X-ray observations.*

Cluster	H_0 (km/s/Mpc)	Reference
A478	32^{+19}_{15}	Myers et al. (1995)
A665	40 ± 9	Birkinshaw et al. (1991)
A1656	74^{+29}_{24}	Herbig et al. (1994)
A2142	57^{+61}_{39}	Myers et al. (1995)
A2163	78^{+59}_{40}	Holzapfel et al. (1996)
A2218	24 ± 11	McHardy et al. (1990)
A2218	65 ± 25	Birkinshaw & Hughes (1994)
A2218	38^{+18}_{16}	Jones (1993)
A2256	76^{+22}_{19}	Myers et al. (1995)

4.3.2. *Cluster structure*

The combination of the SZ effect measurement and X-ray observations, can give the physical size and the distance to the cluster, as well as the Hubble constant. It can also give information on the temperature and matter distributions in the cluster. The linear dependence in n_e of the SZ effect allows to explore the external regions of the gas distribution better than it is possible with the X-ray emission (proportional to n_e^2). Therefore, it would be possible to test some hypotheses such as the sphericity and homogeneity of the electronic distribution.

Another application of the SZ effect is related to the dark matter distribution in the galaxy clusters. An image of the gravitational lensing effect of a cluster gives the total mass of the cluster (Soucail 1992). A comparison between this mass and the one derived from X-ray observations allows to probe the baryon content of the structure and consequently its dark matter content. As the X-ray emission is highly peaked towards the centre of the cluster, the contribution to the mass of the wings of the gas distribution is not taken into account accurately. In this context, the SZ effect seems again an appropriate tool to complete the X-ray observations in the external parts of the structure.

4.3.3. *The large scale structures and the SZ effect*

Up to now, we have given the principal applications of the SZ effect measurements, when we compare them to other observations (X-ray emission, gravitational lenses), for studies of the individual clusters. In the following, we stress on the interest of the SZ effect for the study of large scale structures.

The spatial distribution of the galaxy clusters provides valuable information on the large scale structures of the Universe. As the average distance between two clusters is about 10 to 25 Mpc, a homogeneous catalogue of galaxy clusters would be a very useful complementary information to the galaxy surveys for the study of the structures at scales larger than 10 Mpc.

Such catalogues of galaxy clusters could be used to compute the two points correlation functions. Using the actual catalogues of rich clusters constituted with X-ray and optical observations, studies have been made on the galaxy clusters correlation function (Bahcall 1988 and references therein, Sutherland 1994). They indicate that the correlation scale lies between approximately 15 and 25 Mpc and that there is a strong correlation up to about 50 Mpc. This strong correlation suggests that there is more power at large scales than the power predicted by the standard CDM model normalised to COBE data (Sutherland 1994). This result indicates that this approach is very promising to probe the cosmological models. It leads again to a questioning of the validity of the standard CDM model. Nevertheless, the obtained results have to be taken with cautiousness as they are very sensitive to the quality of the catalogues. It is thus clear that a refined analysis to probe the cosmological models needs more complete and homogeneous catalogues. In this context, the detection of galaxy clusters through their SZ effect is very appropriate in order to constitute homogeneous and complete samples of clusters.

The COBE/FIRAS instrument gave an estimate of the mean Compton parameter characterising the overall SZ distortion of the CMB. It has given an upper limit of $\overline{y} > 1.5\,10^{-5}$ (Fixsen et al. 1996). This value is associated with the sources of hot gas in the whole Universe. In the hypothesis that the average parameter $\overline{y}$ is produced by the SZ effect integrated over the whole population of galaxy cluster, the value of $\overline{y}$ tightly constrains the structure formation and evolution theories, as well as the cosmological parameters such as Ω_0, or n (Markevitch et al. 1992, Barbosa et al. 1996).

It has long been suggested that combining the measurement of the SZ thermal and kinetic effect would allow the measure of the radial peculiar velocity of galaxy clusters (Sunyaev & Zeldovich 1980). As a matter of fact, the radial peculiar velocity of a galaxy cluster can be written as a function of the relative temperature variation $(\Delta T/T)_{SZ}$, associated with the kinetic SZ effect, and of the Compton parameter y, of the thermal effect. We replace the Thomson optical depth τ, in equation (10), by its expression as a function of y (Eq. 9). We then obtain:

$$\frac{v_r}{c} = -\frac{k_b T_e}{m_e c^2}\frac{(\delta T/T)_{SZ}}{y}. \tag{14}$$

The measurement of the radial peculiar velocity with the SZ effect is limited by the spectral confusion between the primary anisotropies and the ones generated by the kinetic effect (Heanelt & Tegmark 1996; Aghanim et al. 1998). This confusion is the main source of errors in the measurement. It results in a *rms* error of about 500 km/s for rich clusters ($y \simeq 10^{-4}$) with core radii of about 1 to 2 arcminutes. The error can be even larger for weaker clusters or larger core radii. In the last case, the primary fluctuations have larger amplitudes. The measurement of the velocity is limited for individual clusters, but if made on a large number of cluster it can give valuable statistical information on the cosmic velocity field. It allows more specifically to measure the overall bulk velocities in defined volumes of the Universe (of about 100 Mpc typical scale) up to a redshift of unity with a precision between 30 and 100 km/s. The direct measurement of the galaxy cluster peculiar velocities, combining SZ effect measurements, is a promising method to obtain the large scales cosmic velocity fields ($\simeq$ 100 Mpc). Under the hypothesis of linear perturbation theory and potential flow for the Universe, one can carry out the reconstruction of the underlying density field and once again probe the cosmological models.

5. Inhomogeneous re-ionisation

The observations of the absorption spectra towards distant quasars ($z \simeq 5$) indicate the quasi absence of neutral hydrogen on their lines of sight (Webb et al. 1992, Jackobsen et al. 1994, Madau & Meiksin 1994). This test, called the Gunn-Peterson test (Gunn & Peterson 1965), suggests that the Universe is highly ionised, in the interval $0 < z < 5$. It is thus necessary to invoke a phenomenon that re-ionised totally the Universe between the recombination ($z \simeq 1000$) and the redshift of the most distant quasars ($z \simeq 5$).

The re-ionisation has two main effects on the CMB. On the one hand, it damps the power associated with the primary fluctuations, and on the other hand, it generates secondary fluctuations at the degree scale through the SZ effect or at very small scales through the Ostriker-Vishniac effect. A detailed description of the thermal and of the re-ionisation history of the Universe represents a crucial point for the interpretation of primary fluctuation observations. In this context, several aspects of the re-ionisation have been, and are, studied. They raise many questions regarding the re-ionisation as the the process(es) at the origin of the re-ionisation, its redshift, its homogeneity or inhomogeneity.

Several physical processes have been invoked to account for the re-ionisation of the Universe. A majority of the models are based on the emission of the first objects which formed then. Some of these models imply the photoionisation (Donahue & Shull 1987, Shapiro & Giroux 1987, Miralda-Escudè & Ostriker 1990, 1992), the explosive galaxy formation (Ikeushi 1981, Vishniac & Ostriker 1985) or the hierarchical structure formation (Couchman & Rees 1986). In these descriptions, the re-ionisation is often inhomogeneous, because it begins in the vicinity of point sources emitting in the neutral medium. Other models based on the decay of exotic particles present in the Universe (Sciama 1982) have been also studied. In this case, the homogeneous distribution of the exotic particles induces a spatially homogeneous re-ionisation.

The effects of the re-ionisation on the CMB have been recently revisited for different cosmological scenarios: CDM (Dodelson & Jubas 1995) and PIB models (*Primordial Isocurvature Baryon*) (Hu et al. 1994). In the standard CDM model, the re-ionisation of the Universe for ($z \simeq 5$) results in a global Thomson optical depth $\tau = 0.01$. As the re-ionisation damps the power of the temperature fluctuations by a factor $e^{-2\tau}$, the C_l coefficients of the angular power spectrum are decreased by a factor of about 2% with respect to their initial values.

The recent study of Dodelson & Jubas (1995) treated in particular two extreme cases. A first case with no re-ionisation and a second case where the recombination never took place. This study shows that the principal effect of an early re-ionisation is to shift the acoustic peaks towards large angular scales. In the extreme case of no recombination, the first peak is located at $l \simeq 50$. The re-ionisation thus produces a more significant signal, which can be explained by the fact that the scattering effects are more important; the velocities are larger and the Doppler interactions are thus more significant. In a re-ionised Universe the damping effects take place at larger angular scales (small l). Other studies have put constraints on the epoch of the re-ionisation based on observations of the Compton parameter y (Tegmark et al. 1994). They indicate that, in a standard CDM model, the re-ionisation is likely to happen at redshifts lower than 50.

As mentioned previously, the main effect of the re-ionisation is to wash out some of the CMB temperature fluctuations on small scales. We present here a new kind of detectable secondary fluctuations that might appear due the kinetic SZ effect in the context of an inhomogeneous re-ionisation by early quasars (Aghanim et al. 1996). Indeed, we have developed a model in which the re-ionisation of the Universe is due to the photoionisation by emitting sources. We study in this class of models, a quasar population present at $z < 10$. We bare in mind that this assumption is likely to maximise the induced secondary anisotropies. The early quasars could constitute the dominant source of UV background radiation (Meiksin & Madau 1993), but their formation epoch remains to be established. Many models of structure formation predict the formation of objects early in the history of the Universe. In PIB and CDM models the emitting sources are formed at respectively $z < 100$ and between $z = 10$ and 20 (e.g. Ostriker & Gnedin 1996). In this context, the basic hypothesis of our model which consists in assuming that the quasars are present at high redshifts is justified. Moreover if the first quasars appear when the Universe is still essentially neutral, a significant fraction of the effects will be due to the quasars even though other sources dominate the re-ionisation later.

The schematic scenario of the model is the following: before the full re-ionisation, quasars ionise the surrounding gas producing ionised bubbles which move together with the quasars in the large scale gravitational potential wells. An effect equivalent to the kinetic SZ effect for rich clusters creates temperature fluctuations which are proportional to the product of the Thomson optical depth of the ionised cloud τ and the ratio of the quasar peculiar radial velocity to the speed of light (v_r/c). Contrary to galaxy clusters, the corresponding thermal SZ effect is negligible in the case of ionised clouds.

This work has no strong dependence on the quasar formation theory, the presence of any luminous ionising point source will have the same kind of effect on the CMB, as long as it can roughly account for the total re-ionisation of the Universe at the present time.

5.1. TEMPERATURE OF THE IONISED GAS

If the intergalactic matter is ionised by the emitted radiation, the ionisation of a hydrogen atom produces a photoelectron with an average kinetic energy defined by $\overline{E_c} = \overline{E_{UV}} - 13.6\,\mathrm{eV}$, where $\overline{E_{UV}}$ is the average energy of the ionising photon. Electrons and protons rapidly thermalise owing to the short time scale of the Coulomb interactions, therefore, the temperature of the plasma is given by $T = \overline{E_c}/3k$ as long as the cooling by recombination is negligible. Hereafter, we neglect the contributions other than those induced by hydrogen and helium atoms.

We compute the average energy of the ionising photon $\overline{E_{UV}} = h_{pl}\,\overline{\nu}$ using a power law spectrum for the quasar emission $L \propto \nu^{-1.5}$ ($\lambda < 1216\,\mathring{A}$) (Bechtold et al. 1987, Bechtold 1994), where L is the luminosity of the quasar and ν the frequency, and assuming that every photon is absorbed sooner or later by the intergalactic medium. We find a temperature $T \approx 10^5\,\mathrm{K}$. In the bubble, the gas is fully ionised, therefore there is no collisional cooling. We find that the bremsstrahlung cooling time is longer than the age of the Universe. Likewise, we investigate the effect of the inverse Compton cooling and find that it becomes important for redshifts greater than $10\,h^{2/5}$, which gives $z \simeq 8$ for $h = 0.5$. In this case, we check that the temperature of the bubble decreases by a factor 2 at $z = 10$. The Compton cooling does not change drastically the temperature. The exact temperature being not critical for the model, we thus assume that it remains constant at $T \approx 10^5\,\mathrm{K}$. This approximation breaks down for redshifts significantly larger than 10.

5.2. RADIUS OF THE IONISED REGION

Let us assume that a quasar turns on at a redshift z_{on} and turns off at z_{off}, in the assumption of a lifetime t_q short compared to the expansion time, we have $1+z_{on} \approx 1+z_{off}$. The proper radius R of an ionised bubble at a given time t (when it has not reached the stationary Strömgren sphere state) is computed by writing that every UV photon emitted by the quasar is absorbed through the photoionisation of a neutral atom. The main quantity that rules the size of the bubble is the total number of ionising photons emitted by a quasar during its lifetime. It is proportional to the product Lt_q. Given a point-like emitting source of luminosity, L, we compute its

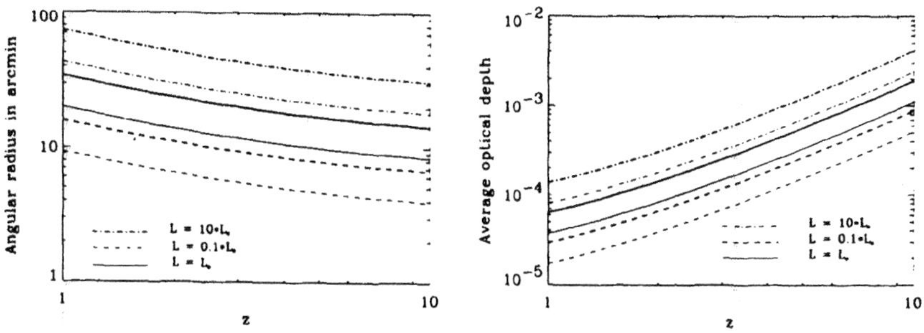

Figure 3. Left: Angular radius Right: optical depth of an ionised bubble as functions of the redshift for three values of the quasar luminosity. The series of thin lines are obtained with quasar lifetime $t_q = 10^7$ *yrs and the thick ones with* $t_q = 5\,10^7$ *yrs.*

emission rate in terms of number of ionising photons per unit of time:

$$S = \int_{\nu_{ion}}^{\infty} \frac{L}{h\nu}\, d\nu,$$

where ν_{ion} is the ionising frequency for hydrogen. If we assume an isotropic emission of the photons, the ionised region produced around the quasar (or the emitting source) is spherical. Other possible geometrical effects are neglected. If every photon emitted by the source is absorbed through the photoionisation of a neutral hydrogen atom, the size of the ionised region is thus given by the number of emitted photons using the following relation:

$$\frac{4\pi}{3} R^3 n_e = S\, t_q,$$

where $n_e = n_{e0}(1+z)^3$ a nd $n_0 = 1.13\,10^{-5}\,\Omega_b\, h^2\,\mathrm{cm}^{-3}$ is the present day electronic density

The proper radius R reachs its maximum value at the redshift z_{off}, and then it undergoes the average Hubble expansion. At a given redshift z smaller than z_{off}, the expansion leads to the following relation:

$$R(l_Q, z) \;=\; R_c\, l_Q^{1/3} (1+z)^{-1}. \qquad (15)$$

where $R_c = 10 \left(\frac{\Omega_b h^2}{0.015}\right)^{-1/3} \left(\frac{\Omega_0}{1}\right)^{-1/6}$ Mpc is a characteristic radius for $t_q = 10^8$ yrs associated with a characteristic luminosity L_c^* and $l_Q = L/L_c^*$. The comparison between the proper and the Strömgren radii shows that recombinations can be neglected. Figure 3, left, show the angular scale of the bubble versus the redshift different values of t_q and L.

5.3. OPTICAL DEPTH

Using the proper radius of the ionised sphere, we compute the average Thomson optical depth over the spherical ionised bubble, it is given by:

$$\tau_a = \frac{2}{3}\tau_c \, l_Q^{1/3}(1+z)^2 \tag{16}$$

where $\tau_c = \sigma_T \, n_{e0} \, R_c$ the characteristic optical depth. Figure 3, right, shows the dependence of the bubble's optical depth with both the lifetime of the quasar and its luminosity, as a function of the redshift. We note that typical optical depths of 10^{-3} (comparable to those of rich clusters) are reached for our redshift range of interest.

5.4. INDUCED SECONDARY EFFECTS

Similarly to galaxy clusters, one can compute the thermal SZ effect associated with an ionised bubble. The amplitude of this distortion is given by $y = \frac{kT_e}{m_e c^2}\tau$. The optical depths are comprised between 10^{-4} and a few 10^{-3}, and the temperature of the ionised bubbles is of about 10^5 K. The Compton parameter hardly reachs 10^{-7}. This effect is thus negligible, especially if compared with the typical Compton parameter of 10^{-4} for the galaxy clusters.

The ionised bubbles generated between $z \sim 5$ and $z \sim 1000$ are assumed random substructures in the large scale structures. They move in the large scale potential wells. Therefore, they induce secondary anisotropies through the kinetic SZ effect. The amplitude of the anisotropies are given by $\Delta T/T = -(v_r/c)\tau$. The distribution of the quasars (and thus bubbles) peculiar radial velocities is given by the Gaussian distribution of the large scale velocity field. Its velocity dispersion $\sigma_v(z)$ varies with the redshift and is normalised to the present day value of the velocity dispersion. We take a semi empirical approach using the z dependence given by the linear theory $\sigma_v(z) \propto (1+z)^{-1/2}$ and a conservative value of $\sigma_{0v} = 300$ km/s (Faber et al. 1993). For the typical ionised bubbles, we find that the average temperature fluctuations produced by the kinetic SZ effect are of a few 10^{-6}.

5.5. GENERALISATION TO QUASARS DISTRIBUTION

The distribution of ionised bubbles is derived from the quasars luminosity function $\Phi(L, z)$ which gives for a redshift z the comoving space density of quasars as a function of the luminosity. The quasars luminosity function is not well determined yet, in particular at high redshifts ($z > 3$), owing to the small number of observed sources. We use the luminosity function given by Boyle (1991). For $z > 3$, following Warren et al. (1991) and Meiksin &

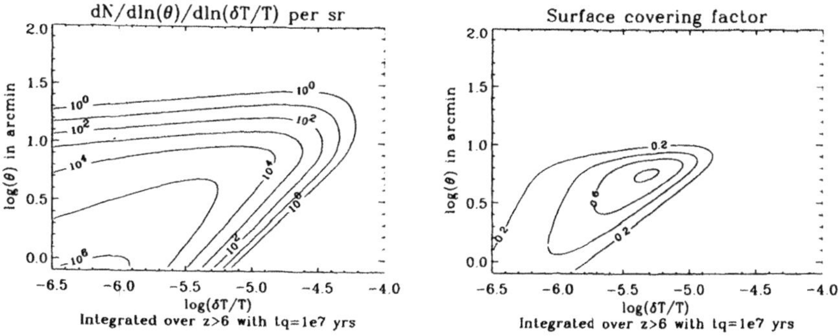

Figure 4. Left: Number of ionised bubbles per unit solid angle, logarithmic interval of $|\delta_T|$ logarithmic interval of the angular radius θ plotted as a function of $\log\delta_T$ and $\log\theta$ integrated over $z > 6$ and computed for lifetime $t_q = 10^7$ yrs. Right: The surface covering factor as a function ($\log\delta_T$) and $\log\theta$ integrated over the redshifts ($z > 6$) for lifetime $t_q = 10^7$ yrs.

Madau (1993) we introduce an exponential decay by a factor of 2 per unit redshift in the comoving number density.

In order to evaluate the redshift z_{ion} of the re-ionisation of the Universe for hydrogen, we compute the porosity parameter. The total re-ionisation of the intergalactic medium occurs when the porosity parameter reaches unity, at $z_{ion} \approx 5.7$, which is compatible with the observed re-ionisation of the Universe.

Using expressions of the angular size, optical depth and temperature fluctuations of the ionised bubbles around quasars, we compute the number of bubbles at redshift z with angular size θ and temperature fluctuation δ_T in the Universe, per unit of solid angle. See Aghanim et al. (1996) for detailed computations. The results plotted, in Fig. 4 left, are integrated over the redshifts ($z > 6$) for $t_q = 10^7$ yrs.

We also compute the covering surface parameter which is the fraction of the sky covered by ionised bubbles inducing a distortion of amplitude δ_T and angular size θ at redshift z. The results plotted Fig. 4 right, are also integrated over z and given for $t_q = 10^7$ yrs.

5.6. DISCUSSION AND CONCLUSIONS

We have discussed the effects of an inhomogeneous re-ionisation by early formed quasars on the CMB. We have shown the thermal SZ effect was negligible compared to the kinetic one. The amplitude of the largest temperature fluctuations induced by the kinetic effect are of the the order of the primordial fluctuations at scales comparable with the scales of the Doppler peaks (10 to 1°). These secondary fluctuations have no spectral signature to distinguish them from the primordial ones. Figure 3, left, shows that

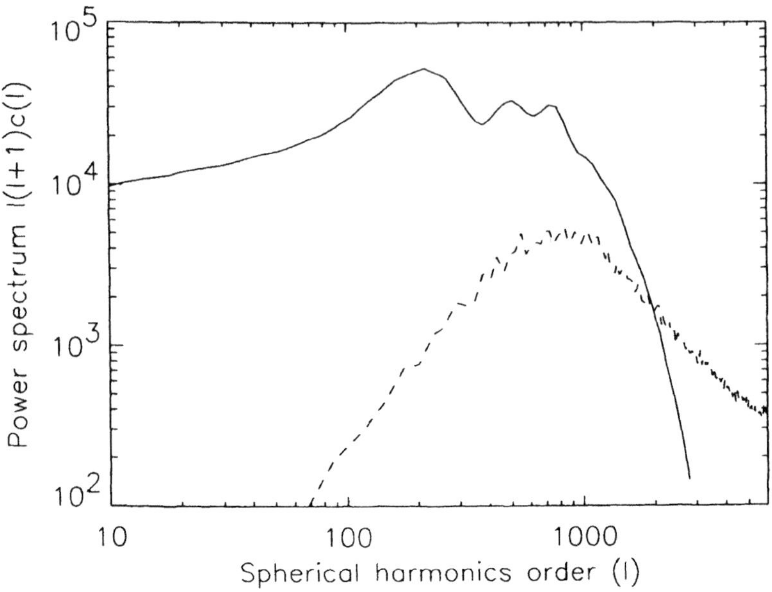

Figure 5. Power spectra of the anisotropies in the spherical harmonics versus the multipoles. Solid curve for the primary CMB and dashed curve for the fluctuations induced by the ionised bubbles for $t_q = 10^7$ yrs

there are approximately 100 objects per unit solid angle with associated temperature fluctuations some 10^{-5} and angular radius 5 to 10 arcminutes. The counts depend significantly on the quasar lifetime value: decreasing the lifetime t_q decreases the size of the bubbles and increases the number of bubbles. Figure 4 , right, representing the surface covering factor, shows that the ionised bubbles around quasars are close to cover the whole sky although they do not fill the space until $z < z_{ion} = 5.7$.

Cosmological models show that there is cut off in the angular power spectrum of the primordial anisotropies. The bubble contribution to the observed CMB could be important, or may be dominant, in particular at scales where the primordial fluctuations are "washed out" because of the damping, as shown Fig. 5. Furthermore, Fig. 5 indicates that the secondary anisotropies due to ionised bubbles add a significant power ($\simeq$ 10%) at scales corresponding to the secondary Doppler peaks.

In the hypothesis of an ionization due to individual sources of radiation, the whole formalism we develop for early quasars (Aghanim et al. 1996) remains valid whatever the kind of sources, early galaxies, first generation of stars.

References

Aghanim, N., Désert, F.X, Puget, J.L., Gispert, R. 1996, *A&A*, **311**, 1
Aghanim, N., De Luca, A., Bouchet, F.R, Gispert, R., Puget, J.L. 1997, *A&A*, **325**, 9
Bahcall, N. A. 1988, *ARA&A*, **26**, 631
Barbosa, D., Bartlett, J.G., Blanchard, A., Oukbir, J. 1996, *A&A*, **314**, 13
Bechtold, J., et al. 1987, *ApJ*, **315**, 180
Bechtold, J. 1994, *ApJS*, **91**, 1
Birkinshaw, M. 1990, in *The cosmic microwave background 25 Year Later*, eds. N. Mandolesi and N. Vittorio (Kluwer, Dordrecht)
Birkinshaw, M. 1998, astro-ph/9808050
Binkinshaw, M., Gull, S. F., & Hardbeck, H. E. 1984, *Nature*, **309**, 34
Binkinshaw, M., Hughes, J. P., & Arnaud K. A. 1991, *ApJ*, **379**, 466
Birkinshaw, M., & Hughes, J. P. 1994, *ApJ*, **420**, 33
Blanchard, A., & Schneider, J. 1987, *A&A*, **184**, 1
Bond, J. R., & Efstathiou, G. 1987, *M.N.R.A.S.*, **226**, 655
Boyle, B. J 1991, in *Texas/ESO-CERN Symposium on Relativistic Astrophysics, Cosmology and Fundamental Physics*
Bunn, E. F. 1996, in Proc. Strasbourg Summer School on Comology, astro-ph/9607088
Carlstrom, J. E., Joy, M., Grego, L. 1996, *ApJ*, **456**, L75
Cavaliere, A., & Fusco-Femiano, R. 1978, *A&A*, **70**, 667
Cayón, L., Martinez-Gonzalez, E., Sanz, J. L. 1993, *ApJ*, **403**, 471
Chase, S. T., Joseph, R. D., Robertson, N. A., Ade, P. A. R. 1987, *M.N.R. A.S.*, **225**, 171
Couchman, h. M. P., & Rees, M. J. 1986, *M.N.R.A.S.*, **221**,53
David, et al. 1990, *ApJ*, **359**, 29
Désert, F. X., et al. 1998, *New Astron.*, **3**, 655
Dodelson, S., & Jubas, J. M. 1995, *ApJ*, **439**, 503
Donahue, M. J., & Shull J. M. 1987, *ApJ*, **323**, L13
Faber, S. M., Courteau, S. J. A., Yahil, A. 1993, in Proc. 9th *IAP Astrophysics Meeting, Cosmic Velocity Fields*, M. Lachièze–Rey & F. R. Bouchet (eds.) (Editions Frontières)
Fixsen, D. J., et al. 1996, *ApJ*, **470**, 63
Grainge, K., Jones, M.,Pooley, C., Saunders, K. 1993, *M.N.R.A.S.*, **265**, L57
Gunn, J. E., & Peterson, B. A. 1965, *ApJ*, **142**, 1633
Heanelt, M.G., & Tegmark, M. 1996, *M.N.R.A.S.*, **279**, 545
Henriksen, M. J., & Mushotzky, R. F. 1985, *ApJ*, **292**, 441
Herbig, T., Lawrence, C. R., Readhead, A. C. S., Gulkis, S. 1995, *ApJ*, **449**, L5
Holzapfel, W. L., Arnaud, M., Ade, P. A. R., Church, S. E., Fisher, M., Mauskopf, P. D., Rephaeli, Y., Wilbanks, T. M., Lange, A. E. 1996, *ApJ*, **480**, 449
Hu, W., Scott, D., Sugiyama, N., White, M. 1995, *Phys. Rev. D*, **52**, 5498
Hu, W. & Sugiyama, N. 1995a, *ApJ*, **444**, 489
Hu, W. & Sugiyama, N. 1995b, *Phys. Rev. D*, **51**, 2599
Hu, W., & White, M. 1996a, *ApJ*, **471**, 30
Hu, W., & White, M. 1996b, *ApJ*, **471**, 30
Hu, W., Scott, D., & Silk, J. 1994, *Phys. Rev. D*, **49**, 648
Ikeuchi, S. 1981, *Pub. Astr. Soc. Japan*, **33**, 211
Jakobsen, P., Boksenberg, A., Deharveng, J. M., Greenfield, P., Jedrzejewski, R., & Paresce, F. 1994, *Nature*, **370**, 35
Jones, C., & Forman, W. 1984, *ApJ*, **276**, 38
Jones, M., et al. 1993, *Nature*, **365**, 320
Jungman, G., Kamionkowski, M., Kosowsky, A., Spergel, D. N. 1996, *Phys. Rev. D*, **54**, 1332
Kaiser, N. 1986, *M.N.R.A.S.*, **222**, 323
Katz, N., & White, S. D. M. 1993, *ApJ*, **412**, 455
King, I. R. 1966, *Astr.J.*, **71**, 64

Klein, U., Rephaeli, Y., Schlickeiser, R., Weilebinski, R. 1991, *A&A*, 244, 43
Kolb, E.W., & Turner, M. S. 1990, in *The Early Universe*, Frontiers in Physics Series
Kompaneets, A. S. 1957, *Soviet Phys. JETP*, **4**, 730
Lachièze-Rey, M., & Gunzig, E. 1999, Cambrige University Press, in press
Lamarre, J.-M., et al. 1998, *ApJ Lett.*, **507**, 5
Lasenby, A. N. 1992, in *Clusters and Superclusters of Galaxies*, A. C. Fabian (ed.), Kluwer Academic Publishers
Lasenby, A. N. 1996, in *Cosmic Microwave Background*, Proceedings des Rencontres de Moriond
Loeb, A., McKee, C. F., Lahav, O. 1991, *ApJ*, **374**, 44
Madau, P., & Meiksin, A. 1994, *ApJ*, **433**, L53
Markevitch, M., Blumenthal, G. R., Forman, W., & Jones, C., and Sunyaev, R. A. 1992, *ApJ*, **395**, 326
Martinez-González, E., Sanz, J. L., & Silk, J. 1990, *ApJ*, **355**, L5
Mather, J. C., et al. 1994, *ApJ*, **420**, 439
Meiksin, A., & Madau. A. 1993, *ApJ*, **412**, 34
Meyer, S. S., Jeffries, A. D. & Weiss, R. 1983, *ApJ*, **271**, L1
Miralda-Escudé, J., & Ostriker, J. P. 1990, *ApJ*, **350**, 1
Miralda-Escudé, J., & Ostriker, J. P. 1992, *ApJ*, **392**, 15
Mushotzky, R. 1994, in *New horizon of X-ray astronomy, first results from ASCA*, Universal Academy press, Inc
Ostriker, J. P. & Vishniac, E. T. 1986, *ApJ*, **306**, L51
Ostriker, J. P., & Gnedin, N. Y. 1996, *ApJ*, **472**, 73
Padmanhaban, T. 1993, in *Structure formation in the Universe*, Cambridge University Press.
Pariskii, Yu. N. 1973, *Sov. Astron.*, **16**, 1048
Partridge, R. B., et al. 1986, *ApJ*, **308**, 46
Partridge, R. B. 1996, in *The Cosmic Microwave Background*, Proc. Strasbourg summer school on Cosmology
Peebles, P. J. E. 1980, in *The Large Scale Structure of the Universe*, Princeton University Press
Peebles, P. J. E. 1993, in *Principles of Physial Cosmology*, Princeton University Press
Penzias, A. A., & Wilson, R. W. 1965, *ApJ*, **142**, 419
Press, W. H., & Schechter, P. 1974, *ApJ*, **187**, 425
Radford, S. J. E., Boynton, P. E., Ulich, B. L., Partridge, R. B., Scommer, R. A., et al. 1986, *ApJ*, **300**, 159
Readhead, A. C. S., Lawrence, C. R., Myers, S. T., Sargent, W. L. W. 1989, *ApJ*, **346**, 566
Rees, M. J., & Sciama, D. W. 1968, *Nature*, **511**, 611
Rephaeli, Y. 1993, *ApJ*, **418**, 1
Rephaeli, Y. 1995, *ARA&A*, **33**, 541
Sachs, R.K., & Wolfe, A.M. 1967, *ApJ*, **147**, 73
Salati, P. 1992, in *Clusters and Superclusters of Galaxies*, A. C. Fabian (ed.), Kluwer Academic Publishers
Salati, P. 1996, in *Cosmic Microwave Background*, Proc. Strasbourg summer school on Cosmology
Sanz, J. L., Martínez-González, E., Cayón, L., Silk, J., Sugiyama, N. 1996, *ApJ*, **467**, 485
Sarazin, C. L. 1988, in *X-ray Emissions from Clusters of Galaxies*
Sarazin, C. L. 1992, in *Clusters and Superclusters of Galaxies*, A. C. Fabian (ed.), Kluwer Academic Publishers, Dordrecht
Sciama, D. W. 1982, *M.N.R.A.S.*, **200**, 13
Shapiro, P. R. & Giroux, M. L. 1987, *ApJ*, **321**, 107
Slejak, U. 1996a, *ApJ*, **436**, 509

Slejak, U. 1996b, *ApJ*, **460**, 549
Slejak, U. 1996c, *ApJ*, **463**, 1
Smoot, G., et al. 1992, *ApJ*, **371**, L1
Soucail, G. 1992, in in *Clusters and Superclusters of Galaxies*, A. C. Fabian (ed.), Kluwer Academic Publishers
Soucail, G., Fort, B., Mellier, Y., Picat, J. P. 1987, *A&A*, **172**, L14 A. C. Fabian (ed.), Kluwer Academic Publishers
Soucail, G., Fort, B., Mellier, Y., Picat, J. P. 1987, *A&A*, **172**, L14
Sunyaev R. A., & Zeldovich, Ya. B. 1980a, *M.N.R.A.S.* , **190**, 413
Sunyaev, R. A., & Zeldovich, Ya. B. 1972, *A&A*, **20**, 189
Sunyaev, R. A., & Zeldovich, Ya. B. 1980b, *ARA&A*, **18**, 537
Sutherland, W., & Dalton, G. 1994, in *Clusters of galaxies*, Proceedings des Rencontres de Moriond, F. Durret, A. Mazure, J. Trân Thanh Vân (eds.) (Editions Frontières)
Taylor, G. B., & Wright, E. L. 1989, *ApJ*, **339**, 619
Tegmark, M., Silk, J., & Blanchard, A. 1994, *ApJ*, **420**, 484
Uson, J. M. 1986, in *Radio Continuum Processes in Clusters of Galaxies*, Proceedings, Green Bank NRAO
Vishniac, E. T., & Ostriker, J. P. 1985, in Proc. 3rd *Rome Meeting on Astrophysics, The Cosmic Background Radiation and Fundamental Physics*, ed. F. Melchiorri (BolognaeItalian Physical Society)
Vishniac, E. T. 1987, *ApJ*, **322**, 597
Walker, T. P., Steigman, G., Schramm, D. N., Olive, K. A., Kang, H. S. 1991, *ApJ*, **376**, 51
Warren, S. J., Hewett, P. C., & Osmer, P. S. 1991, in *The Space Distribution of Quasars*, ed. D. Crampton (ASP Conf. Ser 21), 139
Webb,J. K., Barcons, X., Carswell, R. F., Parnell, H. C. 1992, *M.N.R.A. S.*, **255**, 319
West, M. J. 1994, in *Clusters of galaxies*, Proc. des Rencontres de Moriond, F. Durret, A. Mazure, J. Trân Thanh Vân (eds.) (Editions Frontières)
Wilbanks, T. M., Ade, P. A. R., Fisher, M. L., Holzapfel, W. L., Lange, A. E. 1994, *ApJ*, **427**, L75
Wright, E. L. 1979, *ApJ*, **232**, 348
Zel'dovich, Ya. B., & Sunyaev, R. A. 1969, *Ap and S.S.*, **4**, 301
Zwicky, F. 1933, *Helv. Phys. Acta*, **6**, 110

PRIMORDIAL GALAXIES: FROM LOCAL HII GALAXIES TO HIGH REDSHIFT GALAXIES

D. KUNTH
Institut d'Astrophysique de Paris
98 bis Boulevard Arago
75014 Paris, France

Abstract. Primordial galaxies at early epoch may be strong Lyα emitters. I review the results of recent studies of local HII galaxies on their Lyα emission that are relevant to the search of distant young galaxies. The interdependance of the Lyα escape with the properties of the ISM in starburst galaxies is outlined. A new model explains Lyα profiles in starburst galaxies from the hydrodynamics of superbubbles powered by massive stars. Since Lyα is primarely a diagnostic of the ISM, it is mandatory to understand how the ISM and Lyα are related to firmly relate Lyα to the cosmic star–formation rate. Lyα search of distant dust–free young galaxies is now a reachable possibility.

1. Introduction

The search for high-redshift galaxies has progressed rapidly over the last ten years by using color techniques which basically select galaxies with massive ongoing star formation and little extinction while more evolved objects that could be heavily dust-reddened are picked out at millimeter wavelengths (see [12] for references). It has been conjectured that primeval galaxies at their very early stage would be nearly dust-free hence easily detected from their Lyα emission ([19]; [21]). Assuming that protogalaxies would undergo a rapid collapse on a short dynamical timescale, [7], the expectation was that an L* galaxy, [25], could be detected from its redshifted Lyα emission line. Indeed in a dust–free case and a normal IMF such a galaxy could pop up with a star formation rate of nearly 1000 $M_{\odot}$/yr producing a Lyα luminosity of 10^{45} erg/sec (this amounts few percent of the total luminosity). At a redshift of 3 this translates to 10^{-15} erg/cm^2/sec, in

M. Lachièze-Rey (ed.), Theoretical and Observational Cosmology, 353–364.

principle easily detectable even from deep spectrocopic searches or narrow band imaging using panoramic CCDs. The failure to detect such a strong Lyα is an additional piece of evidence that the overall process of galaxy formation is more complex: galaxies form from smaller building blocks and/or luminous protogalaxies have to be found at much larger redshift and/or are enshrouded in large dusty cocoons.

Indeed our recent view of the distant Universe opens the possibility that small gaseous systems gradually merge into large ones. Galaxy formation appears more gradual that had been anticipated and one would expect to see some left over debris in our local Universe after a priviledged epoch of star formation at around z of 1-2 .

2. The HII galaxies

The major motivation for follow-up studies of unevolved and nearly dust-free HII galaxies has been their possible ressemblance with distant protogalaxies in their early phase of star-formation.

Local star-forming galaxies are often called HII galaxies because their ongoing star-formation rate is large with respect to their integrated past star formation rate so that their overall spectroscopic appearance is that of a Giant HII region. HII galaxies have absolute luminosities dominated by the young stellar population and range from M_B=-13 to -20 at the brighter end. They are bluer than ordinary galaxies (with typically U-B=-0.6 and B-V=0.0) and are richer in ionized gas. The equivalent width of the Hβ line ranges from 30 to 200 Å implying a burst star formation no older than a few Myrs. The role of massive stars is to energize the ISM and enrich it with heavy metals and some helium. These galaxies are on average chemically unevolved and their metal distribution peaks at 1/10th the solar value, with the most deficient galaxy IZw18 at only 1/50th. Their neutral gas content is relatively high, amounting a few 10^7 $M_\odot$ in general, *i.e.* a large fraction of the luminous mass. On the other hand there are some indications that the dynamical mass is much larger [2]. UV to IR studies clearly indicate that massive star formation is a discontinuous process concentrated to short bursts intervened by longer passive periods. Their UV spectra show small amount of dust that makes them ideal targets to study locally.

3. The Early IUE observations and interpretations

While galaxies at their very early stage could be nearly dust-free hence easily detectable from their Lyα emission [21]; early ultraviolet observations of nearby HII galaxies, have revealed a much weaker Lyα emission than predicted by starburst models and simple models of galaxy formation. In some other galaxies, Lyα was non-existent or even appeared as a broad absorption profile ([5, 10, 11, 15, 16, 20, 28]). For young star-forming galaxies without so much dust as to suppress Lyα , large equivalent widths are expected in the range of 100-200 (1+z) Å , [3]. However it was early realized that pure extinction by dust would be unable to explain the low observed Lyα/Hβ (although Calzetti and Kinney [1] tentatively proposed that proper extinction laws would correctly match the predicted recombination value). Valls-Gabaud [30] on the other hand suggested that ageing starbursts could reduce Lyα equivalent widths because they are affected by strong underlying stellar atmospheric absorptions. Early IUE data have provided evidence for an anticorrelation between the Lyα/Hβ ratio and the HII galaxy metallicity. The key explanation of all these effects stems from realizing that Lyα is a resonant line. Hence effect of resonant scattering of Lyα photons into a gas-rich medium dramatically enhances their mean free path therefore the probability that they could be absorbed by dust, even in minute amounts ([3, 4, 9, 20] and references therein).

Finally [3] advocated that the structure of the interstellar medium (porosity and multi-phase structure) is most probably an important factor for the visibility of the Lyα line. Other factors such as the geometry, the galaxy morphology, the strength of the starburst, etc... may play a role.

4. HST observations

New observations performed with the HST have outlined the key role of the velocity structure of the interstellar medium to the transfer and escape of Lyα photons. Kunth et al. [14] and Lequeux et al. [18] have used the Goddard High Resolution Spectrograph (GHRS) onboard the Hubble Space Telescope (HST) to observe Lyα and the interstellar lines O I 1302.2 Å and Si II 1304.4 Å . The O I 1302 Å and Si II 1304 Å allow to measure with reasonable accuracy the mean velocity at which the absorbing material lies with respect to the star-forming region of a given galaxy. Surprisingly, Lyα was observed only in absorption in IZw 18, the most metal–poor starburst galaxy known, [14]. Meanwhile to add to the confusion, a Lyα emission line showing a clear P–Cygni component, has been detected in the star-forming galaxy Haro 2, dustier than IZw 18 and with $Z = 1/3\ Z_{\odot}$, [18]. The detection of such a profile in the Lyα emission line led us to postulate that the line was visible because the absorbing neutral gas was velocity–shifted with respect to the ionized gas. This was confirmed by the analysis of the UV O I and Si II absorption lines (blue–shifted by 200 km s^{-1} with respect to the optical emission lines) and that of the profile of the Hα line [17]. Observations were subsequently made on additional galaxies by Kunth et al. [16] while Thuan and Isotov [29] have obtained GHRS spectra of two more starburst galaxies, namely Tol65 and T1214-277. Tol65 reveals a

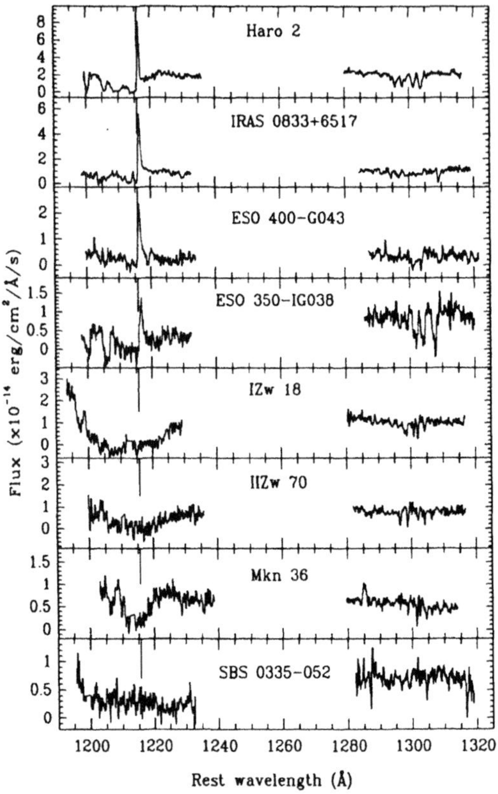

Figure 1. GHRS spectra of all the galaxies in [16]. The spectra have been shifted to rest velocity assuming the redshift derived from optical emission lines. Vertical bars indicate the wavelength at which the Lyα emission line should be located. The geocoronal emission profile has been truncated for the sake of clarity. The spectra have been plotted after rebinning to 0.1 Å per pixel and smoothed by a 3 pixel box filter.

broad damped Lyα absorption while T1214-277 shows a pure Lyα emission profile with an equivalent width of 70 Å and with no blue absorption. The individual spectra of galaxies observed in [16] are shown in Fig. 1.

5. The role of the velocity structure of the ISM

Three types of observed lines have been identified so far: pure Lyα emission; broad damped Lyα absorption centered at the wavelength corresponding to the redshift of the HII emitting gas and Lyα emission with blue shifted absorption features, leading in some cases to P–Cygni profiles.

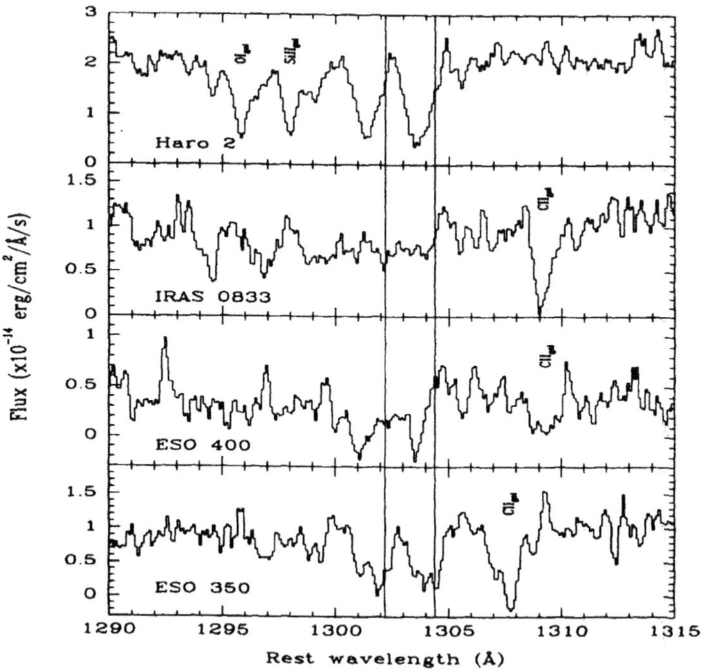

Figure 2. Detail of the O I and Si II region for the galaxies with Lyα emission. The vertical bars indicate the wavelength at which the O I and Si II absorption lines should be located, according to the redshift derived from optical emission lines. Some Galactic absorption lines have been marked. Note that the metallic lines appear systematically blueshifted in these galaxies with respect to the systemic velocity. In some cases there is no significant absorption at all at zero velocity.

Lyα emission with deep blueward absorption troughs evidence a wide velocity field. The equivalent widths of the Lyα emission range between 10 and 37 Å hence much below the value predicted by Charlot and Fall, [3] for a dust–free starburst model. In all cases, interstellar absorption lines (OI, SiII) are significantly blueshifted with respect to the HII gas (see Fig. 2). On the other hand, if the HI is static with respect to HII, the destroyed Lyα photons are those emitted by the HII region and the interstellar lines are not displaced (see Fig. 3).

In galaxies with low dust content - IZw 18 has a dust-to-gas ratio at least 50 times smaller than the Galactic value, [14] -, it remains possible to weaken observationally Lyα by simple multiple resonant scattering from a static neutral gas and produce an absorption feature. The H I cloud surrounding these galaxies might be leaking Lyα photons through its external

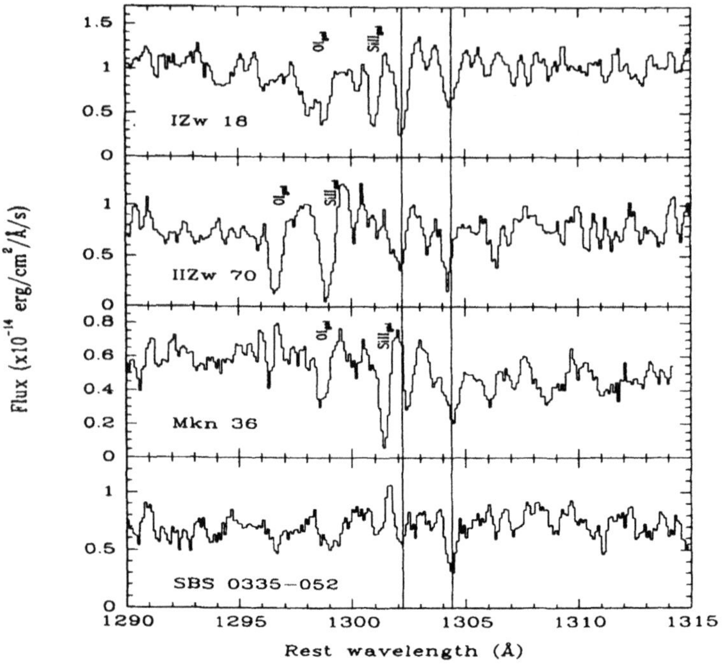

Figure 3. O I and Si II region for the galaxies showing damped Lyα absorptions. Details as in Fig. 2. Note that in these galaxies the metallic lines are essentially at the same redshift as the ionized gas, indicating the presence of static clouds of neutral gas, as discussed in the text.

surface. The Lyα line would then become very hard to detect because of its low surface brightness. On the other hand, the situation in Haro 2 is quite different. The Lyα P Cygni profile suggests a rather modest amount of neutral gas of the order of N (H I) = 7.7x10^{19} atoms cm^{-2}. The crucial point here is that the neutral gas responsible for the absorption is not at the velocity at which the Lyα photons were emitted but is being pushed out by an expanding envelope around the H II region, at velocities close to 200 km s^{-1}. This interpretation is strengthened by the presence of the blueshifted absorptions of O I, Si II and Si III precisely due to outflowing gas in front of the ionizing hot stars, (see also Legrand et al., [17]).

Data on other H II galaxies with detected Lyα emission confirm that Haro 2 is not an isolated case. Most spectra show Lyα emission with a broad absorption on their blue side except for ESO 350-IG038 in which the emission is seen atop of a broad structure requiring several filaments. When metallic lines are detected, they are always blueshifted with respect to the ionized gas, further supporting the interpretation. In the case of ESO 350-IG038 the velocity structure seems to be more complicated and several components at different velocities are identified on the metallic lines. Note that there might be a secondary peak emission in the blue side of the main line in the spectrum of ESO-400-G043.

The main conclusion drawn from this set of data is that complex velocity structures are determining the Lyα emission line detectability, showing the strong energetic impact of the star-forming regions onto their surrounding ISM. We want to stress that this effect seems to be almost independent of the dust and metal abundance of the gas.

Thuan and Isotov [29] have detected strong Lyα emission in T1214-277, with no clear evidence of blueshifted Lyα absorption. They argue that a significant fraction of the area covered by the slit along the line of sight is essentially free from neutral gas, suggesting a patchy or filamentary structure of the neutral clouds. Such a geometry would be unlikely in galaxies surrounded by enormous H I clouds, as in IZw 18 and similar objects.

6. The evolution of superbubbles in extended HI halos

To account naturally for the variety of Lyα line detections in star-forming galaxies, Tenorio-Tagle et al. [27] have proposed a scenario based on the hydrodynamics of superbubbles powered by massive starbursts. This scenario is visually depicted in Fig. 4. The overpowering mechanical luminosity ($E_0 \geq 10^{41}$erg s^{-1}) from massive starbursts ($M_{stars} = 10^5$ - 10^6 M$_\odot$) is known to lead to a rapidly evolving superbubble able to blowout the gaseous central disk configuration into their extended HI haloes.

As shown by [13], the superbubble will blowout and thus massive starbursts will lead to superbubble blowout phenomena even in massive galaxies such as the Milky Way. One can then predict the venting of the hot superbubble interior gas through the Rayleigh-Taylor fragmented shell, into the extended HI halo where it would push once again the outer shock, quite early in the evolution of the starburst ($T_{blowout} \leq$ 2Myr), allowing to build a new shell of swept up halo matter (see Fig. 4b). The blowout or the fragmentation of the shell allows the expansion of the hot interior gas into the extended low density HI halo, and the leakage of the ultraviolet photons emitted by the starburst. These photons establish an ionized conical HII region with its apex at the starburst. The density of the halo steadily decreases with radius, and one can show that the conical sector of the HII region will extend all the way to the galaxy outer edge, i.e. several kpc. Furthermore, the low halo density implies a long recombination time ($t_{rec} = 1/(\beta n_{halo}) \geq 10^7$ yr), with the implication that the ionized conical sector of the galaxy halo becomes, and remains, transparent to the ionizing flux pro-

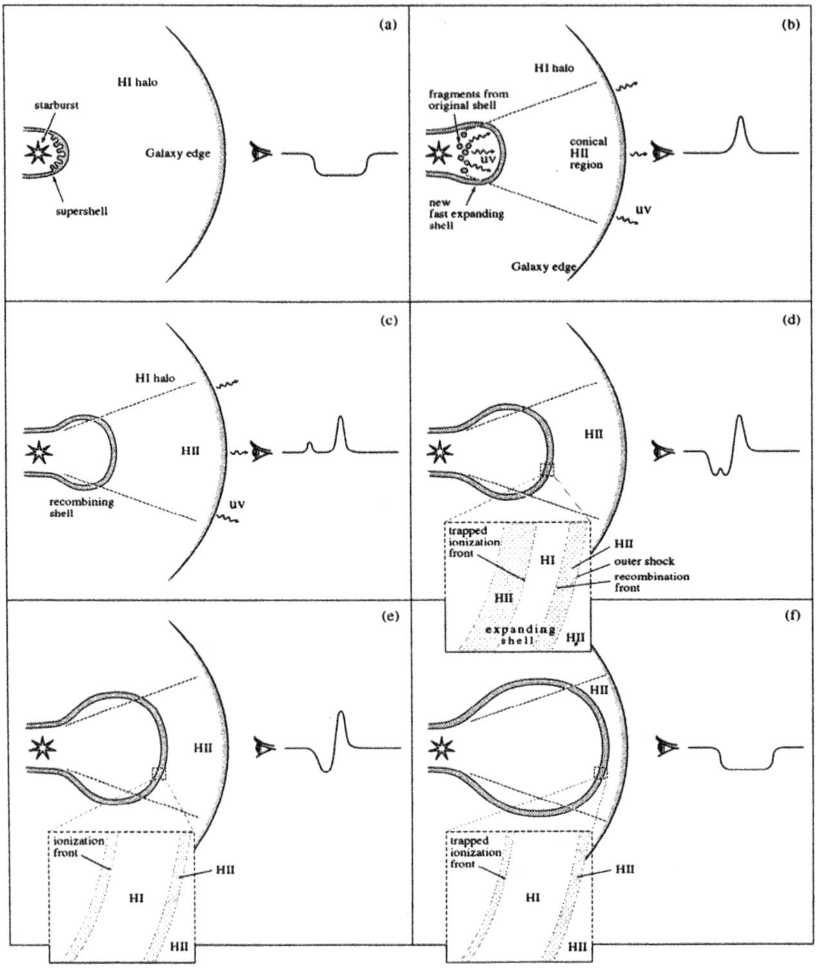

Figure 4. The evolution of a superbubble as a function of time.

duced during the ensuing lifetime of the ionizing massive stars ($\leq$ 10 Myr). Then clearly, the Lyα photons produced at the central HII region will also be able to travel freely in such directions. The escape of UV photons from the galaxy would be particularly important during the early stages after blowout and until the new shell of swept up halo matter condenses enough material as to allow its recombination. If the shock progresses with speeds of a few hundred (≤ 400) km s^{-1}, (i.e. with a Mach number $M \leq 40$) it will promote the rapid cooling of the shocked gas that will cool down to the HII region temperature making the shock isothermal, and thus causing compression factors of several hundreds. Compression leads to recombination in time-scales ($t_{rec} = 1/\beta n_{shock}$) of less than 10^5yr, and this immediately and steadily will reduce the number of stellar ultraviolet photons leaking out of the galaxy. At the same time recombination in the fast expanding shell will lead to a correspondingly blueshifted Lyα emission, as depicted in Fig. 4c. Once the shell presents a large column density ($\sim 10^{19}$ atoms cm^{-2}), as it grows to dimensions of a few kpc, it will trap the ionization front. Note that from then onwards, recombinations in the shell will inhibit the further escape of ionizing photons from the galaxy (compare Fig. 4b, c, and d).

The trapping of the ionization front makes the shell acquire a multiple structure with a photoionized inner edge, a steadily growing central zone of HI, and an outer ionized sector where the recently shocked ionized halo gas is steadily incorporated. The growth of the central layer eventually will cause sufficient scattering and absorption of the Lyα photons emitted by the central HII region, leading to a blueshifted Lyα absorption. As long as recombinations continue to occur at the leading edge of the shell, a blueshifted Lyα in emission will appear superposed to the blueshifted absorption feature (see Fig. 4d). Recombination at the leading edge will become steadily less frequent, depleting the blueshifted Lyα in emission. This is due when the shell and its leading shock move into the outer less dense regions of the halo, and the shell recombination time, despite the compression at the shock, becomes larger than the dynamical time. At this stage, an observer looking along the conical sector of the HII region will detect a P-Cygni-like Lyα line profile as shown in Fig. 4e.

Geometrical dilution of the ultraviolet flux will begin to make an impact as the superbubble grows large. This and the drop in the UV photon production rate, caused by the death of the most massive stars after $t = t_{ms}$, will enhance the column density of neutral material in the central zone of the recombined shell to eventually cause the full saturated absorption of the Lyα line (see Fig. 4f). Full saturated absorption has usually been accounted for by the large column density of the extended HI envelope of these galaxies and thus, as in all models, many different orientations will match the observations.

7. Discussion

7.1. THE RELEVANCE OF THE SUPERBUBBLE MODEL

The main implication of the evolution depicted in Fig. 4 is that it is the feedback from the massive stars that leads to the large variety of Lyα emission profiles. P-Cygni Lyα profiles are predicted when observing along the angle subtended by the conical HII region but only once the ionization front is trapped by the sector of the superbubble shell evolving into the extended halo. This will produce the fast moving layer of HI at the superbubble shell, here thought to be responsible for the partial absorption observed in sources such as Haro 2, ESO 400-G043 (which probably exhibits a secondary blueshifted Lyα emission) and ESO 350-IG038 [16]. Damped Lyα absorption is seen in several galaxies. We note that these objects are all gas–rich dwarf galaxies whereas in most cases but Haro 2, the HII galaxies that exhibit Lyα in emission or with a P-Cygni profile, are on the higher luminosity side of the distribution ($M \leq -18$). Pure Lyα emission is observed in C0840+1201 and T1247-232 ([28]; IUE) or T1214-277 ([29]; HST). Such a line implies no absorption and thus no HI gas between the starburst HII region and the observer, as when observing the central HII region after the superbubble blowout, within the conical HII region carved in the extended HI halo. It is not a straigthforward issue to estimate what is the fraction of Lyman continuum radiation that leaks out from galaxies. This scenario however predicts a short but significant evolutionary phase (between blowout and the trapping of the ionization front by the fast expanding shell) during which a large amount of UV radiation could leak out of a galaxy into the intergalactic medium. Detailed numerical calculations of the scenario proposed here are currently underway. These results and further implications of the model will be reported in a forthcoming communication.

7.2. THE GALAXIES AT HIGH-REDSHIFT

The effect of neutral gas flows helps to understand why luminous high-redshift objects have only been found up to now with linewidths larger than 1000 km s^{-1}. High–redshift galaxies with very strong equivalent widths (EWs > 500 Å) extended Lyα emission are characterized by strong velocity shears and turbulence (v > 1000 km s^{-1}); this suggests an AGN activity, in the sense that other ISM energising mechanism than photoionization by young stars may be operating. On the other hand Steidel et al. [26] have recently discovered a substantial population of star–forming galaxies at $3.0<z<3.5$ that were selected but from the presence of a very blue far UV continuum and a break below 912 Å in the rest frame. These Lyman–break

galaxies (LBG) are much more luminous but similar to our local starbursts in the sense that 50% show no Lyα emission whereas the rest does, but with weak EWs no larger than 20 Å at rest. The Lyα profiles of this population looks also very similar to those of our local starburst galaxies ([8, 23]; see also [6] for the z=5.34 galaxy).
New Lyα emitters are now found at high-redshift from surveys using large telescopes with narrow-band filters ([12, 22]). Limits down to a few 10^{-18} erg/cm^2/sec are now reachable and give access, in principle, to galaxies with fainter continuum magnitudes than the LBG. It is not clear whether these Lyman–alpha–galaxies (LAG) represent the earliest stages of the galaxy formation process as suggested by Hu et al. [12]. One may think that they are less dusty than the LBGs although Lyman-break selected galaxies should not be much more obscured than the LAGs.
Is there such an intrinsically fainter population of dust-free star–forming galaxies, possibly connected to the bright-end of the star-forming dwarfs population we see today? What are their masses and spatial distribution (see also [22] for a recent HST study)? Near-IR spectroscopic studies with large ground based telescopes will in the future help to settle some of these questions [24].
The possible increase of the star formation rates as one moves at higher redshift (between 3 and 6) reported by Hu et al. [12] must be taken with caution since from the preceeding discussion, it is clear that a significant, yet unknown fraction of the youngest galaxies may not be Lyα emitters. Indeed, since Lyα is primarely a diagnostic of the ISM, it is only when we fully understand how the ISM and Lyα are related can we hope to firmly relate Lyα to the SFR.

References

1. Calzetti D. & Kinney A.L. 1992, ApJ 399, L39
2. Carignan C. & Beaulieu S. 1993 ApJ 347, 760
3. Charlot S. & Fall S.M. 1993, ApJ 415, 580
4. Chen W.L. & Neufeld N.A. 1994, ApJ 432, 567
5. Deharveng J.M., Joubert M. , Kunth D. 1986, in the First IAP workshop: "Star-Forming Dwarf Galaxies and related objects", edited by. D. Kunth, T.X. Thuan and J. Tran Thanh Van, Editions Frontières, p.431
6. Dey A., Spinrad H., Stern D., Graham J., Chaffee F.H. 1998, ApJ 498, L93
7. Eggen O.J., Lynden–Bell D. & Sandage A. 1962, ApJ 136, 748
8. Franx M., Illingworth G.D., Kelson D.D., van Dokkum P.G., Kim-Vy T. 1997,ApJ 486, L75
9. Giavalisco M., Koratkar A. & Calzetti D., 1996, ApJ 466, 831
10. Hartmann L.W., Huchra J.P. , Geller M.J., 1984, ApJ 287, 487
11. Hartmann L.W., Huchra J.P., Geller M.J., O'Brien P. & Wilson R., 1988, ApJ 326, 101
12. Hu E.M., Cowie L.L. & McMahon R. G. 1998,AJ in press, astro-ph/9803011

13. Koo, B-C. & McKee, C. F.: 1992, ApJ 388, 93
14. Kunth D., Lequeux J., Sargent W.L.W. , Viallefond F. 1994, A & A 282, 709
15. Kunth D., Lequeux J., Mas-Hesse J.M., Terlevich E. & Terlevich R. 1997, Rev. Mex. Astr. Astrofis. 6, 61
16. Kunth D., Mas-Hesse J.M., Terlevich E., Terlevich R., Lequeux J. & Fall M. 1998, A & A 334, 11
17. Legrand F., Kunth D., Mas–Hesse J.M. , Lequeux J. 1997 A & A 326, 929
18. Lequeux J., Kunth D., Mas–Hesse J.M. & Sargent W.L.W. 1995, A & A 301, 18
19. Meier D.L. 1976, ApJ 207, 343
20. Meier D.L. & Terlevich R. 1981, ApJ 246, L10
21. Partridge R. & Peebles P.J.E. 1967, ApJ 146, 868
22. Pascarelle S.M., Windhorst R.A., Keel W.C.,1998, astro-ph 9809181
23. Pettini M., Steidel C.C, Adelberger K.L. et al. 1997, to appear in 'ORIGINS', ed. J.M. Shull, C.E. Woodward, and H. Thronson, (ASP Conference Series)
24. Pettini M., et al., 1998, astro-ph 9806219
25. Schechter P. S., 1976, ApJ 203, 297
26. Steidel C.C., Giavalisco M., Pettini M., Dickinson M. & Adelberger K. 1996, ApJ 462, 17
27. Tenorio-Tagle G., Silich S.A., Kunth D., Terlevich E., & Terlevich R., 1999, in preparation
28. Terlevich E., Diaz A.I.,Terlevich R. & Garcia-Vargas M.L. 1993, MNRAS 260, 3
29. Thuan T.X. & Izotov Y.I. 1997, ApJ 489, 623
30. Valls-Gabaud D. 1993, ApJ 419, 7

ACTIVE GALACTIC NUCLEI

GUY MATHEZ
Laboratoire d'Astrophysique, Observatoire Midi-Pyrénées

1. Short Historic, Basic Properties

1.1. ACTIVE GALACTIC NUCLEI

In what is defined as Active Galactic Nuclei, the stellar contribution to the emitted light is generally less than (or exceptionally equal to) half the total emitted power. Even if a spectrum of an active galaxy was taken as early as 1908, Seyfert was the first astronomer to understand that *active galaxies* do form a class. Seyfert's original work (1943) listed 12 galaxies, defined at that time as *normal galaxies with high surface brightness nuclei.* Spectroscopy then revealed most characteristic properties of Seyfert galaxies, the standard of AGN: a *continuum with both a UV and an IR excess* (with respect to usual galaxy continua), together with numerous broad emission lines with high degree of ionization and excitation. Other characteristics of AGN are the following: they have in average large redshifts, some are radio loud, almost all are X-ray emitters, a majority seems to emit in the high-energy range, γ-ray and TeV bands, and many of them present time variability. The discrimination between active galaxies and quasars is mostly historical, due to the very different morphologies of the first Seyferts and quasars. Furthermore, the status of AGN remained unclear for a long time, due to the mixture of morphological and spectroscopic criteria used in many of their first classifications. Since that time, and essentially based on the similarity of their spectra and of most of their properties which vary continuously from the low luminosity less active galaxies to the most luminous quasars, there has been progressive understanding of the fact that the very same physical processes are at work in both AGN and quasars. However the connection is not yet completely understood. In particular it is not yet established whether or not there is either some evolutive sequence from active galaxies to quasars, or gradual differences in the degree of activity.

M. Lachièze-Rey (ed.), Theoretical and Observational Cosmology, 365–379.

1.2. SEYFERT GALAXIES

Seyfert Galaxies are now defined as a galaxy clearly visible, with an active nucleus (non-stellar continuum plus emission lines). A criterion is $[OIII]/H_\beta > 3$. There are 2 classes, according to their line spectra. The Seyfert 1 galaxies show two systems of lines. The first one consists of *broad lines*, with typical widths $\simeq 10^3 - 10^4\ km/s$, and it concerns *permitted lines* only. In the second set of *narrow* lines there are both permitted and forbidden lines, with typical widths characterized by velocities around $\simeq 10^2 - 10^3\ km/s$. (Forbidden lines indicate low density plasma: $n_e \simeq 10^2 - 10^5\ cm^{-3}$). Seyfert galaxies often show stellar absorption lines. Intermediate spectra do exist, such as Seyfert 1.1 to 1.9. Recent HST imaging of 256 nearby Seyfert galaxies [24] leads to the following conclusions: little evidence for recent interaction ($<$ 10% galaxies with tidal features or multiple nuclei), most Seyfert galaxies are spirals ($\simeq$ 5% ellipticals), with the same rate of occurrence of barrs in Seyferts 1, Seyferts 2 and non-Seyferts. Seyfert 2 nuclei show more absorption and dust, and Seyfert 1 nuclei reside in significantly earlier type galaxies. Furthermore, significant differences have been evidenced in the stellar populations of the nuclear region of AGN following the type of activity. *Note that some of this results are difficult to reconcile with the unified models (see below).*

1.3. QUASARS

A stellar contribution around 1/100 of the total emission is typical of quasars. In the late 50's, a great work was done for the identification of optical counterparts of the 3CR catalog (9 Jy at 178 MHz). In some cases this led to the so-called *radio stars*, characterized by unusual spectra. The latter were understood in the early 60's as highly redshifted broad emission lines coming from 'quasi stellar radio sources'. A large redshift implies a large distance and a high luminosity. Quasars are no more required to be radio emitters. Their spectra are roughly identical to Seyfert 1 nuclei spectra except that narrow lines are in general weaker. Quasars are more luminous than Seyferts, the limit being around $M = -21.5$ ($H_0 = 100$). As the prominence of the non stellar component increases with the nucleus luminosity, the host galaxies are in general not seen in quasar images obtained from ground based telescopes. HST imagery has revealed the host galaxy of quasars with redshift up to $z \simeq 0.5$. Extreme care is required for statistical studies, as in most quasar catalogs a stellar shape was required for quasars. Recently, it has been shown by assembling quasar samples without this requirement, that samples with the 'quasi-stellar' selection criterium are incomplete by a large factor, up to 5, at bright magnitudes and/or low redshift.

1.4. THE AGN ZOO

Many acronyms have been or are still used to designate various sub-classes of AGN. **ULIRGs**, Ultra Luminous IRAS Galaxies, are strong IR emitters ($\simeq 10^{12}\ L_\odot$ at $\lambda > 10\mu m$). **ELRGs**, Emission Lines Radio Galaxies, are some of the brightest radio sources associated with quasars and active galaxies. Their spectra and most properties are analogous, the main difference is that their host are generally elliptical galaxies. The **BL Lac** are AGN characterized by the absence of strong emission and absorption lines. They are generally Violent Variable. **OVVs** are Optically Violent Variables, with typical variation rates in optical $\Delta m \simeq 0.1\ mag$ on timescales $\Delta t \simeq 1$ day. They exhibit polarization up to a few percents. The **Blazars** are a class consisting of the reunion of Bl Lac's and OVV's. In the 'unified model', they are thought to be seen along the axis of the accretion disk. **LINERs**, Low Ionization Nuclear Emission-line Regions [19], are quite common. They have spectra similar to Seyfert 2, but where low ionization elements have stronger lines: $0.3 < [OIII]5007/H_\beta < 3$ and $0.6 < [NII]6583/H_\alpha 6563 < 4$. **LLAGNs** (Low Luminosity AGN), **Ultra Soft Narrow Line Seyfert 1 Galaxies, N galaxies** (Nuclear galaxies were selected by Morgan (1958) from their nearly stellar shape on photographic plates), **Blue Compact galaxies** (first selected by Zwicky, many of which are luminous intermediate between quasars and less active galaxies), and the **Markarian galaxies** (or Galaxies with Ultraviolet Continuum which have a moderate to strong ultraviolet continuum), are other residents of the AGN zoo.

To conclude this short Introduction, let us point out two good reviews which may be found on AGN in the libraries: [29,43].

2. Physical Properties

2.1. THE CONTINUUM SPECTRUM FROM RADIO WAVELENGTHS TO IR, UV, X, γ AND TEV

As already mentioned the AGN continuum is non stellar. Roughly speaking, the product νF_ν is constant from 100 μm to 1000 Å, and ¿from 0.1 to 10 keV. There are two main questions concerning the typical spectrum from cm to TeV: 1. which fraction is *thermal* (i.e. due to particles with a Maxwell-Boltzmann kinetic energy distribution), or which fraction may be interpreted in terms of *starburst* (see the contribution of D. Kunth) and: 2. which fraction is *primary* (i.e. due to synchrotron emitting particles coming directly from the central engine) as opposed to secondary emission from a gas heated and ionized by primary emission. The product νF_ν is a factor 10^2 (respectively 10^5) lower in the radio range (cm to dm) for 'radio-loud' (respectively 'radio-quiet') quasars. Radio loud QSOs are typically 10-30 times less common that optically selected QSOs. The emission mechanism is believed to be synchrotron emission, i.e. caused by electrons (positrons) spiraling along magnetic fields. The spectrum is observed as a power law, whose spectral index $\alpha \simeq 0.5$ ($f_\nu \propto \nu^{-\alpha}$) is related to the energy spectrum of the relativistic particles $N(E) \propto E^{-s}$; $\alpha = (s-1)/2$. AGN radio maps generally show extended double lobes plus a central compact (unresolved) source. Double lobes generally have steeper spectra, compact sources flatter ones. Jets are commonly observed around compact sources. As revealed by VLBI data, the radio-structures observed in compact sources and/or jets are often variable in morphology, and the apparent superluminal motions are now understood as caused by relativistic expansions with high Lorentz factor γ. Towards higher energies (far UV, keV, MeV, TeV) the continuum is essentially a power-law without prominent structures. Its origin is most likely Inverse Compton emission due to relativistic particles acting onto lower energy photons. The latter are likely to originate from the hot accretion disk thermal emission, a part of which is observed directly, while another one is *reflected* by the optically thick material in or around the disc, giving then rise to the 6.4 keV fluorescent Fe K line. A part of the high energy emission may also arise in a cascade of Inverse Compton emissions from lower energy primary radiation.

Emission Lines The spectrum of quasars is characterized by *broad (up to 5-10000 km/s) permitted emission lines* with high degree of excitation and/or ionization: main permitted lines are $H_\alpha\lambda6562$, $H_\beta\lambda4861$, $H_\gamma + [OIII]\lambda4363$, $Ly_\alpha\lambda1216$, $CIV\lambda1549$, $MgII\lambda2798$. Studies of line ratios show that ions are photoionized in the BLR *Broad Line Region* where electron density is as high as $10^9\ cm^{-3}$ and temperature is around $10^4\ K$. The dimensions of the BLR region are of the order of 1 to 10 parsecs. The main forbidden lines are $[OII]\lambda$ 3727 , $[OIII]\lambda$ 5007, [Ne III] and [Ne V] lines and the semi-forbidden $CIII]\lambda1909$. The rate of *radiative spontaneous decays* is equal to the Einstein coefficient A_{ij} (expressed in s^{-1}). Forbidden lines are characterized by very long lifetimes, i.e. by very large A_{ij}. The rate of *collisional* decays is proportional to the electron density, and such decays occur *without* emission of photon. A condition for observing forbidden lines is thus that the electron density must be less than a given limit. the dimensions of the NLR (Narrow Line Region) are of the order of 10^2 to 10^3 parsecs.

2.2. ABSORPTION LINES

Normal stellar absorption lines may be visible in LLAGs, but are less visible in quasars. Broad absorption line (BAL) systems are observed in 5-10 % of the QSOs. Asymmetric emission-absorption (P CYGNI) profiles are

relatively common in quasars. All quasars above redshift $z \simeq 2$ show absorption lines which are due to intervening material. Thanks to their high luminosities allowing high Signal to Noise Ratios, very low column densities are detectable. Roughly speaking, there are 2 classes of absorbers. The first one is believed to reside in the quasar vicinity, with $z_{abs} \simeq z_{em}$ systems, corresponding in general to clouds around the quasar. The other class of absorbers has significantly lower redshifts, they may be either *metal* or *Ly α absorbers* constituting the so-called *Ly α forests*. Strongest absorption in the forest constitute *damped Ly α* systems.
Light Variations A majority of AGN show 0.3-0.5 magnitude variations on timescales of a few months, some do on a few days, indicating that much of their radiation comes from a region less than 1 light-day in size. These variations concern continua and lines as well, with timescales increasing along the sequence: continuum, lines. In Seyfert 1, broad line variation timescales are shorter than narrow line ones. In a sample of Seyfert 1 recently observed during more than 8 years, a combination of measurements of lags (sizes R) and widths (velocities V) allows one to estimate virial AGN masses ($M \propto RV^2/G$) in the range $10^7 - 10^8\ M_\odot$ [28]. In Seyfert 1, there is a marked trend to increasing variability with decreasing luminosity. However LINERS and LLAGS show little or no significant variability [30].

2.3. MODELS

From observed variations, the size of the central emitting region may be constrained in between 1 light-day and 1 light-year. Energy flows are observed up to 10^{48} ergs/s. 10^{44} ergs/s is typical for quasars, 10^{42} ergs/s for Seyferts 1. 10^{48} ergs/s would represent the conversion of more than $3 \times 10^3\ M_\odot/yr$ if it was radiated ¿from stars (efficiency: $\eta_* < 0.01$). Furthermore, there is about 1 AGN per 100 spirals in the nearby universe, so even if they are short-lived, they must radiate during at least 1/100 of a Hubble time in total. The presence of a hot accretion disk surrounding a MBH is the most generally admitted model to understand how as much energy can flow out for such a time and for such a small region. Furthermore, MBH are being found in most normal galaxies, (see e.g. [40]): firstly, HST observations of surface brightness profiles of early-type galaxies show cusps, either 'core' profiles or 'power-law' profiles. Secondly, spectroscopic data in nuclei of normal galaxies confirm the presence of a highly condensed object. There seems to be a proportionality between the BH mass and the bulge mass.

2.4. TYPICAL RELATIONS

The BLR is spatially resolved and shows variations on timescales of the order of months to years: $0.1\, light - year < R_{BLR} < 10\, light - years$. The observed line widths show typical velocities up to $V \simeq 10^4\ km/s$ in the BLR. ¿From the virial theorem we get an estimate of the mass enclosed at the level of the BLR: $4 \times 10^8\ M_\odot < M < 4 \times 10^{10}\ M_\odot$. Such masses are typical of a small to normal galaxy.
Consider the accretion disk surrounding a MBH of mass M. Let $\dot{M} = dM/dt$ be the mass accretion rate at a distance $R = \alpha R_S$, R_S being the Schwartzschild radius. Let η be the global efficiency of the conversion of gravitational energy to radiated energy, $L = \eta \dot{M} c^2$ the bolometric luminosity. L is also equal to the loss of potential energy: $\dot{U} \simeq GM\dot{M}/R$. This implies $R = GM/\eta c^2 = \alpha R_S = \alpha \times 2GM/c^2$ whence $\alpha = 1/2\eta$. $\eta = 0.1$ is typical, leading to $\alpha \simeq 5$. From the light variations we induce: $R \simeq$ 1-30 light-days, and from the expression of the Schwartzschild radius we derive: $M \simeq 2\times10^9 - 6\times10^{10} M_\odot$. On the other hand if $\eta \simeq 0.1$ and $L_{bol} \simeq 10^{44} ergs/s$, then $\dot{M} \simeq 0.02 M_\odot/yr$.

2.5. EDDINGTON LIMIT

The photons escaping the central engine undergo Thompson scattering by crossing the emitting plasma in the BLR. This is accompanied by a transfer of momentum from radiation to matter, and translates into an outwards radiative force. The Eddington condition is obtained by writing the balance between the gravitational and radiative forces in terms of the central mass and the bolometric luminosity. The momentum of a photon of energy E is $p = h\nu/c = E/c$. To the (ISOTROPIC!) flux of radiative energy, $L/4\pi R^2$, corresponds a flux of momentum: $L/4\pi R^2 c$, and the radiative pressure is obtained after multiplication by the Thompson cross section σ_T, which corresponds to the scattering of a photon by an electron. Due to electric neutrality, the radiative force outwards is obtained by introducing the proton mass m_p: $\sigma_T L/4\pi R^2 c = GMm_p/R^2$. This expresses the Eddington bolometric luminosity, which is the maximum luminosity of an AGN, in terms of its mass: $L_{edd} \simeq 10^{44}\ ergs/s \times M/10^6\ M_\odot$.

2.6. UNIFIED MODEL

Recent models assume that the accretion disk is surrounded by a dust torus responsible for absorption of high energy photons. Energetic particles escape along the magnetic field, i.e. along the disk axis. Seyfert 2 are viewed from the equator. They are 2 populations of radio-sources, corresponding to Fanaroff-Riley morphological classes I and II. These represent 1. the unbeamed, 'side-on' parent population of steep spectra, and 2. the 'beam' sources, including flat spectrum quasars and BL Lac objects, which arise through random alignment of their radio-axis with line of sight [21]. Remember, however, that some observational results do not favor the unified model (see Section 1).

3. Statistics

3.1. DETECTION, SURVEYS, AND CATALOGUES OF AGN

The General Catalogue of Quasars 42] contains 11358 (+2759) quasars (defined as brighter than absolute B magnitude -23), 3334 (+501) AGNs (defined as fainter than absolute B magnitude -23) and 357 (+137) BL Lac objects compiled from 1863 (+201) references.
Radio quasars are the optical counterparts Radio-Sources found in surveys such as 3CR, Bologna, PKS, 4C, 5C, 6C, AO, Ohio, VLA, FIRST and others. Radio emission from optical quasars has also been searched for, with typical success rate 20% ([1],[20]).
Braccesi [5] was the first to detect QSOs by a **purely optical method**, the UV/IR Excess. A little more than a dozen of such samples have been published. Below I focus to some of the largest ones. The **UVX** technique consists in the spectroscopy of candidates, selected as stellar objects which exhibit UV excess. It is quite efficient, as *radio-quiet* quasars are far more common than *radio-loud* ones. However, it suffers from redshift limitations, since Ly_α must fall in the UV band, which corresponds to $z_{UVX} < 2.2$. [3] found a total of 420 QSOs among 1400 faint stellar objects identified in a complete, ultraviolet-excess (UVX) survey. This sample is still one of the largest available today. [10] select quasar candidates in the range $15 < B < 18.75$ and $0.3 < z < 2.2$, using Schmidt plates, scanned with COSMOS and searched for UVX objects, and follow-up spectroscopy: there are already 285 QSOs in this initial sample. A systematic search for quasar candidates was conducted [27] in 40 square degrees at the North Galactic Pole, resulting in 1681 quasar candidates down to V=20.0, selected on the basis of UV excess and, when possible, absence of any measurable proper motion.

To avoid both the redshift limitation ($z < 2.2$) of UVX quasars, and the bias towards blue objects, the method of **Multicolor Surveys** consists in increasing the number of filters, especially towards the red. [4] give positions, magnitudes and identifications for a complete sample of 66 faint ($B \leq 22$) quasars based on spectroscopic observations of multicolour selected samples from deep photographic plates on an effective area of 0.85 square degrees The remainder of this survey is principally composed of galactic subdwarf

TABLE 1. QSO surface density to various faint limiting magnitudes.

Ref.	*mag*	redshift	Surface Density ($deg.^{2}$)	Error ($1\ \sigma$)
[39]	$R_{AB} < 18$	> 3.6	0.12	$\pm.03$
[4]	$B < 22$	< 2.2	68	± 9
[45]	$m_J < 22$	< 2.2	86	± 17
[45]	$m_J < 22$	∀	115	± 17
[17]	$B < 22.5$	< 2.2	130	$^{+60}_{40}$
[23]	$B < 22.6$	< 3.2	100	± 20
[16]	$B < 22.6$	< 3.3	140	$^{+30}_{15}$
[31]	$I < 22.5$	∀	200	$^{+120}_{80}$
[9]	$b_J < 23.5$	∀	160	$^{+390}_{100}$
[13]	$22.5 < B < 23.5$	∀	270	$^{+290}_{110}$
[8]	$K < 20$	∀	740	± 400
[22]	$B < 30$	$3.5 < z < 5$	8600	$\pm$ 2500

stars and compact narrow emission line galaxies. [16] describe a multicolor survey for quasars over 0.83 square degrees of sky imaged at KPNO 4m Mayall telescope in six filters (3000-10000A, magnitude limits 22 to 24). Quasars candidates are identified at all redshifts from their colors alone, special care is paid to high-z quasars.

QSOs Palomar Transit **Grism Survey** catalog [35] was designed to produce a sample of z>2.7 quasars identified by well-defined selection criteria (essentially based on their emission lines). 90 high-z quasars have already been found. [11,12] describe a wide-angle objective prism survey for finding bright QSOs in the northern sky, to be subsequently verified by slit spectroscopy. The magnitudes are in the range $15 \leq B \leq 19.5$ and redshifts $z \leq 2.8$.

In some samples, the **absence of proper motion** has been added to the selection criteria. Furthermore, quasars are used to achieve the connection between, for example, the HIPPARCOS catalogue and an absolute astrometric referential. [18] selected QSOs through their **light variations**, examining plates taken with the UK Schmidt telescope at 15 years interval. More than 300 objects, which are brighter than B=21.0 on a reference plate and which display a peak to peak amplitude of variability larger than B=0.35 mag are selected.

3.2. HIGH ENERGY SAMPLES

These samples consist of the optical counterparts of X-ray and γ-ray sources detected by Ginga, ROSAT, CGRO, ASCA. They are specially interesting, as firstly, about $\simeq$50 % of all X-ray sources brighter than $10^{-13}\, erg\, cm^{-2}\, s^{-1}$ are AGN, and secondly, X-ray emission is not affected by dust, these samples provide an efficient way of estimating the population of those red (absorbed) AGN which do *not* emit in radio bands. [6] and [44] compiled all the quasars from [42] detected by ROSAT (all sky and pointed). The total numbers of ROSAT detected quasars are 654 radio-loud and 846 radio-quiet.

3.3. OBSERVED EVOLUTION OF THE LUMINOSITY FUNCTION

Let $d^2N(M,z)$ be the *total* number of quasars with redshift $\in [z, z+dz]$ and absolute magnitude $\in [M, M+dM]$. The luminosity function $\Phi(M,z)$ is defined by the relation: $d^2N(M,z) = \Phi(M,z)\ dM\ \frac{dV}{dz}\ dz$. where $V(z)$ is the volume of the sphere out to redshift z. Those with *apparent* magnitude *brighter* than m_{lim}, a given completeness limit amount to: $d^2N_s(M,z) = d^2N(M,z) \times H(m_{lim} - m) = d^2N(M,z) \times H(z_{max} - z)$. Here H is the Heaviside function and z_{max} is the maximum redshift at which a quasar with absolute magnitude M may have an apparent magnitude *brighter* than m_{lim}. If Φ is independent of z, the total number of quasars *in the sample* is: $N_s = \int_{-\infty}^{+\infty} dM \Phi(M) \int_0^{z_{max}} dz \frac{dV}{dz} = \int_{-\infty}^{+\infty} dM \Phi(M) V_{max}(M)$, where $V_{max} = V(z_{max})$.
Applying the $< V/V_{max} >$ test, [32,33] found the first evidence for evolution of quasars in 2 available complete samples, 1 radio, 1 optical, with $\simeq$20 quasars each. Let us compute the average $< V/V_{max} >$ in a complete sample of quasars, each characterized by apparent magnitude $m < m_{lim}$ and redshift z.

$$< V/V_{max} >= (1/N_s) \int_{-\infty}^{+\infty} dM\Phi(M) \int_0^{V_{max}(M)} dV V/V_{max}$$

$$^{z_{max}} dz V(z)/N_s = \int_{-\infty}^{+\infty} dM\Phi(M) \frac{1}{2} V_{max}(M)/N_s = \frac{1}{2}$$

Provided the spatial distribution is *uniform* (Φ independent of z), this relation applies whatever the luminosity function $\Phi(M)$, in any range of absolute magnitude. In euclidean space, the flux density f is proportional to z^{-2}, the limiting flux density: $f_{min} \propto z_{max}^{-2}$, the volume is $V \propto z^3$, the maximum volume is $V_{max} \propto z_{max}^3$, and finally: $V/V_{max} = (f_{min}/f)^{3/2}$. The computation is *more difficult* in FRW cosmology! In its first application Schmidt [32] computed $< V/V_{max} >= 0.7$ in two complete samples, whatever the cosmology, and interpreted this result as a proof of evolution, because in cosmology the distance (redshift) corresponds to the cosmic time. Two main simplistic phenomenological models have been often considered, either Pure Density Evolution (PDE) [32,33,34], i.e. quasars more numerous in the past, or Pure Luminosity Evolution (PLE) [25,26], i.e. quasars in average brighter in the past. In both cases V and V_{max} must be computed according to the assumed evolution, leading to $< V/V_{max} >= 1/2$. In PDE, the quasar density increases with redshift z, a dimensionless factor $\rho(z)$ appears in the expression of the number of stars $d^2N(M,z)$: $d^2N(M,z) = \Phi(M)dM\rho(z)\frac{dV}{dz}dz$, the maximum redshift z_{max} is unchanged, the volume to be used, however, is no more $V(z)$ but $W(z) = \int_0^z \rho(z')\ \frac{dV}{dz}\ dz'$, with $< W/W_{max} >= 1/2$. In PLE, the *average* quasar luminosity evolves with epoch z: the luminosity function must be defined at some fixed epoch (usually $z = 0$). The expression of the number $d^2N(M,z)$ is unchanged, but, because each QSO turns out to be brighter as z increases, the true maximum redshift is $z'_{max} > z_{max}$, and $< V'/V'_{max} >= 1/2$.

To check directly whether data favour PDE or PLE, one needs to study the QSO joint distribution of magnitude *and* redshift. The sample size must be large enough to allow this. [2] compute the luminosity function at various z to study the evolution in a complete sample of 400 UVX QSOs, finding that *PLE* is a slightly better fit than *PDE*. All the results described about evolution concern mainly UVX quasars with $z < 2$. At larger redshift, the anlysis of both radio and optical data shows that the observed evolution reverses at $2 < z < 3$ and that the density of quasar light decreases with increasing redshift at $z > 3$ [37].
A lot of recent theoretical progress has been made in understanding the evolution of quasars in terms of accretion onto massive black holes (MBH) lying in galactic nuclei. The latter are in turn embedded in growing Dark Matter Halos (see the lectures of E. Bertschinger, and also see e.g. [15,37,14]. Eddington-limited accretion governs the individual light-curves. In most models, the corresponding timescales are determined by fitting the observed luminosity function. These timescales of emission are generally short (10^6 to 10^7 years), but recurrent emission is predicted in many models, due to the refueling of the black hole which may be expected through either tidal encounters in groups or viscous instabilities in the accretion disk[38]. Luminosity functions are derived by convolving the BH mass function with the average individual light-curve, which leads to predictions of the evolution of the AGN population.

4. AGN and Cosmology

4.1. GUNN-PETERSON TEST, REIONIZATION

Observing quasars allows one to put constraints on the physical properties of the high redshift intergalactic medium. The observed small width of the Ly_α lines in absorption in quasar spectra shows in fact that the intergalactic medium is ionized at $z > 2$. Furthermore, it can be shown that this ionization is precisely due to quasars themselves.

4.2. PRIMORDIAL GALAXIES

It is only recently that direct observational limits on the epoch of galaxy formation may be found from distant galaxies themselves, as dedicated techniques (such as UV drop out) have been successfully applied to the detection of galaxies out to redshifts larger than 3. Quasars however lie in the nuclei of active galaxies. As they are (at least to some extent) easier to find than galaxies, they may allow limits to be put on the formation redshift of galaxies.

4.3. PRIMORDIAL CHEMISTRY

As underlined in the lectures on primordial nucleosynthesis (K. Olive), the Deuterium lines observed in quasar spectra may have a crucial importance to constrain Big Bang Nucleosynthesis.

4.4. AGN AND LENSING

The statistics on observed separations on the sky of multiply lensed quasars may be used to put limits to the cosmological constant Λ, as this parameter plays important role in the variation of the angular diameter distance. Limits of the order of $\Lambda < 0.6$ to $\Lambda < 0.8$ have been found in this way.
On the other hand, the increase with redshift of the proportion of lensed (thus amplified) quasars in flux limited samples may mimic luminosity evolution.

4.5. GEOMETRICAL TESTS

are allowed with large quasar (and AGN) samples *provided that the observed evolution is well understood* (according to theorem 2 of G. Ellis). Many recent such tests have been concerned with quasars: angular diameter-redshift relation [7], clustering of quasars, and V/Vmax statistics [41] have been used to constrain cosmological parameters.

4.6. LARGE SCALE STRUCTURE

The clustering of AGN themselves, as well as of their absorbers has been used to put constraints to the cosmological parameters.

4.7. HIGH ENERGY BACKGROUNDS

Finally, AGN are the main contributors to the **UV, X-ray, Gamma-ray extragalactic backgrounds,** *(See the lectures by J.-L. Puget).*

5. Prospects

Even if galaxies have now larger z, quasars still offer the opportunity of (relatively) easy assembling of large-z samples, thanks to their high luminosity, quasi-stellar shape (easy photometry), emission lines. Many large surveys are in progress or in project:

The AAT 2dF QSO Survey will obtain redshifts for more than 25000 $B < 21$ QSOs through UVX selection $z < 2.2$ in two strips in the South Galactic Pole and in an equatorial region at the North Galactic cap. Already several thousands QSO redshifts have been obtained. The primary scientific aims of the QSO survey are: (1) to obtain the primordial fluctuation power spectrum out to COBE scales, (2) to determine the rate of QSO clustering evolution in the non-linear and linear regimes and hence obtain new limits on the value of Ω and of the bias parameter, (3) to apply a powerful geometric method to measure Λ: with an inflationary, non-zero Λ, the scale-length of QSO clustering at high redshift becomes several (4-5) times larger than the scale-length of either galaxy or Seyfert clustering at low redshift [36]. This result is difficult to explain in terms of clustering evolution because in gravitational instability models the scale-length of clustering tends to decrease at higher redshifts rather than increase. With the 30000 QSOs in the 2dF survey it should be possible to reject the allowed range of inflation models with Λ at a much higher significance (10σ) than the existing level obtained from current QSO redshift surveys [36].

CADIS: Calar Alto Deep Imaging Survey, multicolor, classification algorithm for distinguishing stars, galaxies and quasars, specifically high z quasars.

Chile-UK quasar survey: a total of $\simeq$thousand quasars expected.

DPOSS (Digital POSS II) quasars z>4.

The LZT (Large Zenith (6-meter liquid-mirror) Telescope), operational next year, will be dedicated to cosmological surveys, using a drift-scan CCD with a series of 40 narrow filters. This project is expected to provide redshifts with an accuracy of 0.01 for more than 1 million of galaxies and for quasars.

The Sloan Digital Sky Survey (SDSS,[39]) of the North Galactic hemisphere (10,000 square degrees) will include $\simeq$ 100,000 color-selected AGN and quasar candidates. The prime scientific goals of the SDSS concerning AGN will be their large scale distribution and evolution as well as systematic surveys of absorption line systems.
Most of the surveys above, however, are based on the photometric preselection of quasar candidates, in order to optimize the efficiency of telescope time. Preselection makes the completeness questionable both at low and at high redshift (even specialized color-color diagrams of [16] still reveal a strong bias in the range $2.2 \leq z \leq 3$, exactly where a reversing of evolution is observed). The observational strategy of the VIRMOS Survey should avoid the main drawbacks of preselection. To improve the efficiency of all quasar studies, one obviously has to increase both the size and the upper redshift limit of the sample, and to achieve a maximum completeness over the largest redshift range.
The **VIRMOS** spectroscopic large survey of *faint galaxies* will provide the opportunity of assembling such a sample. Thanks to the fact that stars and stellar objects are only a marginal population of the deep sky, the spectroscopy of a large fraction of stellar objects may be done. A shallow (I<22.5) and a deep (I<24) survey are foreseen. What is important is that two samples will be assembled without any preselection of quasar candidates based on color or morphology. There will be only some random selection for the shallow sample (1300^{+800}_{-500} quasars expected) and no selection at all for the deep one ($700?^{+600?}_{-300?}$ quasars expected). Furthermore we are examining the possibility to get a third, color-selected, sample of 4500^{+500}_{-500} quasars, which would be far larger, but biased because it will be obtained by favouring the spectroscopy of *candidates* based both on color preselection in the associated photometric survey and on counterparts of sources in the XMM Deep Survey, which will have at least some common fields.

In brief, the VIRMOS quasar sample will offer the opportunity to test the completeness of current surveys, to compare the usual modes of selection of quasar candidates, to study numerous absorbers, to study jointly active galaxies and quasars in a complete sample containing both with essentially common selection criteria, and, last but not least, it will be partly free of biases in the redshift range $2 < z < 3$ where evolution is particularly complex.

6. Bibliography

[1] Bischof O.B., Becker R.H. 1997, Astron. J. 113, 2000;
[2] Boyle et al. 1988, MNRAS 235, 935
[3] Boyle et al. 1990, MNRAS 243, 1
[4] Boyle et al. 1991 MNRAS 251, 482
[5] Braccesi, A. et al. 1968 ApJ 152, L105
[6] Brinkmann et al. 1997 A&A 319, 413,
[7] Buchalter et al. 1998 ApJ 494, 503
[8] Cohen et al. 1996 ApJ 462, L9
[9] Colless et al. 1991 MNRAS 253, 686
[10] Cristiani S. et al. A&A Sup. 112, 347 (1995)
[11] Engels et al. 1998 A&A Sup 128, 507
[12] Engels et al. 1998 A&A Sup 131, 196
[13] Glazebrook et al. 1995 MNRAS 273, 157
[14] Haehnelt & Rees, 1993, MNRAS 263, 168)
[15] Haiman & Loeb, 1997, ApJ 503, 505,
[16] Hall et al. 1996 ApJ 462, 614
[17] Hartwick F. & Schade D. 1990 ARA&A 28, 437
[18] Hawkins M. & Véron P. A&A 296, 665 1995
[19] Heckmann, T.M. 1980 A&A 87, 152
[20] Hooper et al. 1995 ApJ 445, 62
[21] Jackson, C.A., Wall, J.V. astro-ph/9808054
[22] Jarvis, R.M., MacAlpine, G.M. 1998 AJ 116, 2624
[23] Koo, D. & Kron, R. 1988 ApJ 325, 92
[24] Malkan M. et al. 1998, ApJ Sup 117, 25
[25] Mathez G. 1976 A&A 53, 15
[26] Mathez G. 1978 A&A 68, 17
[27] Moreau O. and Reboul H. 1995, A&A Sup 111, 169
[28] Peterson B.M. et al. 1998, ApJ 501 82
[29] Peterson B.M. 1997 *'Active Galactic Nuclei'*, *Cambridge University Press*
[30] Ptak et al. 1998, ApJ 501 L37
[31] Schade D. et al. 1996 MNRAS 278, 95
[32] Schmidt M. 1968 ApJ 151, 393
[33] Schmidt M. 1970 ApJ 162, 371
[34] Schmidt M. 1972 ApJ 176, 273
[35] Schneider D. et al. 1994 AJ 107, 1245
[36] Shanks T. & Boyle B. 1994 MNRAS 271, 753
[37] Shaver P. et al. 1996 Nature 384, 439
[38] Siemiginowska A. & Elvis M., 1997, ApJ 482, L9
[39] SDSS, http://www.sdss.org/news/releases/19981208.qso.html

[40] Van der Marel, R.P. astro-ph/9806365
[41] Van Waerbeke L. et al. 1996 A&A 316, 1
[42] Véron-Cetty M.P. and Véron P. 1998, ESO Scientific Report 18,
[43] D. W. Weedman 1986 *'Quasar Astronomy'*, *Cambridge Astrophysics Series*
[44] Yuan W. et al. A&A 330, 108, 1998
[45] Zitelli V. et al. 1992 MNRAS 256, 349

COSMIC VELOCITY FIELDS

J.F.L. SIMMONS
Department of Physics and Astronomy
University of Glasgow, Glasgow G12 8QQ , U.K.

1. Introduction

The velocity of the Local Group with respect to the CMBR has been determined to be about 600 km s^{-1}. A significant difference between the actual direction of motion and the direction to the nearest large cluster of galaxies, the Virgo cluster, led astronomers to believe in the existence of a mass concentration beyond, the so called Great Attractor, and to recognise the importance of large scale bulk motions.

Although these results were initially treated with scepticism in the cosmological community, with hindsight they are not so surprising. Density fluctuations are necessary for structure formation. Necessarily they also imply deviations from ideal Hubble flow. The measurement of peculiar velocities of galaxies should provide information about the distribution of all gravitating matter, including dark matter.

As we shall see, linear perturbation theory provides a simple relationship between the cosmic velocity field and density fluctuations. Measurement of the radial component of the peculiar velocities of galaxies is possible from simultaneous redshift and distance measurements. Thus if there is some way of constructing the vectorial velocity from measurement of the radial component alone, then we should be able to determine the matter density field. The methods we describe below largely depend on the linear theory of gravitational instability, and assume the validity of the Newtonian approximation. For observable cosmic structures today such as clusters and superclusters, such an approach is justified since the corresponding density fluctuations still fall in the linear regime and have remained within the horizon during their evolution.

These lectures are not in any way intended to provide a detailed or up to date review of the subject (see for instance chapter 7 of (Padmanabhan, 1993), (Dekel, 1997) and (Strauss, 1997)). We concentrate on outlining

M. Lachièze-Rey (ed.), Theoretical and Observational Cosmology, 381–396.

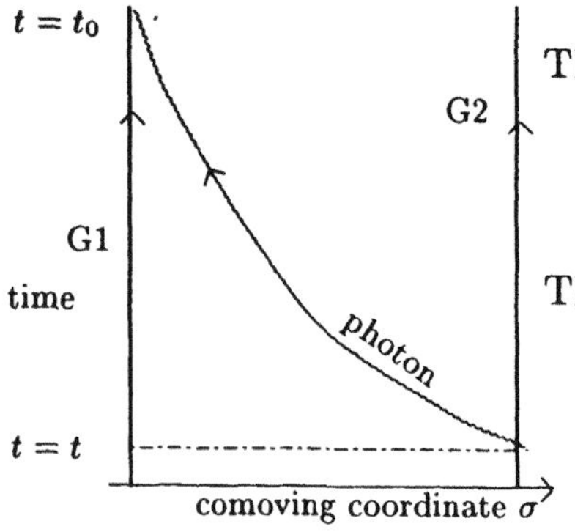

The proper distance between G_1 and G_2 is

$$d(t) = a(t)\int_{\sigma_1}^{\sigma_2} \frac{d\sigma}{\sqrt{1-k\sigma^2}}.$$

The comoving distance is constant in time.

Figure 1. Universal expansion.

of the main ideas and approaches. In the course of the lectures I suggest a number of simple exercises for the reader to solve. Most of these yield important results which are important for the rest of the text. We shall necessarily touch on the subject of Friedmann models and large scale structure formation, although these are dealt with more thoroughly by G. Ellis and E. Bertschinger in their respective lectures.

The metric of an homogeneous and isotropic universe may be written in the form

$$ds^2 = -dt^2 + a^2(t)(\frac{d\sigma^2}{1-k\sigma^2} + \sigma^2(d\theta^2 + \sin^2\theta d\phi^2). \quad (1)$$

$k = 1$, $k = 0$, and $k = -1$ correspond to closed, flat, and open universes respectively. σ, θ and ϕ, are comoving coordinates, i.e. worldlines with $\sigma = \theta = \phi =$ constant are geodesics. The picture we have of universal expansion is given by figure(1). The proper distance between G_1 and G_2 is scales with $a(t)$. With the simplistic interpretation of the relative velocity as the rate of change of the proper distance we regain Hubble's law with $H = \dot{a}/a$. [1] If we take galaxy G_1 to be our own, we see the redshift of a photon emitted at cosmic time t by galaxy G_2 is given by $1+z = a_0/a(t)$. In referring to length scales for fluctuations we shall always use the comoving length normalised to the present epoch. (The comoving distance is necessarily constant in time (see figure(1)). For convenience we shall take $a_0 = 1$. The comoving distance between G_1 and G_2 is $\int_{\sigma_1}^{\sigma_2} d\sigma/\sqrt{1-k\sigma^2}$.

[1]Hubble's law is usually expressed in the form $v = H_0 d$ where v is the radial component of the velocity of the galaxy, d its distance, and H_0 is Hubble's constant, expressed $100\,h$ km s^{-1} Mpc^{-1}. More generally it is expressed as a relationship between apparent magnitude and redshift. The above form is valid for $z << 1$. Any quantity with subscript 0 denotes its value at present epoch.

Can one measure the cosmic velocity field? One immediate problem here is that redshifts only give us the radial part of the peculiar velocity of galaxies via the relationship $v_{\rm pec} = cz - H_0 d$, and not the vectorial velocity. For irrotational flow this is sufficient to construct a velocity potential, and hence the vector velocity field. We still have the problem that we need to measure accurately the distances to galaxies. Moreover, galaxies provide a rather sparse sampling of the cosmic velocity field, and the velocities of galaxies might not represent the flow of non-visible matter. Galaxy surveys now give redshifts and distance estimates for a few thousand galaxies within 100 Mpc or so (Strauss, 1997) (Strauss and Willick, 1995).

Under a number of assumptions, linear perturbation theory of the Newtonian fluid equations provides a relationship between the divergence of the velocity field and the matter density contrast, defined by the fractional variation of matter density, i.e. $\delta(\mathbf{x}, t) = (\rho(\mathbf{x}, t) - \bar{\rho})/\bar{\rho}$, where $\bar{\rho}$ is the average density at a given cosmic time. Specifically we have the relation $\nabla \cdot \mathbf{v} = -H_0 f(\Omega)\delta$ where $f = \Omega^{0.6}/b$ and $\Omega = \bar{\rho}_0/\rho_c$ and ρ_c is the critical density given by $3H_0^2/8\pi G$. Galaxy counts indicate that $\delta < 1$ for scales greater than 10 Mpc. Hence for given H_0, the value of Ω can be obtained using this relation if *both* $\mathbf{v}$ and δ are known. Usually it is assumed that there is a linear relationship between the number density contrast for galaxies, δ_G, and the density contrast for all matter i.e. $\delta_G \equiv \delta n_G/n_G = b\,\delta\rho/\rho = b\,\delta$, where n_G is the number density of galaxies and b is the biasing parameter. There is no physical justification of such a relationship, and it is really only adopted for expediency. The above relation now takes the form

$$\nabla \cdot v = -H_0 \frac{f(\Omega)}{b} \delta_G = -H_0 \beta \delta_G. \tag{2}$$

We see that only β can be obtained, and not Ω and b separately.

2. Friedmann models and horizons

A more complete study of the evolution of cosmic fluctuations requires a general relativistic analysis, in which one considers fluctuations in both the metric and the energy momentum tensor. However on scales that are smaller than horizon size Newtonian theory yields essentially the same results. For this reason I shall introduce general relativistic notions only in order to provide a conceptual framework, and to establish the limits imposed on the validity of the Newtonian treatment. The horizon size will be of order c/H, although will vary for different models, depending on the behaviour of the scale factor $a(t)$. Friedmann's equations for the evolution of the scale factor are obtained by substituting the metric (1) into Einstein's field equations. Taking, for convenience, the cosmological constant to be zero, the field

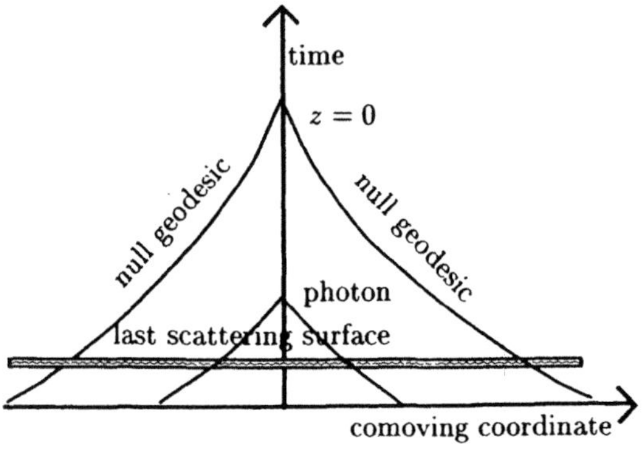

Figure 2. Horizons at different redshifts.

equations yields just two independent equations

$$(a)\quad \frac{\dot{a}^2 + k}{a^2} = \frac{8\pi}{3} G\rho \qquad (b)\quad \frac{\ddot{a}}{a} = -\frac{4\pi}{3} G(\rho + 3p) \tag{3}$$

For matter ($p = 0$) equations (3 a, b) yield $\rho a^3 = \text{const}$ and for radiation ($p = \rho/3$) $\rho a^4 = \text{const}$.

For a given model the horizon size is obtained by considering the null geodesics (worldlines (of photons) for which $ds = 0$ (see figure(2)). Evidently the horizon size will vary with cosmic time. The observable (visible) universe will be further limited to epochs after recombination.

EXERCISE: Consider the case where $a = t^{2/3}$ corresponding to a matter dominated universe ($p = 0$) with $\Omega = 1$ ($k = 0$). Show that (i) $H \equiv \dot{a}/a = 2/3t = 2/3(1+z)^{3/2}$ (ii) $\sigma(z) = 3(1 - (1+z)^{-1/2}$ and hence that as $z \to \infty$ $\sigma(z) \to 3$. Show that the comoving radius of the light sphere on the surface of last scattering is given by $\sigma_{\rm rec} \approx 3$. Show that comoving horizon size, $\sigma_{\rm H}(z)$ for a galaxy at redshift z is given by $\sigma_{\rm H}(z) = 3(1+z)^{-1/2}$. (iv) *See figure(2)* : Calculate the angle subtended by a fluctuation in the surface of last scattering of a given length scale $\lambda_{\rm CM}$ at an observer at z for a universe in which $a = t^{2/3}$.

Note that in the above model the horizon size today is twice as large as c/H_0, the speed of light times the Hubble time. The horizon size at $z = 1000$ is about $1/30$ of the horizon size today. If we take the horizon size now to be c/H_0 or $3 \times 10^3 h^{-1}$ Mpc, then the comoving horizon at recombination is about $100h^{-1}$ Mpc. The value at matter radiation equality would be about the same, and since we should expect structures to form only after this epoch, we see that Newtonian theory should be adequate for

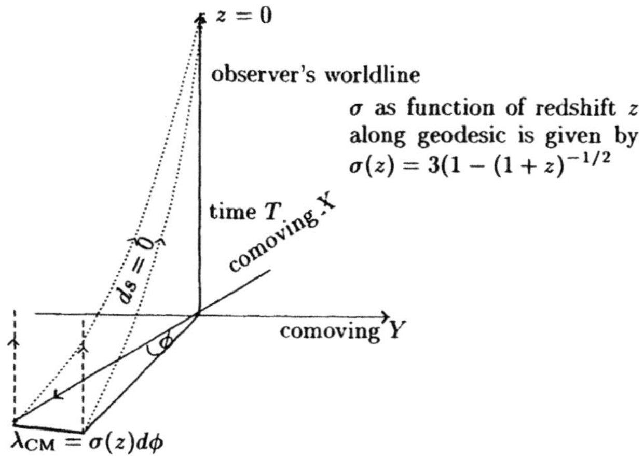

Figure 3. Angular size of fluctuation.

the description of structures such as superclusters, which are smaller than 100^{-1} Mpc.

Observed anisotropies in the CMBR reflect the density or potential inhomogeneities in the surface of last scattering. Fluctuations of comoving scalelength $\lambda_{\rm CM}$ subtend today an angle given by $\lambda_{\rm comoving} = \sigma(z)\ d\phi$ (see figure(2)). If we take $\lambda_{\rm CM}$ to be the horizon size at recombination, then $\lambda_{\rm CM} = \sigma_{\rm H}(z = 1000) \approx 1/10$, and hence with $\sigma(z = 1000) \approx 3$ we obtain $d\phi \approx 1/30$. Thus the horizon at recombination will subtend an angle of about 2 degrees.

3. The Sunyaev-Zeldovich effect

The question of horizons is of considerable importance when we come to discuss the Sunyaev-Zeldovich (S-Z) effect. In the S-Z effect the CMBR photons are scattered by hot intergalactic gas in a cluster. The contribution of Compton scattered CMBR photons at redshift z_c produces a small distortion in the black body spectrum of the CMBR we see today ($z = 0$). Introducing the conformal time given by $d\eta = dt/a$ reduces the line element to

$$ds^2 = -a^2(d\eta^2 - dl^2) \qquad (4)$$

and the spacetime diagram has the simple form given in figure(3)

The light sphere seen by the cluster is of course different from the light sphere we see, and evidently has a smaller radius. Furthermore the anisotropies in the CMBR seen by clusters at different redshifts will have a different angular dependency, and hence the polarization signal should

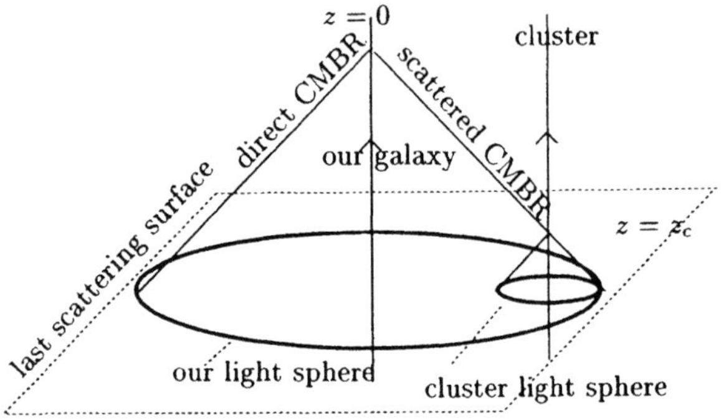

Figure 4. Sunyaev Zeldovich effect.

depend on the redshift of the cluster, and of course on the power spectrum of the fluctuations in the last scattering surface.

The kinematic S-Z effect is somewhat similar to the (thermal) S-Z effect, and was first discussed in Sunyaev and Zeldovich 1980. It provides in principle a direct method of determining the velocity of a cluster with respect to the CMBR frame. Ionized intergalactic gas in the cluster will see an anisotropic CMBR by virtue of the velocity of the cluster relative to the CMBR frame. The variation in intensity of the scattered radiation towards the cluster will be given by $\Delta T/T \approx -\beta_r \tau/c$, where β is the radial component of the relative velocity, and τ is the optical depth for Thomson scattering. In principle this would allow the determination of the radial velocity if some value for τ can be assumed. Of course the ordinary S-Z effect will tend to obscure the kinematic effect, but it can be used to estimate τ. Interestingly, there will also be a polarimetric signal from the quadrupole anisotropy induced by the aberration, which depends on the transverse velocity, $p \approx 0.1\beta_t^2 \tau/c^2$. The direction of polarization will be perpendicular to the plane defined by the vector velocity of the cluster and the line of sight. These are very small effects, but with the refinement of the measurement of the CMBR, it is not out of the question that vectorial peculiar velocities be determined in the future.

4. Statistics and Fourier Analysis

From the considerations of the last section, we see that for density fluctuations on comoving scales less than a few hundred Mpc, Newtonian analysis should be valid. Before deriving the equations governing the evolution of the density fluctuations it would be useful to define a few quantities that are essential to the analysis.

Firstly it is convenient to introduce Newtonian comoving coordinates $\mathbf{x} = \mathbf{r}/a$. It then follows that

$$\dot{\mathbf{r}} = a\dot{\mathbf{x}} + \frac{\dot{a}}{a}\mathbf{x}a \tag{5}$$

Noting that $\dot{a}/a = H$, and that $a\dot{\mathbf{x}} = d\mathbf{x}/d\eta$ equation 5 takes the form

$$\dot{\mathbf{r}} - H\mathbf{r} = \frac{d\mathbf{x}}{d\eta} \tag{6}$$

The left hand side is just the peculiar velocity.

In carrying out the perturbation analysis is convenient to express all the dynamical variables in terms of their Fourier harmonics. We shall adopt the simplest approach (see Kolb and Turner 1990) and do the expansion in a cube of side L and volume V, which has the effect of introducing periodic boundary conditions. Provided the volume is large enough, this procedure is justified. (More sophisticated approaches use the theory of random functions, treating $\delta(\mathbf{x},t)$ as a homogeneous stochastic process.) Thus expanding $\delta(\mathbf{x},t)$ in Fourier modes we obtain

$$\delta(\mathbf{x},t) = \frac{1}{V}\sum_{l,m,n=-\infty}^{l,m,n=\infty} \delta_{\mathbf{k}}(t)e^{-i\mathbf{k}\cdot\mathbf{x}} \tag{7}$$

where $\mathbf{k} = 2\pi(l,m,n)/L$ is the comoving wavevector. Rather than work with discrete sums it is more convenient to approximate them by Fourier transforms.

EXERCISE: Show from equation (7) that

$$\delta_{\mathbf{k}}(t) = \int_V \delta(\mathbf{x},t)e^{i\mathbf{k}\cdot\mathbf{x}}\, d^3x \text{ and } \delta(\mathbf{x},t) \approx (\frac{1}{2\pi})^3 \int \delta_{\mathbf{k}}e^{-i\mathbf{k}\cdot\mathbf{x}}\, d^3k. \tag{8}$$

For any functional of the density contrast we define an average by taking the integral over the volume V

$$\langle f(\delta)\rangle \equiv \frac{1}{V}\int_V f(\delta(\mathbf{x},t)d^3x. \tag{9}$$

In particular we have for the variance

$$\langle \delta^2(\mathbf{x},t)\rangle = \frac{1}{V}\int_V \delta^2(\mathbf{x},t))\, d^3x. \tag{10}$$

Using equation (8) and the relation $\int_{-\infty}^{\infty} \exp ikx dx = 2\pi\delta_D(k)$, where $\delta_D(k)$ is the Dirac delta function, one obtains

$$\langle\delta^2(\mathbf{x},t)\rangle \approx (\frac{1}{2\pi})^3 \int P(\mathbf{k},t) d^3k \;\text{ where } P(\mathbf{k},t) = \frac{1}{V}|\delta_{\mathbf{k}}|^2. \qquad (11)$$

$P(\mathbf{k},t)$ is the power spectrum, and plays a fundamental role in the theory of structure formation. It also turns out to be the Fourier transform of the autocorrelation function of $\delta(\mathbf{x},t)$ (see equation (14)).

We can also define the density contrast averaged over a window function, $W(\mathbf{x}',\mathbf{x})$, that is

$$\delta_W(\mathbf{x},t) = \frac{1}{V_W}\int \delta(\mathbf{x}',t)W(\mathbf{x}',\mathbf{x})\, d^3x' \;\text{ where }\; V_W = \int W(\mathbf{x}',\mathbf{x})\, d^3x' .$$

If we take $W = W(\mathbf{x}'-\mathbf{x})$ and define $W(\mathbf{k}) = \int_V W(\mathbf{x}) \exp i\mathbf{k}\cdot\mathbf{x}\, d^3x$, then we obtain

$$\langle\delta_W^2\rangle \approx (\frac{1}{2\pi})^3 \frac{1}{V_W^2}\int |\delta_{\mathbf{k}}|^2|W(\mathbf{k})|^2 d^3k. \qquad (12)$$

EXERCISE: Prove the above relation.

Defining the mass, M_W within the window as $\int W(\mathbf{x}',\mathbf{x})(\rho(\mathbf{x}',t)d^3x'$, and noting that the average mass within the window is $M_{0W} = V_W\rho_0$, it is evident that $\delta M_W/M_{0W} = \delta_W$. This means that the variance in the fractional mass contained in spheres of a given volume is the same as the variance in the density contrast averaged over similar volumes.

Simply taking a top-hat window function of diameter 8 Mpc, the relative mass fluctuation, σ_8, within the corresponding volume is about 1, and so $\delta_W \approx 1$. In other words scales of 8 Mpc and less have gone non-linear.

We the correlation function is defined in terms of the density contrast as

$$\xi(\mathbf{r}) \equiv \langle\delta(\mathbf{x}+\mathbf{r})\delta^*(\mathbf{x})\rangle. \qquad (13)$$

Using equation (8) in equation (13) we obtain the important result

$$\xi(\mathbf{r}) = (\frac{1}{2\pi})^3\frac{1}{V}\int |\delta_{\mathbf{k}}|^2 \exp -i\mathbf{k}\cdot\mathbf{r}\, d^3k, \qquad (14)$$

that is $P(k) = |\delta_{\mathbf{k}}|^2/V = \int \xi(\mathbf{r}) \exp i\mathbf{k}\cdot\mathbf{r} d^3r$. In other words the power spectrum is the Fourier transform of the autocorrelation function.

The autocorrelation function of the density contrast is related to the galaxy - galaxy correlation function, but the precise relationship will depend on galaxy biasing. Thus if we denote the probability that a galaxy lies in small volume ΔV_1 and another in ΔV_2 by P_{12}, then $P_{12} = n_1 n_2 \Delta V_1 \Delta V_2$. Averaging this probability over the entire volume gives

$$\langle P_{12}\rangle = \bar{n}^2(1+\xi_G(\mathbf{r})).$$

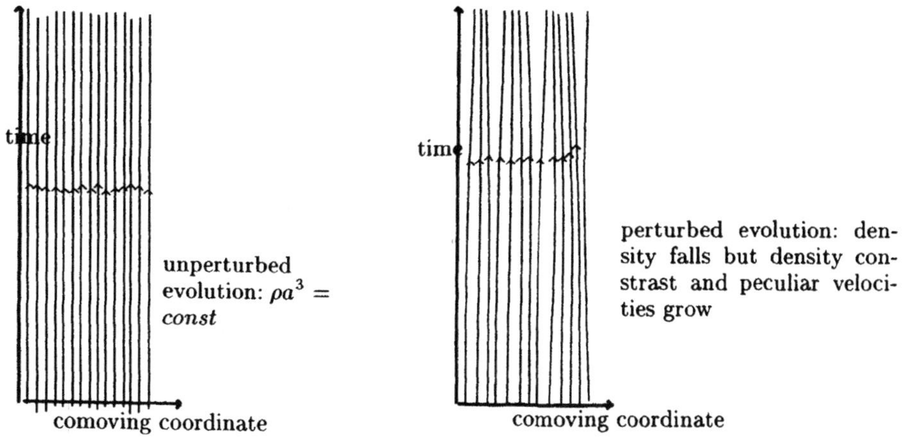

Figure 5. Development of fluctuations.

If galaxy biasing takes the fairly simple form

$$\frac{\delta n_G}{n_G} = b\frac{\delta\rho}{\rho} = b\delta \tag{15}$$

then the correlation function for matter may be obtained from the galaxy galaxy correlation function by dividing by b^2.

5. Newtonian fluid equations

In order to analyse the evolution of fluctuations in the density contrast after matter domination, it is adequate to treat matter and radiation as essentially a gas, assume an equation of state, and apply the fluid equations. (For CDM we take the pressure to be zero.) The equation of continuity, momentum, and Poisson's equation may be written

$$(a)\ \frac{\partial\rho}{\partial t} + \nabla_{\mathbf{r}}\cdot(\rho\mathbf{u}) = 0 \quad (b)\ \frac{D\mathbf{u}}{Dt} = -\frac{\nabla_{\mathbf{r}}p}{\rho} - \nabla_{\mathbf{r}}\phi \quad (c)\ \nabla^2_{\mathbf{r}}\phi = 4\pi G\rho, \tag{16}$$

where $\mathbf{r}$ are the physical space coordinates, D/Dt is the material derivative and $\mathbf{u} = d\mathbf{r}/dt$. Rewriting these equations in comoving coordinates, $\mathbf{x} = \mathbf{r}/a$, noting that

$$(\frac{\partial}{\partial t})_{\mathbf{r}} = (\frac{\partial}{\partial t})_{\mathbf{x}} + (\frac{\partial x^i}{\partial t})_{\mathbf{r}}\frac{\partial}{\partial x^i} \quad \text{and} \quad \nabla_{\mathbf{r}} = a^{-1}\nabla_{\mathbf{x}} \tag{17}$$

equations (16(a),(b) and (c)) become

$$(a)\ \frac{\partial\delta}{\partial\eta} + \nabla\cdot\mathbf{v}(1+\delta) = 0 \quad (b)\ \frac{\partial\mathbf{v}}{\partial\eta} + (\mathbf{v}\cdot\nabla)\mathbf{v} = -\frac{\nabla p}{\rho} - \nabla\tilde{\phi} - \frac{a'}{a}\mathbf{v}$$

$$(c)\ \nabla^2\tilde{\phi} = 4\pi G a^2\bar{\rho}\delta \tag{18}$$

where $a' \equiv da/d\eta$, $\nabla_{\mathbf{x}} \equiv \nabla$, and $\tilde{\phi} \equiv \phi + x^2\ddot{a}a/2$. Note that $\mathbf{v} \equiv d\mathbf{x}/d\eta$ is the peculiar velocity of a fluid element.

EXERCISE: Derive equations (18 a, b, c).

By linearizing these equations (i.e. ignoring all terms of order greater than 1 in δ and $\mathbf{v}$) one obtains

$$(a)\ \frac{\partial\delta}{\partial\eta} + \nabla\cdot\mathbf{v} = 0\ (b)\ \frac{\partial\mathbf{v}}{\partial\eta} = -\frac{\nabla p}{\rho} - \nabla\tilde{\phi} - \frac{a'}{a}\mathbf{v}\ (c)\ \nabla^2\tilde{\phi} = 4\pi G a^2\bar{\rho}\delta. \tag{19}$$

Later we shall see that for $\Omega = 1$ matter-dominated universe, $\delta \propto a$, so $\tilde{\phi}$ is a function only of space coordinates, and not of time. The gravitational potential is 'frozen in'. We shall see the importance of this later on.

EXERCISE: Since at later cosmic times $\mathbf{v}$ is irrotational, we may write $\mathbf{v} = \nabla\phi_{\mathbf{v}}$. Show that in the linear approximation $\tilde{\phi} = 4\pi G a\bar{\rho}\phi_{\mathbf{v}}/f(\Omega)H$ (modulo a function, χ say, satisfying $\nabla^2\chi = 0$.)

From equations (19 a, b, c) it follows that for isentropic processes one obtains

$$\frac{\partial^2\delta}{\partial\eta^2} + \frac{1}{a}\frac{da}{d\eta}\frac{\partial\delta}{\partial\eta} - c_s^2\nabla^2\delta - 4\pi G a^2\bar{\rho}\delta = 0, \tag{20}$$

where $c_s^2 = (\partial p/\partial\rho)_s$ is the speed of sound. $\delta_{\mathbf{k}}$ will grow only for modes larger than a certain scale length, i.e. $\lambda > \lambda_J$ where $\lambda_J = c_s\sqrt{\pi/Ga^2\bar{\rho}}$. One associates a mass, the Jeans mass ,M_J, with this scalelength. Only fluctuations of mass greater than the Jeans mass will grow. By decomposing the velocity field into rotational and irrotational parts one can show (see e.g. Kolb and Turner 1990) that the rotational part will decay. For modes with mass greater than M_J the pressure term will be negligible, and the general solution of equation (20) for δ may be written in terms of a growing part (subscript +)and decaying part (subscript −).

$$\delta(\mathbf{x},t) = \delta_+(\mathbf{x})D_+(\eta) + \delta_-(\mathbf{x})D_-(\eta)$$

and

$$\nabla\cdot\mathbf{v} = -\delta_+(\mathbf{x})D'_+(\eta) - \delta_-(\mathbf{x})D'_-(\eta)$$

EXERCISE: Show that $\delta = \delta_0 a$ is a solution of equation (20) for a matter dominated universe in which $a = t^{2/3}$, $\Omega = 1$. (Ignore pressure term.)

Thus, since we may ignore the decaying mode, we arrive at the key result

$$\nabla\cdot\mathbf{v} = -\frac{D'_+}{D_+}\delta. \tag{21}$$

$D'_+/D_+ = Haf(\Omega)$ where $f(\Omega) = d\ln D_+/d\ln a \approx \Omega^{0.6}$ (see Padmanabhan 1993). The above analysis is valid for any mode which is smaller than horizon size, and for which $|\delta_k|$ remains less than 1. Given the initial values of the density fluctuations, δ_i, at some cosmic time, t_i, linear theory predicts the values at later time, t, and indeed at the present epoch t_0. Furthermore, given an initial power spectrum $P(\mathbf{k}, t_i)$, it is straight forward to obtain the power spectrum at a later time, since in the linear regime each mode evolves independently . Of course this will not hold for all time, and depending on the initial amplitude of the mode, it will sooner or later go nonlinear. For $\delta > 1$ gravitational collapse and virialization will take place. This appears to be the case for structures of size 10 Mpc and less.

6. Methods and Data

Equation (21) tells us that the matter density distribution is related to the divergence of the velocity field on scales for which linear analysis is valid. Thus in principle if $\mathbf{v}$ and δ are measured the parameter β can be determined, or if we believe galaxy biasing is as simple as described by the relation 15, then measurement of $\mathbf{v}$ gives us the density distribution of all gravitating matter.

There are however problems with this simple programme, which can be summarized as follows:

- At the present time we have redshifts and distances to only about 3000 galaxies within 100 Mpc (See Strauss 1997).
- Measurement of the redshifts only gives us the radial component of the peculiar velocities.
- There is considerable uncertainty in the measured distances to galaxies.
- Selection effects lead to uncertainty in the reconstruction of the density contrast from survey.

DISTANCE INDICATORS: The difference between the apparent magnitude and absolute magnitude (the distance modulus) is related to the distance via the expression

$$m - M = 5\log_{10} r/\text{Mpc} + 25. \tag{22}$$

(A similar relation to equation will hold for the absolute size, and the following analysis is easily adapted to this case.) However in equation (22) m is only determined up to an observational error (which we henceforth assume to be zero) and M is selected from the luminosity distribution of galaxies of its type. We can estimate the distance of the galaxy by estimating its absolute magnitude. We might take the absolute magnitude to simply be the average absolute magnitude of galaxies of this class, but such a procedure would give us unsatisfactory estimate. In particular for a distant sample

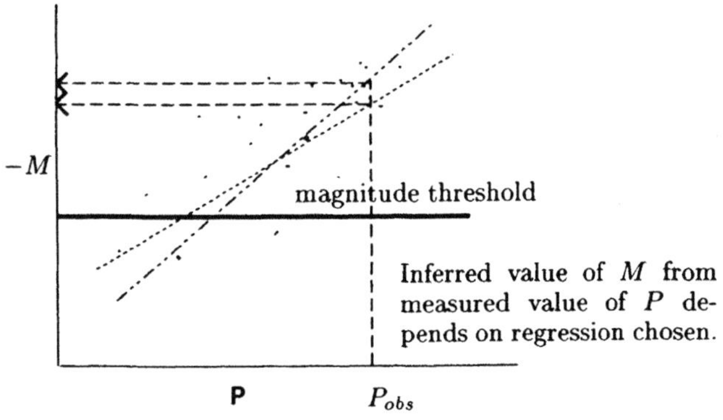

Figure 6. Regression of M on P and P on M.

of galaxies it is more probable that an intrinsically bright galaxy would be selected whose luminosity would be greater than the mean luminosity of the class of galaxies. In such a case the distance to the galaxy would be underestimated. This is called the Malmquist effect.

In order to estimate the absolute magnitude of a galaxy one usually uses an observable, P say, which is correlated with absolute magnitude. The two most commonly used indicators are (i) the Tully-Fisher (TF) relation for spirals, which gives a correlation between line width and absolute magnitude, and (ii) the so called $D_n - \sigma$ relation between absolute diameter or magnitude and velocity dispersion for elliptical galaxies. (See (Jacoby et al., 1990) for a review of distance indicators.)

By measuring the correlated quantity, one estimates the absolute magnitude, and then, in conjunction with the measured apparent magnitude, obtains an estimate of the distance. Introducing $w = \log_{10} r/\mathrm{Mpc}$ and a hat to represent the estimator of a particular quantity, we can write

$$\hat{w} = (m - \hat{M})/5 + w_0 - 5 \tag{23}$$

where w_0 is the true distance. Equation (23) can also be written

$$\hat{w} = (M - \hat{M})/5 + w_0 \tag{24}$$

Evidently, if the difference between the value of $\hat{M}$ and the actual value, M_0 is $\Delta M \approx 1$ then $\Delta w_0 \approx 1$ and the fractional error in the distance estimate is $\Delta r/r \approx 0.2$, or about 20% .

We could take a linear estimator for M, i.e. $\hat{M} = a + b\mathbf{P}$, and upon substitution in equation (23) or 24 obtain an estimator $\hat{w}$. Suppose for simplicity that $\mathbf{P}$ and $\mathbf{M}$ have a bivariate normal distribution. Two obvious choices for a and b are the values obtained from either regressing $\mathbf{M}$ on $\mathbf{P}$, or $\mathbf{P}$ on $\mathbf{M}$ (see figure(6). In the former case we obtain what is referred to in the literature as the direct Tully Fisher (TF) and the latter the Inverse Tully Fisher (ITF) estimator. A simple but tedious calculation gives (Hendry and Simmons, 1994)

$$\hat{w}_{\mathrm{TF}} = 0.2((m - M_0) + (P - P_0)\frac{\rho\sigma_M}{\sigma_P} - 25) \tag{25}$$

and

$$\hat{w}_{\mathrm{ITF}} = 0.2((m - M_0) + (P - P_0)\frac{\sigma_M}{\rho\sigma_P} - 25), \tag{26}$$

where σ_M and σ_P are the standard deviation of $\mathbf{M}$ and $\mathbf{P}$, and ρ is the correlation coefficient. Use of one or other will generally give significantly different results, depending on the scatter.

If we are carrying out repeated measurements or using a large set of galaxy distance data, the statistical bias of an estimator will be important. Suppose $S(\mathbf{M}; w_0)$ is the probability of a galaxy of absolute magnitude M at distance w_0 being included in the catalogue. Then the expectation value, $E(\hat{w}|w_0)$, of the estimator $\hat{w}$ over the observable distribution of $\mathbf{M}$ and $\mathbf{P}$ will be given by

$$E(\hat{w}|w_0) = \frac{\int \hat{w}S(M; w_0)\Psi(M,P)\mathrm{d}M\mathrm{d}P}{\int S(M; w_0)\Psi(M,P)\mathrm{d}M\mathrm{d}P}. \tag{27}$$

An estimator $\hat{w}$ is called unbiased if $E(\hat{w}|w_0) = w_0$. Generally this will not be the case. A propos, one can show that $\hat{w}_{\mathrm{TF}}$ is biased and $\hat{w}_{\mathrm{ITF}}$ unbiased (Hendry and Simmons, 1994).

Various attempts have been made to incorporate prior information in order to reduce the error in the distance estimator. Such a technique is essentially 'Bayesian'. In the context of galaxy distance estimation let us suppose that the prior information consists of knowing the intrinsic distribution of galaxy distances, $\Phi_G(w_0)$. Let $\hat{w}_U$ be an unbiased estimator, which we assume to be gaussian, of w_0. For argument's sake let us take $\Phi_G(w_0)$ to be gaussian also. We can construct an estimator, $\hat{w}_B$, which minimises the square deviation of the estimator from the true value averaged over the prior distribution. $\int(\hat{w}_B - w_0)^2\Psi_{\mathrm{obs}}(\hat{w}_B|w_0)\Phi_G(w_0)\mathrm{d}P\mathrm{d}M\mathrm{d}w_0$.

It can be shown that if both Ψ_{obs} and Φ_G are gaussian, then a gaussian Bayesian estimator is given by

$$\hat{w}_B = \frac{\hat{w}_U + \beta w_0}{1+\beta} \quad \text{where } \beta = \frac{\sigma_U^2}{\sigma_G^2} \text{ and has variance } \sigma_B^2 = \frac{\sigma_U^2}{1+\beta}. \tag{28}$$

Evidently, from equation (28) $\hat{w}_B \to \hat{w}_U$ as $\beta \to 0$: prior information is unimportant, and the two estimators agree. Notice also that the variance of $\hat{w}_B$ is always less than $\hat{w}_U$. On the other hand we can not necessarily be sure that the prior information is reliable.

7. Reconstruction of velocity fields

POTENT: According to linear theory, the velocity field should be potential. Thus if the radial component of the peculiar velocity can be obtained by measuring redshifts on the one hand and from galaxy distances that are determined independenty of redshifts, then the velocity potential at any point can be determined by taking a radial path integral. Thus

$$\mathbf{v} = -\nabla\phi_v \quad \text{and} \quad \phi_v = -\oint \mathbf{v}\cdot d\mathbf{r} = -\int v_r dr. \tag{29}$$

This method was devised by Bertschinger and Dekel and later refined in an attempt to cope with problems of sparse, inhomogeneous data, and uncertain distance estimates (Dekel, 1997). Comparison of the velocity field with the galaxy number density distribution inferred from various galaxy surveys has led to fairly crude estimates of the β parameter. Attempts have also been made to obtain the power spectrum of the velocity potential, and use this to determine the power spectrum of density fluctuations from $k^2\phi_{\mathbf{k}} = f(\Omega)H_0\delta_{\mathbf{k}}$.

VOIDS: A similar method was suggested (Dekel and Rees, 1994) using regions where there were very few galaxies. Using the Zeldovich approximation, an upper bound, δ_c, to the density contrast can be found. δ is necessarily greater than -1. By considering the voids the range of values of Ω for which δ_c is negative can be rejected. Using this method Dekel and Rees found that $\Omega > 0.3$ There is evidently a problem in this method since the velocity field is necessarily poorly determined in such regions. Furthermore linear theory cannot really be expected to hold for $\delta \approx -1$.

REDSHIFT DISTORTION: Various work on this stemmed from the idea that the galaxy number density in redshift space differed from the number density in real space. Using the small angle approximation, Kaiser ((Kaiser, 1987)) showed that the power spectrum, $P_S(\mathbf{k})$, calculated from the number density in redshift space, is related to the power spectrum in real space, $P_R(\mathbf{k})$, by $P_S(\mathbf{k}) = P_R(\mathbf{k})(1 + f(\Omega)\mu^2))^2$, where $\mu = \hat{\mathbf{k}}\cdot\hat{\mathbf{r}}$ and $s = r + (\mathbf{v} - \mathbf{v}_{\mathrm{LG}})\cdot\hat{\mathbf{r}}/H_0$. (There are two very apparent observed effects. Firstly virialisation in clusters brings about a spread in redshift values due to the spread in random peculiar velocities at a given distance. This is manifested in the so called "fingers of God" in the plot of redshift number density of galaxies. On the other hand, bulk infall into a region of mass concentration will produce a concentration of number density in redshift space in the radial direction.) Various more refined methods (Fisher et al., 1994) (Heavens and Taylor, 1995) of exploiting this distortion have been used in order to determine the β parameter. These methods depend on the expansion in terms of spherical harmonics of the density in redshift space and the mean density in real space , viz

$$\rho_s(\mathbf{s}(\mathbf{r})) = \sum_{lm} \rho_{lm}(r) Y_{lm}(\theta, \phi)$$

$$\bar{\rho}^0(\mathbf{r}) = \sum_{lm} \rho^0_{lm}(r) Y_{lm}(\theta, \phi)$$

$$\delta(\mathbf{r}) = \sum_{lm} \delta_{lm}(r) Y_{lm}(\theta, \phi)$$

The quantity $D = \langle\langle \rho_{lm}(r) - \rho^0_{lm}(r)/\bar{\rho} \rangle\rangle$, where the averaging is over the radial coordinate, has a statistical distribution which depends on β and $\langle\langle \delta_{lm}(r) \rangle\rangle$. Likelihood analysis in the various papers using this method have yielded a value of $\beta \approx 1$.

BEYOND THE LINEAR REGIME: Higher order perturbation theory has been used to try to break the degeneracy in Ω and b in the β parameter by using the non-gaussian character of the density contrast and velocity field, (Bernardeau *et al.*, 1995) i.e.

$$\frac{\langle \delta^3 \rangle}{\langle \delta^2 \rangle} \approx (34/7 - 3 - n) \quad \frac{\langle (\nabla \cdot \mathbf{v})^3 \rangle}{\langle (\nabla \cdot \mathbf{v})^2 \rangle} \approx (26/7 - 3 - n)/f(\Omega). \tag{30}$$

For such an approach to be successful, better data will undoubtedly be required.

In summary, the study of the cosmic velocity field can yield very valuable information about the distribution of matter in the universe, and be used for the testing of different cosmological scenarios. The more accurate determination of the cosmic velocity field will in turn depend on uncovering more reliable distance indicators, and possibly on the better modelling of galaxies, and of course a better understanding of nonlinear effects and galaxy formation.

References

Bernardeau, F., Juszkiewiecz, R., Dekel, A., and Bouchet, F.R. Omega from the skewness of the cosmic velocity divergence. *Mon. Not.Ast.Soc.* **Vol. 274.** pp. 20-26.

Bertschinger, E. and Dekel, A. (1989)*ApJ.* **Vol. 336**, L5

Dekel, A. (1997) Cosmological Implications of Large Scale Flows, *Galaxy Scaling Relations.* Springer.

Dekel, A. and Rees, M.J. (1994) Ω from velocities in voids. *ApJ.* **Vol. 422**, L1-L4.

Fisher, K.B. , Scharf, C.A., Lahav, O. (1994) A spherical harmonic approach to redshift distortion and a measurement of Ω_0 from the 1.2-Jy IRAS Redshift Rurvey*Mon. Not.Ast.Soc.* **Vol. 266**, pp. 219-226.

Heavens, A.F. and Taylor, A.N. (1995) A spherical harmonic analysis of redshift space *Mon. Not.ast.Soc.* **Vol. 275**, pp. 483-497.

Hendry, M.A. and Simmons, J.F.L. (1994) Optimal galaxy distance estimators *ApJ.* **Vol. 435**, pp.515 -227.

Kaiser, N. (1987) Clustering in real space and in redshift space. *Mon. Not.ast.Soc.***Vol. 227**, pp. 1-21.

Jacoby, G. H. et al. (1992) A critical review of selected techniques for measuring extragalactic distances. *PASP* **Vol. 104.**, pp. 599-662.

Kolb, W.K. and Turner, M.S.(1990) *The Early Universe.* Addison Wesley.

Padmanabhan, T. (1993) *Structure Formation in the Universe.* C.U.P.

Strauss, M.A. (1997) Recent advances in redshift surveys of the local universe, in Formation of Structure in the Universe eds. A. Dekel and J.P. Ostriker (C.U.P)

Strauss, M.A. and Willick, J.A. (1995) The density and peculiar velocity fields of nearby galaxies. *Physics Reports,* **Vol. 261**, pp. 271-431.

Sunyaev, R.A. and Zeldovich, Ya. B. (1980) The velocity of clusters of galaxies relative to the microwave background. The possibility of its measurement. *Mon. Not.ast.Soc.***Vol. 190**, pp. 413-420.

GPSR Compliance

The European Union's (EU) General Product Safety Regulation (GPSR) is a set of rules that requires consumer products to be safe and our obligations to ensure this.

If you have any concerns about our products, you can contact us on ProductSafety@springernature.com

In case Publisher is established outside the EU, the EU authorized representative is:

Springer Nature Customer Service Center GmbH
Europaplatz 3
69115 Heidelberg, Germany

Batch number: 10755284

Printed by Printforce, the Netherlands